Lecture Notes of the Institute for Computer Sciences, Social Informatics and Telecommunications Engineering 392

More information about this series at http://www.springer.com/series/8197

Mian Ahmad Jan · Fazlullah Khan (Eds.)

Application of Big Data, Blockchain, and Internet of Things for Education Informatization

First EAI International Conference, BigIoT-EDU 2021
Virtual Event, August 1–3, 2021
Proceedings, Part II

 Springer

Editors
Mian Ahmad Jan 🆔
Abdul Wali Khan University Mardan
Mardan, Pakistan

Fazlullah Khan 🆔
Abdul Wali Khan University Mardan
Mardan, Pakistan

ISSN 1867-8211 ISSN 1867-822X (electronic)
Lecture Notes of the Institute for Computer Sciences, Social Informatics
and Telecommunications Engineering
ISBN 978-3-030-87902-0 ISBN 978-3-030-87903-7 (eBook)
https://doi.org/10.1007/978-3-030-87903-7

This Springer imprint is published by the registered company Springer Nature Switzerland AG
The registered company address is: Gewerbestrasse 11, 6330 Cham, Switzerland

Preface

We are delighted to introduce the proceedings of the first edition of the European Alliance for Innovation (EAI) International Conference on Application of BigData, Blockchain, and Internet of Things for Education Informatization (BigIoT-EDU 2021), which was held virtually during August 1–3, 2021. The conference aims to provide an international cooperation and exchange platform for big data and information education experts, scholars, and enterprise developers to share research results, discuss existing problems and challenges, and explore cutting-edge science and technology. The conference focuses on research fields such as "big data" and "information education". The use of artificial intelligence (AI), blockchain, and network security lies at the heart of this conference as we focused on these emerging technologies to excel the progress of big data and information education.

EAI BigIoT-EDU has three tracks: the main track, a late track, and a workshop track. In total, EAI BigIoT-EDU 2021 attracted 500+ submissions. Upon rigorous review, only 144 papers were accepted. The keynote speaker was Paul A. Watters from LaTrobe University, Australia. The workshop was organized on "Information Retrieval and Algorithms in the Era of Information and Communication Technologies" by Ryan Alturki from Umm Al-Qura University, Saudi Arabia. The workshop aimed to focus on advanced techniques and algorithms to retrieve useful information from ICT and connected technologies. Coordination with the steering chair, Imrich Chlamtac, was essential for the success of the conference. We sincerely appreciate his constant support and guidance. It was also a great pleasure to work with such an excellent organizing committee team for their hard work in organizing and supporting the conference. In particular, the Technical Program Committee (TCP), led by our TPC co-chairs, Yinjun Zhang, Yar Muhammad, and Muhammad Imran Khan, who completed the peer-review process for the technical papers and put together a high-quality technical program. We are also grateful to Conference Manager Natasha Onofrei for her constant support and to all the authors who submitted their papers to the EAI BigIoT-EDU 2021 conference, late track, and workshop.

We strongly believe that BigIoT-EDU provides a good forum for all researchers, developers, and practitioners to discuss all science and technology aspects that are relevant to big data and information education. We also expect that the future BigIoT-EDU conferences will be as successful and stimulating, as indicated by the contributions presented in this volume.

October 2021

Mian Ahmad Jan
Fazlullah Khan
Mengji Chen

Organization

Steering Committee

Imrich Chlamtac	University of Trento, Italy
Mian Ahmad Jan	Abdul Wali Khan University Mardan, Pakistan
Fazlullah Khan	Abdul Wali Khan University Mardan, Pakistan

Organizing Committee

General Chairs

Mian Ahmad Jan	Abdul Wali Khan University Mardan, Pakistan
Fazlullah Khan	Abdul Wali Khan University Mardan, Pakistan
Mengji Chen	Guangxi Science and Technology Normal University, China

Technical Program Committee Chair and Co-chairs

Lu Zhengjie	Hechi University, China
Yinjun Zhang	Hechi University, China
Yar Muhammad	Abdul Wali Khan University Mardan, Pakistan

Sponsorship and Exhibit Chairs

Lan Zimian	Harbin Institute of Technology, China
Zhang Mianlin	Guangxi University, China

Local Chairs

Huang Yufei	Hechi Normal University, China
Wan Haoran	Shanghai University, China

Workshops Chair

Abid Yahya	Botswana International University of Science and Technology, Botswana

Publicity and Social Media Chairs

Wang Bo   Guangxi Science and Technology Normal
                                 University, China
Aamir Akbar   Abdul Wali Khan University Mardan, Pakistan

Publications Chairs

Fazlullah Khan   Abdul Wali Khan University Mardan, Pakistan
Mian Ahmad Jan   Abdul Wali Khan University Mardan, Pakistan

Web Chair

Shah Nazir   University of Swabi, Pakistan

Posters and PhD Track Chairs

Mengji Chen   Guangxi Science and Technology Normal
                                 University, China
Ateeq ur Rehman   University of Haripur, Pakistan

Panels Chairs

Kong Linxiang   Hefei University of Technology, China
Muhammad Usman   Federation University, Australia

Demos Chairs

Ryan Alturki   Umm Al-Qura University, Saudi Arabia
Muhammad Imran   Abdul Wali Khan University Mardan, Pakistan

Tutorials Chair

Wei Rongchang   Guangxi Science and Technology Normal
                                 University, China

Session Chairs

Ryan Alturki   Umm Al-Qura University, Saudi Arabia
Aamir Akbar   Abdul Wali Khan University Mardan, Pakistan
Mengji Chen   Hechi University, China
Vinh Troung Hoang   Ho Chi Minh City Open University, Vietnam
Muhammad Zakarya   Abdul Wali Khan University Mardan, Pakistan
Yu Uunshi   Shanxi Normal University, China

Ateeq ur Rehman	University of Haripur, Pakistan
Su Linna	Guangxi University, China
Shah Nazir	University of Swabi, Pakistan
Mohammad Dahman Alshehri	Taif University, Saudi Arabia
Chen Zhi	Shanghai University, China
Syed Roohullah Jan	Abdul Wali Khan University Mardan, Pakistan
Qin Shitian	Guangxi Normal University, China
Sara Kareem	Abdul Wali Khan University Mardan, Pakistan
Mohammad Wedyan	Al-Balqa Applied University, Jordan
Lin Hang	Beijing Forestry University, China
Arjumand Yar Khan	Abdul Wali Khan University Mardan, Pakistan
Liu Cheng	Wuxi Institute of Technology, China
Rahim Khan	Abdul Wali Khan University Mardan, Pakistan
Muhammad Tahir	Saudi Electronic University, Saudi Arabia
Tan Zhide	Anhui University, China

Technical Program Committee

Muhammad Usman	Federation University, Australia
Abid Yahya	Botswana International University of Science and Technology, Botswana
Noor Zaman Jhanjhi	Taylor's University, Malaysia
Muhammad Bilal	Hankuk University of Foreign Studies, South Korea
Muhammad Babar	Iqra University, Pakistan
Mamoun Alazab	Charles Darwin University, Australia
Tao Liao	Anhui University of Science and Technology, China
Ryan Alturki	Umm Al-Qura University, Saudi Arabia
Dinh-Thuan Do	Asia University, Taiwan
Huan Du	Shanghai University, China
Sahil Verma	Chandigarh University, India
Abusufyan Sher	Abdul Wali Khan University Mardan, Pakistan
Mohammad S. Khan	East Tennessee State University, USA
Ali Kashif Bashir	Manchester Metropolitan University, UK
Nadir Shah	COMSATS University Islamabad, Pakistan
Aamir Akbar	Abdul Wali Khan University Mardan, Pakistan
Vinh Troung Hoang	Ho Chi Minh City Open University, Vietnam
Shunxiang Zhang	Anhui University of Science and Technology, China
Guangli Zhu	Anhui University of Science and Technology, China
Kuien Liu	Pivotal Inc., USA

Kinan Sher	Abdul Wali Khan University Mardan, Pakistan
Feng Lu	Chinese Academy of Sciences, China
Ateeq ur Rehman	University of Haripur, Pakistan
Wei Xu	Renmin University of China, China
Ming Hu	Shanghai University, China
Abbas K. Zaidi	George Mason University, USA
Amine Chohra	Université Paris-Est Créteil, France
Davood Izadi	Deakin University, Australia
Sara Kareem	Abdul Wali Khan University Mardan, Pakistan
Xiaobo Yin	Anhui University of Science and Technology, China
Mohammad Dahman Alshehri	Taif University, Saudi Arabia
Filip Zavoral	Charles University in Prague, Czech Republic
Zhiguo Yan	Fudan University, China
Florin Pop	Politehnica University of Bucharest, Romania
Gustavo Rossi	Universidad Nacional de La Plata, Argentina
Habib Shah	Islamic University of Medina, Saudi Arabia
Hocine Cherifi	University of Burgundy, France
Yinjun Zhang	Guangxi Science and Technology Normal University, China
Irina Mocanu	University Politehnica of Bucharest, Romania
Jakub Yaghob	Charles University in Prague, Czech Republic
Ke Gong	Chongqing Jiaotong University, China
Roohullah Jan	Abdul Wali Khan University Mardan, Pakistan
Kun-Ming Yu	Chung Hua University, China
Laxmisha Rai	Shandong University of Science and Technology, China
Lena Wiese	University of Göttingen, Germany
Ma Xiuqin	Northwest Normal University, China
Oguz Kaynar	Sivas Republic University, Turkey
Qin Hongwu	Northwest Normal University, China
Pit Pichapan	Al-Imam Muhammad Ibn Saud Islamic University, Saudi Arabia
Prima Vitasari	National Institute of Technology, Indonesia
Simon Fong	University of Macau, China
Shah Rukh	Abdul Wali Khan University Mardan, Pakistan
Somjit Arch-int	Khon Kaen University, Thailand
Sud Sudirman	Liverpool John Moores University, UK
Tuncay Ercan	Yasar University, Turkey
Wang Bo	Hechi University, China
Ibrahim Kamel	University of Sharjah, UAE
Muhamamd Wedyan	Albalqa University, Jordan

Mohammed Elhoseny	American University in the Emirates, UAE
Muhammad Tahir	Saudi Electronic University, Saudi Arabia
Marwa Ibrahim	University of Technology Sydney, Australia
Amna Khan	Abdul Wali Khan University Mardan, Pakistan
Xiao Wei	Shanghai University, China
Zhiming Ding	Beijing University of Technology, China
Jianhui Li	Chinese Academy of Sciences, China
Yi Liu	Tsinghua University, China
Wuli Wang	SMC Guangxi Electric Power Company Limited, China
Duan Hu	Shanghai University, China
Sheng Li	Xinjiang University, China
Mahnoor Inam	Abdul Wali Khan University Mardan, Pakistan
Yuhong Chen	Inner Mongolia University of Technology, China
Mengji Chen	Guangxi Science and Technology Normal University, China
Jun Zhang	Hechi University, China
Ji Meng	Hechi University, China

Contents – Part II

Applications of Information Technology

Contents – Part I

Applications of Big Data and Cloud Computing

Research on the Development Direction of Computer Software Testing Methods in the Era of Big Data

Ning Wu$^{(\boxtimes)}$

Anhui Sanlian University, Anhui 230601, China

Abstract. With the development of Internet, Internet of things and cloud computing, the era of big data is accelerating. In this background, higher requirements are put forward for the testing of computer software. Based on the data background, the author of the database system software testing to start a brief discussion, hope this research can play a role in promoting the development of computer software testing.

Keywords: Big data era · Software testing · Computer technology

1 Introduction

Big data refers to the data set that cannot be captured, managed and processed by conventional software in a certain period of time. It has the characteristics of large amount of data, many types of data, low value density of data and fast data processing speed. The era of big data originated from McKinsey, which pointed out that data has spread to various industries and fields. Later, Alibaba founder Jack Ma also proposed that people will enter the era of big data in the future. With the continuous progress of the times and the continuous development of science and technology, in the era of big data, hardware products are becoming more and more complex, and the scope of application is also expanding. In this context, the scale of software system is expanding, and the complexity is increasing. In order to ensure the quality and operation safety of software, the development and application of software testing technology is very important [1].

2 Challenges of Software Testing in the Era of Big Data

2.1 Oracle Problems Become More and More Prominent

The purpose of software testing is to find out the wrong operation of the software, and Oracle is to make a special judgment on whether the testing process has passed the verifiability. In the era of big data, whether it is trend analysis or graph computing, software testing has become more difficult. Big data processing mode is divided into physical mode and chemical mode [2]. Big data processing in physical mode refers to

M. A. Jan and F. Khan (Eds.): BigIoT-EDU 2021, LNICST 392, pp. 3–12, 2021.
https://doi.org/10.1007/978-3-030-87903-7_1

continuously reducing the scale of big data and fully cleaning some fixed basic attributes of data on the premise of ensuring the value of big data. This process includes many data processing methods, which can effectively realize the physical processing of big data. It can also be seen that in the physical mode, there is no problem in the big data processing test oace itself. However, in the chemical mode, there are two problems to deal with big data: one is the most important prediction; the other is fast algorithm. These two problems increase the difficulty of Oracle to a great extent, which makes Oracle extremely difficult. For example, when calculating personalized recommendation statistics, through the analysis of personalized data, we can recommend products or commodities that meet the needs of users, but it also means that half of users may not like this commodity. The chemical processing of big data can only calculate the user's liking for a certain kind of commodity, and cannot further analyze it. The appearance of this problem shows that the accuracy and correctness of the results have an essential deviation, which makes Oracle more difficult to determine [3].

2.2   The Traditional Test Platform Cannot Meet the Needs of Big Data Processing

The traditional method of software testing is to use the controller to coordinate the local, and to test the server pressure by sending the service request to the server. This method is very practical for the system with few servers, but after entering the era of big data, especially the emergence and wide application of cloud computing technology, the number of users of application server is increasing, its demand is also increasing, and the number of concurrent users of the system is constantly rising, which makes the system access increase rapidly. At this time, in order to ensure that the system can carry this huge amount of user access and run normally, it is necessary to test the server system. The server system test can start before the system goes online, and the test content can be fully tested first [4]. However, the traditional LAN testing method is difficult to meet the testing requirements of the server, resulting in a lot of software testing problems: first, the number of physical machines of the load generator is difficult to achieve dynamic expansion; second, big data drives cloud computing, now cloud computing adopts a large number of distributed clients, increasing the workload of software testing; third, due to the promotion of big data, the number of physical machines of the load generator is difficult to achieve dynamic expansion, The test of load generator state limits the test of system performance and increases the risk of test failure. Fourth, at present, the synchronization problem between the controller and load generator used in software testing in China is becoming more and more complex, which seriously affects the test effect of load generator [5].

It is defined as the sum of the effective sample values. These effective sample values are extracted in the corresponding interval of the maximum linear density. The estimated normal values are shown in Eq. 1:

$$R_0 = \frac{1}{m} \left(\sum_{i=i_0}^{i_0+v} \sum_i rs \right) \tag{1}$$

Because of the equal interval sampling, the sampling time is equal, and the change trend can be determined by the increment of R. Therefore, there are:

$$K_{ij-ij+1} \times K_{0j-j+1} \geq 0 \tag{2}$$

It can be transformed into the following formula, as shown in Eq. 3:

$$(R_{ij} - R_{ij+1}) \times (R_{0j} - R_{0j+1}) \geq 0 \tag{3}$$

When one of them is 0, the further determination depends on the relative increment of amplitude [6].

3 Development of Software Testing in the Era of Big Data

3.1 Adjust and Optimize Oracle Memory Area

Oracle's memory area can be divided into SGA and PGA. SGA provides buffer for Oracle database, and can realize resource sharing and data log buffering. Whether the allocation of SGA regions is reasonable directly affects the performance of database system, which is very important for the performance of database system. The database buffer can store the searched data. If the data request sent by the database user enters the data buffer, the database will directly return the received data to the user, so as to reduce the user's retrieval time [7]. If the data request sent by the database user does not enter the data buffer, it needs to use a special server to read from the data file, then convert it to the data buffer, and feed it back to the user through the data buffer, which obviously prolongs the data retrieval time. In order to ensure that users can quickly receive the data they need, it is necessary to improve the performance of the database system. Resource sharing includes two parts: database buffer and data dictionary buffer. Database buffer is used to store executed code or execution plan information, and data dictionary buffer is used to store data objects and database user permissions of database related systems. Reasonable allocation of space shared by data in these two parts can effectively improve program execution efficiency [8]. The data log buffer is used to store the modification information of the database system. If there are many log write failures in the data log buffer, it indicates that the capacity of the data log buffer is insufficient, which affects the storage of the data log and ultimately affects the formation of the database. Therefore, we must constantly adjust and optimize [9].

3.2 Regularly Defragment the Database

In the actual operation, the operation of the database is uninterrupted, and the operation of the relevant data is changing all the time, resulting in disk fragments in the database. Disk fragmentation is divided into three levels: table space level, index level and table level. The steps to clean up the disk fragmentation of table space level are as follows: first, export the data in the database by using table space reorganization and command operation, then delete the data in the table space by using runcate, and finally import the valid data by using mport program import mode, so as to clean up the disk fragmentation of table space level. There are two ways to clean up the disk fragments at the index level: the first is to minimize the related index data at the table space level; the second is to re create the index by transforming the relatively low frequency columns, so as to clean up the disk fragments in the index. The method of cleaning up disk fragments at the table level is simpler than that at the table space level and index level. It only needs to reasonably configure and set the size of the system data block, and at the same time, it can effectively clean up the disk fragments by applying the relevant pcefree data parameters (see Fig. 1).

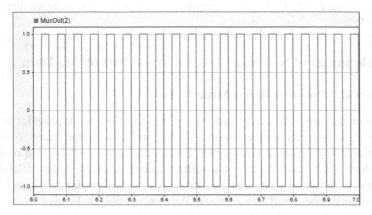

Fig. 1. Simulation for Regularly defragment the database

3.3 Improve the Accuracy of Software Test Data

The main reason for the poor effect of software testing is that the fuzzy understanding of data storage and the sending of useless duplicate requests cause the appearance of error information, which makes the accuracy of data information decline. Therefore, if we want to apply the database, we should first test the software used in the database several times, ensure the quality of software testing and the accuracy of data, and try to reduce the impact caused by data errors. In addition, relevant researchers should deeply study software testing technology. In the past, most software testers used a single software testing technology to test the software performance, and the test results were limited. Nowadays, software products have been widely used in people's production and life [10]. If the traditional single software testing technology or method is still used, the result will be too one-sided to ensure the safety and reliability of software operation. Therefore, software testing staff should further study software testing technology, flexibly use various testing technologies and methods, and appropriately use intelligent data processing technology, so as to continuously improve and perfect software testing system, gradually eliminate one sidedness of software testing, and apply diversified testing technologies in all aspects to ensure the accuracy of software testing data and information, So as to improve the efficiency of software testing and effectively avoid the "pesticide" phenomenon [11].

The software meets the requirements of relevant platforms; the software functions meet the requirements of product design and operation. Based on this, in computer software testing, we should adhere to the principle of scientific and practical. In terms of scientificity, testers should realize that there are differences in software functions and development platforms [12]. To test different software, they need to adopt more appropriate methods according to the specific situation of the software. For example, in order to detect the software running environment and structure, white box testing method can be used to quickly locate the problems in the software structure (see Fig. 2).

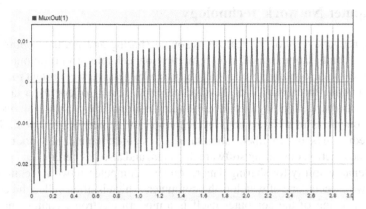

Fig. 2. Simulation for Improve the accuracy of software test data

4 Artificial Intelligence

Artificial intelligence technology is the product of the development of science and technology in our country. The main purpose of creation is to replace human by computer, use computer program to imitate human thinking, so as to replace human, reduce the use of human resources, and improve work efficiency and accuracy. Artificial intelligence is mainly the use of technical means to make electronic devices have the basic skills and characteristics of human beings [13]. In practical application, it is generally used to improve the efficiency and quality of work, or some work that can not be completed by human beings. This technology has appeared, which has brought a good role in promoting the social development of our country. Taking some large-scale production work as an example, the traditional production work requires a lot of staff to operate the machinery, overhaul the equipment, and check the qualified products. Artificial intelligence technology can replace these artificial, realize the mechanical automation in the real sense, improve the work efficiency to a great extent, and reduce the allocation of human resources, Further reduce the cost of human resources. With the passage of time, China's artificial intelligence technology has been improved and improved, it can not only replace human to complete the task, after the program system is determined, it can also control the machinery to carry out production work independently, and operate according to the design instructions [14].

The characteristic of artificial intelligence technology is that it has very high information processing ability. After getting the information, it can analyze the information content. Artificial intelligence can choose the working mode independently through information instructions, so as to carry out automatic operation. In the face of some wrong information, artificial intelligence can judge the information, prevent the impact on the work due to information problems, and send feedback information to the operator to determine the authenticity of the instructions, so as to further ensure the quality and efficiency of work.

5 Computer Network Technology

Computer network technology realizes resource sharing and mutual communication [15]. It is the combination of modern communication technology and computer technology. It has certain information processing ability and plays a very important role in social economy. In order to encourage the development and training of computer talents, now computer network technology has corresponding majors. The computer network itself connects computers in different regions according to the network protocol. The connecting media can be optical fiber, cable, satellite, etc. the computer network technology itself has the function of sharing software and data, and also has certain data processing and maintenance ability for sharing data resources. Computer network technology first includes two aspects, one of which is the computer, which is often called the computer. The system setting of the computer itself is a modern electronic equipment that can automatically and efficiently process data according to the program operation. Now it has been widely used in people's daily life. The computer itself is composed of software and hardware, The common computer styles in life are notebook, desktop and so on. For network technology, it is through physical connection to connect multiple computers to form a data line to form a LAN, so as to achieve the purpose of resource sharing and communication between computers.

6 Advantages of Artificial Intelligence in Computer Network Technology in the Era of Big Data

Now our country's science and technology level has developed mature, each technology has also made good achievements, the development of artificial intelligence technology has also reached a certain level, if the computer network technology and artificial intelligence are combined, we can design the robot system, and realize "artificial intelligence" in the real sense. At the present stage, artificial intelligence technology in China can also be called mechanical intelligence technology, which is more applied in industrial production. Through intelligent technology to control machinery, reduce labor force, make mechanical equipment intelligent, and replace employees to do some dangerous and complex work, it can not only effectively ensure the safety of employees, It can also replace human resources and reduce production expenditure. At the same time, in the era of big data, artificial intelligence can also improve work efficiency. Artificial intelligence technology can instruct machines and robots to ensure that robots and equipment can work according to the wishes of employees. In the process of development, artificial intelligence technology is a combination of various technologies. Artificial intelligence technology involves many aspects of knowledge, The same is true of computer network technology. In the process of combining the two, the effect of the two promotes the development of computer technology in China to a certain extent. Computer network technology with the help of artificial intelligence technology, the work content that computer network technology needs to be responsible for will become complex, at the same time, the processing ability of computer network technology will also become faster and stronger, on the contrary, the improvement of computer network technology will also help the application of artificial intelligence. First of all, after the combination of AI technology with computer network technology, AI technology will have better analysis ability, especially for those uncertain data, and the analysis process will be more stable. Secondly, AI technology has characteristics, and its integration with other devices can play a better work effect. AI technology also has excellent learning ability, Can keep up with the development of the times, conform to the trend of the times, in dealing with nonlinear ability also has a good effect. The advantages of network technology in the era of big data are shown in Fig. 3.

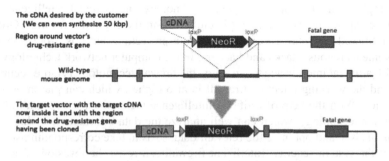

Fig. 3. Advantages of network technology in the era of big data

7 The Specific Application of Artificial Intelligence in Computer Network Technology in the Era of Big Data

Under the background of big data era, the development direction of computer network technology is constantly changing. It is in the position that computer network technology can better serve the society and the masses. Integrating artificial intelligence technology into computer network technology can achieve better application effect. Computer network technology in the process of integration can use its own expert knowledge base to build a more perfect management system, through the management system to achieve the role of management computer. Through the artificial intelligence technology, we can combine the two. When the computer network technology encounters problems, we can let the expert system automatically analyze and solve problems, so as to realize the role of automatic protection of computer network security. In this process, the computer internal system is mainly used for network management, and the corresponding evaluation system can be set to rate the management work, so as to continuously improve the application effect of computer network technology and help people solve the corresponding problems. After learning the relevant management information, artificial intelligence will save, describe and use the learning results, and set different technical methods according to different learning contents. When external factors invade the host, artificial intelligence technology can take the initiative to detect and scan the external factors, protect the system security and improve the work efficiency.

The so-called data mining work refers to a series of data information sharing and dialogue on the network, which can be analyzed by artificial intelligence technology. Artificial intelligence technology can make good use of its own technology to set corresponding standards for data application, and store the standards in its own database at the same time. When the computer system is invaded, artificial intelligence technology can extract the data information in the database, and compare the intrusion information with the stored information, so as to achieve the effect of efficient processing of information. The working principle of data mining is similar to this, its main working principle is to find the characteristics of information, so as to realize the monitoring of information, and then effectively learn the information data, and retain the learning results, so as to provide reference for data analysis.

The application of computer network technology and artificial intelligence technology in computer network security is embodied in the computer firewall. Ordinary computer firewall can play a very good role in protection, but this protection is not against some malicious attacks and hackers. In the computer network technology combined with artificial intelligence technology, the firewall technology is more secure than before, and the working effect of firewall is also higher, which can play a good role in protection. With the help of artificial intelligence technology, the firewall also has identification technology, which can well analyze the data information in the external intrusion, analyze and optimize the relevant data, so as to take corresponding strategies to protect the system security. Intelligent firewall can resist the invasion of advanced virus, ensure the security of computer operation, protect the file information and data information inside the computer system, prevent the problem of information loss caused by virus invasion, and reduce the economic loss caused by information loss. The intelligent firewall has its own intrusion detection system. When it is attacked by the outside world, it can form a report of the intrusion information, so that users can accurately understand the security level of the computer. Artificial intelligence will also specify a professional database according to the intrusion information to record the information data.

In terms of computer network security, the management work is mainly carried out from the internal and external aspects. In the internal aspect, it mainly controls the operation form of computer equipment, enhances the safety awareness and network skills of operators, ensures that people can set network specifications in the process of practical operation, and realizes the issuance of equipment instructions, so as to improve the standardization of system operation. Especially in the current era of big data, users must have a solid theoretical foundation, It has the ability of independent risk identification, and further restricts and controls its own operation behavior. In the external aspect, it is mainly to set the security protection for the operating environment of the computer equipment, such as regularly cleaning the working environment of the computer equipment, line maintenance treatment, moisture-proof treatment, etc., so as to ensure the continuity and integrity of the equipment in the process of operation. At the same time, in order to prevent the data loss problem caused by the device in an emergency, the data information in the physical server of the computer device should be synchronized in the cloud environment to establish a virtual data repository. Through the real-time synchronous data storage, a data backup mechanism should be established, so as to effectively avoid the data damage problem during the operation of the computer device.

As a kind of information encryption technology, digital signature itself presents the characteristics of data docking, which can further prevent the problem of data mistransmission in the process of system processing. Generally speaking, the information transmission mechanism with digital signature technology as the main body can be divided into line encryption and endpoint encryption. Through the transmission of different channels and information sources, the system has the point-to-point transmission attribute of information. In the process of data encryption, the attribute of the file can be further confirmed to ensure that the ciphertext can be transmitted in the form of encryption transformation and encryption, It is based on the user's legal mechanism to define the user's current information instruction behavior and ensure the user's current data reading behavior through permission audit, so as to avoid the phenomenon of ultra vires, match the digital information with identity authentication and enhance the transmission quality of data information.

8 Conclusions

In a word, the coming of big data era brings many challenges to software testing, which affects the development of software testing technology to a certain extent. In order to better solve the problems of software testing in the era of big data, as software testing staff, on the one hand, they should constantly adjust and optimize Oracle memory area; on the other hand, they should regularly clean up database fragments; in addition, they should improve the accuracy of software testing data to ensure the quality of software testing, ensure the safe operation of software, and promote the further development of software testing technology, Promote the rapid progress of China's software industry.

References

1. Li, N., Zhuang, L., Shi, L., et al.: Challenges faced by software testing in the era of big data cloud computing. Educ. Teach. Forum **51**, 275–276 (2017)
2. Wang, J.: Research on software testing technology under the background of big data. Inf. Comp. **7**, 26–27, 30 (2018)
3. Wu, Z.: Challenges and prospects of software testing in the context of big data. Jiangsu Sci. Technol. Inf. **35**(19), 69–71 (2018)
4. Zhang, Z.: Challenges and prospects of software testing in the context of big data. Electron. Technol. Software Eng. **09**(06), 61 (2016)
5. Guan, Y.: Application of artificial intelligence in computer network technology in the era of big data. China Sci. Technol. Inf. **20**, 5354 (2020)
6. Ren, P.: Application of artificial intelligence in computer network technology in the era of big data. Comput. Prod. Circ. **11**, 23 (2020)
7. Bao, Y.: Application analysis of artificial intelligence in computer network technology based on big data era. Bonding **43**(09), 98–101 (2020)
8. Li, Z.: Research on the application of artificial intelligence in computer network technology in the era of big data. Netw. Sec. Technol. Appl. **09**, 103–104 (2020)

9. Xu, S., Sun, Y., Liu, J., Gao, Y.: Research on the application of artificial intelligence in computer network technology in the era of big data. Netw. Sec. Technol. Appl. **08**, 112–113 (2020)
10. Yang, B.: Application of artificial intelligence in computer network technology based on big data era. Commun. World **27**(07), 213–214 (2020)
11. Dong, Z.: Discussion on the application of artificial intelligence in computer network technology in the era of big data. Comput. Knowl. Technol. **16**(21), 169-170+174 (2020)
12. Dai, C.: Computer network information security and protection measures under the background of big data. Netw. Sec. Technol. Appl. **09**, 4–5 (2020)
13. Sun, K.: Problems and countermeasures of computer network security in the era of big data. Public Stand. **17**, 145–146 (2020)
14. Li, C.: Discussion on computer network security, hidden dangers and management strategies. Digital World **08**, 253–254 (2020)
15. Li, H.: Research Review on the Development of China's Short Video Industry Under MCN Mode. Harbin Normal University (2019)

A Study on the Construction of Behavioral Medicine ESP Corpus Resources in the Context of Big Data

Lei Wang, Guoting Yuan, and He Ren[✉]

School of Foreign Languages, Jining Medical University, Jining 276826, China

Abstract. ESP (English for specific purposes) refers to the English based on the theory of Applied Linguistics, which is related to a specific occupation, subject or purpose. It can also refer to the English courses offered according to the learners' specific purposes and special needs. In teaching, esp mode attaches importance to the cultivation of practical language skills, which can solve the professional problems in the future. It is pointed out that "modern vocational education system construction and international vocational education system construction in 2014" should play an important role, In recent years, higher vocational colleges have made some achievements in ESP teaching mode in response to the call of national vocational education, and also carried out some effective reforms. However, due to the existing objective conditions, ESP English teaching mode has not yet developed mature.

Keywords: ESP · Teaching · Teacher · Student · Textbook · Strategy

1 Introduction

Talent training must be carried out around the future career needs to meet the actual needs of students' knowledge and skills for their jobs after graduation. Higher Vocational English education also needs to implement teaching around students' majors, serve for professional teaching, and improve students' core literacy and professional knowledge application skills [1]. Although higher vocational English course is a humanities course and basic course, English is also a language communication tool, which needs to meet the foreign communication needs of future jobs. Therefore, English Teaching in higher vocational colleges should not only follow the law of language teaching, but also follow the development trend of professional requirements of the society, reflecting the real demand of employment for English. In recent years, China's vocational education has made great achievements, but the effect of English teaching reform in higher vocational colleges is not significant. At present, the teaching mode still follows the EGP mode, and there is a lack of high-quality unified English teaching materials suitable for higher vocational education. Even some colleges are still using their own teaching materials, and the teaching methods and means are mostly cramming teaching, emphasizing vocabulary and grammar learning, Not paying attention to the combination of post needs to

M. A. Jan and F. Khan (Eds.): BigIoT-EDU 2021, LNICST 392, pp. 13–18, 2021.
https://doi.org/10.1007/978-3-030-87903-7_2

cultivate students' practical application ability, resulting in the students can not skillfully use the English knowledge to solve the professional problems encountered in the post after graduation. Through a careful analysis of senior high school English and higher vocational English teaching, senior high school students have a good English foundation through the middle school English learning, which even exceeds the basic requirements. As a result, there are more repetitions in Higher Vocational English knowledge compared with middle school, students are not fresh, learning interest and enthusiasm are gradually lost, which directly affects the effect of classroom teaching.

2 Investigation on the Current Situation of ESP Teaching in Medical Vocational Colleges

ESP (English for specific purposes) focuses on the practical use of language. It combines language training with the future job needs and the development trend of social development for post ability requirements, so as to cultivate students' ability to solve practical problems in future jobs. In recent years, Heze Medical College has responded to the call of the state and Shandong Province to develop the internationalization of vocational education, actively selected teachers to visit or study abroad, and invited foreign experts to teach regularly [2]. We will cooperate with relevant institutions to provide financial support, select outstanding students to work abroad, and jointly organize Sino foreign cooperative education projects with British universities. The ultimate goal is to absorb foreign high-quality teaching resources, promote the connotation construction of schools, comprehensively improve the teaching level of vocational education, realize all-round education, and provide excellent professional talents for the national economic and social development. In the face of the promotion of international communication, the school proposed to carry out medical ESP teaching measures, in order to adapt to the actual needs of the school's foreign exchange and students' overseas employment practice for English language. Up to now, although we have made some achievements in teaching, we are facing "bottleneck period" and many problems. In order to further promote the medical ESP teaching in Heze Medical College, this study carried out a questionnaire survey among students and interviews with relevant teaching administrators, English teachers and professional teachers, hoping to find the deep-seated problems that plagued the ESP teaching in Heze Medical College, sort out and summarize them, and explore the strategies to solve the problems.

3 Analysis of ESP Teaching Problems and Causes in Medical Vocational Colleges

3.1 The Curriculum is not Scientific Enough

Medical English is a link between medicine and English. The setting and teaching arrangement of medical English course are decided by each department independently, and the teaching of medical subject has its unique rules. However, the current medical English teaching arrangement can not be combined with the overall medical curriculum arrangement to play the role of a link. Moreover, the focus of medical English is different with different majors. Nursing majors emphasize communication, while medical device majors emphasize reading. The teaching of medical English should be based on needs analysis, and the teaching progress and content should be adjusted [3].

There is a lack of in-depth discussion on the teaching objectives, curriculum setting, teacher training, teaching assessment and other issues of ESP course, and there is no scientific and unified curriculum planning and evaluation method, so it is difficult to effectively ensure the teaching quality and effect. Different majors have different demands. As a discipline closely practicing with medicine, each medical course has its common parts. On the whole, the teaching contents of the common parts of medicine can be arranged in a unified way, and each department will focus on the medical English education of each specialty according to the needs. At the same time, with the rapid development of the world's medical, medical knowledge should be updated at any time, can not pursue "unification", schools should establish a standardized assessment system, so that the teaching objectives can be implemented.

3.2 Attribution of ESP Teaching Problems

Due to historical reasons and the lagging development of Higher Vocational English education, there is a lack of in-depth research on ESP teaching theory and practice in China, and ESP teaching in domestic higher vocational colleges is still in its infancy. "Basic requirements" only briefly mentions that only after the completion of basic English can specialized English courses be carried out. However, there is no specific definition of the essence of ESP teaching or whether it is a foreign language course. ESP teaching is more than 10 years later than that of European and American countries, and its theoretical development is not perfect, and it did not begin to develop until this century. ESP teaching is not guided by mature theory and lacks empirical research. However, higher vocational colleges can try to carry out teaching. The reform and opening up has promoted the rapid development of China's economy, and the development of China's vocational education lags behind the economic development. There is no real requirement for the necessary quality of technical personnel. Until recent years, with the promotion of the "belt and road" strategy, the country has gradually liberalized the access of foreign-funded medical institutions, and the demand for medical talents with professional background and solid English skills has gradually increased. Based on the above historical reasons, the orientation of medical ESP teaching is not clear, and its teaching objectives are not clear. Many people classify it as English class.

It is determined by many unknown influencing factors, and this part of the influence is considered as a small influence, which is called random error, and is recorded as μ. Therefore, the linear regression equation is as follows:

$$Y = f(x_1, x_2, \ldots x_{p-1}) + \mu \tag{1}$$

The standard deviation between the output value of each input sample and the predicted value of the output should not exceed ε, and the regression function should be as smooth as possible. The above problems can be described by the following formula:

$$\min_{w,b,e} J(w, e) = \frac{1}{2}\omega^T\omega + \frac{1}{2}\gamma \sum_{k=1}^{N} e_k^2 \tag{2}$$

It is impossible to guarantee the appropriateness of the selected teaching materials. All the teachers who participated in the interview said that most of the medical English textbooks on the market are original books and textbooks compiled by the school itself, and there is a serious shortage of high-quality textbooks. The arrangement of teaching materials is too poor, either too difficult or boring, which makes it difficult for teachers to organize teaching. The self compiled teaching materials are lack of clear teaching objectives, the content is difficult to form a system, and can not find a complete exercise, and need to spend a lot of time and energy.

4 Strategies of ESP Teaching Reform in Medical Vocational Colleges

The essence of higher vocational education is to enable learners to obtain a certain professional practical ability and meet the needs of their future jobs. It will be an inevitable trend to carry out ESP teaching in Higher Vocational Education in the future. Due to the late development and immature development of ESP teaching in China, it is necessary to conduct in-depth research and discussion in theory and other aspects, and carry out empirical research. The relevant education departments of the government need to formulate policies and provide financial support at the macro level. Higher vocational colleges are faced with the problems of changing concepts, organizing teaching and building teaching staff. Teachers also need to change the existing backward teaching concepts and update teaching methods and means. Therefore, to carry out EP teaching is a systematic education project full of difficulties, which is related to the government, higher vocational colleges, teachers and students. ESP teaching in foreign countries developed earlier and has rich experience. We should learn from each other to promote the development of ESP teaching in Higher Vocational Colleges in China. The change of concept is the premise. Only by improving the understanding and deeply realizing that English is a tool for communication and problem solving, can we attach importance to ESP teaching and promote its implementation [4].

4.1 Overall Promotion and Local Implementation

The change of concept is the premise. Only by raising awareness can we really pay attention to ESP and face up to the current situation. ESP teaching involves all levels, all levels need to strengthen the understanding of the importance of ESP teaching. At the beginning of 2019, the State Council promulgated the implementation plan of international vocational reform, and the golden period of the development of vocational education has come. As a special educational value, ESP teaching in higher vocational education will be the inevitable trend of the future development of vocational education, which is determined by the essence of higher vocational education. Teachers and students need to have a clear understanding of the orientation of ESP teaching and the importance of future career for students. Foreign language is not only a communication tool, but also a tool to solve the problems in the future vocational field. It emphasizes practicality and conforms to the talent training goal of higher vocational education. At the same time of strengthening management and raising awareness, teachers should face the problems and difficulties in current ESP teaching, actively deepen the theoretical and empirical research of existing ESP teaching, promote practice with theory, and improve ESP teaching effect. The teaching effect and understanding promote each other. Good teaching effect can promote people's understanding of ESP teaching.

4.2 Training Design

The Department in charge of teaching should, according to the school's characteristics and social needs, select teachers with high quality and good foreign language skills for further study and training outside the school. At present, ESP teaching in some national key vocational colleges or high-quality vocational colleges has been carried out earlier, and students' satisfaction is high. The school can send teachers to the above-mentioned colleges for long-term and short-term study according to its own professional construction, and expatriate teachers can participate in the course teaching, so as to strengthen the cultivation of professional foreign language ability, In depth study of the basic principles of foreign language teaching and teaching methods. Learn from others, learn from each other, improve their own professional composition and level, improve their own teaching level and quality. According to the actual situation of the college, the school should formulate the ESP teacher qualification standards and improve the teacher training system and methods. According to the characteristics of the school, the relevant departments of the school should focus on the professional needs, closely follow the pace of the development of the times, formulate the teacher promotion plan, give policy and financial support, and implement a variety of training based on the post needs. They should have foresight for the future professional construction, and the formulation of the content should also be based on the demand analysis. Only by meeting the learning needs and requirements of ESP teachers, can they mobilize the initiative of participating teachers and achieve the expected goals.

4.3 Training Classification

In order to promote the professional development of ESP teachers and improve the quality and level of teaching, education management departments and schools can provide support in terms of policies and funds. According to the classification of professional courses, relevant education management departments can encourage colleges and universities with better ESP teaching or relevant academic associations to carry out ESP teacher training from time to time, and invite well-known experts to give face-to-face lectures. For colleges lacking ESP teachers, the school selects teachers with relatively high English level who are willing to engage in ESP teaching to conduct research and training in high-level colleges; for professional course teachers, considering the particularity of language teaching, they should pay attention to the training of linguistic principles and various knowledge of English teaching. Generally speaking, European and American countries carry out ESP teaching earlier, and the best training time is about three months. Teachers can make full use of the language environment to better integrate the curriculum teaching, exchange experience with foreign teachers, pay attention to induction and summary, learn from each other's strong points and retrain other teachers in the school, It is conducive to the overall development of teachers. Under the background of highly developed Internet and increasingly perfect education informatization, we should also make full use of online training, provide various forms of personalized training for teachers' different needs, and carry out offline experience sharing and case study, so as to make the training system full of sense of hierarchy, which is conducive to the overall improvement of training quality.

5 Conclusion

Combined with the three elements of teachers and students' teaching materials, this paper summarizes the problems existing in the teaching of ESP in medical vocational colleges, and points out that the teaching materials, teachers, teaching mode and other aspects need to be improved. The main problems are the lack of high-quality teaching materials, the lack of qualified ESP teachers, the unreasonable and scientific curriculum, the weak English foundation and low learning enthusiasm of higher vocational students, the rigid and conservative teaching methods and means, and the imperfect and standardized assessment system. In order to promote the effective implementation of medical ESP teaching, this study attempts to put forward the following methods: overall promotion, implementation of ESP teaching involves the government, higher vocational colleges, teachers and students and other levels, all levels need to strengthen the understanding of the importance of ESP teaching, fully realize the significance of higher vocational ESP teaching for the realization of higher vocational talent training objectives. After understanding, all levels should actively create a favorable environment for ESP teaching. The government and higher vocational colleges provide policy and financial support. Facing the current problems and difficulties in ESP teaching, teachers actively deepen the theoretical and empirical research of existing ESP teaching, promote practice with theory, and improve ESP teaching effect.

As an effective carrier of classroom teaching content, teaching materials are the three components of language teaching. Higher vocational colleges should make a scientific analysis based on the needs, formulate and design the corresponding syllabus and courses, scientifically set up the ESP courses in higher vocational colleges, apply the appropriate teaching methods and means, and carry out situational teaching to meet the needs of students. As an effective carrier of classroom teaching content, textbook is one of the three components of language teaching. According to the steps of choosing first and then reasonable, we should reasonably choose high-quality teaching materials with "authenticity, effectiveness, interest and scientificity".

References

1. Bai, M.: Research on ESP Curriculum Reform in Higher Vocational Colleges Based on Demand Analysis. Minnan Normal University (2016)
2. Zhang, X.: Application of ESP syllabus design procedure and teaching method. North China Coal Med. J. **02**, 271 (2007)
3. Liu, F.G.: On the attributes of ESP and corresponding teaching methods. Primary Foreign Lang. Foreign Lang. Teach. **12**, 25 (2001)
4. Gao, J.: Construction of ESP textbooks in colleges and universities. Foreign Lang. Circles **06**, 84 (2000)

A Study on the Measurement of POI Data in Shandong Urban Governance Under Big Data

Changjuan Shi[✉]

Shandong Xiehe University, Jinan, China

Abstract. This paper takes the public participation in the field of urban planning in Shandong Province as the research object, introduces the meta governance theory and takes it as the theoretical basis. Through the case study of public participation in urban planning, this paper analyzes the achievements and problems of public participation in urban planning in China and the reasons for these problems; Based on the perspective of meta governance theory, this paper puts forward the basic conception of public participation in urban planning, defines the roles of the public, the third-party organizations, especially the government in the process of urban planning preparation and implementation; optimizes the workflow of urban planning implementation stage, and invites the public to participate in the urban planning work in advance, In order to achieve the purpose of early and continuous participation in urban planning; at the same time, put forward suggestions on improving public participation in urban planning and other related systems, through the improvement of relevant systems to ensure that public participation in urban and rural planning can be carried out orderly and efficiently.

Keywords: Meta governance · Public participation · Urban planning

1 Meta Governance Concept

Since the beginning of the 21st century, "governance theory" and "new public service theory" have become more characteristic of "public participation" after "new public management". They all emphasize that public managers should be committed to serving the public and devolving power to citizens when they manage public organizations and implement public policies, It also emphasizes that the government should take citizens as the service object, take the civil rights as the core, and strive to realize the public interests through more democratic participation [1]. However, the "new public service theory" and "governance theory" overemphasize the power of the public. If there is no certain foundation of civil society and perfect legal protection system, it may lead to the rise of extreme populist forces. Relevant scholars have seen that there are certain defects in "governance" and the governance process itself, so many scholars have made new exploration.

The broad sense of urban governance mainly involves the problems of urban positioning, urban planning, urban sustainable development, and mainly dealing with various

M. A. Jan and F. Khan (Eds.): BigIoT-EDU 2021, LNICST 392, pp. 19–24, 2021.
https://doi.org/10.1007/978-3-030-87903-7_3

elements of urban development; The narrow sense of urban governance mainly involves the organizational form, conflict of interest and integration of interests of the main body of governance, focusing on the provision of urban public services. The broad sense of urban governance refers to the social process that the decision-making of cities and cities can be made and implemented. It is marked that the traditional urban management mode led by the government is replaced by the "urban enterprise" governance mode led by the cooperation of government and enterprises. This is because the traditional policy mode under the guidance of monopoly supply of government can not effectively meet the needs of regional development, so industrial and commercial capital and social funds should be introduced, and then the market pressure in global market competition forces municipal authorities to abandon some autonomy and play an enterprise role. The governance mode of urban enterprise introduces the market spirit and the means of enterprise management into the process of urban governance. Local government integrates the marketing strategies and marketing methods of enterprises such as market mechanism, competition, innovation, public-private partnership and risk bearing into urban environment development, and forms a partnership mechanism of growth alliance or alliance between government and enterprises to strongly promote local development. At the regional level, more and more cross regional affairs that cross regional boundaries and beyond the authority of a single government force the local government to change its governance model, forming a "multi center, one and multi-level" regional governance pattern.

The city has a high complexity, so it is an extremely difficult task to effectively manage a big city. Therefore, setting an effective governance framework in advance will help to maintain the normal operation of the city and improve the management efficiency of the city.

In the era of globalization, the competition between cities is becoming more and more intense, which directly promotes the rise of urban governance research. Because of the different time and space and background, the urban governance model has the form diversity. Different cities and different urban departments in the same country background show different governance modes, and different governance modes have different theoretical basis. There is no fixed pattern of urban governance, and different countries, different regions, even the same city in the same country can show their own characteristics at different times. More precisely, each established urban governance model will bring some new problems, and these problems will lead to a new reform plan, which will help scholars interested in the process of change to study. At present, the practice of urban governance mode selection is undergoing a wide range of changes, its breadth and depth are unprecedented. This change is the process of reviewing and reflecting on the practice of the choice of early governance model, and also the process of adapting to the new era and environment. In a broad sense, urban governance includes two aspects: external governance and internal governance.

The external governance of a city mainly investigates the relationship between the city and the central government and the surrounding cities. As for the internal governance of a city, from the perspective of the main body, it refers to the division and interaction of the interest boundaries between the stakeholders in the city; From the specific content, including social, economic, environmental, emergency management and other aspects.

Among them, the urban economic operation governance is carried out by the city government according to the economic control objectives of the central government, aiming at the problems in the urban economic operation, including four main tasks: determining the appropriate economic development speed, forming a reasonable economic structure and spatial layout, maintaining the normal market economic order and tamping the city finance. Urban social governance is to reduce the social problems of cities by adjusting and improving the relationship between urban distribution, rectifying and maintaining the public security, and reconstructing and innovating urban community organizations. The urban environmental governance mainly includes the treatment of urban pollution, urban public transport and other hard environment and the construction of a good soft environment for urban development. Urban emergency management is to take timely and effective measures to prevent the occurrence of crisis or reduce the damage degree of the crisis and protect the public interest of the city, mainly including four stages: disaster reduction, disaster preparation, response and recovery.

2   Quantitative Identification of Urban Functional Areas Based on POI Data

For each functional area unit, frequency density (FD) and category ratio (CR) are constructed to identify functional properties:

$$F_i = \frac{n_i}{N_i}(i = 1, 2, \ldots 6) \tag{1}$$

$$C_i = \frac{F_i}{\sum_{i=1}^{6} F_i} \times 100\%(i = 1, 2, \ldots 6) \tag{2}$$

Where, denotes the type of Po i; n_i represents the number of type i POI in the unit; n represents the total number of type i POI; F_i represents the frequency density of the type i POI in the total number of this type of POI; C_i; represents the ratio of the frequency density of the type i POI to the frequency density of all types of POI in the unit.

According to the formula, the frequency density and type proportion of each unit are calculated, and the type proportion value of 50% is determined as the standard to judge the functional properties of the unit [2]. When a certain type of POI accounts for 50% or more in a unit, it is determined that the unit is a single functional area, and the nature of the functional area is determined by the type of POI; when the proportion of all types of POI in the unit does not reach 50%, it is determined that the functional area is a mixed functional area, and the mixed type depends on the three main POI types in the unit; when the POI is not included in the unit, the unit is determined as a single functional area, That is, the type proportion is null, and the type unit is called no data area.

3   Basic Conception of Public Participation in Urban Planning from the Perspective of Meta Governance

3.1   Guiding Ideology of Public Participation in Urban Planning System

First, according to the idea of "separating government from enterprise, separating government assets, separating government affairs and social affairs", we should promote the separation of public participation in various urban and rural planning plans from the work of organization, preparation and approval. In the case of public participation in urban planning in the third chapter, the public participation part of the planning and reconstruction of Enning road in Guangzhou and the overall planning of Shenzhen city are organized or participated by the third party, and have achieved certain results. In the future public participation work, we can learn from this kind of experience, under the premise of refining the relevant provisions, actively introduce the third-party institutions to carry out relevant public services through PPP mode, and weaken the government's "interest subject" image in public participation.

Second, public participation in urban planning should be as early as possible and sustainable. Through the analysis of the experience of foreign representative countries, it is found that the introduction of public participation at the beginning of the scheme is a very important method. In the current relevant legislation, public participation is emphasized only at the planning level of the master plan, while in the detailed planning, opinions are solicited after the scheme is determined. The lack of early participation is not conducive to the public's deep understanding of the program. Although it is a kind of right of objection to a certain extent, it is undoubtedly a disguised deprivation of rights compared with the right of participation that should be enjoyed. For the stakeholders around the specific project, due to whether the planning of the construction project is legal or not, and whether the scheme is excellent or not has a great impact on the value of their own assets [3]. A successful or unsuccessful construction project may bring sunshine, fire protection, parking and other impacts, and traffic, noise, light, smell and even the groups attracted will cause assets devaluation. The public participation should be carried out as early as possible, so that the interested parties can fully express their right of objection.

3.2 The Principle of Constructing the System of Public Participation in Urban Planning

Through the interpretation and extension of the "ladder of citizen participation", it can be found that in the embryonic period of political democratization, civil rights and consciousness began to awaken gradually, and the ability and degree of organization of citizen participation were improved. At this time, the corresponding forms of citizen participation should be based on citizens' knowledge, consultation with citizens and interactive dialogue, Government decision-making power begins to be shared, citizens gradually identify with their own citizenship, and citizen participation is gradually organized and institutionalized, which has a certain influence on policy. At this time, the degree of public participation is "moderate participation". In the stage of high participation, the government began to authorize citizens, and the community managed independently. The awareness of citizenship was mature, and the knowledge and ability of participation were greatly improved. At this time, the characteristics of public participation form are: citizens become the master of community governance, active participation policy, and citizens' opinions have substantial influence.

4 Countermeasures of Public Participation in Urban Planning from the Perspective of Meta Governance

4.1 Improve the System of Public Participation in Urban and Rural Planning

The urban and rural planning law only requires public participation at the legal level. There are no specific provisions on how to carry out public participation, when to start and end public participation, the effectiveness of public participation, and what kind of punishment will be imposed for failing to carry out public participation according to the requirements, which leaves certain discretion for the planning department in the specific operation of public participation. Therefore, in the formulation of the provisions on public participation in urban planning, we should focus on how to refine the protection of citizens' right to know, to participate and to supervise, to formulate and promulgate relevant rules to make public participation in urban planning concrete and legal. According to meta governance theory, the government should be the main body of establishing public participation system and management cooperation network, platform and channel. While participating in negotiation and dialogue, it should encourage governance forces to play games fairly, so as to ensure that the final decision-making is the common interest pursuit of all governance subjects.

4.2 Improve Urban Planning and Other Related Systems

As for the negative impact of construction projects, government departments need to establish a sound mechanism and quantify them through specific values. For public welfare projects, the government should not only bear the normal operation cost of the project, but also increase the investment in infrastructure of affected residential areas, such as maintenance fund, environmental sanitation facilities, greening and so on, to balance the adverse effects; For quasi public welfare projects or industrial projects with certain profitability, enterprises should bear this part of the cost and establish a mechanism to share benefits with the surrounding affected residents.

People who hold a rational view will not object to the construction project itself, but from the perspective of economic man, they are opposed to the construction project built around themselves, especially those who have bought or even moved into the community before the construction of unfavorable projects [4]. On the one hand, it is the planning department's insufficient publicity for the planning scheme; on the other hand, the developer is suspected of deliberately concealing it. Even if the adverse project has been identified in the upper planning, the stakeholders are not willing to accept the fact from the emotional and self-interest. For the unfavorable projects that have been planned first but the stakeholders are later, they should be publicized to the public as soon as possible, and the developers should be required to clarify the adverse factors in the sales contract, so as to avoid transferring the risk to the government; for the adverse projects that must be added in the planning revision, many site selection schemes and income sharing mechanisms should be made, and decision-making should be made in an open and transparent way.

5 Conclusion and Prospect

The meta governance theory has a more clear and appropriate positioning for the government's functions, has a strong goal for the future social development, and fully considers the stage of civil society development in the realization path, which is more consistent with the development goal of "strong government and strong society" and the current situation of social development of "strong government and weak society". From the perspective of meta governance theory, in order to realize the legal, orderly and efficient public participation in urban planning, planning departments should change their own positioning, on the one hand, they should be separated from the specific and transactional work and become the main body of balancing the interests of all parties; on the other hand, they should be the makers and organizers of public participation in urban planning.

Acknowledgements. Shandong Social Science Planning Research Project:Research on the Path and Countermeasures of public participation in Urban Governance of Shandong from the perspective of multiple Governance, Item number: 20CPYJ27.

References

1. Fu J.: Research on the Right of Public Participation in Urban Planning. Jilin. Jilin University (2013)
2. Yang, G.: Public participation in urban planning in the United States today. Foreign Urban Plan. **2**, 2–5 (2002)
3. Carol. Pettman's theory of participation and democracy, [M]. Chen Yao. Trans. Shanghai: Shanghai People's publishing house, 2006: 8–10
4. Zhang, J.: A historical outline of western urban planning thought. Nanjing. Southeast University Press, pp. 205–206 (2005)

A Study on the Sustainable Participation Behavior of MOOC Learning Educators in the Context of Big Data

Peng Nan and Liu Lu(✉)

East University of HeiLongjiang, Harbin 150060, HeiLongjiang, China

Abstract. MOOC, as a new type of online education course, has the characteristics of initiation, personalization and networking, which provides a new opportunity for educational reform and innovation in the Internet era, and also promotes the development of national learning and lifelong learning. One of the main problems of MOOC is the low passing rate of course learning. There fore, it is of great significance to explore how to promote the continuous participation of MOOC learners from the perspective of research and practice.

Keywords: MOOC · Piper · Learner · Continuous participation · Behavior · Motivation

1 Introduction

This paper aims at "MOOC learners" and "continuous participation behavior", obtains the first-hand data through literature research, personal practice, interview, questionnaire and other means, adopts grounded theory, modal analysis and other qualitative research methods, and SPSS data statistical analysis, Amos structural equation analysis and other quantitative analysis methods, aiming at the mooc learners' persistence We should continue to explore and study the participation behavior [1].

This paper mainly carried out the following work.

Firstly, on the basis of literature research, this paper summarizes the connotation and characteristics of mooc, and systematically combs the related research on learning motivation and participation behavior of mooc at home and abroad.

Second, explore the motivation evolution mechanism of mooc learners continuous participation behave or. It is found that the motivation formation of mooc learners continuous participation behavior can be divided into three stages: initial participation motivation, staying motivation and continuous participation motivation. The stability and maintenance of learner sparticipation behavior ultimately depends on the satisfaction of psychological needs and the generation of motivation for ability realization.

Thirdly, the Expectation Confirmation model is extended by introducing intrinsic motivation, perceived entertainment, self-efficacy and contributing factors. The influencing factors of MOOC learners continuous participation behavior are analyzed, and the research model is verified.

© ICST Institute for Computer Sciences, Social Informatics and Telecommunications Engineering 2021
Published by Springer Nature Switzerland AG 2021. All Rights Reserved
M. A. Jan and F. Khan (Eds.): BigIoT-EDU 2021, LNICST 392, pp. 25–30, 2021.
https://doi.org/10.1007/978-3-030-87903-7_4

Fourthly, it compares the continuous participants and non continuous participants in mooc learning from three perspectives: obstacle analysis, modality analysis and case analysis.

Fifthly, the possible paths to promote the continuous participation of learners are proposed, including MOOC based online and offline mixed learning path, peer mutual learning path and mooc for primary and secondary school teacher training.

2 Introduction

The rapid development of network technology makes the scale of Internet users expand rapidly. As of December 2016, the number of global Internet users has reached 3.42 billion, accounting for 46% of the global population. According to the 39th statistical report on the development of China Internet network issued by China Internet Network Information Center (CNNIC), by the end of December 2016, China Internet users had reached 731 million, with a penetration rate of 53.2%. Compared with the end of 2015, there were 42.99 million new Internet users, with a growth rate of 6.2%. The total number of websites in China was 4.82 million, including 2.59 million websites under the domain name One. In addition, CNNIC 39th statistical report on the development of China Internet also pointed out that the Internet is closely integrated with the traditional economy, its status in the society has been improved, and various applications have a greater impact on the lifestyle of Internet users. The theme of Internet development needs to be changed from "quantity" to "quality". The network eliminates the limitation of time and space, and constructs an extremely large online world of user gathering and mutual communication. In education, society, entertainment, management and other fields, various specific applications constantly bring forth new ones. Such as BBS forum, discussion group, Blog, micro-blog, online game community, online commentary community, online question and answer community, online gambling community, online shopping community, WeChat official account, various APP applications, etc., provide people with a new way of existence, including communication mode, learning mode, working mode, entertainment and leisure mode, and so on. And so on. However, the most important application area of the network world is learning and education 2.Since 2013, major institutions, such as Baidu education, Tao bao and Ten cent, have either built their own platforms or made strategic investments to focus on online education. Lei Jun, the founder of Xiaomi technology, has invested 1 billion yuan to build "100 education" since 2014. In July 2013, Ding Lei of Neteas invested in 91 foreign teachers. On the other hand, according to public information, the revenue of online brand chalk network for civil service examination in 2015 was 56 million yuan, and the revenue in the first two months of 2017 had exceeded 100 million yuan, which focuses on one-to-one online English learning for children aged 4–12, grew more rapidly, with little revenue in 2014 and no revenue in 2016Revenue has exceeded 1 billion yuan [2].

At the national level, China Ministry of education and Ministry of finance support the construction of higher education curriculum resource sharing platform Yi ai curriculum network, including "China University video open course", "China University resource sharing course", "online open course", "China University mooc" and "China Vocational Education mooc".Aiwang is a platform for the sustainable construction and operation of higher education resources. It provides high-quality resource sharing services for anyone at any time and place, and meets the needs of personalized teaching resources.

3 Definition of Core Concepts

3.1 Participate in

In our daily language, "participation" is a very high frequency word. The explanation of "participation" in Chinese dictionary is that "participation" also serves as participation, participation in Henan Province; anticipation and discussion; intervention and participation. The English verb corresponding to "participate" is "participate", which means: to participate in (a business or activity), often with others; to participate in something or have a share in it. The corresponding English noun of "participation" is "participation", which means the action or state of participation. The concept of "participation" can be defined and analyzed from many perspectives.

Generally speaking, as long as all the people present are allowed to "move", everyone has the opportunity to explore, cooperate, experience, express and communicate. This is participation. The concept of participation is widely used in the fields of organizational behavior, modern management and so on. It means that the individual is involved in a certain state, which is the process of the subject initiative to the activity. According to the cognitive strategies used by learners in learning, pintrich (1994) divides participation into surface participation and deep participation. Surface participation is an explicit manifestation of behavior, such as simple exercises and exercises, and cognitive strategies are at a lower level. Deep participation is manifested as creation and reflection, which is a kind of self-control behavior. Marks (2000) believes that participation is a psychological process, which refers to the degree of attention, interest, energy input and effort involved in learning. In fact, emotional participation and behavioral participation are implied in learning experience. Both cognitive participation and emotional participation are realized through behavioral participation (Fig. 1).

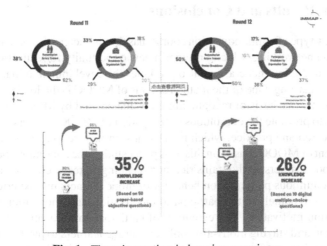

Fig. 1. There is emotion in learning experience

3.2 Types of Learner Participation

Previous studies focus on learners participation behavior in the network environment, which is often divided according to different dimensions. Armstrong and Hagel (1997) divided it into four categories according to the degree and value of participation [3]. Visitors: new learners, usually browsing at will, have the lowest value; divers: stay for a long time, but generally do not actively contribute content, and can collect browsing path, personal data and other information from them, which is of higher value than visitors; contributors: usually the most passionate, most active sharing and dedication of their own creative content, stay for a long time, which is the second most valuable Value members; buyers: actively participate, actively contribute to the content and purchase the corresponding services for the most valuable (Fig. 2).

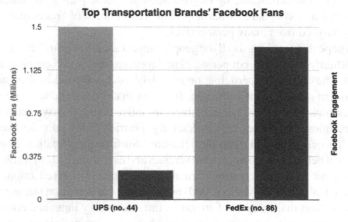

Fig. 2. Purchase the most valuable corresponding services

4   Research Results and Conclusions

MOOC, as a new type of online education course, has the characteristics of openness, personalization and networking, which provides a new opportunity for educational reform and innovation in the Internet era, and also promotes the development of national learning and lifelong learning. One of the main problems of MOOC is the low passing rate of course learning. On the basis of recognizing the value created by MOOC, we still need to explore how to promote the continuous participation of MOOC learners from the perspective of research and practice, which is of great significance to enhance the survival and development of MOOC. Based on this, this paper studies the continuous participation behavior of mooc learners, and obtains rich and meaningful research conclusions.

First, the continuous participation behavior of mooc learners has experienced three stages of evolution, namely initial participation motivation, stay motivation and continuous participation motivation. The evolution of participation motivation indicates that initial motivation and staying motivation only encourage learners to participate in learning activities, but they are not enough to maintain and explain the subsequent continuous participation behavior. The maintenance and stability of learners participation behavior ultimately depends on the satisfaction of psychological needs and the generation of motivation for ability realization.

Secondly, this study introduces intrinsic motivation, perceived entertainment, self-efficacy, contributing factors and so on.

The results show that perceived usefulness, Expectation Confirmation and satisfaction are still the core factors that affect the intention of continuous use in the classical Expectation Confirmation model, and the intention of continuous use has a significant positive impact on the continuous participation behavior; the extended Expectation Confirmation model has a significant positive impact on the continuous participation behavior Confirmatory model is effective in the prediction and explanation of mooc learners continuous participation behavior, and has a complementary role to expectation confirmatory model; intrinsic motivation and perceived entertainment factors have a significant positive impact on satisfaction; self efficacy and contributing factors have a significant positive impact on continuous participation intention, while contributing factors have no significant positive impact on continuous participation behavior; perceived ability has a significant positive impact on continuous participation behavior The perception ability is similar to self-efficacy, that is to say, they think that they can do it, and they are competent for the learning of mooc course. This inner potential mode of thinking, positive self affirmation and suggestion can significantly and positively affect the continuous participation [4].

Thirdly, for persistent participants and unsustainable participants, from the analysis of hindrance factors, modal analysis, case analysis.

This paper makes a comparison from three aspects. Whether continuous participants or non continuous participants, the biggest obstacle encountered.

5 Conclusion

Applied statistics major in Local Application-oriented Universities tends to cultivate application-oriented talents, which requires students to make full use of the theoretical methods of statistics according to the actual digital resources, carry out empirical research and give corresponding decision-making schemes or reference opinions. The "project progressive" teaching method can connect students learning process through multiple progressive projects. Students can master the relevant statistical theory knowledge by completing each course project. According to the characteristics of Local Application-oriented Universities, the practice of "project-based progressive" teaching method will provide a very meaningful guidance for the cultivation of professional talents to a certain extent.

References

1. Qing, C.: Analysis of the significance and ways to enhance customer loyalty. J. Petrochem. Manag. Cadres **4**, 23–34 (2005)
2. Chen, T.: empirical research on improving marketing efficiency by using improved RFM. In: Proceedings of the 7th Symposium on Artificial Intelligence and Application (tai2002) (2002)

3. Su, J., Zhang, J.: Research and application of RFM in customer relationship management. In:The First Symposium on Circulation and Global Operations Research of Taichung Institute of Technology, Taiwan. Taichung Institute of Technology (2003)
4. Qiu, H., Su, J.: A Fuzzy RFM Model with Flexible Support for Customer Relationship Management and Data Mining Marketing. Business press, pp. 149–173 (2004)

Analysis and Design of Human Resource Management Model of Retail Enterprises Under the Background of Big Data

Yuan Yuan[✉]

Guangdong Polytechnic of Industry & Commerce, Guangzhou 510515, Guangdong, China

Abstract. In this paper, ASP technology and SQL are used to realize the three-tier architecture of browser/Web/server for human resource management of retail enterprises in the web environment, and some key technologies in the system development are mainly studied. Through the effective combination of ASP and SQL/server, the standardization, standardization and informatization of human resource management in retail enterprises are realized.

Keywords: Human resource management in retail enterprises·ASP · SQL

1 Introduction

With the development of economy and the improvement of management level, human resource management system has become a very important part of many enterprise managements. As a computer tool of human resource management, human resource management in retail enterprises can manage almost all the information related to the most important asset in an enterprise, such as employee recruitment, post and organization setup, training, skills, salary and benefits, performance and resignation management, with a unified database, It effectively avoids the problems of information incompatibility, updating and sharing caused by the discrete storage of human resources related data, and makes the management of human resources in enterprises move towards standardization, scientization, digitalization and networking.

The traditional human resource management in retail enterprises is based on C/S (client/server) structure, but it has its shortcomings for large and complex enterprise applications: it can not be controlled centrally, its security performance is poor, the client load is heavy, its maintainability and reusability are poor. The system is based on B/S (Browser/server) structure, which can solve the above problems well, make full use of the company's network resources, give full play to the network efficiency and improve labor productivity.

2 System Analysis

A Japanese owned enterprise is a large-scale enterprise with scientific research and production as a whole, multiple varieties and complex institutions, with more than 4000

© ICST Institute for Computer Sciences, Social Informatics and Telecommunications Engineering 2021
Published by Springer Nature Switzerland AG 2021. All Rights Reserved
M. A. Jan and F. Khan (Eds.): BigIoT-EDU 2021, LNICST 392, pp. 31–37, 2021.
https://doi.org/10.1007/978-3-030-87903-7_5

employees. The traditional human resource management in retail enterprises based on client/server was originally adopted in the enterprise. With the continuous expansion of the enterprise scale and the rapid development of computer technology and network technology, it has been unable to meet the requirements of various aspects of the growing enterprise.

2.1 Functional Requirements of the System

(1) Realize dynamic management
    As an information management system, human resource management system can reflect the actual parameters of affairs and society timely and accurately only by using dynamic management technology to manage the database dynamically. Therefore, the timeliness of data is the life of information, and only the flowing and constantly collecting information source has value.
(2) Auxiliary decision function
    With the help of enterprise internal information network, a multi-level, fully functional and intelligent human resource management information and decision support system with computer and communication as the main means is established to collect, store, retrieve, process, analyze and output human resources, so as to serve the management of personnel departments at all levels and the scientific decision-making of leaders at all levels.

2.2 Data Requirements of the System

In the process of database design, the main and foreign key relationships between tables should be established with human as the object and the staff number as the clue, and other tables should be established based on the basic information of staff, along the three main lines of staff entering, changing and leaving the factory, so as to ensure the integrity, accuracy and uniqueness of the data of the whole system.

2.3 Database and Programming Selection of the System

Based on the function and data requirement of the system, we choose SQL Server 2000 which is suitable for large database development as the background database of the program, and use ASP (active server pages) to design and program the web page.

(1) ASP Technology
    ASP (active server pages) is a server-side scripting environment, which is a text file, composed of HTML identifier and active server source program. It can send client-side source programs such as VB script and JavaScript to create and run dynamic and interactive web server applications. Using ASP, you can combine HTML pages, script commands and active components to create interactive web pages and powerful web-based applications, and applications written with ASP code are easy to develop and modify.
    ASP not only depends on the request of client to produce dynamic HTML, but also can detect the ability of existing system, such as database, file detection

and COM based information server. The source code of ASP is the HTML code generated after the source code is interpreted by the server. Therefore, the source program of ASP will not be transferred to the client browser, which can avoid the source program being plagiarized by others and improve the security of the program.

(2)   SOL Server 2000

Microsoft SQL Server 2000 is a relational data management system based on client and server. It is a reliable and easy to manage database and analysis system. SQL (Structured Query Language) is a structured query language, which is used to define, add, delete, modify and manage data, and to control the database with tables, indexes, keywords, rows and columns storing data, as well as the control of database access rights. Microsoft SQL server uses sq statement to transfer request and reply between client and server, and uses client/server structure to decompose workload to execute task on server and client respectively. A client application can run on one or more clients, or on a server, providing data to users. The server is responsible for the management and allocation of server resources, and the client does not need to add the function of managing data locally; at the same time, the server does not need to spend the processing capacity on the display data, only return the data required by the application program, thus optimizing the network traffic.

3   Design and Implementation of the System

The system structure of human resource management system should not only conform to the management system of enterprise production and operation, but also conform to the characteristics of computer software itself. Therefore, it should be comprehensively considered from the following aspects:

Considering the requirements of various functional departments under the current management system of the enterprise, some functions that are closely related, have the shortest path of data collection, exchange, processing and analysis, and have relatively independent business are classified into one subsystem.

Fully consider the business division of the enterprise management department, and try to make each functional subsystem belong to the jurisdiction of one function, so as to facilitate the management and maintenance of each subsystem in the future.

It is conducive to the development, design and maintenance of subsystems, which should be relatively independent and stable.

3.1   Overall Structure of the System

This system adopts the B/S Browser/Web/server) three-tier architecture model (Fig. 1), forming the overall network structure of the system (Fig. 2).

The process of ASP running on the server is that when the browser requests the ASP file from the WCB server, the web server calls ASP, ASP reads the requested file completely, executes all script commands, and transmits the web page to the browser. Because the script runs on the server rather than the client, the web page delivered to the browser is generated on the web server; do not worry about whether the browser can handle the script, the web server has completed the processing of all scripts and transferred the standard himl to the browser. Because only the result of the script is returned to the browser, the user can not see the script command being browsed, so the server-side script command is not easy to copy and view, ensuring the security of the system.

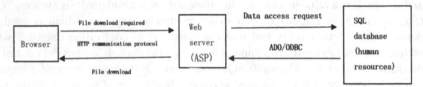

Fig. 1. B/S model

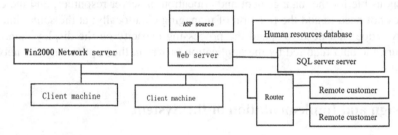

Fig. 2. Overall system model

3.2 Functional Module Structure of the System

Considering the system structure and division of business of human resource management, the human resource management system is divided into:

(1) Human planning subsystem. According to the needs of the company's business, make a period plan according to the future demand development of human resources, including human resource demand forecast, personnel recruitment plan, personnel training plan, and human resource utilization plan.
(2) Performance test subsystem. It is the basic work of enterprise management and the important basis of salary and bonus, labor insurance and welfare.
(3) Employee reward and punishment subsystem. Formulate reward and punishment standards according to the needs of the company, and display the rewards and punishments of employees in a timely manner.
(4) Personnel transfer subsystem. Transfer and promotion, appointment, dismissal and resignation of employees
(5) Salary award gold system. Set up salary and bonus standards at all levels of the company.

(6) Employee insurance subsystem. Employee endowment insurance, medical insurance, unemployment insurance, housing fund, enterprise accident compensation, etc.

(7) Personnel file subsystem. Personnel information, work performance files and training files.

4 Key Technologies of the System

ASP accesses the database through a set of object modules called ADO (ActiveX data objects). No matter what database the server uses, as long as the database has the corresponding ODBC or OLE DB driver, ADO objects can be accessed, as shown in Fig. 3.

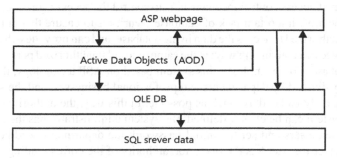

Fig. 3. The process of ASP accessing SQL Server

The specific connection commands are:

```
<%
set RS = Server . Createobiect ( " ADODB . Recodset " )
RS Activeconnection = " DSN dsname;UID = uidname;PWD = pass-
word;"
RS . SoRS Open=str
SQL
RS.Open()
%>
```

Where, dsname: data source name; uidname: user name; password: password; strsol; sol operation command.

This allows you to connect to the database and share resources in the database using the appropriate table action commands.

5 System Maintenance and Safety

5.1 User Management Mark and Identification

This system is a multi-user system, and the identification of authorized user mark is the most important part of the security control mechanism and the first link of the security defense line. The identification here refers to the user's own identification presented to the system. The easiest way is to enter userId and password. In the HUMAN RESOURCE MANAGEMENT IN RETAIL ENTERPRISES system based on ASP technology, the form form can be used to submit the user's account and password, and match with the corresponding fields in the user logo database.

5.2 Access Control

The definition of access control mechanism is to control the access right of one object to another 8. The most important task of database security is to ensure that only qualified users can be authorized to access the data in the database, and can prove/test the reliability of this guarantee in a convincing way; at the same time, all unauthorized personnel cannot open the database. For example, in this system based on ASP technology, the database can be strictly divided by departments and professional businesses, and the subsystems should be logically isolated as much as possible. In this way, the authority of different roles in the system can be clearly defined: the system administrator has the authority to add users, delete users and set the initial password; the department head controls and manages whether the user has the operation authority of the corresponding database in the system authority setting module of each subsystem.

The implementation method is to use session technology of ASP for system isolation protection and user authentication mark keeping. The process is as follows: the user's identity information in this system is composed of employee number, name, login name, password, authority type, identity description, locking and other fields. These fields are a data table in the database. Each time the user logs in, he accesses the data table through the database to identify the identity mark.

Pages in the system are attached with page types, such as <% page type = "HR"% >. When users enter the system for the first time, they are required to enter their own user account and login password. After submitting, the system will open the user database for data matching. If the data matches, the authentication tag is given to a session object, and the user needs to compare whether the authentication tag matches the preset value every time the page is converted. The procedure is as follows:

```
<%
If session ( " MM_userName " ) = " " then
Response . Redirect " Errorinfo . asp "
Elseif
trim(session("MM_userType "))< > "Administrator "then
If
instr(trim(session("MM_userType ")),pageType)=0 then
Response.Redirect "ErrorInfo.asp "
End if
End if
End if
% >
```

6 Conclusion

This paper puts forward the application of ASP technology and SQL to realize the establishment of human resource management in retail enterprises with browser/Web/server three-tier architecture under the environment of WB, and focuses on some key technologies in the development of human resource management in retail enterprises. The application of this system can grasp the situation of human resources and organization resources in time, accurately and comprehensively. It can exchange data with other business systems, realize resource sharing, and provide decision support for senior management of enterprises. The data and operation functions of the system are clear, convenient and flexible. By using this system, the human resource department can get rid of the tedious and repetitive daily work of human resource management, such as files and attendance, improve the working efficiency, and ensure the correctness and reliability.

References

1. Lloyd, B., Leslie, L.: Human Resource Management. Yekun, L., et al. Beijing, Huaxia press (2002)
2. Xu, X., Yuxin, B.: Comparison between BS mode and C/S mode. J. Yanbian Univ. **28**(2), 126–129 (2002)
3. Weng, W.: Programming Practice of Active Server Pages. Higher Education Press, Beijing (1999)
4. Yu, J., et al.: Sol Server 2000 Programming Guide. Beijing Hope Electronic Publishing House, Beijing (2001)
5. Ye, H.: ERP Integrated Resource Management of Enterprise Resource Planning Song Xianglin. Electronic Industry Press, Beijing (2002)

Analysis and Research on Application of Data Mining Algorithm Technology in E-Commerce Platform

Guihong Zhang[✉]

School of Artificial Intelligence, Leshan Normal University, Leshan 614000, Sichuan, China

Abstract. In today's era, the effective use of data technology can not only benefit the development of enterprises, but also help to promote the progress and development of society. This requires enterprise technical personnel to strengthen the understanding of data technology, in order to make better use of data technology and play the role of data mining technology. This paper expounds the data mining technology and discusses its application in e-commerce, hoping to provide reference for relevant enterprises.

Keywords: Data mining technology · E-commerce · Enterprise

1 Introduction

With the development of the times, the development of various industries are constantly combined with data technology, and has made great progress. The application of data technology usually needs the support of a large amount of data information to achieve the purpose of effective use of big data technology. In order to obtain more data information, we can use data mining technology [1–3]. Data mining technology is based on the characteristics of information in the network, to achieve the collection of similar information, and then make targeted analysis. Technicians can combine data mining technology with e-commerce to promote the development of e-commerce enterprises in China.

2 Overview of Data Mining Technology

In e-commerce, customer browsing information is automatically collected by web server and saved in access log, reference log and proxy log. The log file can be divided into service logs, errors logs and cookielog. In e-commerce, data mining technology is used to analyze user's access behavior, And use data mining ideas and related methods to process the data to get the relevant information of the user groups. According to the relevant information, we can launch targeted marketing measures, make use of the customer characteristics with group similarity, and control more targeted business activities, It will achieve better economic benefits. Through data mining technology, relevant enterprises

M. A. Jan and F. Khan (Eds.): BigIoT-EDU 2021, LNICST 392, pp. 38–42, 2021.
https://doi.org/10.1007/978-3-030-87903-7_6

can accurately locate the market, and according to the characteristics of the market, it can not only improve the market objectives, but also effectively enhance the competitiveness of enterprises in the industry.

For example, through data mining technology, we can more systematically analyze the customer group's preference for a website or a product, and we can also get the publicity channel results through the frequency of customer visits [4–8]. In this way, we can better optimize the publicity channel, and optimize the focus of publicity according to the user's transaction situation, so as to strengthen the marketing strategy and achieve better economic benefits.

3   E-Commerce Recommendation System

Sub commerce recommendation system, that is, using e-commerce website to provide customers with help to understand more content and provide customers with more choice reference, can act as a simulation salesperson to help customers make the decision whether to buy or not. In the specific application process, in the process of customers visiting the network site, the business recommender system will analyze customers' purchase behavior habits, and then give users the promotion of similar products. General users will be interested in the recommended products, and enhance their purchase intention. At present, business recommendation system mainly includes analysis recommendation, attribute recommendation and item relevance recommendation.

3.1   E-Commerce Push Technology

Recommendation technology of e-commerce system mainly refers to the operation method of data mining technology. The accuracy and efficiency of system transportation will directly affect the service quality of recommendation system. At present, in order to effectively ensure the reliable application of the recommender system and the real-time performance of the system, data mining researchers have studied the clustering, association rules, collaborative filtering and other technologies. Methods based on e-commerce calculation can be divided into two kinds of technologies, namely model-based recommendation algorithm and memory based recommendation algorithm. Model based recommendation algorithm is to build a model according to user data. In the specific application process, the model needs to be transferred into memory. Model recommendation algorithm can build models through various technologies, such as Bayesian technology, clustering technology, etc.; recommendation algorithm with memory, that is, in the process of running, the whole user's data needs to be transferred into the memory, and then personalized data recommendation is given according to the data left by the user browsing the commodity website. The computing materials of this technology can be collaborative filtering algorithm, association recommendation algorithm and so on.

3.2   Application of NEB Data Mining Technology in E-Commerce

As one of the important ways to find access path, path analysis realizes the investigation and data analysis of user access under the function of veb server. By mining user access path, it obtains user access information, so as to grasp user interests, realize targeted design, and provide better services to users. The knowledge process of browser acquisition is path pattern mining. In the process of subgrade pattern mining, there are three contents. First, browse the information of each station, and obtain the specific access path, including the previous reference and the subsequent reference. By mining the forward reference information, we can delete the worthless information in the back reference in time. Secondly, the maximum reference sequence is obtained. If the information in all forward references is greater than the threshold sequence [9, 10]. If each site is regarded as an item, when searching for the maximum reference sequence, it is necessary to find out the maximum item set according to the association rules. Both of them should meet the requirements that the disposal of the occurrence should be greater than the specific value. According to the rules, the value of a "B" with a certain confidence in transaction set D is C. if D contains a transaction and includes the percentage C of B, the specific formula is:

$$support(A''B) = P(A \cup B) \tag{1}$$

$$confidence(A''B = P(A|B)) \tag{2}$$

4 Application of Data Mining Technology in E-Commerce

4.1 E-Finance

In today's era, electronic finance is a product of very advanced development, for example, people can realize consumption, financial management and other financial realization through WeChat, Alipay and other software. Based on the records of people's consumption using electronic money, data mining technology can effectively predict people's consumption trend. Through the prediction of consumption records, relevant enterprises can provide more perfect services, stimulate people's consumption and promote the development of enterprises. In addition, as the bank credit card, insurance industry, etc., we can also use data mining technology to predict people's deposit and loan trends through the consumption situation of consumers in the network, and then optimize the deposit and loan strategy, so as to enhance the business volume of enterprises and make enterprises obtain more profits.

4.2 Customer Relationship Management

In the process of e-commerce operation, we need to keep close contact with customers. Only by strengthening the understanding of customers, can we deal with the relationship with customers, provide better services for customers and increase the business quota. Specifically, based on the e-commerce model, staff can use data mining technology to analyze the characteristics of customers using electronic software and understand customers' behavior, which is convenient for enterprises to recommend targeted products for customers, so that customers can pay more attention to the products they are interested in, so that customers can improve their consumption intention [11, 12]. The correct analysis of customers' life rules is conducive to the enterprise to deal with the relationship with customers, and ultimately promote the growth of the enterprise.

4.3 E-Commerce Marketing

Under the mode of e-commerce, enterprises generally promote their products by means of web page or software app, which not only facilitates people's life, but also becomes a generally recognized shopping channel at this stage. For better marketing and development of enterprises, when using software app channel marketing, enterprises can use data mining technology to analyze customers' purchase intention. For example, data mining technology can be used to analyze the items in customers' shopping carts, collect information about the products concerned by customers, and then recommend similar products to customers. This can not only help customers compare the same products, but also help enterprises make profits. In addition, because e-commerce enterprises will launch some activities from time to time, at this time, the marketing department of e-commerce enterprises can also timely recommend some activities to consumers, which is bound to increase the number of transactions in the process of e-commerce.

4.4 Business Data Analysis

In the process of e-commerce marketing, consumers generally focus on the marketing volume of a single commodity, and are willing to buy the products of enterprises with more sales. For example, when consumers buy a product, they will visit the online store of the seller according to the sales volume, and then decide whether to buy it or not. At this time, the responsible unit of e-commerce software needs to use data mining technology to realize the sales data analysis of similar products. Only when e-commerce enterprises provide enough good services, can they satisfy Gu Jiao, increase business volume, and finally make e-commerce enterprises better and more competitive.

5 Concluding Remarks

Nowadays, information technology is very developed. In the development of enterprises, the application of information technology is of great significance. Managers should master enough data information, and then use big data mining technology to analyze, so as to help managers make scientific decisions and make enterprises develop healthily. In order to obtain a large amount of data information, enterprise managers need to recognize the important role of data mining technology, especially in the context of e-commerce model becoming more and more popular. If enterprises can effectively use data mining technology to mine and analyze market information, it is bound to help enterprises make better marketing plans, and finally make the development of e-commerce enterprises in line with the development of the times.

Acknowledgements. 1. A Project Supported by Key Lab of Internet Natural Language Processing of Sichuan Provincial Education Department (No. INLP201908).

2. A Project Supported by Scientific Reserch Fund of Sichuan Proincial Education Department - Research on the propagation law and the Countermeasures of Negative Public Opinion in Tourist Spots - (No. LYC19-39);

3. Sichuan Science and Technology Program (No. 2018JY0523),

4. The Scientific Research Fund of Sichuan Provincial Education Department (No. 18ZA0233).

References

1. Xin, X.: Discussion on data mining technology and its application in e-commerce. Manage. Technol. Small Medium Enterp. (zhongxunjiao) **8**, 174–175 (2020)
2. Yang, Y., Zheng, G.: Application of web data mining technology in e-commerce. Inf. Technol. Inf. **06**, 104–106 (2020)
3. Feng, L., Wei, W.: Research on the application of data mining technology in e-commerce. China Market **13**, 189–190 (2020)
4. Liu, S., Liu, Y.: Application of data mining technology in campus e-commerce. Southern Agricult. Mach. **51**(08), 211–220 (2020)
5. Guo, X., Wang, L., Zhao, X.: The linux-based e-commerce platform. Comput. Appl. (2000)
6. Schummer, E.: Internet communications and e-commerce platform: US, US20010032154 A1 (2001)
7. Yang, X., Pan, T., Shen, J.: On 3G mobile E-commerce platform based on Cloud Computing (2010)
8. Yang, X., Pan, T., Shen, J.: On 3G mobile E-commerce platform based on Cloud Computing. In: IEEE International Conference on Ubi-Media Computing. IEEE (2010)
9. Da Lvit, L., Muyingi, H., Terzoli, A., et al.: The Deployment of an E-Commerce Platform and Related Projects in a Rural Area in South Africa. Fountain Publishers, Kampala (2007)
10. Fan, Y., Ju, J., Xiao, M.: Reputation premium and reputation management: evidence from the largest e-commerce platform in China. Int. J. Ind. Organ. **46**, 63–76 (2016)
11. Huang, H.: Research of the Internet Finance with the Core as E-commerce Platform. Shanghai Finance (2013)
12. Luo, R.C., Lan, C.C., Tzou, J.H., et al.: The development of WEB based e-commerce platform for rapid prototyping system. In: IEEE International Conference on Networking. IEEE (2004)

Analysis and Research on the Development of Illustration Design Under the Background of Data Analysis

Hui Wang[✉]

Gongqing College of Nanchang University, Nanchang 332020, China

Abstract. Generally speaking, in the field of art design, music and picture are two different forms of art design. It is very difficult to transform them. But with the rapid development of artificial intelligence in recent years, it is easy to realize the transformation of the two. Not only that, because of the wide application of artificial intelligence technology, the field of art design and creation is also assisted by artificial intelligence. Designers should learn to combine artificial intelligence technology in the process of learning design creative programming. This paper mainly discusses the research and development of art design under the background of artificial intelligence. By introducing the concept of illustration, the development of illustration and the forms of illustration, this paper briefly analyzes the changes in the performance style of illustration in different periods, and further analyzes the characteristics of the performance style of illustration under the background of visual culture era, which helps to improve the artistic charm and practical value of illustration design.

Keywords: Art design · Artificial intelligence · Scheme

1 Introduction

AI has been widely concerned after alpha go defeated Li Shishi. After that, AI has gradually entered our life and work, and some mechanized work will be gradually replaced by AI. At the same time, the relationship between the field of art creation and artificial intelligence also began to receive attention, people began to seek new development of art design creation under the background of artificial intelligence [1].

"The function of illustration to the content of the text is to supplement explanation or artistic appreciation, and to insert it in the middle of the book by printing it in the middle of the text or in the form of inserts." This is the definition of illustration in Cihai. In the traditional sense, as the visual explanation and supplement of the text content, illustration plays a more important role in the visual expression of the text narrative content. Therefore, the most basic meaning of illustration is a kind of picture inserted in the middle of the page to express the meaning of the content of the article. Images are indispensable in human daily life. As early as in ancient China, people had a deep understanding of the meaning of "casting tripod like objects to make people know God"

© ICST Institute for Computer Sciences, Social Informatics and Telecommunications Engineering 2021
Published by Springer Nature Switzerland AG 2021. All Rights Reserved
M. A. Jan and F. Khan (Eds.): BigIoT-EDU 2021, LNICST 392, pp. 43–47, 2021.
https://doi.org/10.1007/978-3-030-87903-7_7

(Zuo Zhuan); at the same time, the proverb "a picture is better than a thousand words" was also widely spread in the West. "The effect of painting on illiteracy is the same as that of writing on literate people." This is Archbishop Gregory's exposition of the role of painting. After the industrial revolution, human civilization has developed to a new height. At the same time, the proportion of illiteracy in the social population has also dropped to a very low level. However, the reality is that people do not reduce their dependence on images because of the improvement of literacy. People's demand for images is more intense due to the development of new image technology.

Commodity image is the expansion of animal personification in the field of commodity, and the personified commodity gives people a sense of intimacy. Personalized modeling, refreshing feeling, so as to deepen people's direct impression of goods.

In terms of the concept of commodity personification, it can be roughly divided into two categories.

The first type is the complete personification, that is, exaggerating the commodity, using the commodity's own characteristics and modeling structure to perform personification.

The second type is semi personification, that is, adding hands, feet and heads as the characteristic elements of personification.

The above two personification molding techniques make the goods full of humanity and personalization. Through the form of animation, emphasis on the characteristics of goods, its action, language and goods directly linked, publicity effect is more obvious.

Illustrators often draw illustrations for graphic designers or directly for magazines, newspapers and other media. They are usually professional illustrators or free artists, like photographers, with their own themes and painting styles. Being sensitive and eager to new forms and tools, many of them began to use computer graphic design tools to create illustrations. The function of computer graphics software makes their creative ability get greater play, whether simple or complex, whether traditional media effects, such as oil painting, watercolor, printmaking style or endless new changes and new interests of digital graphics, can be more convenient and faster. Digital photography is the latest development of photography. Photographers use digital cameras or scanners to scan traditional positive films into the computer, then adjust, combine and create new visual images on the computer screen, and finally output positive or negative films through the film recorder. This new photography technology has completely changed the creative concept of optical imaging in photography, and takes digital graphics processing as the core, also known as "photography without darkroom". It blurs the boundaries between photographers, illustrators and graphic designers. Nowadays, as long as you have the ability, you can complete these three kinds of work on the same computer.

2 Artistic Expression of Artificial Intelligence

The so-called art is actually the embodiment of human thought, is to express our sense of experience through various art forms. Art is not only a physical form of expression, but also a transmission of people's thoughts and spirit. Artists pass their spiritual world to the world through art forms, so that we can understand the unknown fields. With the advent of the information age, art should also change with the change of the times. We can't separate the connection between art creation and the times. Therefore, it is an important task of current art creation to connect art with science and technology. Nowadays, art design needs different space and materials. With the continuous development and popularization of artificial intelligence, the field of art design has also been greatly affected. Therefore, in this context, we should seize the opportunity to realize the combination of art and artificial intelligence, and promote the new development of art design [2].

Usually, the title page painting of King Kong Prajna Sutra in Tang Dynasty is the first illustration work in China. This is also the first illustration found in books. Later, the development of illustration in China was mostly influenced by it. Although scholars have found that there are similar images in the silk books before the Tang Dynasty, there are obvious differences between the books that people usually say and the ancient silk books. Then we can get the characteristics of illustration creation in the era of non visual culture:

First, the carrier of illustration creation is single. It only appears in the form of illustrations in paper books, excluding the form of bamboo and silk books. The reason why they are called "illustrations" is that these pictures appear in the book, and the book is the main body. When we only consider pictures and ignore books, pictures are called "illustrations". At this time, although the picture is presented in the book and serves the book, it is the subject. Because we need to consider both the nature and the source of "painting", we call it "illustration".

Second, illustration creation appears in the form of woodcut. Movable type printing not only facilitates the typesetting of words, but also facilitates the flexible typesetting of all kinds of pictures. Since then, the presentation of illustration in books has become more flexible. The presentation of illustration has also developed from the first page to the whole book, and experienced the change from the presentation of the whole page to the more flexible and flexible. Illustration in books is also more diversified.

Gaussian smoothing. In order to avoid the influence of noise in the image on the effectiveness of edge detection, it is necessary to filter the image. Gaussian blur is a way to smooth the image. Convolution of the image with Gaussian filter can filter out the noise details in the image. Gauss blur is to assign different weights according to the distance of pixels in the neighborhood, and the distribution mode is normal distribution [3]. According to the two-dimensional Gaussian function to calculate the weight of each point around, the calculation formula is as follows:

$$G(x, y) = \frac{1}{2\pi\sigma^2} e^{-(x^2+y^2)/2\sigma^2} \tag{1}$$

The gradient intensity and direction of all pixels in the image are calculated. The edge points of the image can point to different directions. Calculating the direction of each pixel can find out the related pixels and combine them into a contour. The amplitude g and direction α of the gradient can be changed.

$$G = \sqrt{G_x^2 + G_y^2} \tag{2}$$

$$\alpha = \arctan(G_y, G_x) \tag{3}$$

According to the direction of gradient, the pixels whose amplitude is not the maximum are suppressed. This step is to refine the coarse edges calculated by SOBE operator in the previous step. The outline with thick edge is fuzzy. Here we think that only one pixel is the best edge width.

3 Practical Application of Artificial Intelligence in Art

Artificial intelligence has more and more influence on the field of art. Through artificial intelligence, artists can more completely express their thinking. For example, deep learning algorithm can generate paintings. We can also use artificial intelligence to create music and poetry, which effectively link art design and creation with artificial intelligence. Specific applications are as follows: (1) we can use artificial intelligence to create posters. For example, Alibaba's "Luban" artificial intelligence system can make posters, which not only takes a short time, but also has good quality, which effectively reduces the work pressure of Taobao designers; (2) artificial intelligence technology can help designers select materials. For example, Sensei system, which can store all kinds of pictures and videos, can select appropriate materials in the system when designers carry out art design. Sensei system can analyze and screen the materials stored in the system according to the requirements of designers, and users only need to select one key to generate the scheme that meets the requirements; (3) artificial intelligence can generate architectural design scheme. Xkool is a platform for building users. This platform is easy to operate. Users only need to input design requirements and parameters, and the platform can generate multiple schemes for customers to choose according to their design requirements. This process can be completed in a few seconds. This fully proves that artificial intelligence can greatly shorten the design time and reduce the work pressure of designers; (4) artificial intelligence can provide color matching schemes for art design. George Hastings system can automatically generate color matching schemes according to the colors selected by users, which not only takes a short time, but also realizes high-precision color combination and improves the efficiency of design work; (5) Artificial intelligence can also help people to design clothes. In 2017, Amazon launched the artificial intelligence algorithm for designing clothes, which can generate the design drawings of clothes according to the pictures provided by users. The system can also analyze the market big data, predict the future trend, and design new clothes in line with the trend [4].

In the development of printing technology, CI technology, photography and other new media technologies, images are gradually integrated into all aspects of people's lives. Jin Yuanpu once said: "there is a close relationship between the great changes of modern society and the emergence and development of the era of visual culture". From the perspective of the mode of communication, electronic media and digital media have gradually replaced the original print media, which has produced many phenomena of the visual culture era in the real society. In the past, the main way to acquire knowledge was words. The appearance of visual images has changed our way of life and thinking. At first, images were spread through computers. There is no doubt that the emergence of a wealth of electronic products (tablet computers, MP4, etc.) in recent years is also conducive to the spread of images. In the era of visual culture, in order to adapt to the requirements of the development of the times, people also need to constantly change and develop a new way of image communication which is different from the traditional way.

Artificial intelligence technology can promote the development of art design, not only in the application of design, but also in the application of art education. Teachers can use artificial intelligence technology to teach students art and design knowledge faster and better, and then students can use more time to practice, accumulate more design experience and lay a solid foundation for their future work. At present, many schools will use computer aided instruction system for auxiliary teaching, through the multimedia class for students to intuitively display art knowledge. For example, teachers can use virtual technology to create a good learning environment for students, and use 3D printing technology to show students' thinking, so as to improve students' interest and enthusiasm in learning art and design. However, students must actively participate in the teacher's teaching, so that the teacher can carry out targeted guidance according to the students' personal learning situation. It can be seen that the application of artificial intelligence to art education can cultivate students' divergent thinking ability, help them develop their imagination and better complete the learning content.

Illustration creation has changed a lot under the background of visual culture era. In the era of visual culture, people have a deeper understanding of the definition and extension of the concept of illustration, the carrier of illustration, the way of creation and the function and application of illustration. With the development of new technology, illustration is no longer limited to printing. The development of multimedia technology, network technology and digital technology has enriched the presentation and dissemination of illustration. The creation way of illustration also changes with the change of presentation way and communication way of illustration. For example, through the artist's direct or indirect woodcut creation printed on books, this is the traditional way of illustration creation. However, illustration creation has become more diverse in the era of visual culture.

4 Conclusion

Artificial intelligence has a great role in promoting the development of art design. Its application in art design can not only reduce the work pressure of designers, but also shorten their working hours, and promote the good development of art education. However, works of art are always created based on human subjective thoughts and feelings. Therefore, no matter what achievements artificial intelligence has made in art design, there is one thing that can not be changed. Artificial intelligence is defined by human beings and is a tool to assist the development of human art design. In order to create shocking works, designers' emotional injection is indispensable.

References

1. Zhou, H.: Research on the relationship between art design and artificial intelligence. Art Design (Theory) (2019) (z1)
2. Chen, Q.: Future design, walking with AI. Art Observ. **10** (2017)
3. Li, S.: Painting Art Form. Jilin Fine Arts Publishing House, Changchun (2007)
4. Chang, R.: Painting Composition. People's Fine Arts Publishing House, Beijing (2008)

Analysis of Big Data's Influence on the Development of Chinese Language and Literature

Xiaoyan Yin[✉]

Ganzhou Teachers College, Ganzhou, Jiangxi Province, China

Abstract. Chinese culture has a long history, broad and profound. Chinese language and literature have witnessed the development of our country. With the development of history, Chinese language and literature has reached a relatively perfect and mature stage. With the progress of the times and the development of Internet technology, China has entered the information age, and many new network languages are gradually added to the Chinese language. The rise of network language is in line with the development of the times. On the one hand, it increases the fun of people's life and learning, on the other hand, it also affects the development of Chinese language and literature in China. Based on this, this paper will take the overview of network language as the starting point, combined with the application of network language in Chinese language and literature, and deeply explore the influence of network language on Chinese language and literature.

Keywords: New era · Network language · Chinese language and literature · Influence

1 Introduction

Today's society is an information-based society, people have invested a lot of energy in the field of scientific exploration of information, and the text still uses its basic functions to some extent. After the emergence of the Internet, the tools and media of writing information and ideological exchange have undoubtedly provided a lot of convenience for people's life, and at the same time, they have also created convenient conditions for the continuation and development of Chinese language and literature. With the development of network, a new language form appears, which is network language. Most people think that network language is a form of transformation of traditional Chinese language and literature, which enriches traditional Chinese language and literature and brings different development to traditional Chinese language and literature [1]. Network language has added new elements and vitality to the traditional Chinese language and literature. Most of the network language has been recognized and accepted by the public. However, from a macro point of view, this new language form lacks a unified and standardized standard, which has a positive and negative impact on people's life.

M. A. Jan and F. Khan (Eds.): BigIoT-EDU 2021, LNICST 392, pp. 48–52, 2021.
https://doi.org/10.1007/978-3-030-87903-7_8

Chinese language and literature, including language and literature, is one of the earliest majors in the history of Chinese universities, which appeared in the late 19th century. Since the 1980s, the major of Chinese language and literature has developed greatly. For more than a century, the major of Chinese language and literature has cultivated a large number of well-known scholars, professors, writers, journalists, playwrights, etc., which have made great contributions to Chinese humanities.

According to the division of disciplines by the Ministry of education, it mainly includes eight secondary disciplines: Linguistics and applied linguistics, Chinese philology, literature and art, Chinese classical philology, Chinese ancient literature, Chinese modern and contemporary literature, comparative literature and world literature, and Chinese minority language and literature, These eight secondary disciplines will not necessarily set up master's program; The division of research direction, different departments are also based on their own scientific research conditions and faculty to establish, can be said to be "independent."

2 Artificial Intelligence and Deep Learning

2.1 Overview of Artificial Intelligence

Artificial intelligence is a science based on computer technology. It is a comprehensive discipline formed by the interdisciplinary infiltration of cybernetics, neurology, mathematics, information theory, game theory and other disciplines. Since the 20th century, scientists have been looking for ways to endow machines with the same intelligence as human brains, hoping that computer systems can learn human thinking, perception and reasoning abilities. Therefore, artificial intelligence has been widely studied [2]. In 1956, the Symposium on artificial intelligence held in Dartmouth University defined "artificial intelligence" for the first time, which enabled people to have a certain understanding of artificial intelligence, and then a large number of research results came out one after another. BM engineer Samuel designed a checkers machine and defeated the checkers champion at that time. The machine used heuristic search theory to simulate the human solving process. In the 1960s, the emergence of expert system simulating human experts to solve special problems further promoted the development of artificial intelligence technology. In the 1980s, Japanese scientists successfully developed the fifth generation computer system kis, which makes the logical reasoning level of the machine comparable to the speed of numerical operation. In 1997, the chess computer developed by BM defeated the chess master Kasparov by using the theory of mixed decision-making, and let human beings see the wisdom and creativity of the machine.

2.2 Deep Learning

Machine learning and deep learning are artificial intelligence algorithm models which have developed rapidly in recent years. Deep learning is a deep neural network algorithm model which is gradually developed from the traditional neural network. It belongs to the branch of machine learning. It is an algorithm model which uses multiple nonlinear transformations to process the data to obtain the abstract characteristics of the data. It was proposed by Hinton et al. In 2006. Deep learning was originally proposed to solve the problem that shallow networks such as BP neural network can not effectively extract the features of complex data and it is difficult to train. Deep learning can simulate the powerful cognitive principle of human brain's nervous system, and extract the information features input into the neural structure layer by layer by connecting multi-layer neural structures similar to those in the cortex of the brain, Each layer of neurons will extract deeper abstract features of object information until the object information can be distinguished. Figure 1 shows the model diagram of a neuron structure in deep learning.

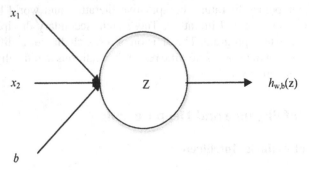

Fig. 1. Neuron model

As the input object of the neuron, the neuron transforms it linearly:

$$Z = w^T x + b \tag{1}$$

Where Z is the input of neuron after weighted sum. Then the output expression is:

$$h_{w,b}(Z) = f(Z) = f(w^T x + b) \tag{2}$$

There are many layers of such neuron structure in deep learning network. Each layer of neuron transfers the abstract features of data to the next layer through nonlinear mapping of activation function, and then learns more abstract data features. With the deepening of the number of layers, the final features can be revealed. Finally, the deep data abstract features learned are sent to the classifier, which can achieve the purpose of data classification and recognition [3].

3   The Influence of Network Language on the Development of Chinese Language and Literature

For the majority of netizens, the network language has a unique language charm, for experts and scholars, the network language has gradually been taken seriously. The rapid development of network language and network culture has also been popularized in the field of education and linguistics. With the in-depth exploration of network language, network linguistics is a new academic derived from network language, which is not difficult to develop. Network language has certain social practical significance. Network language and social culture are closely related, interact and influence each other. As one of the hottest topics in language research, network language is not perfect in theoretical system and research methods, but in a deeper sense, network language is the innovation and reform of Chinese language and literature, and has injected a steady stream of vitality into the steady development of Chinese language and literature. In addition, with the combination of virtual network and network external environment, the network language at this stage is still in the situation of good and bad, but it is also actively improving. The network language is imperceptibly forming a sound language system. Relying on the network, the majority of Internet users are the main body. In the network language system, no one can guide the network language as an authority. All people can express their own suggestions and opinions, express their real ideas, and create new words and languages by themselves, Everyone is the creator of network language. Once the new words and languages are discussed by netizens on the Internet, they will be easily spread on the Internet and then enter people's real life, which greatly enriches the structure of Chinese language and literature [4].

3.1 It Disturbs the Normalization Standard of Chinese Language and Literature

In the network environment, the good and the bad are intermingled, so when people use the network language, they will deliberately use wrong characters, breaking the standard of Chinese language and literature, which leads to the misunderstanding of teenagers in the understanding of Chinese language and literature, unable to understand the essential meaning of the characters in Chinese language and literature. For example, the Internet language "qiafan" refers to eating, because it is a daily language, so it has a great impact on Teenagers' life and learning, because when people are in the adolescent stage, their mind is not yet mature, they like and are willing to accept some new things, but because of their limited ability to distinguish right from wrong, their experience level and self-restraint are poor, And because the network environment is good and bad, it is easy for teenagers to be misled by some bad factors in the network language, resulting in biased understanding of Chinese language and literature. These informal network language make students unable to use the formal Chinese language for self-expression, reading ability and expression ability will decline, and then unable to use the standard Chinese language for writing.

3.2 It Disturbs the Normal Interpretation of Chinese Language and Literature

The wrong use of network language has been gradually accepted and recognized by the public, and no one has put forward contradictory remarks. Even in daily communication with others, they will inadvertently use these wrong network language to express their ideas. The most fundamental Chinese word "person" has been rewritten as "silver" in the Internet language, which is a blasphemy to traditional culture. However, among the young people who have not yet fully developed their minds, they will still use these internet languages for daily communication, and as time goes by, they will forget the standardized use of Chinese language. At the same time, in today's increasingly internationalized society, many foreigners love and begin to learn Chinese language and literature.

4 Conclusion

Generally speaking, with the emergence of network technology, shopping will become more convenient. Online shopping has a great influence on the development of society and people's consumption concept. Even to lead the consumption concept of the whole society. Although it also has personal information security risks, as long as it is well controlled, it can solve such problems and promote the healthy development of online shopping. Only in this way can society progress and people's way of life be fully optimized and gradually improved.

References

1. Wu, L.: Research on influencing factors of consumers' online shopping behavior. Shang Xun **6**, 164–165 (2020)
2. Wang, X.: Analysis of the influence of network language on the development of Chinese language and literature in the new era. Cons. Guide **15**, 93 (2019)
3. Li, P.: On the influence of network language on the development of Chinese language and literature in the new era. Youth **26**, 60–61 (2018)
4. Yao, J.: Research on the Curriculum Reform of Chinese Language and Literature Education Major Facing the New Curriculum Reform. Guangxi Normal University (2015)
5. Ying, J.: A teaching mode of the thesis on graduates of Universities at the department of chinese language and literature suited for the newsituation-mode in the "Three-in-One". J. Guangdong Polytech. Normal Univ. (2007)
6. Library, U.M.: Chinese Language and Literature
7. Qiu, J., Xiong, Z.: The Information communication research based on academic BBS——Take the chinese language and literature section of chinese forum of Peking University as example. Lib. Work Study (2008)
8. Ning, W.: (15)Chinese linguistics and the teaching of chinese language and literature. Soc. Sci. China (2000)
9. Liu, Z., et al.: Modernity: a century's value pursuit of chinese language and literature education. Educ. Res. (2008)
10. Zhao, R., Guo, F., Tan, J.: Evaluation of academic papers impact based on altmetrics: a case study of chinese language and literature. J. Lib. Sci. China (2016)
11. Pan, X.H.: A rustic opinion about curriculum reform of chinese language and literature specialty in higher teacher education. J. Fujian Normal Univ. (Philosophy and Social Sciences Edtion) (2004)
12. Wang, C.W.: A research on the reformation of graduation thesis and design of specialty of chinese language and literature in applied undergraduate colleges——Taking Chongqing University of Arts and Sciences as an example. J. Chongqing Univ. Arts Sci. (Social ences Edition) (2012)
13. Pan, X., Qiu, X.G.: Bias and mismatch of college chinese language and literature research. J. Zhejiang Indust. Trade Vocat. Coll.
14. Xu, L.: English abstracts writing for the thesis in chinese language and literature. J. Wuyi Univ. (2009)

Auxiliary Design of Oil Painting by Computer Technology Under the Background of Big Data

Xiaofei Jia(✉)

XianYang Normal University, Xianyang 712000, China

Abstract. At present, China's oil painting is moving towards a new field of development, but there are few monographs and discussions on the artistic research of oil painting creation. Most of the research on oil painting creation in academic circles is the introduction of oil painting skills and critical articles. In order to make the contemporary oil painting creation have a good development space in China's academic circles, it is necessary to conduct in-depth research on the artistry of oil painting creation, follow the development principle of keeping pace with the times, and apply the advanced Internet technology in the research process of China's oil painting creation, so as to promote the development of China's oil painting creation.

Keywords: Internet plus · Background · Oil painting · Art of creation

1 Introduction

In the new era, people's life is more affluent and their living standard is also greatly improved. The use of modern advanced Internet technology can transfer all kinds of information to people. Internet technology not only provides a lot of convenience for people's life, but also promotes the rapid development of China's culture and art field. In the field of oil painting creation, we need to break through the previous research on oil painting creation, follow the artistic rules of oil painting creation, and combine with modern advanced Internet technology to promote the emergence of new oil painting creation mode in China [1].

Image based oil painting stylized rendering is one of the hot topics in the field of computer graphics. In order to further improve the quality of image oil painting stylized rendering, a hierarchical image oil painting stylized rendering algorithm based on multi-scale brush is proposed. The algorithm simulates the oil painting process of artists, In each layer of brush painting, the incremental Voronoi sequence sampling points and image tangent direction field are used to determine the brush streamline, and then the brush shape and brush height field are combined for texture mapping to obtain the final image oil painting stylized rendering results, The algorithm can not only simulate the real process of oil painting, but also generate a more hierarchical oil painting effect, which fully reflects the structural characteristics of the image and oil painting details.

M. A. Jan and F. Khan (Eds.): BigIoT-EDU 2021, LNICST 392, pp. 53–62, 2021.
https://doi.org/10.1007/978-3-030-87903-7_9

2   The Important Role of Real Emotion in Oil Painting Creation

The art of oil painting is a creative form of labor. Only through careful drawing, careful consideration and continuous improvement can painters create fine works. In today's era, oil painting creation needs a breakthrough to meet the needs of modern people's pursuit of personality. Oil painters need to express their feelings in the process of creation [2]. An oil painting can be called perfect only when it presents the rich inner feelings of oil painters. For creation, the most important thing is the injection of emotional factors, to meet people's wishes and needs. The creation of oil painters' works is to bring pleasure to people and help people form a positive, healthy and optimistic mental state in life. For oil painters, good mood will directly affect the creation of oil paintings.

Compared with the urbanization process of western countries, China's historical agricultural civilization and agricultural social tradition make China's urbanization process difficult. However, in the process of accelerating the development of urbanization, the development and changes of rural areas are obvious. At the same time, social imbalance brings more and more problems and contradictions. Like the history of urbanization in the west, painters agree with the inevitable trend of modernization in concept, but they are still full of nostalgia and longing for the tradition of local social civilization (see Fig. 1).

Fig. 1. Painting assistant system

3   The Artistic Expression of Rhythm and Rhythm in the Process of Oil Painting Creation

3.1   The Influence of Rhythm and Rhythm in Creation

Art is interlinked, and different works of art have different artistic characteristics. Rhythm and rhythm are not only used in the field of music, but also in oil painting. In the development history of oil painting in China, rhythm and rhythm play an indispensable role in the process of artistic creation. With the help of artistic rhythm and rhythmic features, creators can enhance the expressiveness of oil painting works, so as to create more excellent oil painting works for the world. In the new era, in order to promote the continuous development of China's oil painting art creation, borrowing modern advanced Internet technology, selecting modern advanced painting materials and integrating 3D printing technology can create better oil painting works for the world. Oil painting materials provide conditions for creating the rhythm and rhythm of oil painting works of art, thus enhancing the ability of image processing and expression of oil painting works of art. For example, the cold and warm color, the virtual and real performance of the picture, the dry and wet of the media, the density of the object, the black and white gray of the picture and the thickness of the color, etc., add the inherent artistic expression to the oil painting art works.

3.2 The Expression form of Rhythm and Rhythm in Artistic Creation

Rhythm and rhythm are very important in music creation, and rhythm and rhythm are also very important for oil painting creation. Oil painting creation also requires quality, and is related to personality. The quality of oil painting is reflected in the outside and inside. The appearance of an oil painting reflects the external character of oil painting, while the internal character of oil painting mainly refers to the internal meaning of oil painting. In order to make an oil painting have a profound connotation, the creator needs to use rhythm and rhythm to add color and charm to the work in the process of oil painting creation. An excellent oil painting can set off the painter's psychological quality, aesthetic taste and comprehensive quality. The artistry of oil painting is mainly reflected in rhythm and rhythm. Rhythm and rhythm can bring different visual effects to people and make people bright in front of their eyes, so as to improve the quality of oil painting works. Therefore, if an oil painting lacks rhythm and rhythm, it will not reflect the real connotation of the oil painting, will affect people's aesthetic appreciation of the oil painting, and will reduce the aesthetic value of the oil painting.

The most common activation functions in CNN structure are sigmoid, tanh and relu, and their corresponding mathematical expressions are as follows.

$$sigmoid : f(x) = \frac{1}{1 + e^{-x}} \tag{1}$$

$$tanh : f(x) = \tanh(x) \tag{2}$$

$$ReLU : f(x) = \max(x, 0) \tag{3}$$

Through the mathematical expression and image of the activation function, we can see that the value range of sigmoid function is $(0,1)$, and the characteristic of the derivative function is that it is 0 at the beginning, and it tends to be 0 soon [3].

4   The Influence of Visual Culture on Contemporary Oil Painting

With the advent of the era of visual culture, great changes have taken place in our lives. The brand-new visual experience has opened a mysterious door to the development of contemporary art. Contemporary art is in an era of diversified development, with diversified creative ideas and ways. Nowadays, many of our artists are innovating for the sake of "innovation" and working hard to highlight "personality". It is driven by this mentality that their works are becoming more and more empty, artists are becoming more and more utilitarian, and the artistic creation environment is becoming more and more market-oriented. The "contemporaneity" of art has become a problem worthy of consideration for our creators, which has distinctive characteristics of the times. "Innovation" is not just building a brand new thing and abandoning the traditional excellent things. It should, as the predecessors say, make the best use of the fine tradition and make the past serve the present, and carry forward and expand it on this basis. Especially for the artist's own artistic language service, will eventually form their own completely unique artistic personality. The development of art should be a process of inheritance and continuation. Artists should first pay attention to the political, economic and cultural characteristics of the times they live in, and then express their feelings and feelings in such a background, rather than blindly following the trend and catering to the market (see Fig. 2).

The creation of Chinese contemporary artists once again makes people feel the great power of thought. It is precisely in the repeated encounter of the current problems of Chinese people's survival that has become a basic problem. It is repeatedly shown in their works and forms a special system of cultural symbols, which further shows that the creators have the problem of thinking about the meaning of life, which is worthy of our happiness. The development and progress of new technology is bound to promote the development of art, but the key problem is still related to human imagination, aesthetic sensibility and creativity. That is to say, the essence of art concern is to pay attention to people's problems, that is, to pay attention to people's life state, ideology, living situation, spiritual dilemma and so on. The development of art is eternal in paying attention to people's problems. It will not change because of the development of modern scientific and technological means. The development of technological means can only improve the expression form of creators' attention to people's problems. As the most sensitive nerve in society, artists constantly experience the creation by using modern scientific and technological means between the double feelings of life and art. Therefore, we should be more clearly aware that the creation of modern images and contemporary art is always centered on "human problems".

4.1 The Lack of Spiritual Connotation Under the Influence of Mass Consumption Culture

Since the end of the 20th century, China has entered an era dominated by mass consumption and dominated by mass media. At the same time, it is a cultural era of diversified development oriented by practical spirit. Artists try their best to express their own feelings while paying more attention to the aesthetic taste of the public. In the creative activities aimed at the aesthetic taste of the masses, the artists of later generations use different forms and means of expression, trying to find new breakthrough opportunities from both conforming to the aesthetic taste of the masses and rich cultural connotation [4].

4.2 The Confusion of Aesthetic Pursuit and Artistic Standard Under the Background of Visual Culture

With the deepening of China's reform and opening up, the rapid development of economy has greatly improved the people's material living standards, the rapid improvement of material living standards and quality, and further emancipated the mind [5]. People also gradually bid farewell to the Puritan ideology and life concept under the idealistic system before the reform and opening up. With the advent of the consumption era, consumerism, as the inevitable ideology of the current society, has been growing and spreading rapidly. Under the influence of consumerism, Chinese aesthetic culture has entered an unprecedented period of secularization [6]. One of the remarkable characteristics of secularization is that metaphysical idealism and aestheticism are replaced by a kind of metaphysical pragmatism and realism. The auxiliary system of oil painting is shown in Fig. 2.

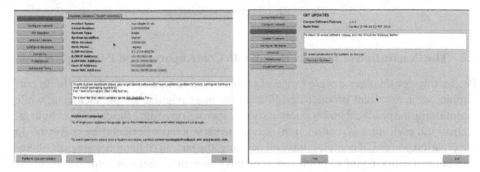

Fig. 2. Oil painting assistant system

5 The Penetration of Design Factors into Contemporary Oil Painting Creation

5.1 The Penetration of Design Elements in Visual Communication Design into Contemporary Oil Painting Creation

Eye catching signs, beautiful packaging, moving illustrations and shocking posters in life affect the expression of oil painting creators' thoughts and feelings in painting in different visual forms. Music art is to express the inner emotional ups and downs through the pleasant melody, and mobilize our auditory organs; while painting is to solidify the music, express through shaping different shapes and rich colors, and mobilize our visual organs [7].

The penetration of many design elements in visual communication design into contemporary oil painting makes its sense of form more intense and tends to be flat. In terms of picture effect, it pursues the rule of formal beauty, emphasizing change and unity, rhythm and rhythm, personality and order, etc. [8]. First of all, the creation process reinterprets the point, line and surface, such as Zhao Haibo's oil paintings (Fig. 3). We can see that the painter uses the simplest point, line and surface to exaggerate the graphics, and uses flat painting and gradient in the bright colors. The intersection of pictures makes the picture form an order, an overall rhythm, and the picture is composed and the color is harmonious.

Fig. 3. Zhao Haibo's oil paintings

5.2   The Penetration of Product Design into Contemporary Oil Painting Creation

Product design involves a wide range of fields, covering most of our daily life. In particular, some excellent product design uses the most popular simple elements, while maintaining the practical and economic style, it focuses on adding the concept of humanization and environmental protection, striving to show a more perfect life experience. In the creation process of contemporary oil painting, the pursuit of natural, pure, self liberation of personality, trying to product design, environmental design of some factors of the real feelings, incisively and vividly show in the painting. In product design, the function, structure, material, shape, color, surface treatment, decoration process and mechanical properties of the product, such as strength, rigidity and stability, will bring users high efficiency, comfort and aesthetic enjoyment [9]. At the same time, oil painting creators get inspiration from these elements to promote the contemporary oil painting creation to pursue a more up-to-date visual texture, Strong mechanical modeling, philosophical and profound picture. We can't help feeling that this is the product of the industrial age [10].

6   The Infiltration of Multimedia into Contemporary Oil Painting Creation

Randall Parker and Kim Jordan said in the "prelude" to multi media: "multimedia (may, will blur the boundaries between life and art, personal and media, real and virtual [11]. And we are only now beginning to understand what these trends mean." I think that as we gradually "understand what these trends mean", our art and artistic creation gradually become clear. It is multimedia that exists in our daily life and presents it to us in a brand new and exciting form, which makes us find the road of painting innovation. The main way for multimedia to infiltrate contemporary painting creation is to let different images infiltrate with the help of network platform, reflecting various forms of interaction.

Because art comes from the experience and observation of life, the starting point of oil painting creation is also based on this origin. We will find that in the daily busy life, the network has changed our society, and mobile phones have enriched our life. No matter where you go, you will feel the convenience of your life by tapping the keyboard. It is precisely because of the new lifestyle of the Internet that our oil painting creators have changed their previous observation and creation methods, found new elements in the new media, and created better oil painting works in a unique way. It is precisely because of the rich and changing cyberspace that we have, Different image culture guides our visual appreciation and changes our creative thinking. I would think that perhaps our creation at first as a virtual symbol involved in the process of creation, behind the symbol of people's emotions, emotions, emotions, creators through the canvas platform combined with their own feelings of life to show this illusory effect, and even amplified or strengthened, is the process of artistic refining and creation [12].

The infiltration of multimedia into contemporary painting creation, on the one hand, is based on the background of the times, the way of oil painting creation has been changed, and the painting theme has become diverse. When we look at the history of Western painting, we will feel that many excellent western paintings choose religious themes, such as Renaissance painting, which has the tendency of secularization of religious themes. In Raphael's painting, there is obviously pagan spirit. When Raphael created "the virgin of Sistine", he has a strong sense of religious theme, The painter depicts the beautiful virgin Mary in a sweet and leisurely lyric style. The composition is solemn and balanced, and the content and form reflect the characteristics of the Renaissance. In today's society, magic video games, animation, these dynamic images, bring us perfect visual experience and super impact, constitute an era of digital art.

Computer games and animation in our life are the most interesting areas in multimedia. A large number of them are emerging to denounce our life, and promote many oil painting themes to be displayed in the form of cartoon animation. For example, I recently saw Liu Yinmeng's work call (Fig. 4).

Fig. 4. "Call" oil painting

7 Design Factors in the Works of Western Painting Masters

7.1 The Relationship Between Animation and Painting

Animation comes from life and is higher than life; animation comes from painting and is higher than painting [13]. There are various forms of artistic expression in painting, such as oil painting, watercolor, gouache, pen painting, pencil painting, etc. Painting works choose a certain space to express the content, and coagulate it with material media to form an artistic image with perfect spiritual outlook and essential significance. Animation art captures motion through a series of pictures, showing the state of the subject in space, resulting in a sense of motion of time continuity. Painting works of art can bring animation art designers rich source of inspiration, both in the ocean of art are interlinked and integrated. Animation art and painting art, in a certain sense, are to show the various forms of things, flowing and changeable actions, people's inner emotions and mysterious scenes in the plane space side by side through light and shadow, color, points, lines and surfaces, creating a harmonious and perfect whole. Both of them have some common characteristics in emotional expression and artistic expression, thus producing resonance [14].

Painting is to use the author's subjective feelings to allocate aesthetic factors to depict objects. It pays more attention to color and emphasizes the independent performance function and artistic language of color. In the creation of animation design, learning from the artistic language of painting will bring unique artistic charm and vitality to animation works. The composition, color, light and shadow, material and texture of pictures in animation works will bring different psychological hint to the audience, so as to convey the author's creative intention and emotional expression. Drawing what we need from paintings is also an expansion of the edge of different art fields.

7.2 Painting is the Foundation of Animation

With the rapid development of computer technology, in the process of animation design and painting creation have had a huge impact, it not only greatly improves the efficiency of people in the creative process, but also provides people with a greater space for thinking. In the CG overwhelming today, the computer is almost omnipotent [15]. However, animation is very important because of its painting. Computer can not replace the traditional role of paper and pen, the simplest and most direct way is often the most effective and important way. Even in the process of using computer to create animation, painting as its basis is still an indispensable part. Painting is of realistic significance to animation. Because painting does not rely on real things, not limited by time and space, it can easily follow their own ideals and wishes to depict images, as long as you want to, you can achieve it through painting.

7.3 Animation has a Unique Painting Style

Animation is not only very different from painting in the two aspects of "personalization" and "attachment", but also in the pursuit of formal beauty. Animation, divided into realistic and non realistic decoration, the former image close to life. In painting, they have different meanings and degrees. In the process of animation production, the image of animation is required to be simpler and more restrictive. The latter is more interesting [16].

In today's world, the social division of labor is more and more clear, the pressure of work and life makes people more and more nervous, people want to work more relaxed, animation is developing rapidly in this social environment. Therefore, animation is easy to be accepted by the public, with a broad customer base and market development space. In the spiritual field, material consumption and many other aspects have had a great impact, become part of people's entertainment, consumption. Beginners should not mistakenly think that computer is omnipotent. Although computer technology provides a very important and practical technology for drawing animation, technology will always serve the purpose of achieving the art of animation. The computer is just a programming machine, which can only provide technology and efficiency, but can not cause the change of ideas. It can never replace the creative foundation, modeling ability and artistic behavior ability of hand painting. We can only use it as a tool. This tool must serve the artistic concept of creating animation. It can't be confused by it, and can't lose the artistic concept, so as to ignore its own hand-painted ability.

8 Animation Masters' Reference for Artistic Ideas

Modern art creation is closely related to the development of science and technology, such as industrial technology thought and futurism, style school, structuralism and Cubism, Freud's dream interpretation theory and surrealism, etc., which can be said to be the combination of science and art, and art creation is a kind of regular creative activity. As an art form, animation not only emphasizes art, but also emphasizes the flexible use of visual language and performance technology. To be exact, it is not only art research but also design application. The beginning of any kind of aesthetic consciousness in the world is always accompanied by the concept innovation in the field of philosophy, and modern technology provides the possibility for us to explore the means of expression. Learn some art theories and understand some art schools, learn from the works and experience of art masters to improve the cultural quality and life experience of their works.

In the process of painting, color plays a more important role in expressing the painter's feelings, especially in the creation of oil painting. In the light and shade changes of colors, it reflects people's social environment, and also reflects the relationship between people and society. The more harmonious the light and shade of colors in oil paintings under the action of light, the more harmonious they are, the more harmonious they are between people and society. In a specific work, if there is a lack of harmonious relationship, it will be impossible to achieve artistic success. In the process of oil painting creation, the colors used in oil painting can touch the natural hearts of the appreciators. They act on the depths of people's hearts through the eyes, nose and other sensory organs of the appreciators.

9 Conclusion

In today's era, the art field also needs to follow the pace of the times, using modern advanced Internet technology to innovate and break through in oil painting creation. We should always follow the law of artistic creation, integrate artistic rhythm and rhythm into the creation process of oil painting works, and add the artist's real feelings, so as to express the strong color and internal meaning of oil painting works. Selecting appropriate oil painting materials and giving full play to the language symbols of abstract beauty in the process of oil painting artistic creation can add artistic sense to oil painting works and promote the development of oil painting art in the Internet era.

References

1. Zhou, X.: The Turn of Visual Culture (Nanjing University). Peking University Press (2008)
2. Zhang, S., et al.: Introduction to Visual Culture. Jiangsu People's Publishing House (2003)
3. Zhan, J.: The Situation and Choice of Chinese Oil Painting, vol. 3. Chinese Oil Painting (2001)
4. Wang, B.: The Influence of Popular Culture on the Youth Generation, vol. 1. Youth Guide (2001)
5. Wang, C.: Introduction to Television Image Communication, pp. 140–174. Sun Yat sen University Press, Guangzhou (2006)
6. Xue, Y.: History of Animation Development, pp. 2027–2036. Nanjing Southeast University Press
7. Feng, W., Sun, L.: Introduction to Animation Art. Ocean Publishing House, Beijing (2012–2014)
8. Liang, D., Park, K.: Pencil drawing animation from a video, computer. Anim. Virtual Worlds 24(3–4), 307–316 (2013)
9. Zhou, C.: A Brief History of Europe, pp. 306–308. Jilin University Press, Changchun (2010). Research status and reasons of Tang Zhonghui Gong Wen's animation ontology. J. Chongqing Normal Univ. 1, 80–84 (2011)
10. Cholodenko, A.: The illusion of the beginning: a theory of drawing and animation. Afterimage 28(1), 9–10 (2000)
11. Birgitta Hosea drawing animation. Animation 5(3), 353–367 (2010)
12. Written by Clive bell, translated by Xue Hua, pp. 1–56. Jiangsu Education Press (2004)
13. Su, X.: The successful application of "thick" expressive language in contemporary oil painters' works. J. Art College PLA 3, 91–93 (2011)
14. Hegel, translated by Zhu Guangqian, Aesthetics, vol. I, p. 362. People's Publishing House, Beijing (1959)
15. Feng, M.: Tour of Color Symphony, Interpretation of European Famous Paintings, pp. 26–30. Zhengzhou University Press, Zhengzhou (2006)
16. You, Z.: Monet, pp. 39–42. Shanghai People's fine arts Publishing House, Shanghai (1998)

Based on the Big Data Analysis of the Task-Driven Teaching Model of Electronic Circuit CAD Teaching Content Planning

Hailan Wang[✉]

Bayingol Vocational and Technical College, XinJiang 841000, China

Abstract. W In view of the problems and disadvantages of electronic circuit CAD in teaching mode, based on the circuit schematic diagram and PCB design tool Altium designer as the support platform, relying on the task driven teaching mode, the teaching objectives, corresponding teaching contents and specific tasks of electronic circuit CAD are planned, and the task driven method is adopted to mobilize students to actively learn electronic circuit CAD. The enthusiasm of the teachers has improved the malpractice of the students who did not study in the previous examination course, and achieved good teaching effect.

Keywords: Electronic circuit CAD · Task driven · Electrical schematic · PCB

1 Introduction

The basic guiding ideology of task driven teaching mode "task driven" teaching is the most fundamental "task as the main line, teachers as the leading, students as the main body", which is a kind of teaching mode based on constructivism learning theory.

Learning method, change the traditional teaching idea of imparting knowledge to the teaching idea of solving problems and completing tasks, change the "cramming" teaching mode to inquiry teaching, change the previous "teachers speak, students listen", change the passive teaching mode of teaching learning, change students into initiative, teachers as guidance. Students in the teacher's guidance, around some specific tasks and teaching activities, because students received a specific targeted teaching task, so in the problem motivation driven, students will actively carry out research or exploration, students received specific teaching tasks, targeted learning, in the learning process, through the completion of the task [1]. And problem solving, students will get the feeling of success, will stimulate students' desire for knowledge, and gradually form a virtuous circle of perceptual mental activities, which is conducive to cultivating students' self-learning ability of independent research and exploration.

2 Research Objective

Based on the teaching goal module of task driven teaching mode, planning the implementation of task driven teaching mode, the most important thing is the planning of

M. A. Jan and F. Khan (Eds.): BigIoT-EDU 2021, LNICST 392, pp. 63–68, 2021.
https://doi.org/10.1007/978-3-030-87903-7_10

teaching goal. According to the engineering education certification standard, determine the teaching goal of the course, and then according to the teaching goal, reasonably plan the teaching content, according to the teaching content, extract the specific teaching task. The teaching goal of electronic circuit CAD comes from two levels: one is to master the drawing of electrical schematic diagram, the other is to master the drawing of PCB; on the drawing level of electrical schematic diagram, we need to master the skills of adding component library, searching and placing components, editing and drawing; on the drawing level of PCB, we need to master the setting of circuit board layer, the layout of PCB, etc. Therefore, the teaching goal of electronic circuit CAD is to enable students to master the method of drawing circuit schematic diagram and printed circuit board diagram by using computer tool software [2]. According to this overall goal, three teaching modules are planned: basic skills of electrical schematic diagram drawing, basic skills of PCB drawing, and practical engineering case training. Among them, seven modules and PCB drawing are planned for basic skills of electrical schematic diagram drawing. Three modules are planned for drawing basic skills, and five modules are planned for practical engineering case training. The planned teaching module is shown in Fig. 1.

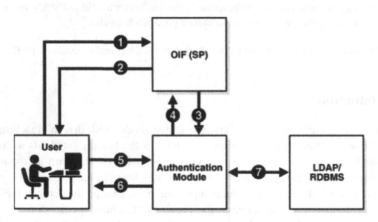

Fig. 1. Target module planning

3 The Planning of the Basic Skills Teaching Module of Electrical Schematic Drawing

This module mainly enables students to master the basic skills of electrical schematic drawing, and lays the foundation for students to draw electrical schematic diagram in the follow-up courses. It is divided into seven functional modules. The specific functional module planning is as follows: module 1: module 1. The teaching task is to enable students to master the engineering establishment, schematic diagram establishment, drawing attribute setting, drawing environment setting, viewfinder application and component library management based on engineering project management. The specific task flow is shown in Fig. 2.

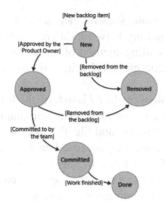

Fig. 2. Asks and processes

Module 2: the teaching task of module 2 is to enable students to master the component placement, component orientation adjustment, component removal, component attribute modification, wiring operation, electrical node placement, power and ground placement.

Module 3: the teaching task of module 3 is to make students master the bus placement, bus branch placement, network label placement, IO port placement and attribute modification, component model management, graphics and text addition and other methods by drawing the schematic diagram of AT89C51 and 2764 program memory expansion circuit.

Module 4: the teaching task of module 4 is to give a single tube AC signal amplification circuit, so that students can master the operation, editing and quick marking (selection) of single and multiple objects through the drawing of the schematic diagram. The editing skills include operation object, batch modification of device attributes, text information modification, automatic alignment of components and drawings, rapid drawing of multiple parallel lines, array pasting, system default settings, navigator use, automatic numbering of components, etc.

Module 5: the teaching task of this module is to enable students to master the editing and creation of electrical graphic symbols of components [3]. The specific task is to give a new package diode to enable students to master the starting operation method of schematic component library file editor, the modification operation skill of schematic component library file, and the Sch library panel. Use and modify the components in the schematic component library, batch update the components in the schematic, and create the schematic primitives.

Part library file, create component electrical graphics symbols, create schematic project component library. The modification of components in schematic component library includes two skills: one component modification and batch modification; there are four step-by-step tasks to create electrical graphic symbols of components, which are creating electrical graphic symbols of components from scratch, copying electrical graphic symbols of components from current library files (the premise is to open the existing schematic Component Library), copying electrical graphic symbols of components from existing library files, and making electrical graphic symbols containing multiple unit circuit components.

Module 6: through this module, students can master the concept and characteristics of multi-level circuit, the switching between different schematic files in the design of hierarchical circuit, and the editing method of hierarchical circuit. The specific task planning is to give two specific electrical schematic diagrams and draw hierarchical circuit on this basis.

Module 7: the teaching task of module 7 is to form the list of components on the basis of independently completing the electrical schematic diagram, and master the output and pasting skills of schematic report and file; the specific task planning is to draw the corresponding electrical schematic diagram based on the power circuit diagram. On this basis, master the device list of the schematic diagram, copy the schematic diagram (with template information) and report to word text, copy and paste only the selected part of the schematic diagram (without template information) to word textText, the operation method of adding the components or circuits in other engineering schematic diagram to the current design project, and the printing method of schematic diagram (using word document printing method and direct printing method).

4 PCB Drawing Teaching Module Planning

4.1 Main Process

This module mainly enables students to master PCB drawing and lay the foundation for students to participate in innovation training and research and development of electronic products.

Module 1: the task of module 1 is to master the creation of PCB file, the application of PCB editor interface and PCB editor interface, and the manual design of single panel method this task includes the setting of working layer, the method of manually adding board layer, the setting of visual grid, the setting of PCB environment, the loading of component packaging library, and the use of drawing tools (it includes component placement, pad placement, circuit board size setting, component serial number of silk screen layer and annotation information editing.

4.2 Part II

Module 2: the task of module 2 planning is to automatically draw PCB Based on power supply schematic diagram. Through this training, students can master the task PCB design process, preparation before PCB design, PCB layout, pad selection and wiring, ground wire/power line layout rules. Pad selection includes through element (THC). The routing tasks include parasitic parameters and crosstalk of printed wires, selection of minimum line width, selection of minimum wiring spacing, routing control of printed wires, setting principle of jumper in single panel, etc. the layout rules of ground wire/power wire include ground wire classification, common impedance interference and elimination of ground wire and power wire, grounding mode, and some basic principles of ground wire distribution and so on.

4.2.1 Specific Task Planning

The specific task planning is to give a new PCB packaging component, so that students can master the creation of component library file of PCB packaging drawing, create PCB packaging drawing of component in PCB library file, understand component management and maintenance of PCB packaging drawing library and 3D model addition. In order to create PCB package drawing component library file, we should master three methods: creating PCB package drawing component library file in user integration library file, creating PCB package drawing component library file in design project, and creating project PCB package drawing library; in order to create the PCB package drawings of components in the PCB package library file, we should master the following methods: creating the PCB package drawings of components manually in the PCB package library file, making the component package drawings by using the component wizard, making the surface mount component package drawings by using the IPC footprint wizard, and making the component package drawings by using the component copy function. Practical engineering case training task planning the main teaching task of this module is to enable students to master the basic skills of electrical schematic diagram and PCB drawing. Through this teaching module, students can master DC regulated power supply, minimum CPU system, single tube AC signal amplification circuit and AC contactless switch control circuit.

4.2.2 Practical Circuit Principle

Module 1 is the practical circuit schematic drawing, the training goal of this module is to complete the most systematic schematic drawing based on 89C51 single chip microcomputer, and the selected content is the drawing of the most systematic schematic circuit board based on 89C51 single chip microcomputer;

Module 2 is the drawing of circuit schematic diagram of self-made components. The training goal of this module is to complete the drawing of silicon controlled voltage regulating circuit based on 89C51 single chip microcomputer, which is divided into two specific tasks. Task 1 is to make electrical graphic symbols of light controlled silicon controlled moc3062 device, and task 2 is to complete the drawing of silicon controlled voltage regulating circuit based on 89C51 single chip microcomputer; Module 3 is the basic drawing of practical circuit PCB board. The training task of this module is to draw the PCB board of warning light, draw the electrical schematic diagram of warning light and PCB board drawing. The specific task includes two parts [4]. Task 1 is to draw the electrical schematic diagram of warning light, and task 2 is to complete the drawing of PCB board of warning light and give PCB Board size, power line width, signal line width, device packaging, etc.; Module 4 is the standard drawing of practical circuit PCB board. The main task is to draw the warning lamp electrical schematic diagram and PCB board based on 555 oscillator control. Three tasks are specifically planned. Task 1 is to draw the warning lamp electrical schematic diagram based on 555 oscillator control, and task 2 is to complete the drawing of warning lamp PCB board based on 555 oscillator control, and give PCB. The specific task is to form the width of the device line and the width of the signal line. Module 5 is practical circuit special structure PCB drawing, the main task of the module is to draw special structure PCB, specifically planning two tasks, task 1 is to draw warning light electrical schematic based on 555 oscillator control, task 2 is to complete the drawing of warning light special structure PCB Based on 555 oscillator control.

5 Conclusion

Based on the planning of teaching objectives and teaching modules, teachers first assign teaching objectives and teaching tasks, and briefly explain the specific implementation methods and measures. Then students learn computer operation on their own initiative. Relying on the specific tasks given by teachers in each class, they carry out skill learning and research. Then teachers summarize, make the finishing point, and point out the students in the process of practice. Finally, students practice according to the teacher's skills, so that students can thoroughly grasp the teaching content and achieve the teaching objectives. After a semester of practical teaching, this teaching method has received good teaching effect.

References

1. Zhang, X.: Research on the Application of "Task Driven" in the "Comprehensive and Exploratory" Learning Field of Junior High School Fine Arts. Guangdong Normal University of Technology (2019)
2. Long, Y.: Research on the Application of Task Driven Teaching Method in High School Ideological and Political Course. Guangxi Normal University (2019)
3. Wang, D.: Experimental Research on Task Driven Teaching Mode in Tennis Course of Physical Education Major in Colleges and Universities. Liaoning Normal University (2019)
4. Ma, Y.: Research on Effective Questioning Strategies in Classroom Teaching of High School Ancient Poetry. Guangdong Normal University of Technology (2019)

Construction of Management Model of Vocational Education Based on Data Analysis

Yunli Dao[(✉)]

School of Civil Engineering and Architecture, Chongqing Vocational College of Transportation, Chongqing 402247, China

Abstract. Due to the strong self-organization and self-adaptive ability of DEEP LEARNING network, this paper constructs a prediction model of management deep learning level of vocational education based on DEEP LEARNING network, using nsse-china 2013 questionnaire as the source of data, taking the five comparable indicators between vocational colleges as the input of the network, and the management level of vocational education as the output of the network, and simulating in MATLAB. The experimental results show that the prediction model overcomes the complexity and subjectivity of the traditional evaluation of deep learning vocational education management, has the characteristics of fast convergence speed and high prediction accuracy, and has good applicability.

Keywords: DEEP LEARNING network · Vocational education management deep learning level: NSSE China · Prediction model

1 Introduction

Vocational education refers to the education that enables the educated to obtain the professional knowledge, skills and professional ethics needed by a certain occupation or production labor, including primary vocational education, secondary vocational education and higher vocational education (college level vocational education, undergraduate level vocational education and graduate level vocational education). Vocational education and general education are two different types of education, which have the same important position. Vocational education is a type of education, not a level of education.

Vocational education includes vocational school education and vocational training. Vocational school education includes various vocational and technical schools, technical schools, vocational high schools (vocational middle schools), etc. Vocational school education is academic education, which is divided into primary, secondary and higher vocational school education. Vocational training is non academic education, including pre employment training for workers, re employment training for laid-off workers and other vocational training [1–4].

Vocational education is an important part of the national education system and human resource development. It is an important way for the majority of young people to open the door to success. It shoulders the important responsibilities of cultivating diversified

© ICST Institute for Computer Sciences, Social Informatics and Telecommunications Engineering 2021
Published by Springer Nature Switzerland AG 2021. All Rights Reserved
M. A. Jan and F. Khan (Eds.): BigIoT-EDU 2021, LNICST 392, pp. 69–78, 2021.
https://doi.org/10.1007/978-3-030-87903-7_11

talents, inheriting technical skills, and promoting employment and entrepreneurship. Since the eighteen Party's Congress, general secretary Xi Jinping has made important instructions on developing occupation education for many times, demanding that "we must attach great importance to and accelerate development". In recent years, a series of major measures have been taken to promote vocational education to a new level, such as the "national vocational education reform implementation plan", the pilot project of jointly building innovative development highland of vocational education by Ministry and province, and the implementation of the East West cooperation action plan of vocational education [5–8].

According to the statistics, there are 11300 vocational schools and 30.88 million students in China, which has built the largest vocational education system in the world. From the perspective of industry distribution, in the fields of modern manufacturing, strategic emerging industries and modern service industry, more than 70% of the new frontline employees come from vocational colleges. It can be said that since the 18th National Congress of the Communist Party of China [9], vocational college graduates have become the main source of China's industrial forces, and the main force supporting the aggregation and development of small and medium-sized enterprises, the transformation and upgrading of regional industries, and the development of urbanization. At the same time, whether it is through vocational education to cut off the root of the intergenerational transmission of poverty, or a large number of ex servicemen, laid-off workers and migrant workers have acquired skills, vocational education has also played an increasingly important role in serving employment and improving people's livelihood [10–15].

At present, the 14th five year plan starts. In the new journey of building a socialist modern country in an all-round way, vocational education has a bright future. With China's entering a new stage of development, industrial upgrading and economic restructuring continue to accelerate, the demand for technical and skilled personnel from all walks of life is becoming more and more urgent. Especially in the new round of scientific and technological revolution and industrial change, the deep application of artificial intelligence, Internet of things, big data and other technologies puts forward higher requirements for the quality of workers. To adapt to the high-quality development stage of our country, it is urgent to have a high-quality labor force and build a grand industrial workers team.

According to the public data, in 2020 [16–18], the gap of skilled talents in key fields in China will exceed 19 million, and the data is still expanding, and it is expected to be close to 30 million in 2025. How to make vocational education better light up the life of workers and serve the needs of national development? How to increase the supply of vocational education and build a high-level and high-level talent training system? The key is to thoroughly implement general instructions of general secretary Xi Jinping, uphold the occupation of virtue, optimize the type and speed up the construction of modern vocational education system. We should not only further promote the reform of education mode, school running mode, management system and guarantee mechanism, but also increase institutional innovation, policy supply and investment, improve the quality of education by relying on reform, and consolidate the foundation of development through system.

The outline of the 14th five year plan proposes to "enhance the adaptability of Vocational and technical education". In recent years, the undergraduate level vocational education has taken a substantial step forward. 27 vocational colleges have independently held undergraduate level vocational education, and the "ceiling" of vocational education has been gradually broken. Facing the future, aiming at the direction of technological change and industrial optimization and upgrading, steadily improving the quality and level of vocational education, and constantly enhancing the recognition and attraction of vocational education, will be able to provide strong talents and skills support for the comprehensive construction of a socialist modern country.

2 The Connotation of Deep Learning

Generally speaking, the typical deep learning model refers to the neural network with "multiple hidden layers", where "multiple hidden layers" represent more than three hidden layers, and the deep learning model usually has eight or nine or even more hidden layers. When there are more hidden layers, there will be more parameters such as neuron connection weight and threshold. This means that deep learning model can automatically extract many complex features. In the past, when designing complex models, we would encounter the problem of low training efficiency and easy to fall into over fitting. After obtaining a better feature representation, we need to design a corresponding classifier, and use the corresponding features to classify the problem. Deep learning is a learning algorithm of automatic feature extraction. So to sum up, deep learning has the following three advantages compared with traditional machine learning.

(1) High efficiency: for example, using traditional algorithms to evaluate the merits of a chess game may require professional players to spend a lot of time studying every factor affecting the chess game, and it is not necessarily accurate. Using deep learning technology, as long as we design a good network framework, we don't need to consider the tedious feature extraction process. This is also the reason why alphago of deep mind company is so powerful that it can easily beat professional human chess players. It saves a lot of feature extraction time and makes things that are not feasible become feasible.
(2) Plasticity: when using traditional algorithms to solve a problem, the cost of adjusting the model may be to rewrite the code, which makes the cost of improvement huge. Deep learning can change the model only by adjusting the parameters. This makes it have a strong flexibility and growth, a program can continue to improve, and then close to perfect.
(3) Universality: neural network is to solve problems through learning, and can automatically build models according to the problems, so it can be applied to all kinds of problems, rather than limited to a fixed problem.

After years of development, deep learning theory includes many different deep network models, such as classic deep neural network (DNN), deep belief network, convolutional neural network (CNN), deep Boltzmann machines (DBM), recurrent neural network, etc., they all belong to artificial neural networks. Different network structures are suitable for different data types, such as convolutional neural network for image processing, recurrent neural network for speech recognition, etc. At the same time, there are a number of different variants of these networks.

3   Building Prediction Model

The data used in this paper comes from the Chinese version of the "vocational education investment survey" (NSSE China) of a vocational college in 2013. The in-depth learning scale is formed by selecting topics. To examine the relevant influencing factors of the level of deep learning in vocational education management and build a prediction model. The prototype of NSSE China questionnaire is the "national survey of student engagement" (hereinafter referred to as NSSE) questionnaire developed by Indiana University. NSSE is an annual survey of high-level learning activities and development of vocational college students nationwide in the United States. The 15lnsse China project was launched in 2007. After a series of cultural adaptation and pre testing, it was first tested nationwide in 2009. The questionnaire specifically measures several indicators, including five comparable indicators of Vocational Colleges (specifically lac of academic challenge, ACL of active cooperative learning level, SFI of student teacher interaction, EEE of educational experience and SCE of campus environment), nine indicators of Vocational Colleges diagnosis, DL indicators and social desirability indicators.

Through the SPSS software, the five comparable indexes and deep learning indexes of each student are calculated. According to the analysis of data, the distribution range of students deep learning index is 9.09. In this paper, the level of deep learning is divided into three levels: excellent, good and unqualified. The classification standard is shown in Table 1:

Table 1. Classification Standard

Comprehensive score DL	Evaluation level
$DL \leq 40$	Unqualified
$40 < DL \leq 70$	Good
$70 < DL \leq 100$	Excellent

3.1   DEEP LEARNING Network Model Structure

Deep learning network is a multilayer feedforward neural network. The most basic deep learning network consists of input layer, hidden layer and output layer. Each layer has many unconnected neuron nodes. The adjacent two layers of nodes are connected by connection weight. Its topology is shown in Fig. 1.

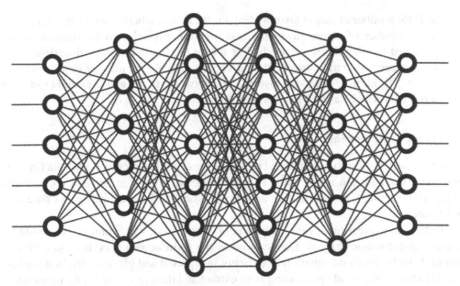

Fig. 1. Structure of deep learning network

(1) Determination of input layer nodes

There are 5 layers in the deep learning system with the vocational colleges, so n = 5, where x_1 means academic challenge, which includes the meaning of students individual and Vocational Colleges organization. It not only reflects students learning behavior performance in their studies, but also indirectly carries out academic requirements, academic standards and academic support through student's behavior performance and self-report Evaluation, to reflect the quality of education in Colleges and universities. x_2 represents the level of active cooperative learning. x_3 refers to students and teachers. It evaluates the frequency of interaction between students and teachers, their initiative and the quality of interaction. x_4 represents educational tasks. It represents teachers teaching practice, including organized teaching activities, clear knowledge explanation. It evaluates the quality of interpersonal communication in campus and the support degree of school for promoting students learning and development.

(2) Determination of output layer nodes

We use the deep learning to evaluate the college students. Therefore, the quantitative value of management deep learning level of vocational education is taken as the output of deep learning.

(3) Determination of the number of hidden layers

At present, there is no certain standard to calculate the number of neurons in the hidden layer, but the approximate range of the number of neurons in the hidden layer can be calculated according to the following formula, and then the best number of neurons in the hidden layer can be determined by trial and error method. Since the number of input layer neurons is 5 and the number of output layer neurons is 1, on the basis of the reference empirical formula $l < n - 1$, $l < \sqrt{(m + n)} + a$, $l = \log_2 n$ (where, is the number of input layer nodes; l is the number of hidden layer nodes:

m is the number of output layer nodes; a is the constant between 0–10), a relatively small number of hidden layer nodes is preliminarily selected for training, if the specified training times are reached or within the limited training timesIf there is convergence, stop training, and then gradually increase the number of hidden layer nodes. The number of hidden layers of the network is two, and the number of each hidden layer is 25 and 20 respectively.

3.2 Data Preprocessing

Data processing refers to the technical process of analyzing and processing data (including numerical and non numerical). That is, the process of data acquisition, storage, retrieval, processing, transformation and transmission, and the conversion of data into information.

Data is an expression of facts, concepts or instructions, which can be processed by manual or automatic devices. Data can be in the form of numbers, text, graphics or sound. Data becomes information after being interpreted and given a certain meaning. The basic purpose of data processing is to extract and deduce valuable and meaningful data for certain people from a large number of data that may be disordered and difficult to understand. Data processing is the basic link of system engineering and automatic control. Data processing runs through all fields of social production and social life. The development of data processing technology and the breadth and depth of its application greatly affect the process of human social development.

Data processing cannot do without the support of software. Data processing software includes various programming languages and Compilers for writing processing programs, file system and database system for managing data, and application software packages for various data processing methods. In order to ensure data security and reliability, there is also a set of data security technology. It includes the analysis, arrangement, calculation and editing of various original data. It is more meaningful than data analysis.

3.3 Content of Data Processing

Computer data processing is a technology that uses computer to collect and record data and produce new forms of information through processing.

Data sorting: arrange data in order according to certain requirements.

The process of data processing is divided into three stages: data preparation, data processing and data output. In the data preparation stage, the data is input offline to punch card, punch tape, tape or disk. This stage can also be called data entry stage. After data entry, it is necessary for the computer to process the data. For this reason, the user should program and input the program into the computer in advance. The computer processes the data according to the instructions and requirements of the program. Processing refers to the combination of one or more of the above eight aspects. The final output is a variety of text and digital tables and reports.

4 Matlab Simulation Implementation

4.1 Learning Process of DEEP LEARNING

According to the network prediction model and the setting of function parameters, the network is trained. The specific training process is as follows: (1) normalize the training data so that it is distributed between [0,1], and normalize the prediction input data in the same way. (2) The normalized data is input into the network model, the network is trained according to the learning algorithm of DEEP LEARNING network, and finally the predicted output is de normalized. (3) The predicted output and expected output are divided into three categories according to the score interval, i.e. excellent, good and unqualified, and the accuracy of the network model is calculated. (4) Draw DEEP LEARNING network prediction output graph and deep learning prediction error percentage graph.

4.2 Result Analysis of DEEP LEARNING

The learning of DEEP LEARNING belongs to supervised learning. A set of learning samples with known target output is needed. Therefore, this paper selects 1000 groups of data required by the prediction model from the NSSE China questionnaire of a vocational college in 2013, and randomly selects 900 groups of data as training samples, the rest 100 groups of data as test samples, input them into the network model, and get the DEEP LEARNING network learning and training process curve as shown in Fig. 2. As can be seen from the figure, when the number of training iterations reaches 5, the network has.

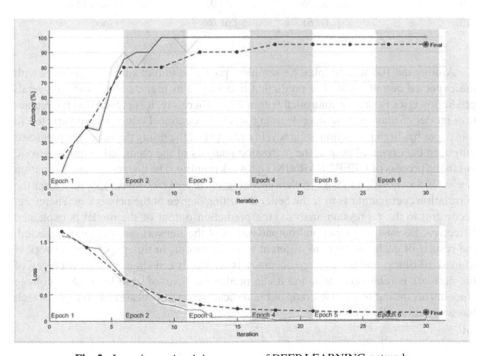

Fig. 2. Learning and training process of DEEP LEARNING network

The prediction accuracy of the network reaches the target value. After the training meets the requirements, 100 groups are selected as test samples. The results of the above network model are consistent with the results of the evaluation level of vocational education management deep learning shown in the questionnaire, and the relative error between the predicted output and the expected output is small. For example, as shown in Table 2, due to the large number of test samples used to test network accuracy, only 10 groups of randomly selected data from the test samples are listed in Table 2 for comparison.

Table 2. Data comparison

Sample serial number	Desired output	Predictive output	Evaluation level
1	45.45	44.9	Good
2	48.48	45.8	Good
3	69.70	63.91	Good
4	63.64	61.43	Good
5	51.52	50.16	Good
6	100.00	99.51	Excellent
7	48.48	48.99	Good
8	54.55	54.64	Good
9	36.36	38.09	Unqualified
10	60.61	61.08	Good

Among the 100 test samples, 88 samples' predicted output value is consistent with the expected output value. The prediction accuracy is as high as 88%, and the overall prediction error is strictly controlled within 20%. Therefore, it is proved that the prediction model of management deep learning level of vocational education constructed in this paper has high prediction accuracy. In order to test the fitting degree of the predicted output to the expected output, the regression analysis of the predicted output is carried out in the process of DEEP LEARNING network training. The linear regression equation is $y = 0.18x + 0.13$, and the correlation coefficient $R = 0.897$. Generally, the closer the correlation coefficient r is to 1, the better the fitting degree of the network is. Therefore, according to the regression analysis, the prediction output of the model is explained Effective. Because the input and output values of the network are randomly selected, the results of each training are different to some extent. In this paper, the network is trained 10 times, and the average accuracy is 85.3%. It can show that the accuracy of the network is relatively ideal, and it can predict the level of deep learning of college students according to the five comparable indicators among vocational colleges, which has a certain application value.

5 Conclusion

DEEP LEARNING network has similar shape input. The prediction accuracy and scientificity of this method depends not only on the number of training samples, but also on the quality of training samples. The more the number of training samples, the higher the quality, the more accurate the prediction of College Students deep learning level. At the same time, DEEP LEARNING network has strong self-organization and self-adaptability. Therefore, using DEEP LEARNING network algorithm to build the prediction model of vocational education management deep learning level can make its prediction results more accurate and reasonable. The prediction model of management deep learning level of vocational education constructed in this paper has the characteristics of high prediction accuracy and fast learning speed. Through this prediction model, not only can college students grasp their learning situation in time, but also can teachers evaluate their learning situation and improve their own teaching strategies more conveniently. At the same time, it can also provide scientific guidance for school teaching reform. Therefore, the prediction model of deep learning level of Vocational Education Management Based on DEEP LEARNING network algorithm has certain practical value and wide application prospect in the field of learning analysis and education reform.

References

1. Marton, F., Saljo, R.: On qualitative differences in learning—outcome and process. Br. J. Educ. Psychol. **46**, 4–11 (1976)
2. Ling, F.Q., Li, J.: Promoting students deep learning. Comput. Teach. Learn. **5**, 29–30 (2005)
3. Wang, J.: Deweys educational thoughts and in-depth learning. Educ. Technol. Guide. **9**, 6–8 (2005)
4. Tang, J., Zhang, X., Cheng, L.: evaluation of teachers educational technology ability training based on DEEP LEARNING neural network. Comput. Technol. Develop. **6**, 249–252 (2013)
5. Shi, H.: The idea of improving the management system of vocational education in China. Educ. Explor. **2**, 68–69 (2009)
6. Feng, W., Xu, Q.: Research on the innovation of vocational education management system. Times Educ. **7**, 281 (2015)
7. Meng, L.: Thoughts on the reform of vocational education management system. China Vocat. Tech. Educ. **17**, 23–24 (2005)
8. Cai, G., Liu, J.: Research on the innovation of Guangxi vocational education management system. Macroecon. Res. **10**, 63–69 (2008)
9. Zhou, C.: Establishing a sound operation mechanism of vocational education management system. China Adult Educ. **6**, 23–24 (2002)
10. Zhang, H.: Thoughts on the reform of vocational education management system. J. Henan Univ. Sci. Technol. **3**, 58–59 (2001)
11. Niu, Y., Ling, Y.: Comparison and enlightenment of vocational education management system in developed countries. World Educ. Inf. **10**, 42-44+66 (2006)
12. Song, N.: Research on the Innovation of Vocational Education Management System. Hunan Normal University (2004)
13. Xiao, Y., Li, M.: Reform and innovation of vocational education management system in China. J. Jiangxi Normal Univ. (Philos. Soc. Sci.) (2005)
14. Shao, H.: Comparison and reference of foreign higher vocational education management system. Scientific Mass Sci. Educ. (2007)

15. Zhou, W., Wu, X.: The reform of vocational education management system in the period of American social transformation. Vocational education communication. J. Jiangsu Normal Univ. Technol. (2009)
16. Li, K., Liu, F.: The reform of vocational education management system in the period of social transformation in Japan. J. Jiangsu Normal Univ. Technol. (Vocat. Educ. Commun.) (2009)
17. Li, D., Zeng, L.: The historical evolution and reform of Guangxi secondary vocational education management system. Guangxi Soc. Sci. **6**, 45–48 (2012)
18. Zhu, X.: Thoughts on the reform of vocational education management system. Contemp. Educ. Pract. Teach. Res. (Electr. J.) **12**, 287 (2018)

Construction of Multi-dimensional Dynamic Innovation and Entrepreneurship Platform for College Students Under the Background of Big Data

Shu Tan[✉]

Yunnan Land and Resources Vocational College, Yunnan 650000, China

Abstract. Based on the school enterprise cooperation, it can bring great benefits to enterprises, opportunities to students and fresh blood to schools. At present, there are many problems in the construction of school enterprise cooperation innovation and entrepreneurship platform under the background of big data. Although the school enterprise cooperation in China is still in its infancy, it is needed by universities, enterprises and the government, especially in the western region. At present, although there are still some problems in school enterprise cooperation, we can't take it as an excuse to be complacent.

Keywords: Big data background · School enterprise cooperation · College students · Innovation and entrepreneurship

1 Introduction

With the continuous enrollment expansion of colleges and universities in China, graduates are facing many difficulties in employment. In order to solve this problem, in addition to policy guidance and creating more employment opportunities, it is of great significance to actively guide and support college students with innovation and entrepreneurship to carry out individual or team entrepreneurship, It is beneficial to create employment opportunities, improve the employment rate and improve the comprehensive quality of talents, Due to the characteristics of College Students' innovation and entrepreneurship, such as wide coverage, certain risks, lack of systematic and consistent policies, and insufficient financial support, it is necessary to conduct a systematic analysis, and then design and develop a college students' innovation and entrepreneurship experience platform [1]. Basic framework of innovation and entrepreneurship platform is shown in Fig. 1.

Tao Xingzhi once said: "the common fault of Chinese education is that people who teach with brain don't use hands, and people who don't teach with hands use brain, so they can't do anything. The strategy of China's educational revolution is the hand brain alliance. As a result, the power of both hands and brain can be incredible." This sentence of Mr. Tao tells the current situation of China's education, and also points out the way out of the dilemma of China's education "hand brain alliance".

M. A. Jan and F. Khan (Eds.): BigIoT-EDU 2021, LNICST 392, pp. 79–88, 2021.
https://doi.org/10.1007/978-3-030-87903-7_12

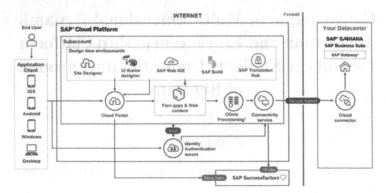

Fig. 1. Basic framework of innovation and entrepreneurship platform

2   Research on Innovation and Entrepreneurship Mechanism of College Students Based on Multidimensional Dynamic Innovation Model

The innovation and entrepreneurship mechanism of college students involves the government level, the social level, the university level and the enterprise level, which is a system engineering with considerable complexity. The multi-dimensional dynamic innovation model (mdmi) with multi factor integration and dynamic adjustment characteristics can be used for system analysis of the innovation and entrepreneurship mechanism of College students, The multi-dimensional dynamic innovation model can be divided into three sub levels: the first level is the entity level, which includes government entity, University entity, enterprise entity and social entity, and is in the dominant position in the whole model; the second level is the influencing factor level, which includes the elements that have great influence on the innovation and entrepreneurship mechanism of college students, and has uncertainty, The third layer is the driving layer, including the external factors that drive the innovation and entrepreneurship mechanism of college students, and the driving layer also has uncertainty.

3   Research on the Importance of Influencing Factors of College Students' Innovation and Entrepreneurship Based on Multivariate Support Vector Machine

The multidimensional dynamic innovation model established above analyzes the innovation and entrepreneurship mechanism of College Students under the influence of multiple factors from a qualitative point of view, An importance model of College Students' innovation and entrepreneurship influencing factors based on multivariate support vector machine is established [2]. The original SVM model is extended to binary classifier to solve the multivariate classification problem, and the multivariate classification problem is decomposed into multiple binary self classification problems. In order to solve the binary sub classification problem between class J and class k and maximize the boundary between the data, the soft marginal objective function is as follows.

$$\min_{w_j, w_k \in R^d} \frac{1}{2} \|w_j - w_k\|_2^2 + C \sum_{y_i \in \{j,k\}} \xi_i^{jk} \tag{1}$$

$$s.t. \quad y_i^{jk} f_{jk}(x_i) \geq 1 - \xi_i^{jk} \tag{2}$$
$$\xi_i^{jk} \geq 0$$

Finally, the prediction category of the sample will get the highest number of votes.

$$\tilde{y}_i = \arg\max w_j^T x_i + b_i \tag{3}$$

4 Data Analysis

4.1 College Students Should Take an Active Part in the Practice of Enterprises when they have Spare Time to Learn

University teachers should participate in the development and practice of enterprises under the condition that they have spare resources to teach, and enterprises should also choose to cultivate talents in Colleges and Universities under the condition that they can guarantee to complete their own production tasks. Only by sticking to their own position, can the cooperation between colleges and enterprises be the icing on the cake, and universities and enterprises can achieve their own goals [3].

4.2 First of All, Enterprises and Schools Should Sign a Confidentiality Treaty, so that Enterprises Can Provide Academic Resources for Colleges and Universities

At the same time, colleges and universities should also respect the operation of enterprises. The research results assisted by enterprises should be published with reservation after discussion with enterprises, so as to ensure the interests of enterprises and carry out school enterprise cooperation on this basis.

4.3 In Order to Improve the Engineering and Pertinence of the Experimental Content, the Actual Project Example is Introduced

In order to improve the efficiency of College Students' project research, we should change the previous model of College Students' scientific research on paper, combine with practical problems, improve the efficiency of College Students' project research, supplement the deficiency of classroom teaching, give full play to students' theoretical advantages, and make up for the lack of social derailment.

4.4 Developing Comprehensive Training Programs

According to the requirements of the quality and ability of applied talents, talent cultivation takes knowledge and ability as the structural standard, and can be divided into four types: academic, engineering, technical and skilled. In addition to academic type, engineering type, technical type and skill type all have a common feature application. According to the law of education and cognition, the cultivation of science practical ability is divided into three levels: basic skill level, professional ability level and comprehensive innovation and application ability level. According to the three different cultivation levels, different practical teaching contents and corresponding practical teaching links are arranged, and teaching is organized step by step. In addition, in the process of achieving the purpose of training, step by step, the use of progressive subsection to achieve the purpose of training. According to the logical sequence of cognition practice on campus comprehensive training on campus production cognition practice of off campus enterprises post training and project training on off campus enterprises, it provides and creates a good environment and conditions for the cultivation of application-oriented talents [4].

4.5 Choose Appropriate Enterprise Projects and Teacher Projects

Not all the problems encountered by enterprises are suitable for college students to study. College students should choose the appropriate difficulty and combine their own interests and advantages to ensure that they have the ability to complete the project and get the correct research results [5]. The research results can be used as graduation design, which can ensure that they have enough time and energy to invest in the research of the topic, and can also write their own research results in the form of papers to ensure that their research results will not cause waste. In this way, on the one hand, we can cultivate our own academic ability, on the other hand, we can encourage ourselves to study the subject more systematically and carefully. At the same time, we should ensure that the resources invested by schools and enterprises will not cause waste while the students are paid for their work [6].

4.6 Establish Resource Sharing Platform

Due to the influence of knowledge, environment and other factors, personal ability is always limited [7]. For example, there are nearly 50 units in manned space engineering, and about 1000 people participate in the research work. The main units and departments participating in the application system of manned space engineering are distributed in all parts of the country and all industries, including universities, scientific research institutes, and application departments of various ministries and commissions [8]. With the continuous development of manned space engineering, more and more scientific research institutes, schools and industrial departments will join the ranks of space application systems. This fully shows that "only width can accommodate people, only thickness can carry things." Tagore said, "when we are greatly humble, we are closest to greatness." Only by integrating all advanced scientific thoughts can we gain the strength and experience to advance. Only when we step on the shoulders of giants can we see and go further(see Fig. 2).

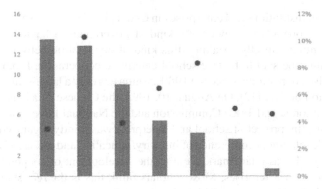

Fig. 2. Demand of students in innovation and entrepreneurship platform

5 The Historical Evolution of School Enterprise Cooperation in Higher Education

5.1 The Background of School Enterprise Cooperation in Higher Education

In the world, school enterprise cooperation is a relatively new mode of production [9]. In the 1950s, California in the United States formed the "Silicon Valley model" of high-tech complex, which combines scientific research institutions and universities, combines production and scientific research, and can quickly transform scientific research achievements into productivity or commodities. Due to the rapid development of "Silicon Valley mode", school enterprise cooperation has opened a new process all over the world. In recent years, school enterprise cooperation has connected the production of enterprises with the research and development of universities, realizing the close combination of science and technology and economy. Especially after the global financial crisis, countries pay more attention to the improvement of independent innovation ability. In recent years, the development of school enterprise cooperation in China is relatively rapid. In the context of domestic economic restructuring and upgrading, especially in the current colleges and universities vigorously advocate entrepreneurship education, through school enterprise cooperation to achieve entrepreneurship education is welcomed by the main body of cooperation. However, as far as the development of entrepreneurship education is concerned, the effect of school enterprise cooperation is not ideal. Therefore, in order to strengthen the mechanism of school enterprise cooperation on entrepreneurship education in Colleges and universities, we must explore the internal law, development process and characteristics of school enterprise cooperation, and put forward countermeasures for the development of entrepreneurship education in China from its development stage and characteristics [10].

5.2 An Overview of the Development of School Enterprise Cooperation in China's Higher Education

Some scholars believe that the state proposed the development of school enterprise cooperation in 1958, which can be regarded as the origin of the cooperation between higher education institutions and enterprises in China [11]. However, some scholars also think that this cooperative behavior is a kind of government behavior in the context of planned economy. Strictly speaking, this kind of cooperation between schools and enterprises cannot be said to be true school enterprise cooperation. From the scale of school enterprise cooperation, since the 1980s, China has only a large-scale school enterprise cooperation project [12]. On August 10, 1992, the Chinese Academy of Sciences, the national economic and Trade Commission and the National Education Commission announced that joint project of school and enterprise was jointly organized at the "first national joint development conference of industry, education and research". This project is generally regarded as a landmark event of the development of cooperation between Chinese schools and enterprises, Most scholars think this is the real starting mark of school enterprise cooperation in China. In recent 20 years, the cooperation between school and enterprise has developed vigorously in China. This chapter draws on the factors of national economy, science and technology, education development strategy, and specific practice mode of school enterprise cooperation [13]. After analyzing and combing the stages of the process of school enterprise cooperation, the development process of the school enterprise cooperation is divided into the following stages. At the same time, combined with the development of entrepreneurship education in different periods, the characteristics and shortcomings of school enterprise cooperation are analyzed [14].

6 Theoretical Discussion on the Relationship Between School Enterprise Cooperation and Entrepreneurship Education in Colleges and Universities

6.1 The Connotation and Characteristics of Entrepreneurship Education in Colleges and Universities

The word "innovation" was first proposed by Schumpeter. In his works, Schumpeter explained the concept of innovation and initiated the research of innovation theory. He thinks that innovation is "the establishment of production function" and "a new combination of production means". And the entrepreneur reorganizes the means of production to achieve new production methods can be called innovation. In Schumpeter's view, entrepreneurs must have certain excellent quality to achieve innovation. First of all, Schumpeter proposed creativity. He believes that there is a clear difference between entrepreneurs and technical experts in traditional society. Entrepreneurs need to break the stereotypes, they need to constantly change their behavior patterns to achieve a new combination of production factors; secondly, practicality is also an indispensable part of entrepreneurs. Entrepreneurs do not engage in invention and creation, but put the existing invention and creation into practice through practice, so as to create social economic value. Third, entrepreneurs need to be opportunistic. Entrepreneurs need to capture opportunities in the changing socio-economic situation. Fourth, excellent will quality. Because the high risk of entrepreneurship itself requires entrepreneurs to have strong willpower [15].

The content of current entrepreneurial knowledge can be described from the following aspects: the excellent entrepreneurial model in today's society, the attitude

of the society towards entrepreneurship, the support of entrepreneurship related policies and legal systems for entrepreneurial behavior, etc. It is not difficult to find that
entrepreneurial knowledge has a characteristic, which is comprehensive. Secondly,
although entrepreneurship is a kind of business behavior from the text, many factors
should be considered to cultivate students' entrepreneurial ability, such as the special mode of entrepreneurial thinking and reasoning, insight, imagination and creative
ability are important components of entrepreneurial ability. Therefore, to understand
entrepreneurship from a deep level and embody innovation education, entrepreneurship
education is the best practice path. Entrepreneurship education can not only optimize the
knowledge structure of college students, but also help them adapt to the future innovative
society, which also reflects the essential characteristics of entrepreneurship education.
The characteristics of entrepreneurship education in Colleges and universities are shown
in Fig. 3.

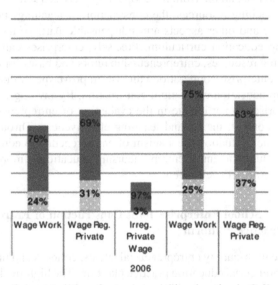

Fig. 3. Distribution of characteristics of entrepreneurship education in Colleges and Universities

6.2 Analysis of the Mechanism of School Enterprise Cooperation on Entrepreneurship Education in Colleges and Universities

To construct the curriculum system of entrepreneurship education based on school enterprise cooperation, the first thing to analyze is the curriculum objectives and characteristics of entrepreneurship education, and sort out the related tasks and abilities of
entrepreneurship education, summarize the corresponding disciplines, and transform
them into the corresponding entrepreneurship education curriculum. School enterprise
cooperation investigates many problems faced by entrepreneurship education, analyzes
the ability and quality requirements of entrepreneurship education with researchers in
the field of entrepreneurship education and experts from various industry associations,

defines the main tasks of entrepreneurship education, formulates the training objectives of entrepreneurship education, and constructs the curriculum system oriented by entrepreneurship education. At the same time, in order to constantly meet the needs of society, enterprises and universities, we should regularly revise the talent training program, and design and develop entrepreneurship education.

Therefore, school enterprise cooperation plays an optimized mechanism in all aspects of entrepreneurship education curriculum construction in Colleges and universities. School enterprise cooperation plays a guiding role in the design of curriculum standards for entrepreneurship education. The tools and equipment needed for entrepreneurship education are inseparable from the help of enterprises. Enterprises can provide corresponding industries to help colleges and universities formulate entrepreneurship education courses, so that colleges and universities are no longer blind in the design of curriculum standards for entrepreneurship education. In the implementation of entrepreneurship curriculum, it is also inseparable from the cooperation between schools and enterprises. The guarantee of teaching resources, the construction of teaching staff, the construction of training base and other aspects are indispensable parts of the implementation of entrepreneurship education curriculum. Precisely, enterprises can provide schools with various teaching resources, entrepreneurship tutors and bases for entrepreneurship education. School enterprise cooperation can also improve the evaluation mechanism of entrepreneurship education curriculum construction. By inviting enterprise experts, schools and enterprises to participate in the evaluation of entrepreneurship education teaching methods, content, means and teaching effectiveness. Through school enterprise cooperation, the evaluation mechanism of entrepreneurship education is diversified, and the curriculum content of entrepreneurship education is more in line with the requirements of enterprise development.

6.3 Mechanism of School Enterprise Joint Construction of Entrepreneurship Education Practice Platform

In order to cultivate high-quality entrepreneurial talents, colleges and universities need a high-quality entrepreneurial education practice platform. The high-quality entrepreneurship education practice platform needs high investment, but it is difficult to solve with the strength of the school itself in the case of the lack of funds in many colleges and universities. Therefore, we must mobilize the enthusiasm of enterprises to participate in the construction of entrepreneurial practice platform, and improve the construction mechanism of entrepreneurial practice platform. A fully functional entrepreneurship education practice platform needs to have the function of entrepreneurship education practice teaching, entrepreneurship skills training, and the training function of "double teacher" quality teachers. The school enterprise joint venture education practice platform has unique advantages in cultivating entrepreneurial talents. Firstly, the platform integrates the resources of schools and enterprises, realizes the reorganization of resources, and jointly cultivates entrepreneurial talents. Secondly, the entrepreneurship education practice platform simulates a genuine entrepreneurial environment for students, and organically integrates entrepreneurial theory and practice.

The training mechanism of entrepreneurship education teachers is not a simple training problem, it is a systematic project around the construction of entrepreneurship discipline, including the selection mechanism of entrepreneurship discipline leaders, the incentive mechanism of entrepreneurship teachers, the sharing mechanism of entrepreneurship teacher resources and so on. The training of entrepreneurship education teachers needs the cooperation of school and enterprise, give full play to the role of enterprise equipment and environment, organize teachers to receive training in entrepreneurship training base, and update entrepreneurship knowledge and skills. The source of entrepreneurship education teachers in Colleges and universities is not only teachers in Colleges and universities, but also enterprise experts or professional skilled talents. Therefore, the cooperation between university teachers and enterprise experts can be adopted to cultivate entrepreneurship education teachers. At the same time, we should improve the incentive mechanism for entrepreneurial teachers, stimulate the enthusiasm of university teachers, and protect the treatment of part-time teachers.

In order to improve the service quality of entrepreneurship education, colleges and universities need to establish a set of effective teacher supervision system and evaluation system. The system needs to achieve a comprehensive evaluation of the work of entrepreneurship education teachers, help entrepreneurship teachers improve their teaching, and put forward relevant suggestions. The evaluation mechanism of entrepreneurship education teachers should be different from other teachers, and its evaluation should have the following three characteristics. First, the "bottom-up" evaluation method still works, and students' evaluation of teachers is still an indispensable part of the evaluation system of entrepreneurship teachers; second, the leading group evaluates entrepreneurship teachers through the "top-down" method; third, in addition to the above evaluation methods, through school enterprise cooperation in entrepreneurship education teaching, The teaching evaluation of entrepreneurial teachers is more important. Enterprise peers use their unique experience to judge the teaching results of entrepreneurship teachers and help teachers improve their teaching.

7   Conclusion

Based on the school enterprise cooperation mode, this paper analyzes the existing problems and development trend of College Students' innovation and entrepreneurship activities from multiple perspectives by using multi-dimensional dynamic innovation model (mdmi), and constructs a college students' innovation and entrepreneurship experience platform, which has good performance. The importance of influencing factors of College Students' innovation and entrepreneurship is analyzed by using multiple support vector machine algorithm, The platform is stable, practical and functional, which can better meet the requirements of university innovation and entrepreneurship experience platform.

References

1. Zhang, H.: Discussion on the optimization of College Students' innovation and entrepreneurship education system under the mode of school enterprise cooperation. J. Xuchang Univ. **34**(4), 144–146 (2015)

2. Luo, Y., Chen, M.: Thinking and exploration of College Students' innovation and entrepreneurship practice based on school enterprise cooperation mode. Sci. Educ. Guide **15**, 178–179 (2016)
3. Wang, H.: Research on the construction of school enterprise cooperation mode for college students' innovation and entrepreneurship in Tibet University. J. Tibet Univ. **31**(4), 134–139 (2016)
4. Zhu, R., Peng, P.: On the design and implementation of ecological innovation and entrepreneurship mobile service platform in Colleges and universities. Modern Comput. **27**, 73–76 (2016)
5. Li, C.: Incentive and Restraint Mechanism of Modern Management. Higher Education Press, Beijing (2002)
6. Jiang, Y., Wang, W.N.: Research on cooperative education mode between University and enterprise. Modern Enterp. Educ. **10**, 25–26 (2008)
7. Pan, M.: Research on several theoretical problems of industry university research cooperation education. China Univ. Educ. **3**, 4–6 (2008)
8. Xu, X.: The concept of education and the strategy of Entrepreneurship. 6405 (2012)
9. Xu, X., Zhang, M.: concept change and strategic choice of entrepreneurship education. Educ. Res. **5**, 6468 (2012)
10. Kuang, W.: School enterprise cooperation in Higher Vocational and technical education under the "triple helix" theory. Explor. High. Educ. **1**, 115–119 (2010)
11. Feng, J.: Research on long term mechanism of school enterprise cooperation in higher vocational education. J. Hubei Univ. Econ. **7**, 125–128 (2008)
12. Yu, Z.: Progress, problems and improvement of long term mechanism of school enterprise cooperation in Vocational Education. Vocat. Tech. Educ. (2008)
13. Xu, X., Mei, W., Ni, H.: Entrepreneurial dilemma and institutional innovation of college students. China High. Educ. Res. **1**, 45–48 (2015)
14. Ding, J., Tong, W., Huang, Z.: Innovation of school enterprise cooperation mechanism in Higher Vocational Education. Res. Educ. Develop. **17**, 67–70 (2008)
15. Huang, Z., Song, Z.: Entrepreneurship education in Colleges and universities faces three major turns. Res. Educ. Develop. **9**, 45–48 (2011)

Design and Implementation of Innovative Entrepreneurial Experience System for College Students Based on Data Analysis Model

Lufeng Li[✉]

North Sichuan Medical College, NanChong 637002, SiChuan, China
lufengli1981@sina.com

Abstract. Aiming at some key problems of College Students' innovation and entrepreneurship under the mode of school enterprise cooperation, this paper uses multidimensional dynamic innovation model (mdmi) to qualitatively analyze the problems and development trend in the process of College Students' innovation and entrepreneurship activities from multiple perspectives, and uses multi factor Logistic regression analysis to quantitatively analyze the probabilistic nonlinear regression relationship between College Students' innovation and entrepreneurship mechanism and multiple influencing factors. This paper designs and implements an innovation and entrepreneurship experience system for college students based on multi-dimensional dynamic innovation model. The system adopts Java EE design framework and Hadoop development mode, and adopts the ASP.NET The collaborative filtering technology and support vector machine (SVM) algorithm are used for personalized recommendation and data mining of innovation and entrepreneurship. After the completion of the system design, the actual operation shows that the overall operation of the system is stable, which has a positive significance to promote the development of College Students' innovation and entrepreneurship activities.

Keywords: College Students' innovation and entrepreneurship · Multidimensional dynamic innovation model · Hadoop technology · SWM algorithm

1 Introduction

The course of modernization. The vigorous development of College Students' innovation and entrepreneurship can effectively optimize the employment problem of colleges and universities, not only relying on the number of resources invested, but also on the employment structure of college students in China. In the traditional employment mode of colleges and universities, it depends on the investment of modern science and technology. Only under the guidance of science and technology, can the dissemination and circulation of employment information be very limited, The lack of timely understanding of the market, the continuous transformation and upgrading of the traditional employment mode of colleges and universities, and the full use of the needs of modern graduates

M. A. Jan and F. Khan (Eds.): BigIoT-EDU 2021, LNICST 392, pp. 89–95, 2021.
https://doi.org/10.1007/978-3-030-87903-7_13

lead to great blindness in the employment of college graduates. Strict information technology and data fusion technology can accelerate the development of College Students' innovation and entrepreneurship mode, and restrict the healthy development of college graduates' employment. Therefore, the integration of information technology innovation and entrepreneurship experience system is the inevitable trend of college graduates' employment. By effectively combining with the needs of the employment market, scientific and reasonable overall planning is carried out in the process of college employment analysis and implementation to ensure timely information communication between college graduates and the employment market. The core purpose of the author is to develop a stable performance, suitable for college full-time employment counselors to use college students' innovation and entrepreneurship experience system, hoping to popularize the modern information advanced management concept and technology to college full-time employment counselors. Therefore, the combination of theory and practice, the current actual situation of college employment into the design and development of software. In the software design architecture of this paper, the multivariate logistic regression analysis is used to quantitatively analyze the probabilistic nonlinear regression relationship between College Students' innovation and entrepreneurship mechanism and multiple influencing factors. The system uses Java EE design framework and Hadoop development mode, uses aspnet language to realize dynamic web pages, and uses collaborative filtering technology and support vector machine (s VI m) algorithm for personalized recommendation and data mining of innovation and entrepreneurship. After the completion of the system design, the actual operation shows that the overall operation of the system is stable, which has a positive significance to promote the development of College Students' innovation and entrepreneurship activities [1].

2   System Requirement Analysis

As a programmatic guidance document for system development, system requirements analysis is very important for the smooth implementation of the system. There are many ways to obtain the system requirements. For example, through the establishment of relevant personnel, including experts and scholars in relevant fields, the establishment of learning groups, through interviews, with the help of market research, etc., and through field investigation, browsing historical information to collect relevant information, with the help of corresponding cases for detailed research, combined with the characteristics of the program itself, Multi level investigation and analysis of system requirements analysis. Based on the general method of forming the above demand analysis documents, this paper obtains the demand analysis of College Students' innovation and entrepreneurship experience system by combining actual research and literature review. On the one hand, it actually visits the current development status of College Students' innovation and entrepreneurship and their needs for future development in a university in East China, and obtains first-hand information, Provide objective and real data guarantee for system requirement analysis [2]. On the other hand, by consulting the relevant literature of College Students' innovation and entrepreneurship at home and abroad, and referring to the latest development trend of College Students' innovation and entrepreneurship at home and abroad, the corresponding demand analysis documents are formulated to ensure that the obtained demand analysis documents are in line with the current development trend. As the innovation and entrepreneurship technology and application place of college students are constantly changing, the innovation and entrepreneurship experience system of college students designed in this paper must be able to adjust according to the update and application place of innovation and entrepreneurship technology of college students, and realize the upgrading and improvement of the system, so as to meet the changing needs and have good scalability.

3 Research on the Importance of Influencing Factors of College Students' Innovation and Entrepreneurship Based on Multivariate Logistic Regression Analysis

The multidimensional dynamic innovation model established above analyzes the mechanism of College Students' innovation and entrepreneurship under the influence of multiple factors from a qualitative perspective. In order to further quantitatively analyze the probabilistic nonlinear regression relationship between College Students' innovation and entrepreneurship mechanism and multiple influencing factors, logistic regression analysis is introduced, This paper establishes the importance model of College Students' innovation and entrepreneurship influencing factors based on multi factor Logistic regression analysis, and the modeling process is as follows [3].

3.1 Sample Pretreatment

Set the test sample, set the two classification observation results (College Students' innovation and entrepreneurship results) as the d-junction, use the LR classifier after classification learning to preprocess the sample, set the processing weight sequence, then the sample and processing weight group can be obtained according to the linear superposition.

$$C = q_0 + q_1 x_1 + q_2 x_2 + \cdots + q_n x_n \tag{1}$$

3.2 Introducing Nonlinearity by Sigmoid Function

On the basis of sample preprocessing by LR classifier, sigmoid function is used to introduce nonlinearity, which is conducive to control the output range and ensure that the data is not easy to diverge in the regression process, as shown in Eq. (2):

$$f(x) = \frac{1}{1 + e^{-x}} \tag{2}$$

From Eq. (2), it can be concluded that the key of logistic regression analysis is to solve the weight of sigmoid function.

4   System Implementation and Test

4.1   Design of System Use Case Diagram

According to the actual needs and combined with the general design ideas of software engineering, the roles of College Students' innovation and entrepreneurship experience system based on multi-dimensional dynamic innovation model mainly include system administrator, full-time employment personnel in Colleges and universities, and general personnel. The use case diagram of each role user is shown in Fig. 1. In order to improve the system performance, the system performance requirements mainly include the following aspects. First, the system operation interface should be unified and friendly. In order to meet the needs of more users, the ease of use of the system should be improved. Second, the basic information of users in the innovation and entrepreneurship experience system must be standardized and complete, so as to ensure the reliability of the whole system. Fourth, ensure the real-time query of students' employment information data in the college students' innovation and entrepreneurship experience system. Fourth, when the employment information data in the college students' innovation and entrepreneurship experience system has the operation of writing and deleting, the system can complete the real-time automatic update and has the function of recording. Fifth, the database of College Students' innovation and entrepreneurship experience system should have good platform expansibility and high security performance [4].

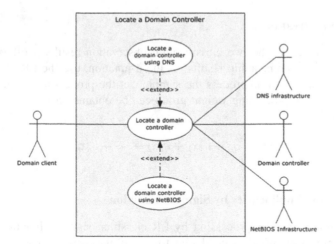

Fig. 1. System analysis use case diagram

4.2   System Implementation

According to the system architecture and overall function module design scheme given above, the system is programmed in vs2012 environment. Considering the universality of the system application, the system operation carrier takes Microsoft's classic stable

Windows 7 operating system as the standard. At the same time, in order to improve the user's access experience, the hardware configuration should be equivalent to the common computer workstation. The first is to design and develop the system, and then to meet the user's requirements of using browser to access the system. In other words, in order to meet the user's requirements of using domain name and IP to enter the system, it is necessary to first arrange and allocate the server of the system. Hardware configuration and environment requirements are the primary contents of system configuration database server. The environment requirements here are windows Server2008 server operating system, and sqlserver2015 is the database version to be installed. The client software environment can be reduced to meet the general Internet needs. In order to improve the access success rate, the browser software is recommended to use the IE browser of Windows 7 operating system.

5 System Software Design

System software design is the guarantee of system function realization. According to the system structure framework, system software can be divided into two parts: monitoring terminal software and monitoring platform software. The embedded software of monitoring terminal is the core content to ensure the function realization of monitoring terminal, and the display software of monitoring platform is the support to ensure the function realization of monitoring system. This chapter focuses on the specific design of embedded software of monitoring terminal and display software of monitoring platform, so as to improve the overall design of system software.

5.1 Overall Structure

The software of logistics monitoring system based on multi-dimensional information synthesis includes two parts: the embedded software of monitoring terminal and the display software of monitoring platform. The designed functions are as follows: (1) the embedded software of monitoring terminal has complete DSP chip control process mechanism, initialization, parameter configuration, interrupt response and program cycle operation; it realizes the data acquisition function of monitoring terminal in the process of logistics transportation, It includes temperature and humidity collection, object status image information collection, positioning information collection and so on: taking TMS320F28335 as the carrier, it processes the collected data, analyzes the data alarm threshold and data changes, analyzes and integrates the multi-dimensional information comprehensive sensing data, and stores the data in the cache; designs the working strategy of DSSP main control chip, Set up a complete sleep and wake-up mechanism, control the power system of the hardware platform for policy management, reduce power consumption, intelligently judge the sensor state, add function verification and restart reply function; prepare for data sending and receiving, establish handshake agreement with the monitoring platform, receive and respond to the command of the monitoring platform, and cache the failed data, And in chronological order. (2) The display software of the monitoring platform can connect data with the monitoring terminal, establish handshake protocol, analyze data and save data information, verify and display current data information, issue commands to the monitoring terminal, configure the parameters of the monitoring terminal and modify the operation strategy, display the positioning information and operation track of goods in the process of logistics transportation, and display the status information of goods, It can store photo information, be compatible with electronic map for location query, and connect with server back end to view historical data.

5.2 Data Processing

(1) Data analysis. The data information of monitoring terminal is divided into three categories: temperature and humidity information, image information and positioning information. In the temperature and humidity information, AD590 can directly get the digital temperature value by controlling the AD conversion module of tms320f2835, while SHT15 also returns the digital temperature and humidity value, which only needs to do simple averaging processing; the image information from the image module ov7670 returns the information byte with the opening 64 bytes of "image name" size, shooting time and image frame number, After decoding according to the protocol, the subsequent image information is framed to complete the image information analysis; the positioning information is extracted according to the keyword structure of BD to complete the data analysis of positioning information.

(2) Data communication. The serval communication of monitoring terminal is divided into three categories: internal data communication of core controller, data communication between core controller and data acquisition unit, data communication between core controller and wireless communication unit. The internal data communication of the core controller means that the control chip TMS320F28335 communicates with the data memory through SPI communication bus, through JTAG and simulation interface, through 232, 422 standard interface and communication interface, and through SP I and timer. The data communication between the core controller and the data acquisition unit is connected by the way of communication bus. AD590 converts data directly through AD conversion, SHT15 communicates data through 12C, image module communicates data through the cached UART, and Beidou positioning module communicates data through UART. The data communication between the core controller and the wireless communication unit is carried out through UART, and the prepared data is finally sent to the monitoring platform.

6   Conclusion

This paper first analyzes the important role of school enterprise cooperation mode in promoting students' innovation and entrepreneurship mechanism, and then analyzes that the university students' innovation and entrepreneurship mechanism involves many levels, which belongs to the system engineering with considerable complexity, and must be analyzed by using the systematic method, This paper analyzes the existing problems and development trend of College Students' innovation and entrepreneurship activities from multiple perspectives by using multi-dimensional dynamic innovation model (mdmi), and constructs a college students' innovation and entrepreneurship experience system. The system adopts Java EE design framework and Hadoop development mode, and adopts a variety of data storage and data mining technologies. The system has good performance. The actual test shows that the system runs stably, has strong practicability and functionality, and can better meet the requirements of university innovation and entrepreneurship experience system.

Acknowledgement. The Education Department of Sichuan Province:First-class courses in Sichuan Province (2020242), The third and fourth batch of innovation and entrepreneurship demonstration courses in Sichuan Province (2019145; 2020020), Sichuan Province Curriculum Ideological and Political Model Course Funding Project (2019057).

References

1. Zhang, H.: On the optimization of College Students' innovation and entrepreneurship education system under the school enterprise cooperation mode. J. Xuchang Univ. **34**(4), 144–146 (2015)
2. Luo, Y., Chen, M.: Thinking and exploration of College Students' innovation and entrepreneurship practice based on school enterprise cooperation mode. Sci. Educ. Guide **15**, 178–179 (2016)
3. Wang, H.: Research on the construction of school enterprise cooperation mode for college students' innovation and entrepreneurship in Tibet. J. Tibet Univ. **31**(4), 134–139 (2016)
4. Liu, J., Hao, F., Wu, G.: Design and implementation of College Students' innovation and entrepreneurship project management system based on J2EE. J. Luliang Univ. **7**(2), 37–42 (2017)

Design and Implementation of Piano Performance Grading System Based on Data Analysis

Jing Zhao$^{(\boxtimes)}$

Baoshan University, Baoshan 678000, Yunnan, China

Abstract. In order to meet the needs of piano learners to get feedback on their performance accuracy, this paper designs and implements a piano performance scoring system based on wechat. In the audio preprocessing part, the user's uploaded playing audio can be converted into a file format that is easy to analyze, and the music name can be detected by the music detection algorithm; in the performance scoring part, the music fingerprint based localization algorithm and the non negative matrix factorization based multi pitch detection algorithm are used to detect the correctness of the user's uploaded music. Based on the test data, the system scores the accuracy and the accuracy of the whole play for the subsection. The final result is presented in the form of images. The WeChat interactive part, which develops a public official account through the WeChat public platform development technology, can check the music contained in the system library by passing the official account. And upload their own playing audio, get graphical feedback and get some related services.

Keywords: Piano score · WeChat official account · Multi pitch detection · Music fingerprint location

1 Introduction

As a branch of the Internet, mobile Internet can provide people with high-quality network services through mobile terminals. In mobile Internet products, instant messaging applications have a large number of users. Among the instant messaging applications, the most outstanding one is the wechat released by Tencent in 2011. According to the statistics of professional market research companies, the market share of wechat in mainland China is as high as 93% 0. By June 2015, wechat had 600 million active users worldwide. In August 2012, WeChat launched the official account function, and enterprises and organizations can provide corresponding services to users through developing official account numbers. According to statistics, the official account number of WeChat has exceeded 12 million in 2016, an increase of 46.2% over the same period. In China, 523% of Internet users have the habit of using official account numbers. All this shows the attraction of the official account and the huge market hidden behind it. After a large number of questionnaires, we found that most piano learners want to have the function

© ICST Institute for Computer Sciences, Social Informatics and Telecommunications Engineering 2021
Published by Springer Nature Switzerland AG 2021. All Rights Reserved
M. A. Jan and F. Khan (Eds.): BigIoT-EDU 2021, LNICST 392, pp. 96–102, 2021.
https://doi.org/10.1007/978-3-030-87903-7_14

of piano performance evaluation in piano education products, but the current evaluation methods are very scarce [1]. Therefore, taking the accuracy of piano playing as the breakthrough point, this set up is designed to design a WeChat official account that can evaluate the piano playing audio of users.

2   Introduction of Related Technologies

Including the introduction of sound signal spectrum analysis method and multi pitch detection algorithm based on non negative matrix factorization. Then it introduces the music fingerprint algorithm used in the system, and then introduces the development technology of WeChat official account platform access technology and background development framework Codeigniter.

2.1   Basic Principle of Ant Colony Algorithm

The pitch information of sound signal is reflected in its frequency domain, and the spectrum analysis is the basis of frequency domain research. Most of the spectrum analysis is based on STFT (short time for short Fourier transform). Fourier transform can be applied to stationary random process and periodic transient signal, but sound signal is not a stationary process, so Fourier transform cannot be used to analyze it directly. However, the sound signal after short-time processing can be approximately regarded as a stationary random process, so it can be analyzed by short-time Fourier transform. The spectrum analysis of sound signal can be realized by two steps: windowing and fast Fourier transform.

2.2   Calculation Method

The traditional FFT spectrum generation algorithm is discrete Fourier transform (DFT), DFT is difficult to be widely used in practical applications because of its time complexity of N, and FFT algorithm is an improvement of DFT. FFT decomposes the DFT matrix into the product of sparse factors to achieve the purpose of fast calculation, thus reducing the computational time complexity to the order of nlogn. The definition of DFT is shown in Formula 1. Where x is the plural. If we calculate directly according to this definition, we need to do o (n) operations in total. Among them, XK has n outputs in total, and each output needs.

$$x_k = \sum_{n=0}^{N-1} x_n e, k = 0, \ldots, N - 1 \tag{1}$$

Foot (k) and feel (k) are the odd and even numbers of the sequence {xn −) respectively. So far, the first N2 points of YK can be calculated. For the last N2 points, because both fodd (k) and fven (k) are functions with period N2, and the unit root is symmetric, we can get formulas (2) and (3).

$$y_{k+\frac{N}{2}} = F_{even}(k) - W_N^K F_{odd}(K) \tag{2}$$

$$y_k = F_{even}(k) - W_N^K F_{odd}(K) \tag{3}$$

In this way, a Fourier transform of length n is decomposed into two Fourier transforms of length N2. Decomposition can continue in the same way. By using the method of the main theorem, it can be concluded that the time complexity of the algorithm is O (nlogn).

3 Multi Pitch Detection Algorithm Based on Non Negative Matrix Factorization

3.1 Introduction of Multi Pitch Detection Algorithm

In 1999, hsseung and ddle proposed a new matrix decomposition method in nature. This method is nonnegative matrix factorization (QMF) algorithm, which is a matrix factorization method when all elements in the matrix are nonnegative. This paper has attracted great attention of scholars in various fields, because in scientific research, there are many analysis methods with huge data that need to be processed by matrix, The method of non negative matrix factorization opens up a new direction for dealing with a large number of data. In addition, compared with the traditional algorithm, NMF algorithm is more simple in implementation, its decomposition form is easy to explain, and it occupies less storage space. There are many ways to realize NMF algorithm, among which NMF decomposition based on Euclidean distance and NMF decomposition based on dispersion are the most widely used [2]. The NMF algorithm based on dispersion can be described as decomposing the input matrix V into v wh (the elements of matrix V, W and H are not negative), so as to minimize the dispersion D. In the calculation, two multipliers are generated according to some new rules. When the dispersion of the product of V, W and H satisfies the set size limit, the iteration is stopped.

3.2 Form of Expression

First, Moreover, if the weight coefficient of a note changes from lower than the threshold value to higher than the threshold value in a certain period of time, it means that the piano key corresponding to the note has just been pressed; if the weight coefficient of a note changes from higher than the threshold value to lower than the threshold value, it means that the piano key corresponding to the note has just been released. We need to extract the spectrum template for each single note. For each audio containing a single note, first of all, it needs to detect the silent area, and get the playing section of the note through the detection of the silent area. Then, FFT transform is performed on the playing section of the note to extract the spectrum information, and then nm algorithm is used to extract the spectrum template which can represent the note in the spectrum information. The spectrum templates of each single note are assembled into w matrix. For the performance audio, after frame division and preprocessing, the frequency spectrum V of each frame is obtained by short-time Fourier transform. NMF algorithm is used to decompose h from V so that v = wh, and H is the weight coefficient of each note. After further thresholding, the results of multi pitch detection can be obtained. The principle of multi pitch detection algorithm based on NMF is shown in Fig. 1.

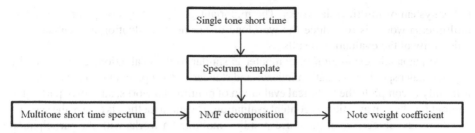

Fig. 1. Schematic diagram of multi pitch detection algorithm

4 Evaluation Index System and Evaluation Strategy

4.1 Main Introduction

Multimedia work evaluation is an important means to measure the practical teaching value and teaching effect of multimedia courses. The evaluation of multimedia works not only involves the teaching design thought, teaching content arrangement, design intention of teaching scheme, but also involves the application of multimedia technology and artistic experience of students. Student multimedia work is a multimedia assignment that students make and complete by using the technology and methods they learn in multimedia courses according to the requirements of teachers' proposition [3]. The evaluation of multimedia works should be multi-objective and individual, which should have mathematical statistics of formative evaluation and language description of summary evaluation. Each index item should be refined as much as possible.

4.2 Index System

According to various teaching evaluation methods at home and abroad, we think that the evaluation index of multimedia teaching can be divided into four levels. The first level is technical evaluation; the second level is artistic evaluation; the third level is content evaluation; the fourth level is creative evaluation.

(1) Technical evaluation. Mainly through the evaluation of the technical content and expression effect of multimedia works to evaluate students' mastery of multimedia technology.
(2) Artistic evaluation. It is mainly to evaluate the interface layout, frame structure and color collocation of multimedia works to evaluate the students' feelings on the rational use of multimedia elements in the works.
(3) Content evaluation. Content evaluation is mainly to judge the integrity, logic and expression effect of students' multimedia works.
(4) Creative evaluation. It is mainly about the evaluation of students' conception of multimedia works and their personality expression.

4.3 Evaluation Strategy

In order to realize the computer automatic evaluation of multimedia works, it is necessary to formulate specific evaluation strategies and methods on the constructed evaluation index system of multimedia works. The important goal of using computer to evaluate multimedia works is to reduce the workload of teachers' evaluation and enhance the objectivity of the evaluation results.

The automatic evaluation of computer is similar to the evaluation results given by experts under specific evaluation indexes and standards. Computer automatic evaluation is mainly to complete the technical evaluation of multimedia works, and solve part of the content involved in artistic and content evaluation. The main method is to give the specific problems and objectives of the corresponding multimedia works with the participation of teachers, and give the technical key points, art, content evaluation points and the final reference effect documents that need to be evaluated. According to the corresponding information provided by teachers, the multimedia works automatic evaluation system gives the specific score of each item according to the scoring standard, and finally summarizes into a final evaluation.

5 Design of Automatic Evaluation System for Works

5.1 System Design Objectives and Basic Ideas

The goal of this system is to explore a scientific and reasonable evaluation strategy, so that under the guidance of this strategy, the computer automatic evaluation of multimedia works can be realized. Therefore, the important criterion of system implementation is operability, which not only reflects the evaluation of each evaluation index to the work, but also enables the evaluation to be realized by computer. As a matter of fact, it is very difficult to form a unified evaluation standard for computer automatic evaluation because there are too many subjective elements in a work completed under an open topic without any constraints [4]. Therefore, it is necessary to have a suitable topic design as an important part of the system and have a suitable topic, In the process of computer automatic evaluation, we mainly solve the following problems: how to obtain the document information of multimedia works. how to use and analyze the obtained information to reflect the index components in the evaluation strategy. This part is mainly guaranteed by scoring strategy and logic.

5.2 System Composition

According to the above analysis and the actual environment and result requirements of the system, I divide the system into the following functional modules Block.

(1) Database module: the module mainly includes two parts, one is the question database, which is composed of examination document, standard answer and scoring strategy; the other is the result database, which stores the examinee information and evaluation results, which is the main data recording part of the evaluation system. Since there is no need for the original document in the evaluation process, the standard answers and scoring strategies in the question bank which affect the marking are called "multimedia works scoring standard management tool".

(2) Preprocessing module: This module is used to unpack and decrypt the recovered test paper, and get the legal marking document of the specified question type.

(3) Document recognition module: This module mainly completes the recognition of multimedia documents. That is to say, using the information export tool of multimedia works, the animation or image files can be converted into ordinary text with key information, which provides the basis for marking and matching. This module is the core module of the whole system.

(4) Marking module: the module extracts and analyzes the information of candidates' documents which have been parsed into text files, and gives the evaluation results according to the scoring strategy.

5.3 Design and Implementation of Evaluation System

The evaluation system consists of preprocessing module, document recognition module and marking module. The preprocessing and marking module are completed by the system platform. The document recognition module is implemented in the form of plug-in, which is the "multimedia work file information export tool" mentioned above. The implementation process of the evaluation is described as follows:

(1) The standard answer files of multimedia works produced by teachers are exported by using the "multimedia works file information export tool" to form the "multimedia works standard answer export information", and the score knowledge points are extracted from the information as the standard answers of system evaluation.

(2) The result file uploaded by the candidates is preprocessed, the result package is decrypted, the candidates' information is read, and the legitimacy of the results is verified. This process is completed by the preprocessing module.

(3) Thirdly, the document information identified in the examinee's multimedia file is stored in the temporary file by using the "multimedia work file information export tool".

(4) Compare the information of the designated knowledge points in the temporary file with the standard answers, and give the score according to the detailed rules.

(5) Record the score into the database.

   Exactly speaking, the second step in the process is the beginning of batch review. The formation of standard answers in the first step should belong to the task of proposition stage, but because there is a process of multimedia document recognition here, it is regarded as a part of the evaluation execution [5]. It can be seen that the "multimedia works file information export tool" used for document recognition is useful in the process of extracting standard answers and batch review. Therefore, we design the script plug-in as a separate tool that can run, and call it where it needs to be used in large platforms.

6 Work Summary

In this paper, the multi pitch detection algorithm and music positioning algorithm are studied. According to the needs of piano learners to obtain the evaluation feedback of their own performance accuracy in practice, the investigation is carried out. After finishing, the demand of piano performance scoring is analyzed, and the system design and implementation are carried out according to the functional requirements obtained from the analysis. The system finally adopts the NMF based multi pitch detection algorithm and music fingerprint based music localization algorithm as the core algorithm of the system, and implements a piano score system that can upload the playing audio to the user's score and overall score, and the system is finally presented in the form of WeChat official account. After testing, the system performs well in function and scoring accuracy.

References

1. Hou, Z.: Design and Implementation of Higher Vocational Learning Evaluation System Based on Web. Shanxi Normal University (2014)
2. Xing, M., Zhang, J., Wang, X., Zheng, Y., Li, M.: Design and implementation of provincial ambient air automatic monitoring system. In: Proceedings of 2013 Annual Meeting of Chinese Society of Environmental Sciences, vol. 5, p. 7. Chinese Society of Environmental Sciences: Chinese Society of Environmental Sciences (2013)
3. Wu, D.: Design and Implementation of Micro Auto Focusing System for Automatic Urine Sediment Analyzer. Wuhan University of Technology (2012)
4. Xiang, J.: Design and Implementation of Multimedia Works Automatic Evaluation System. Beijing University of Posts and Telecommunications (2010)
5. Yi, Q., Zhang, H., Wu, P., Lv, X., Luo, Q., Zeng, S.: Design and implementation of an optical microscope auto focusing system. Comput. Eng. Appl. **27**, 119-120+215 (2007)

Exploration of Computer-Based Course Teaching in Data Thinking

Yun Zhu[(⊠)]

Zhuhai College of Jilin University, Zhuhai 519000, China
juliet_zy@sohu.com

Abstract. The research results of this paper are as follows: put forward suggestions and Countermeasures to improve the construction of university computer basic teaching system and deepen the reform of computer basic teaching: combine the cultivation of "computational thinking ability" with the cultivation of "compound high-quality innovative talents"; promote the integration of computer basic teaching and professional teaching. At present, there are some problems in the teaching of computer basic course in our country, such as "special tool theory", disconnection with high school information technology teaching, "learning" goal, etc. Starting from the same level, improve the specific program of teaching reform of basic computer course.

Keywords: Basic computer course · Teaching exploration · Computational thinking · Teaching model

1 Introduction

Computational thinking is an important topic concerned by the international and domestic computer science, philosophy and education circles. The research and development of CT is of great significance to China's computer education. Academician sun Jiaguang, Dean of the school of software, Tsinghua University, also put forward in "the revolution of computer science" that "Computational Thinking" is the most long-term and basic thought in the field of computer science. In addition, China's computer education sector has also accelerated the pace of research and Discussion on Computational Thinking. In July 2010, the Ministry of education's College Computer Foundation Course Teaching Steering Committee issued the joint statement on the development strategy of nine University Alliance (C9) computer foundation teaching, which clearly requires all colleges and universities in China to carry out the reform of computer foundation teaching with computational thinking as the core. At present, some colleges and universities in China have taken the lead in carrying out the reform and practice of basic computer courses with "computational thinking as the core". For example, Shanghai Jiaotong University and Southern University of science and technology have opened a new basic computer course - "Introduction to Computational Thinking" for 2010 freshmen.

© ICST Institute for Computer Sciences, Social Informatics and Telecommunications Engineering 2021
Published by Springer Nature Switzerland AG 2021. All Rights Reserved
M. A. Jan and F. Khan (Eds.): BigIoT-EDU 2021, LNICST 392, pp. 103–108, 2021.
https://doi.org/10.1007/978-3-030-87903-7_15

At present, Although scholars at home and abroad have carried out some research on Computational Thinking and achieved certain results, there is no consensus on some basic theories of computational thinking, such as the concept, characteristics and principles of computational thinking. In addition, some research fields, such as the basic computer curriculum system, teaching mode and teaching mode with "Computational Thinking Ability Training" as the core, have not yet reached a consensus Methods and teaching evaluation system have not attracted the attention of the majority of scholars [1].

Nowadays, the development of Internet provides more convenient conditions for acquiring knowledge resources and information resources, and also provides better learning tools and broader learning space for lifelong learning. The popularity of computer brings us not only convenience in study and work, but also great improvement in life. Therefore, it is essential to master computer technology. As colleges and universities to cultivate modern talents, the opening of basic computer courses is particularly important.

2   Basic Requirements of Computing in English Vocabulary Query System Based on Struts

Generally speaking, computational thinking refers to the thinking method that well-trained computer science workers are used to in the face of problems. It is embodied in some typical means and ways of analyzing and solving problems in the brilliant development of computer and information technology in the past half century. Computational thinking is recursive thinking and parallel processing, which can translate code into data and data into code. It is a type checking method promoted by multidimensional analysis. Computational thinking is a method that uses abstraction and decomposition to control complex tasks or design huge complex systems. Computational thinking is a way of thinking to choose an appropriate way to state a problem, or to model the related aspects of a problem, making it easy to deal with. Computational thinking is a way of thinking to prevent, protect and recover the system from the worst through redundancy, fault tolerance and error correction. Computational thinking is a way of thinking that uses heuristic reasoning to seek solutions, that is, planning, learning and scheduling in uncertain situations. Computational thinking is a way of thinking that uses massive data to speed up computing, and makes tradeoffs between practice and space, processing capacity and storage capacity. Through in-depth analysis, it can be seen that there is a certain correlation between higher-order ability and computational thinking. The cultivation of Computational Thinking as the core represents the generation of a series of higher-order ability. The corresponding relationship between computational thinking and higher-order ability is shown in Table 1.

Table 1. Advanced capabilities

Ability	Definition	Corresponding points with computational thinking
Innovate	The ability to produce novel, original and socially significant products	a b c f
Problem solving	It refers to any goal oriented cognitive operation procedure	a d
Policy decision	Faced with multiple options. The ability to make judgments through a careful process of thinking	f g
Critical thinking	Have independent, comprehensive and constructive thinking ability to a certain phenomenon/thing	e
Information maintenance	The ability of problem solving and innovation by using various information tools/resources	a b d

The purpose of university computer basic course is to cultivate students' basic quality. Therefore [2], in the process of university education, the goals of quality education, the cultivation of computational thinking ability, and the teaching of university computer basic course are consistent. The relationship between the three is shown in Fig. 1.

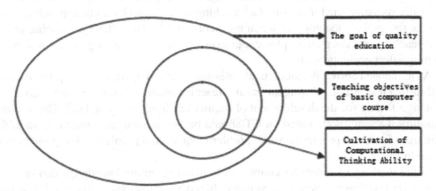

Fig. 1. The relationship between "quality education basic computer course Computational Thinking"

For a long time, our country has been carrying out the reform of quality education, and has achieved fruitful results. Therefore, in the process of teaching reform of College Computer Basic Course Based on Computational Thinking, it is necessary to actively learn from the experience, methods, means and achievements of quality education reform to change the teaching concept of teachers and the learning concept of students in the teaching of college computer basic course.

3   The Design of Teaching Mode of Computing Thinking

With the deepening of the teaching reform of the new curriculum, the teaching reform goal based on the cultivation of CT ability as the core of Cultivating University Learners' scientific thinking and scientific methods has been recognized and concerned by the majority of scholars, experts and teachers. The construction of "CT" teaching and learning mode as the core ability has also become the focus of the realization of this core ability. The reason why the "CT" method is used as a direct means to construct the reform means to cultivate the "CT" ability is to make a more intuitive and complete description of the abstract "CT", so as to help researchers and computing enthusiasts understand and analyze. The construction of a series of teaching mode and learning mode based on "CT" is based on relevant theories and methods. Through the analysis of relevant theories and the summary of practical experience of course teaching, the teaching process based on CT is explained and analyzed. The discussion on the teaching and learning mode of CT series helps to promote the understanding of the process of creating teaching and learning mode for the majority of scholars, and apply the series of modes to the course teaching activities, so as to promote the learners' mastery of CT methods and the improvement of CT ability [3].

As modern educators, especially those who have been fighting in the front line of education for a long time, each of them actually has a set of teaching methods belonging to their own style, which can be said to be their own personal "teaching mode". Once they produce good teaching results in their teaching, they can promote their teaching methods. Generally speaking, the formation of a "teaching mode" must have corresponding links, such as the guiding ideology of teaching theory and learning theory, teaching objectives, teaching process plan, implementation conditions, teaching organization strategy, teaching effect evaluation, etc.

As mentioned above, the theoretical basis and practical operation of inquiry teaching need further research. Exploring the inquiry teaching mode with thinking as the core is of great significance for the development of inquiry teaching theory and CT. The research of inquiry teaching mode based on CT should be constructed from the three variables of Inquiry Teaching: problem posing, problem inquiry and problem solving method, as shown in Fig. 2

In the teaching of computer course, all kinds of scientific knowledge can be used, and inquiry teaching is gradually developed. In order to fully express the whole teaching activity, it is established as the following mathematical function expression [4].

$$Q = F(A_T, A_S, P) \tag{1}$$

In general, A_T and A_S has the following action set:

$$A_T = \{q, i, r, h, c\} \tag{2}$$

$$A_S = \{l, t, a, d, m\} \tag{3}$$

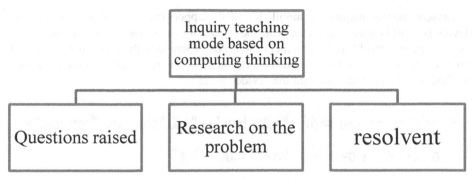

Fig. 2. The construction basis of inquiry teaching mode

In the process of implementing inquiry teaching based on CT, learners are required to use a series of methods of CT to explore and discover the essence of the problem, and to cultivate learners' independent ability to think and answer questions like scientists under the guidance of scientific inquiry methods. To achieve this goal, the scientific CT into the inquiry teaching process, can greatly improve the ability of teachers and learners to quickly find and solve problems.

4 Application of Computational Thinking Ability in Basic Computer Course Teaching

The rise of MOOC curriculum platform is inseparable from the support of learning support technology. Learning support refers to the learning support provided to meet the reasonable needs of individual or group learners in the learning process under the learner centered modern distance education environment, including learning conditions, learning environment, learning materials and other learning resources and comprehensive, high-quality and efficient help and services. Creating environment for learners' learning needs, improving learners' learning efficiency, and promoting learners' interactive learning are the three levels to gradually improve learning support, and the technical support of hardware, software, communication and other aspects provided for these three levels is learning support technology [5].

Teachers will upload the recorded course video resources to MOOC course platform, and learners can watch them freely after landing on MOOC course platform. Learners not only need to watch learning videos, but also need to share and interact videos. When learners discuss in the forum of MOOC course platform or finish practical homework, they can give up monotonous text or picture content and use mobile terminal to record a certain length of micro video. The learners upload the recorded micro video to the homework area, and the teachers make comments and corrections on the video, and then conduct quantitative assessment on the students. This kind of video interaction mode has changed the present situation of one-way presentation of teaching resources, and enriched the interaction between learners and teachers, and between learning peers. This two-way video interactive learning mode needs three aspects of technical support: video recording equipment support, network bandwidth support and platform function support.

At present, the vast majority of mobile terminals support video recording function, and device support is easy to achieve; with the popularity of WFI and 4G communication technology, bandwidth is easy to meet the needs of learners; therefore, the support of MOOC course platform for video technology is the most critical link. Figure 3 shows a method to include videos through the <video> tag.

<video>   <video width="320" height="240" controls="controls">

<source src="movie.ogg" type="video/Ogg">

<source src="movie.mp4" type="video/MPEG4">

Your browser does not support the video tag.   </video>

Fig. 3. Video standard method

5   Conclusion

At present, in order to meet the needs of the development of the times, colleges and universities need to reform the basic computer courses. Of course, in our modern information technology environment, if we want to face the change correctly, our teachers' teaching methods are bound to change. In the process of teaching, no matter what kind of teaching method we use, the final task is the same, that is, in the teaching of basic computer course, in addition to cultivating students' computational thinking ability, we should also improve their innovation ability. In learning, through learning, they can really become the main body of learning.

References

1. Dong, R., Gu, T.: Computational thinking and computer methodology. Comput. Sci. **1**, 1–4 (2009)
2. Feng, B.: Computational thinking and basic computer teaching. In: The 7th Summit Forum on Computer Teaching Reform and Development in Colleges and Universities, p. 7. Xi'an Jiaotong University, Zhangjiajie (2011)
3. Guo, X., **2**:31–2001, on the curriculum model
4. Lan, B.: Construction of learning support service system model for distance education. China Dist. Educ. **5**, 45–48 (2006)
5. Guan, J., Li, Q.: Development status, trend and experience of online education in China. China Audio Vis. Educ. **8**, 62–66 (2014)

Research on Convolutional Neural Network Based on Deep Learning Framework in Big Data Education

Xuan Luo$^{(\boxtimes)}$

Wuchang Shouyi University, Wuhan 430064, Hubei Province, China

Abstract. Deep learning is an important part of the development of artificial intelligence. Deep learning has made breakthroughs in many fields (such as image recognition, speech recognition, natural language processing), and has made gratifying achievements in the application of traditional algorithms which are not easy to solve, It includes automatic driverless vehicle, automatic pattern recognition, automatic simultaneous interpretation, commodity image retrieval, handwritten character recognition, license plate recognition, etc. In recent years, with the continuous improvement of research and development personnel's requirements for deep learning development process, the traditional deep learning programming methods can not meet the current needs. The traditional deep learning programming methods will take researchers and developers months or even years to implement the most basic algorithms. At the same time, the traditional deep learning programming methods can not meet the current needs, In this paper, a variety of in-depth learning frameworks, including CAE, have been developed for some of the world's top research institutions. These deep learning frameworks not only provide efficient and fast development models for scientific research institutions and related developers, but also provide several convolutional neural network models for developers to study and improve on the more advanced and perfect convolutional neural network models.

Keywords: Deep learning · Artificial neural network · Convolutional neural network · Deep learning framework · Caffe

1 Related Concepts of Personalized Recommendation System

Deep learning is a branch of machine learning based on artificial neural network. "Artificial neural network" ("Ann") tries to build "artificial" biological neural cells (i.e. neurons) and neural networks, and realize the functions of human brain nervous system in information processing, learning, memory, knowledge storage and retrieval at different levels. Deep learning comes from "artificial neural network", which combines the bottom features to form abstract high-level attribute categories or high-level features, and then finds the distribution feature representation of data [1].

The earliest deep learning framework is recognized as a new cognitive machine proposed by Fukushima K. Neocognitron in 1980. In 1989, in the research of handwritten

M. A. Jan and F. Khan (Eds.): BigIoT-EDU 2021, LNICST 392, pp. 109–114, 2021.
https://doi.org/10.1007/978-3-030-87903-7_16

character set recognition, Yann lecunn1 used back propagation algorithm, MP algorithm and graphics card acceleration, which is similar to the combination of new cognitive machine, weight sharing and convolution neural layer. Since then, this algorithm has become a basic part of many deep learning models. This is also an important milestone in the development of convolutional neural network. Although the algorithm can be successfully implemented, the cost of calculation has exceeded the normal range, so that it can not be applied in the actual research work. Then in 1991, Sepp Hochreiter proposed a solution to the gradient vanishing problem. However, with the development of computational model, SVM (support vector machine) for unknown problem modeling in many ways is more simple and easy to use than neural network, so the development of neural network in the 1990s compared with the slow progress.

2   Overview and Structure of Artificial Neural Network and Convolutional Neural Network

Artificial neural network is to use the connection between neurons, input and output to model, or to study the relationship between data. At present, the most typical artificial neural networks are BP network, Hopfield network, Boltzmann machine, SOFM network and art network [2].

The function of perceptron is to classify and recognize two kinds of attributes. Although the structure of perceptron unit is simple, it has all the elements of artificial neural network. The perceptron is composed of input part and output part. The input and output parts are directly connected. The perceptron can get an output value f (x) by mapping the input value X.

2.1   Feedforward Back Propagation Algorithm

The iterative process of feedforward back propagation algorithm is mainly divided into the following two steps: 1. Input a series of samples and output a result through forward propagation; 2. Adjust the weights in the neural network by calculating the deviation. In the whole calculation process, feedforward and back propagation are carried out alternately until the deviation is within a threshold or the number of iterations reaches a certain requirement. First, explain what the variance loss function is. For a problem with Class C and N training samples, the loss function is defined as:

$$E^N = \frac{1}{2} \sum_{n=1}^{N} \sum_{k=1}^{c} (t_k^n - y_k^n)^2 \tag{1}$$

Because the error of the whole data set is obtained by summing all the errors in each class, when calculating the feedback function, we need to focus on a single mode. The nth class error of the feedback function is:

$$E_n = \frac{1}{2} \sum_{k=1}^{c} (t_k^n - y_k^n) = \frac{1}{2} \| t^n - y^n \|_2^2 \tag{2}$$

2.2 Convolution Neural Network

Generally speaking, using multi-layer back propagation network to train data can get better results. The traditional multi-layer back propagation network needs to collect the effective information in the data through manual input. It needs to go through a complex preprocessing process to get the feature variables, and then classify and recognize the above processed feature variables through the definition of a trainable classifier. This kind of traditional multi-layer back propagation network can play the role of classification [3].

Compared with the traditional shallow neural network, the structure of convolutional neural network is much more complex. Each layer of neurons uses local connection to connect, neurons share connection weight and use reduced sampling in time or space to make full use of the characteristics of data itself. These characteristics determine that the dimension of convolution neural network is greatly reduced compared with traditional neural network, Thus, the complexity of the calculation is reduced. Convolutional neural network is divided into two processes, namely convolution and sampling. Convolution is mainly to extract the upper data and abstract, while sampling is to reduce the dimension of the data.

Activation function is an important part of neural network. Through the function transformation of the input of neural network, the appropriate output is obtained. Generally, the activation function injects nonlinear factors into the neural network with poor linear expression ability, which makes the data separable in the nonlinear condition, and also can sparse expression of data, and more efficient data processing (Fig. 1).

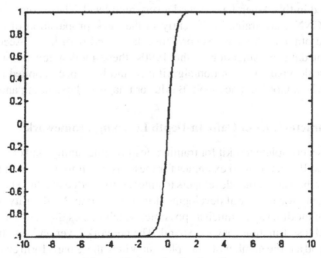

Fig. 1. Tanh activation function

3 The Construction and Principle of Caffe Deep Learning Framework

Cafe deep learning framework provides a pure and changeable framework for multimedia scientists and developers. Cafe framework includes the most advanced deep learning algorithm and a series of reference models. O Cafe framework is written in Python language and MATLAB language, and calls the C + library of Berkeley University, It is used to train and deploy common convolutional neural networks and other effective depth algorithms for common structures. When cafe is applied to commercial and Internet scale media, it needs CUDA GPU for computing. It can process 40 million images (each image is about 25 ms) on independent K40 or Titan GPU every day. The cafe framework realizes the model representation of extraction, can carry out simulation experiments and support more efficient development, and can seamlessly transform with the prototype deployment platform in the cloud environment.

3.1 Overview of Caffe Deep Learning Framework

The key problem of multimedia data analysis is to find the effective representation of perceptual input - image, sound wave, touch and so on. The manual feature extraction method of the input data has entered a stable stage in recent years, but the way of extracting features in the structure of depth composition by convolution has been steadily rising. Depth model has been better than manual feature representation in many fields, and it can be extracted in those areas where manual features are imperfect.

Large scale visual recognition is the most concerned area. Caffe, the deep learning framework used in this paper, has reached a very high level. These convolutional neural networks, or CNNs, are trained differently in the back propagation of convolutional filter layer and other layers such as correction layer and pool layer. According to the early results of data classification in the 1990s, these models are far superior to all known large-scale visual recognition algorithms, and have been applied to large-scale businesses, such as Google, Facebook, Baidu for image understanding and search [4].

3.2 The Characteristics of Caffe In-Depth Learning Framework

Caffe provides a complete toolkit for training, testing, fine tuning, and developing models, and it has well-documented examples for these tasks. Similarly, it is an ideal starting point for researchers and other developers to enter cutting-edge machine learning, which makes it available for industrial development in a short time. Modularity: Cafe follows the principle of modularity as much as possible, which makes the new data format, network layer and loss function easy to expand. The network layer and loss function have been defined, and a large number of examples show how these parts form an identification system for different situations.

Separation of representation and Implementation: the definition of cafe model has been written into configuration file in protocl buffer language. Cafe supports network construction in any directed acyclic graph. According to the instantiation, cafe keeps the memory needed by the network and extracts the memory from the host or the bottom of GPU. Only one function needs to be called to convert between CPU and GPU.

3.3 Training of Convolutional Neural Network Model Based on Caffe

MNIST data set is a large binary handwritten numeral data set, which is a standardized data set in machine learning and deep learning. The MNIST dataset contains 60000 training samples and 10000 test samples. Mnst dataset is a kind of black-and-white image which keeps the aspect ratio of the original 20×20 pixel handwritten character image in NIST dataset and expands the center of NST image to 28×28 pixel. The special database SD-3 and SD-1 of NIST dataset are composed of SD-3 as training set and SD-1 as test set. The SD-3 data set is clearer than the SD-1 data set, because the SD-3 data set is collected from the employees of the census and Census Bureau of the United States, while the SD-1 data set is collected from high school students in the United States. It is concluded that the training results should be independent of the selected data set. Therefore, mnst data set selects 30000 samples from SD-3 and SD-1 data sets respectively. The test set selects 5000 samples from each of the two data sets. A total of 60000 training samples were collected from 250 writers, and it can ensure that the writers of the training set and the test set do not coincide.

In this paper, we use the relu activation function instead of the original sigmoid activation function. Inspired by alexnet, we add the activation function in the convolution layer and the downsampling layer to make the network converge more quickly. There is no normalization level processing, which will reduce the training speed of the network. If mnst is normalized, errors may occur.

4 Conclusion

Today, with the development of artificial intelligence, deep learning has realized many tasks for scientific research and traditional industries that traditional methods can't accomplish. In recent years, deep learning has become the focus of scientific research. Although there are still many problems to be solved, the breakthrough of deep learning in image recognition and speech recognition has brought hope to scientists and related researchers. Google, Baidu and other research centers have developed a number of deep learning products, and have been applied in many civil fields. Many efficient and practical deep learning development tools are updated at an unprecedented speed. Convolutional neural network is the best in the field of deep learning. Most of the deep learning algorithms and network structure are convolutional neural network. Convolution neural network has been developing rapidly in the past twenty years. After 2010, it has developed more rapidly, and has made many achievements in the field of scientific research and business application (such as shopping picture recognition, Google self driving car, voice assistant, etc.).

Convolutional neural network has become a hot topic in deep learning. At present, there are many deep learning frameworks for developers and researchers to use. There are still many problems in the field of convolutional neural network and artificial intelligence that need to be further studied by researchers. The neural network of handwriting recognition data set can do a lot of follow-up work.

References

1. Yu, K., Jia, L., Chen, Y., et al.: Yesterday, today and tomorrow of deep learning. Comput. Res. Develop. **50**(9), 1799–1804 (2013)
2. Ding, S.: Basis of Artificial Neural Network. Harbin Engineering University Press (2008)
3. Ao, D.: Application of Unsupervised Feature Learning Combined with Neural Network in Image Recognition. South China University of Technology (2014)
4. Xuke: Application of Convolutional Neural Network in Image Recognition. Zhejiang University (2012)

Research on Development Status and Innovation Strategy of Broadcasting and Hosting in We Media Era Under Data Mining

Zhehao Li[✉]

School of Literature and Journalism, Leshan Normal University, Leshan 614000, China

Abstract. With the advent of the self butterfly era, we media website broadcasting and hosting industry under data mining has a great impact on this field in China. Therefore, whether it is for this new thing, or for the broadcasting and hosting of traditional media, there are further requirements for their own positioning, media understanding and professional level.

Keywords: Data mining · We media · Broadcasting and hosting

1 Introduction

Although the development of we media communication industry has a continuous impact on broadcasting and hosting, it also brings more development opportunities. The broadcasting and hosting industry, which is at the forefront of the media, has a strong social influence, and it has to operate with a new way of thinking for the new situation. Therefore, the key to the current field is to abandon the traditional inherent mode and use new forms of expression to ensure that its broadcasting and hosting image is more accessible to the public, deeper and more spiritual.

With the continuous popularization of China's Internet, the development of China's Internet and mobile Internet is gradually mature, and even unlimited traffic has begun to appear. At the same time, Internet products are increasingly filling our lives. At the same time, the number of mobile end users is increasing, even more than twice as many as the number of PC end users. People's demand for simplicity, quickness and interest is also increasing. From fragmented reading to short video viewing, China's we media is also developing rapidly.

In July 2003, two Americans, Sherin Bowman and Chris Willis, clearly put forward the concept of "We Media", which is translated into Chinese as "We Media" and has a very strict definition. So far, the concept of "We Media" has really entered the public field of vision.

The development of we media has gone through three stages: the first stage is the initial stage of we media, represented by BBS; the second stage is the embryonic stage of we media, mainly represented by blog, personal website and microblog; the third stage is the era of we media consciousness awakening, mainly represented by wechat

M. A. Jan and F. Khan (Eds.): BigIoT-EDU 2021, LNICST 392, pp. 115–123, 2021.
https://doi.org/10.1007/978-3-030-87903-7_17

public platform and Sohu News client. At present, the development of we media is in the transition period from the embryonic stage to the era of we media awakening. However, since we media has only been born for more than ten years, these three stages actually exist at the same time, but at this stage, microblog and wechat public platform are the main body of we media, and the others are relatively weak.

In China, the development of media has been divided into 4 stages: the 2009 Sina micro-blog launched, which caused the social media to be in the media; in 2012, the WeChat official account was launched on the Internet and moved from the media to the mobile terminals; from 2012 to 2014, portal websites, videos, e-commerce platforms and other media have been involved in the media field, and the platform has been diversified; from 2015 to now, live and short videos have become media content. New hot spots in the industry.

Broadcasting and hosting is a college major in Colleges and universities, which belongs to the major of radio, film and television. The basic length of schooling is three years, formerly known as hosting and broadcasting. In 2015, the Ministry of education changed its name to broadcasting and hosting.

This major cultivates advanced technical application-oriented professionals who have the basic knowledge and technical skills of broadcasting and hosting, and can be engaged in broadcasting and program hosting.

2   Data Mining Algorithm

2.1   Basic Application Algorithm

TF4df algorithm can be used to evaluate the importance of a word to a given document in a corpus or document set. The importance of a word is positively related to the number of times it appears in a document. The more times a word appears in a document, the more important it is. However, it is negatively related to the frequency it appears in the corpus [1]. The more times it appears in the corpus, the less important it is. Therefore, tf1df can be used as the basis for news text classification.

Term frequency. F refers to the total number of times a word appears in a given file. The reverse file pre rate is obtained from the total number of files in the library, followed by the number of pieces containing the word, and then the quotient is "time". If the word frequency in a file is higher than that in the whole corpus, the word is more important to the file. Therefore, the f-4df value can be used to filter common words and only retain the words that can highlight the characteristics of the document.

The traditional tf-df algorithm takes the file after word segmentation as the input (word segmentation by stuttering participator and removing stop words) and the output is the tf-df value of each word. Then the F-C% value is sorted from large to large, and the top-N (top-N) words are taken as the feature words for differentiation. The TF-df algorithm steps are as follows. Count the number of times each word appears in the file, and calculate the TF value of the word. The calculation method is shown in formula (1):

$$TF_{ij} = \frac{n_{ij}}{\sum Tl_j} \tag{1}$$

How many files in the file set do the unified words appear. The calculation method of words is shown in formula (2):

$$IDF_i = \log \frac{|D|}{\sum t_i \in d + 1} \qquad (2)$$

The TF * DF value of each word is calculated as shown in Formula (3):

$$TF - IDF_i = TF_{ij} * IDF_i \qquad (3)$$

2.2 Classification Method Based on Improved t-df and Bayesian Algorithm

News classification based on improved tfdf and Bayesian algorithm firstly uses stutterer to segment news text and remove stop words. Then the improved tfdf algorithm is used to calculate the tfdf value of the remaining words, and according to the statistical results of tf4df value, the highest n valuesv of tfdf value are selected as feature vectors. Finally, the news text is classified by Bayesian algorithm. The flow of news text classification is shown in Fig. 1.

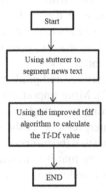

Fig. 1. News text classification process

3 Enhance the Overall Ability of Network Broadcasting and Hosting

(1) Familiar with and proficient in new media technology. Mastering new media technology is one of the first skills that media people need to have in the era of we media. The operation of new media software and hardware is just like the sharp weapon in the hands of local soldiers. With the help of computers and mobile phones, it is an indispensable and important skill for broadcasting and hosting to upload, receive and reprint various kinds of information.

(2) Have sensitive reaction ability. Up to now, any mobile terminal that can meet the needs of Internet access can become a media to release information. For example, the first publishers to report the "7.23" bullet train accident, the death of bin Laden and the fire in CCTV's auxiliary building are all from we media users, and there are countless examples like the above. Therefore, as a successful announcer, he needs to have the ability to control emergencies, and quickly seize the opportunity through his own sensitivity to events. Not only that, he should also have a deep understanding of the current hot news and unpopular topics, so as to transmit the information in time with the fastest efficiency.

4   Strengthen and Improve Their Comprehensive Skills

In the era of we media, every individual can use the Internet to spread information. And these individuals upload their interesting things, events and people to the network in the form of text, pictures or videos, and this series of operations is equivalent to the whole process of traditional media collection, editing and broadcasting. But for traditional media reporters, cameras and radio hosts, in view of the different positions, there are also significant differences in the nature of their work. For example, most of the work of radio hosts is completed in the studio, while such tasks as going out to interview are relatively few. However, nowadays, with the further development of the media, even the traditional broadcasting host also needs to actively go out of the studio, through the all-round forging of their own, with a new mode of work and a new way of thinking to quickly integrate into the current atmosphere. For another example, most radio hosts now actively participate in news interview and production in addition to completing the task of broadcasting and hosting [2]. More importantly, the topic selection information of radio hosts comes from we media, so as to further explain the topic selection and give full, in-depth and thorough interpretation of the information based on the current mainstream values, In order to ensure that the information provided for the public is more authoritative and effective, we can say that it is moving forward and growing towards the omnipotent and comprehensive broadcasting host.

5   The Positive Role of We Media Broadcasting and Hosting

5.1   Improve the Credibility of Your Image

Because we media information release is very random, and the quality of netizens is uneven, the credibility of information is also reduced. Based on the above environment, as a broadcaster, they should not be affected by these bad information, and should always adhere to the principle of disseminating true and effective information. In addition, they should also have the ability to distinguish the authenticity of information, collect and absorb the hot topics from the public, and have their own unique opinions to make their hosting more quality, style and depth.

5.2 Highlight Self Advantages and Show the Charm of Hosting

As an active broadcasting host in the forefront of the media, it makes the dissemination of information more personalized through the integration of their voice, language, posture and body. Especially for today's increasingly serious homogenization of media environment, the key of broadcasting host depends on personalized communication [3]. For example, Cui Yongyuan, who presided over the column of "telling the truth", fully showed his unique temperament and personal charm in the whole hosting process. Witty words combined with quick wit will lead people to a deeper realm successfully. Now, in the face of the noisy we media era, a very attractive and innovative announcer is lacking in this era (see Fig. 2).

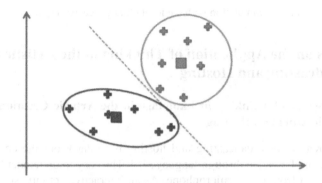

Fig. 2. Simulation with charm of hosting

5.3 Promote the Level of Language Expression

Throughout the information transmission based on the current we media era, due to the great differences in the identity, quality and level of communicators, the phenomenon of nonstandard language also occurs from time to time. Moreover, a series of network words and expressions are emerging one after another, such as "Huichang, Shenma, Fuyun". Not only that, the phenomenon of improper collocation and unreasonable quotation is becoming more and more serious. What's more, even the well-known and influential media have misused and explained the words. Moreover, the relevant departments have made a lot of requirements and norms on the use of language, such as requiring the announcer to use Putonghua in addition to the special festival needs, forbidding to develop or imitate in the form of regional characteristics, eliminating the use of nonstandard network language, and ensuring that the use of grammar always follows the rules of modern Chinese [4]. As the most advanced broadcasting and hosting industry in the media, it has a great influence on the society, especially in the standardized use of language, which is both a responsibility and an obligation. Based on the context of the former fast food culture, the announcer should be fully aware of the responsibility of a media person, and should actively and positively transmit the standard language through the language full of taste, culture and connotation (see Fig. 3).

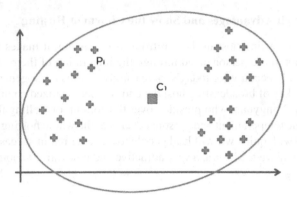

Fig. 3. Simulation with the level of language expression

6 Analysis on the Application of Thinking in the Artistic Creation of Broadcasting and Hosting

6.1 The Function of Thinking in Language in the Artistic Creation of Broadcasting and Hosting

The artistic creation of broadcasting and hosting is to carry out the communication activities of sound language through language [5]. The easy to understand point is that the host speaks in front of the microphone. As an interactive activity, speech not only includes "speech" as the subject and "listening" as the object, but also includes the content and way of speech in this activity. So when we study broadcasting and hosting this activity, what we study is actually the host's "what to say" and "how to say" in front of the microphone. The artistic creation of broadcasting and hosting relies heavily on logical thinking and image thinking, and it is difficult to distinguish between the two kinds of thinking. But we can make an easy to understand analogy. Logical thinking is like the skeleton of broadcasting language, which is the solid backing to support the inner core of language [6]. Image thinking is the flesh and blood of broadcasting language, The flesh and blood growing on the basis of the skeleton is the most appropriate and fresh; the inspiration thinking is not only integrated into the logical thinking and inspiration thinking, but also guides the logical thinking and inspiration thinking, and injects the soul into the broadcasting language, so that it has vitality. Only by combining logic thinking, image thinking and inspiration thinking, can such language be full and accurate [7].

6.2 "Language Skeleton" Derived from Logical Thinking

Logical thinking is often compared to "the grammar of thinking", which fully illustrates the connection between logical thinking and language. Logic, as the long-term experience of human beings in the objective world, is the theory and direction to guide thinking and communication. Logical thinking plays a very important role in real life. As a guide to reveal the internal laws of the objective world, the application of logical thinking can improve people's overall quality, cultivate people's pursuit of rational thinking and identify with the scientific spirit. People's understanding of the world is summarized as thinking, but people's expression depends on language. Thinking is the spiritual guidance of language, language is the material carrier of thinking, and the formation of thinking needs grammar to follow specific grammatical rules [8].

As an important tool of language logical thinking, it is also an important tool of language logical thinking. Logical thinking is to analyze data, infer the whole process, judge the truth and falsehood to understand the internal laws, characteristics of objective things, and the relationship between other things [9]. The analysis of logical thinking is rigorous and progressive. Logical thinking is used as a guide to create rational thinking language in language art creation, such as material selection, sentence structure, vocabulary selection, rhetorical decoration and theme cohesion [10].

7 The Proper Way of Thinking in the Artistic Creation of Broadcasting and Hosting

7.1 Correctly Understand the Importance of Thinking Ability

It is self-evident that the importance of the ability of artistic creation and hosting is whether or not there is the ability of artistic thinking in broadcasting [11].

In the daily teaching of broadcasting and hosting, it is often found that students prefer to broadcast with manuscripts and are better at broadcasting with manuscripts. In contrast, the enthusiasm of training and learning for broadcasting without manuscripts is not high. At present, most of the teaching of broadcasting and hosting focus on the broadcasting with manuscripts, and most of them focus on the students' Putonghua pronunciation, pronunciation, speaking speed and fluency. However, they don't pay much attention to the content of "speaking" in the process of language expression. They only focus on the external characteristics of expression, but ignore its connotation. In teaching, we often encounter such situations: some students have pleasant timbre, good pronunciation, and full language to express their feelings, but they are fluent in repeating redundant words; some students have solid basic skills in broadcasting with manuscripts, but they can't say anything in improvisation, and their sentences are empty and they don't know what to say. Such a language is flashy and lacks logic core. Broadcast with manuscript is to reprocess the original manuscript based on the understanding and experience of image thinking, while broadcast without manuscript is the direct coding and transmission of the announcer and host's own logical thinking products. Broadcasting host should not be full of emotion and eloquence to talk nonsense, but should be a process of using exquisite external language conditions to wrap the heavy thinking core and succinctly deliver these ideological contents to the audience [12]. It can be seen that the application ability of logical thinking, image thinking and inspiration thinking will have a far-reaching impact on the future development and personal achievement of students majoring in broadcasting and hosting.

7.2 Although they have Different Emphases, they Should Not Be Neglected

The broadcasting with articles relies on the integration of image thinking and logical thinking, while the broadcasting without articles emphasizes logical thinking and intersperses image logic. It's true that the two ways of broadcasting and hosting creation have their own emphasis, but we can't think that only this kind of thinking is respected. These two ways of broadcasting and hosting are the product of logical thinking and image thinking [13]. In the creation of broadcasting with manuscripts, image thinking does occupy a dominant position, but without logical thinking, the announcer will not be able to correctly grasp the logical connection between the upper and lower sentences in the manuscript, accurately understand the level of the manuscript, and even more difficult to summarize the core ideas of the manuscript. If we lose the guidance of logical thinking and only use the thinking of images, we will eventually turn into a poor work that moans without illness and expresses feelings without connotation [14]. As far as the creation of non manuscript broadcasting is concerned, broadcasters use more logical thinking, but if they lose the blessing of image thinking and have no reasonable emotional mobilization and expression, it is difficult for broadcasters and hosts to produce emotional resonance with the audience [15]. The unpublished broadcast, which only focuses on the logic of words without emotion, will only become a blunt narration that the audience is unwilling to accept [16].

After deeply understanding the application characteristics of the two kinds of thinking in broadcasting and hosting creation, announcers and hosts can also train their weak thinking ability according to their own situation [17]. For example, announcers and hosts who have been engaged in broadcasting with manuscripts for a long time can increase the practice of broadcasting without manuscripts at ordinary times, and try to organize logical and language idiomatic words and sentences without the prompt of manuscripts, so as to adapt to the active state of logical thinking in the artistic creation of broadcasting without manuscripts. However, some announcers and hosts who have been broadcasting without manuscripts for a long time need to exercise their image thinking ability, improve their adaptability in the face of manuscripts, and strengthen their emotional investment and scene restoration in the creation of existing manuscripts [18]. Only in the two forms of broadcasting and hosting creation, can we achieve the nearly perfect state of announcer and host professionally. The art creation of broadcasting work should pay attention to the coordinated development of logical thinking and image thinking. Although they are two different thinking systems, they are closely related, infiltrating, complementing and supporting each other.

8   Conclusion

Zhang song is to respect people, care for people, guide people, inspire people to give a clear advocate. With the advent of we media era, the channels for the audience to receive information are further broadened, and the optional content is also more and more diverse, while the traditional preaching expression is far from meeting and attracting the needs of the audience. The appearance of new we media broadcasting and hosting, from the perspective of civilians, makes a more appropriate, more real and more effective description of things, so as to build a platform for equal communication with the audience, and effectively disseminate positive, positive and valuable information in full communication with the audience, so as to ensure the positive guidance of social civilization and esoteric theory.

References

1. Peng, X.: Research on the development of broadcasting and hosting in the era of big media. Sci. Technol. Commun. **4**, 169–175 (2015)
2. Cheng, Y.: Development strategy of broadcasting and hosting in we media era. Sound Screen World **8**, 36–37 (2015)
3. Dou, H.: Opportunities and challenges faced by broadcasting and hosting industry in all media era. Commun. Copyright **1**, 107-108+111 (2017)
4. Liu, Q.: Innovation of professional talent training mode of broadcasting and hosting in the era of mass media. Educ. Teach. Forum **21**, 64–65 (2016)
5. Zhang, S.: Foundation of Broadcasting Creation. Communication University of China Press, Beijing (2004)
6. Zhang, S.: Chinese Broadcasting. Communication University of China, Beijing (2003)
7. Wu, Y., Zeng, Z.: Research on the Training of Broadcasting and Hosting Professionals. Communication University of China Press, Beijing (2009)
8. Wu, Y.: The Training Path of Host's Thinking and Language Ability. China Radio and Television Press (2005)
9. Wu, Y.: Research on the Comprehensive Quality of TV Program Hosts. China Radio and Television Press (2007)
10. Song, X.: Guide for On-the-Spot Reporting of on Camera Reporters. China Radio and Television Press (2008)
11. Bi, Y.: Outline of Modern Radio and Television, 1st edn. China Radio and Television Press (2007)
12. Lu, Y., Zhao, M.: Introduction to Contemporary Radio and Television, 1st edn. Fudan University Press (2008)
13. Luo, L.: Performance of Literary Works. Communication University of China Press, Beijing (2003)
14. Wu, Y.: Host Language Skills. China Radio and Television Press, Beijing (2011)
15. Fu, C., Wu, H.: Practical Broadcasting Course. Putonghua Pronunciation and Broadcasting Voice. Communication University of China Press, Beijing (2005)
16. Lu, M., Qian, X.: Thought of Thinking Science. Science Press (2012)
17. Jin, B.: Dictionary of Marxist Philosophy. Shanghai Dictionary Press, Shanghai (2003)
18. Jiang, T.: 23 Kinds of Thinking Wisdom Benefiting from Life. CPC Central Committee Party School Press, Beijing (2003)

Research on Human Resource Management System Model Based on Big Data Management

Lei Zhou[✉]

Nanchang Institute of Science and Technology, Nanchang 330025, Jiangxi, China

Abstract. Human resource management system (HRMs) is an important part of enterprise information management system. With the development of modern personnel management informatization and the increasing demand of leaders for statistical analysis of information data, the optimization of data quality and process has become the development trend of HRMs design and development. In this paper, the idea of cloud computing is introduced. Based on the analysis of the existing problems of human resource management information system, we constructed the cloud computing for the HRMs model is constructed, and the system implementation and system security are discussed.

Keyword: Human resource management · Cloud computing · Management information system

1 Introduction

Human resource management (HRMs), the upgrading of personnel management, refers to the effective use of relevant human resources inside and outside the organization through recruitment, selection, training, remuneration and other management forms under the guidance of economics and humanism, so as to meet the needs of the organization's current and future development, The general term of a series of activities to ensure the realization of organizational goals and the maximization of member development. It is the whole process of forecasting the demand of human resource and making the plan of human resource demand, recruiting and selecting personnel and organizing effectively, assessing performance, paying compensation and motivating effectively, and developing effectively according to the needs of organization and individual so as to realize the optimal organizational performance. It is also an important position in the company.

The academic circles generally divide human resource management into eight modules or six modules: 1. Human resource planning; 2. Recruitment and allocation; 3. Training and development; 4. Performance management; 5. Salary and welfare management; 6. Labor relations management. Explain the core idea of the six modules of human resource management to help business owners grasp the essence of employee management and human resource management.

Cloud computing is the latest application technology in the development of computer science, which has been widely concerned. In a narrow sense, cloud computing refers to

M. A. Jan and F. Khan (Eds.): BigIoT-EDU 2021, LNICST 392, pp. 124–128, 2021.
https://doi.org/10.1007/978-3-030-87903-7_18

that manufacturers build data centers or supercomputers through distributed computing and virtualization technology, and provide data storage, analysis and scientific computing services to technology developers or enterprise customers in a free or on-demand rental way. Generalized cloud computing includes three service forms: SAS, PAS and IAS. It means that manufacturers provide different types of services such as online software services, hardware leasing, data storage, calculation and analysis to different types of customers through the establishment of network server cluster [1]. In the practical application of cloud computing, in addition to the application characteristics of network technology, it also has the following characteristics:

1. Users can apply for computing services from designated service providers according to their personal needs. 2. It supports multiple operating platforms and multiple terminal access. 3. Computing resources are centralized to provide services for multiple customers through a multi customer sharing model. 4. The ability to quickly respond to service applications and provide services in a scalable way. 5. Provide measurable computing services.

2 Current Situation of Human Resource Management System

Human resource theory has always been a hot pot problem in enterprise management. Managers of various industries are more and more aware of the key role of human resource in enterprise development and competitive advantage, and put forward higher requirements for human resource management. At the same time, with the development of information technology, human resource management extends from "transaction processing" and "data management" to "process management" and "decision support". Therefore, from the perspective of human resource management, high performance database system is used to manage human resource related data, and provide friendly interface, process management, information management, statistical analysis, decision support and other functions, so as to improve the work efficiency of human resource management transaction processing [2]. However, many enterprises and institutions in our country still have to stop.

Let G_1 be a bilinear group with prime P as its order, and let G be its generator, bilinear mapping

$$e : G_1 \times G_1 \to G_2 \tag{1}$$

Threshold D, decryption user attribute set u, system extraction attribute set ω, so Lagrange coefficient can be defined.

$$\Delta_{i,s}(x) = \prod_{j \in s} (x - j)(i - j) \tag{2}$$

(1) The idea of management information system is not popular. At present, most of the so-called informatization just stays at the stage of using computer to process documents, data tables, presentation documents, and searching for information on the Internet, without the idea of using information technology to manage resources and optimize processes. (2) The function coverage of human resource management system is narrow. From the function of management information system, the functions most

used by enterprises are "personnel information management" (71%), "salary" (682%), "report" (62.2%), "attendance" (595%), "Recruitment" (56%), "welfare" (538%), etc. According to this analysis, the human resource management information system of most enterprises in our country is still in the "transaction processing level" and "business process level", and the realization of "comprehensive human resource management" is less.

3   Cloud Computing for Construction of HRMs Model

3.1   System Structure Model

The model is a cloud computing model management system based on SAS, which usually relies on SOA architecture and web service technology. SOA is a service-oriented structure, which can extend many web services related to human resource management applications on the basis of SOA. We based on such as database processing, network transmission, interface platform with the system application of software and hardware, and constitute the "HRM cloud" [3]. Its structure is shown in Fig. 1.

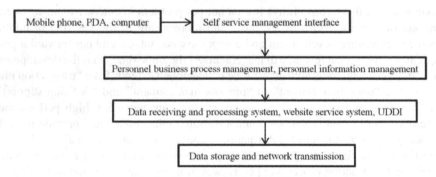

Fig. 1. Cloud Computing for HRMs

From the Fig. 1, we know there are four layers in the system. From the top to the bottom, each layer has the related function in order to ensure the data transmission.

3.2   Data Storage and Data Management Technology

Cloud computing usually uses distributed storage to store data, and uses redundant storage to ensure the reliability of stored data. Cloud computing system also needs to meet the needs of a large number of users at the same time and provide concurrent services, so it also needs to have the characteristics of high throughput and high transmission rate. At present, the mainstream cloud computing data storage technology has Google's GFS and Hadoop development team.

4 Security Measures of Cloud Computing Human Resource System Model

Nowadays, many enterprises are considering adopting cloud computing human resources (HR) software systems with many advantages, but it departments and human resources (HR) must also consider their technical challenges.

The human resource system based on cloud computing has brought a lot of progress to the human resource department. The services they provide help simplify human resources tasks, realize employee experience, improve and reduce costs. However, the human resource system based on cloud computing also brings a lot of technical burden. Everyone in the software procurement team, especially the staff in the IT department, should consider the selection, deployment and promotion of each tool.

The following are nine challenges faced by cloud computing based human resources systems:

1. New security burden
   The newly implemented GDPR regulations are increasing compliance requirements to track these data, which adds an additional burden on employees.

   "Because of these problems, it is important to study and select suppliers that meet all the regulatory needs of the enterprise and protect its sensitive data. Before entering the implementation phase, ensure their professional qualifications in network security. Then, work closely with the enterprise's chief technology officer (CTO), as well as Devops and it functions to ensure the security and maintenance of the system. Don't take anything for granted." Bazar cautioned.

2. Balance security and ease of use
   We are concern the security for the consumerism of human resources.

   We can understand the use must be easy, as once the technology is for nomal HRM, the HR just want to use the technology and the HR doesn't know what is the security, technology and so on. If the system is so complexity, it will cause some problems. So as a HR, we will integrate all workflow, let it be easy.

3. Significantly change the use mode
   Expansion can be a major challenge. There are great differences in the computing load that human resources systems based on cloud computing must support. Menon said that, for example, the average system load of a company with 2000 employees may be 100 page views per hour. But at the end of the month, the peak may be as high as 8000 pages per minute.

   A good strategy is to adopt the human resource system based on cloud computing, which can be dynamically expanded and reduced the workloads. As HRM need to consider expanding services and rights.

4. Keep pace with integration
   Cloud computing can easily add various human resources services. Anup yanamandra, chief product officer of betterworks systems, which focuses on employee engagement and development services, said that the number of individual human resources systems to be maintained can be between 10 and 100, and it can be very difficult to manage all integration and permissions consistently in all these systems.

5. Custom cloud service

   Chris Schaaf, general manager of Human Resources Department of Accenture, said, "there is no general cloud computing human resources solution." Some IT organizations is developing some demands for differents client that focus on user needs, experience and participation, as well as integration and seamless capabilities, enabling users to better complete their work anytime, anywhere on any device they use.

6. Scalability and flexibility testing

   It and human resources departments should work with suppliers to test the scalability and flexibility of human resources cloud computing services in their own organizations, rather than measuring suppliers' statements by surface value. There are many other factors that affect scalability. In addition, it is important to ensure that specific services can adapt to different human resources processes. Chris Schaaf, Accenture human resources portfolio director, said that a key best practice is to outline the organization's operation mode to suppliers, identify potential risk points, and then jointly prevent and repair. For example, Accenture uses servicenow to provide guided vacation services for its employees.

7. Prepare for data flooding

   Cloud computing can more easily access raw data by following the natural path advocated by specific cloud computing services. However, more needs to be done to obtain, understand and share these data, so as to help human resources managers meet unique challenges. Enterprises look for tools that can be set up to calculate meaningful return on investment and indicator information about personal business. Human resources professionals need to make full use of such tools.

5 Conclusions

As a cutting-edge network application technology, cloud computing has carried out application research in many fields. With the development of scientific management, human resource management system, as an important part of management information system of enterprises and institutions, is also facing design ideas and technological innovation. The application of cloud computing technology to human resource management system will inject new vitality, Have a positive and far-reaching impact on it.

Acknowledgements. Project No. YJ201908, Social Science Programme for Nanchang 2019.

References

1. Jia, Y.: Research on human resource management system model based on cloud computing. Human Res. Develop. **2**, 1–2 (2015)
2. Bo, W.: Research on human resource management system model based on cloud computing. Netizen World **Z3**, 3 (2013)
3. Wu, Z.: Research on human resource management system model based on cloud computing. Inf. Comput. (Theor. Ed.) **16**, 142–143 (2012)
4. Feng, D., Zhang, M., Zhang, Y., Xu, Z.: Research on cloud computing security. Acta Softw. Sin. **22**, 71–83 (2011)

Research on Innovation and Practice of School Mode and Operation Mechanism of Open University Under Big Data Environment

Junrong Guo(✉)

Hebei Open University, Hebei 050051, China

Abstract. Based on the open university model and operation mechanism of the relevant theoretical research. This paper expounds that colleges and universities should give full play to their own advantages, explore innovative new models, serve local economic and social development, and study and plan the development path of Open University from multiple dimensions and perspectives.

Keywords: Open University · School running mode · Operation mechanism

1 Introduction

The essence of the transformation from radio and Television University to open university is to change from the extensive scale development mode to the connotative quality development mode. The goal of the reform of talent training mode is directly to the teaching quality, and the teaching mode is the core part of the talent training mode, which directly determines whether the training goal can be achieved, and whether the appropriate and effective teaching mode can be selected and constructed, It is of great significance to the quality of personnel training in open universities [1].

The distance teaching mode is changing and developing with the social demand and technological revolution. For example, the Open University in the UK initially focused on students' independent learning, and the learning activities were carried out in the form of classroom groups. With the rapid development of information technology, it began to use course websites, e-mail and computer conferences. After 2005, it mainly adopts the open distance learning mode characterized by virtual learning environment, and uses the integrated online learning environment to support students' learning activities, information collection and all-round management.

At the beginning of its establishment, China's Radio and television university system mainly followed the traditional classroom teaching mode, transmitting and receiving curriculum teaching programs through radio and television. Since 1999, through the implementation of the "pilot project of talent training mode reform and open education of China Central Radio and Television University", radio and Television University has carried out extensive and in-depth practical exploration on teaching organization form, teacher-student interaction mode, learning platform and resources. In the summative

M. A. Jan and F. Khan (Eds.): BigIoT-EDU 2021, LNICST 392, pp. 129–138, 2021.
https://doi.org/10.1007/978-3-030-87903-7_19

evaluation in 2007, most of the provinces have summed up their own teaching modes, on this basis, the teaching mode of "combination of learning and guidance" with students' self-study as the core has been formed [2–4]. After more than ten years of active exploration, the teaching mode with the core feature of "combination of learning and guidance" emerges in endlessly in the radio and television university system, which has become an important achievement in the reform of the talent training mode of distance open education.

2 The Current Situation of Teachers in Large Scale System

2.1 Main Development Status in China

For most years, he has devoted himself to the exploration and practice of open and distance education, playing a unique role in promoting educational equity in the autonomous region, and has become the main force of distance higher education. As of September 2012, there were 64 Open Education Majors in 100 teaching centers of RTVU, with more than 70000 students. In order to develop RTVU education with regional and national characteristics, the pilot project of Mongolian Chinese bilingual teaching in open education was launched in 2010, offering undergraduate law, finance and accounting majors. By the autumn of 2012, 183 students were enrolled in open education bilingual teaching, which has become the characteristic of RTVU [5–7]. At present, Inner Mongolia RTVU is making every effort to build an ancient Open University. The separation of teachers and students, as well as the diversity of learning subjects and learning needs in Open University make the exploration and practice of teaching mode, the organization and management of teaching process, the construction and sharing of learning resources extremely complicated, which puts forward higher requirements for the construction of teaching staff. With this problem in mind, the course team takes the full-time and part-time teachers engaged in Distance Open Education in Inner Mongolia Radio and TV university system as the research object to conduct questionnaire survey, interview and collect data, and conduct qualitative and quantitative analysis on the survey data, so as to provide the basis for the construction of teaching team.

2.2 All Aspects of the Specific Performance

The number of full-time teachers is insufficient. As of 2011, Inner Mongolia TV University has 680 full-time teachers and 386 part-time and external teachers. Compared with the ever expanding enrollment scale, problems such as insufficient total number of teachers, low teacher-student ratio, poor stability of part-time and external teachers, and insufficient teaching ability of distance education are increasingly prominent, It makes the distance educators question the teaching quality of radio and Television University (the requirement of the Ministry of education for the summative evaluation of the talent training mode reform and open education pilot project of the Central Radio and Television University is 501, which is a structural problem. The unreasonable age structure, more old teachers and less than 6% young teachers under 30 years old affect the sustainable development of TVU education. The structure of professional titles and academic qualifications is unreasonable [8]. The proportion of senior and deputy senior professional titles is seriously unbalanced. Less than 11% of postgraduates have academic qualifications. It is difficult for disciplines and academic echelons to form. The structure of disciplines and specialties is unreasonable. Grammar teachers account for 396%, finance and Economics teachers account for 21.2%, and science and engineering teachers account for 357%. Statistics show that teachers in various disciplines are basically balanced, but specific to each major is very unbalanced. There is a shortage of teachers in new majors, a surplus of teachers in traditional majors and basic courses, and a more serious problem in science and engineering. 59% of the science and engineering teachers are computer application teachers, and about 3800 civil engineering students are in school. Only 6.4% of the teachers are teachers, which is due to the unbalanced construction of the teaching team in each branch. In many branch schools, there are few professional teachers with a large number of students, and the teaching tasks are undertaken by interdisciplinary or external teachers, which leads to the practical problem of low professional coincidence rate. The coincidence rate of civil engineering is only 12%. Fourth, the teaching ability of modern distance education needs to be improved. Although after more than ten years of teaching practice in open education, the teaching ability of full-time teachers in distance education has been significantly improved, and 60% of them can skillfully use modern educational technology, there are still many problems to be solved [9]. Among the surveyed teachers, only 43% can organize online teaching according to learners' needs, only 31% can independently develop multimedia courseware, and only 12% have presided over distance education research.

3   The Significance and Basic Characteristics of Teaching Team Construction

3.1   The Significance of Teaching Team Building

To sum up, the teacher-student ratio, structure and ability of teachers in Inner Mongolia Radio and TV university system are not suitable for the construction of Open University. Only by integrating excellent teacher resources of the whole system and building a high-quality teaching team across time, space and organizational boundaries, which integrates learning resource development, teaching organization and management, learning support service and scientific research, can we help the construction of Inner Mongolia Open University. (1) Realize the co construction and sharing of high-quality education resources of the system. Break the boundaries of region and organization, set up a teaching team composed of excellent teachers, technicians and managers of the same specialty at all levels of the system [10]. Through equal communication, unity and cooperation, and complementary skills and knowledge among members, the team members can make common progress and improvement, and promote the co construction and sharing of high-quality teacher resources in the system. (2) To realize the individual professional development of teachers and improve the overall quality of teachers. In the process of completing the task, team members can learn and communicate with other members to achieve personal goals and promote professional development, and promote the improvement of the overall quality of the whole system of teachers through the training of team members. (3) It is necessary to promote teaching reform and practice and improve teaching quality. Team members based on the network environment of remote cooperation, explore and practice the new teaching mode based on the network environment of autonomous learning, remote support services and face-to-face counseling, effectively solve the main problems in the teaching process, and strive to improve the teaching quality of distance open university education. (4) Build the learning support service system of Open University. The teaching process of Open University is the process of providing diversified learning support services for all kinds of learners at all levels. The learning process of students is the process of receiving and using learning support services [11–15]. A perfect learning support service system needs the support of a high-quality team of teachers.

3.2 Basic Characteristics of Teaching Team

There are many definitions of team by scholars at home and abroad. American scholars kazenbach and Smith (1993) define team as a group composed of a few individuals with complementary skills and intended to take responsibility for the common vision, performance goals and methods. Based on the research of related literature at home and abroad, combined with the characteristics of Open University, the author thinks that the teaching team of neijiagu Open University is a multi-ethnic teachers, administrators and technicians with different knowledge and skills, which is based on the professional construction platform, According to the concept of teaching team, the basic characteristics of teaching team are as follows: (1) having clear team objectives and members' responsibilities. Teaching team is a temporary teaching organization based on a certain specialty or a certain course construction task. A high-performance team must ensure that the team goals are consistent with the individual goals of its members. (2) Team organization structure with appropriate scale and reasonable structure. Team members across time and space and organizational boundaries, no hierarchy, complementary knowledge and skills. (3) Based on the network environment, remote division of labor and cooperation. (4) Through the network information technology and other communication technology to communicate and establish a trust relationship. (5) Task driven management [16]. Team leader overall planning, individual members self-management and self-control, mutual management and cooperation within the team.

4 Obstacles to the Construction and Development of Teaching Teams

4.1 The Influence of Administrative System

First, the team operation mechanism is not perfect. There is no hierarchy among teaching team members, and the task driven team internal management mode increases the difficulty of team management. Third, professional leaders and curriculum responsible teachers have not introduced the competition mechanism, and they are assumed by full-time teachers appointed by the University Department [17]. As shown in Fig. 1. The quality of professional leaders and curriculum responsible teachers is uneven, and even some professional leaders are assumed by interdisciplinary teachers, which also restricts the construction and development of the team to a great extent.

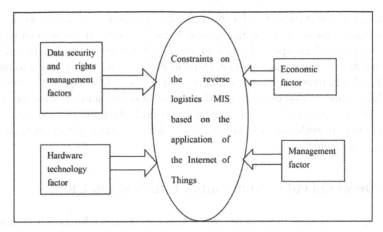

Fig. 1. Main influence of administrative system

4.2   The Influence of Team Members' Cultural Differences

The difference of working environment and scientific research direction adds to the complexity and diversity of communication within the team. It is difficult to establish and maintain the team trust relationship. Teachers at all levels lack effective communication and cooperation. First, teachers at different levels have vague role positioning, unclear division of labor, and all teachers are doing repetitive work, Second, most teachers are lack of team work experience, lack of cooperation and communication among teachers in the system, even teachers of the same course seldom communicate with each other, and the headquarters does not provide professional guidance and training for grass-roots teachers [18]. The survey data shows that in the past two years, only 36% of the branch school and teaching point teachers have participated in the systematic teaching seminar and secondary training, only 32% are satisfied with the teaching and research meeting organized by the University Department, only 26% often communicate with the superior professional head or curriculum responsible teacher, and less than 339% have received the guidance of the superior professional head or curriculum responsible teacher.

4.3   Enhance the Understanding, Strengthen the Construction and Management of Teaching Team

A special organization should be set up in the headquarters of Open University. Take full charge of team building and management, supervise and guide team building regularly or irregularly to coordinate with the construction of Open University. The second is the system construction. According to the characteristics of the teaching team, formulate corresponding rules and regulations, standardize the team application, approval and acceptance procedures. According to the needs of specialty and curriculum construction, the current situation of teaching staff and the situation of students, scientific planning of team building should not be carried out in a rush. Third, reform the administrative mechanism. Establish a team leader responsibility system with unified responsibilities and rights, set up a special fund for team building, give full support to people, finance and materials, and help team leaders solve the difficulties and problems in their work. Fourth, improve the team leader selection mechanism. Introduce competition mechanism, select team leaders within the scope of the system, and effectively allocate teachers with strong professional ability, organizational management and communication skills to the team leaders [19]. Fifth, improve the management mechanism of learning guidance teachers. The teaching team will be used as the whole system of the subject guidance teachers resources, and the guidance teachers will be deployed in the regional radio and TV university system to provide students with hierarchical and diversified guidance, assistance and promotion services.

5   The Development of Community Education in China

Although the rise and development of community education in China has a short history, it has made great contributions to improve the quality of life of community residents and enrich their spiritual and cultural life. After nearly 30 years of development, it has gradually formed a new type of community education with the development orientation and characteristics of "all staff, whole process and all-round". The status of community education in the whole education system has been constantly strengthened, and the relationship between community education and regional economic and social development is becoming closer and closer. Therefore, it is of great significance to accelerate the development of community education to build our country into a learning society. This chapter mainly reflects the development pace of Guangxi community education from the development process of community education in China, observes the development characteristics of Guangxi community education through the basic characteristics of community education in China, and analyzes the development form and direction of Guangxi community education by summarizing the experience of developing community education in developed areas in China [20, 21].

The depth of field can be defined as the distance between the depth planes of the back and front edges:

$$DOF_{geom} = 2\frac{l^2}{g \times p}p_d \tag{1}$$

The optical imaging model of lens can be used to analyze the imaging optical path of camera lens:

$$\frac{1}{f_c} = \frac{1}{g_c} + \frac{1}{L_c} \tag{2}$$

The integrated imaging display mode can be divided into real mode, virtual mode and focus mode, and the display process also meets the Gauss imaging principle:

$$\frac{1}{f_2} = \frac{1}{g_2} + \frac{1}{l_2} \tag{3}$$

6 General Situation of Community Education in China

6.1 The Basic Characteristics of Community Education in China

Community education, which integrates education with management, service and cultural activities, is a breakthrough, expansion and extension of single school education. The role of community education in the community is constantly highlighted, such as the opening of educational resources and sharing with residents, holding various cultural activities to improve the knowledge literacy of community residents, etc. At the same time, community education plays a role in promoting and guiding the formation of residents' values in community development. Therefore, the basic characteristics of Chinese community education can be summarized as follows:

One is human nature, which means that the fundamental value orientation of Chinese community education is to promote people's all-round development. This requires that community education should take "people-oriented" as the starting point and foothold. The second is comprehensiveness, which means that the educational activities and contents of community education cover a wide range. The third is all staff, which can be summarized as community education for all members of the community. The main objects of community education are vulnerable groups, the disabled, the elderly and migrant workers in the community. Due to the change of social structure, community education is more and more widely concerned. With the aggravation of population aging in China, the elderly has become an important object of community education. The floating population has also become a large group of community education objects, so we should improve their cultural quality and professional skills to make them qualified citizens. Enterprise reform has led to more and more laid-off workers need to change jobs, coupled with the influx of new labor force into the human market, community education should play a role, adjust the psychology of the unemployed, strengthen their skills training, and enhance employment competitiveness [22]. As the community education circles say, one of the purposes of community education is to "improve the quality of life of community members and promote the life development of community members the fourth is diversity, which means that the composition of community members is complex and changeable, so the content and form of community education should have the characteristics of diversity.

6.2 The Basic Management System of Community Education in China

The management system of community education is the guarantee of the stable development of community education and a kind of government behavior. System is usually defined as the general term or systematization of fundamental management system, mode, method and form, such as institution setting, subordinate relationship and authority division in organization system. Community education management system is a comprehensive and systematic education management system composed of community development power, interest subject, power structure, operation mechanism and supervision mechanism. At present, the management system of community education in China basically takes the form of Street (town) Community Education Committee. Because of the participation of administrative officials, such organizations have both authority and universality, which is one of the biggest characteristics of our community education management system.

Model is a rational and simplified form to reproduce the existing facts. In the process of the development of community education in China, four basic modes of community education management have been gradually formed. One is the street centered regional community education management mode, which is usually a community education organization mode with the district government or streets as the main body and all sectors of society participating in the management [23]. The second is the interactive community education management mode between school and community, which requires two-way communication between school and community.

7   System Simulation

The research on the innovation of Open University Based on big data mainly focuses on the optimization of its curriculum, because in all the optimization process, the selection of curriculum is an integration and optimization process of the whole system. In the big data environment, we optimized 12 courses, as shown in Figs. 2 and 3. In these 12 courses, we mainly focus on which courses must be opened, which courses are open, which courses are old and old [24]. From these aspects, we can use students to integrate into society faster in innovation. Innovative mechanism, innovative curriculum system and innovative system are the future of Open University.

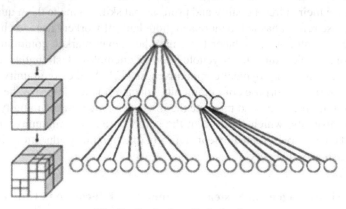

Fig. 2. Curriculum Optimization

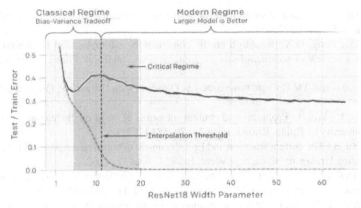

Fig. 3. Optimization Results

8 Conclusion

The teaching and learning of distance education are in the state of "quasi permanent separation". How to adopt the appropriate training mode, fully consider the learning characteristics of learners, actively play the role of teachers and platforms in guiding the learning of distance learners, and realize the talent training goal of distance open education has become an important reform content of the transformation from radio and Television University to open university.

References

1. Shen, J., Wang, W., Qiu, J., Chen, J., Zhong, Z.: Ten expectations of teaching mode innovation in Open University. Dist. Educ. China **7**, 41–47 (2012)
2. National Medium and Long Term Education Reform and Development Plan (2010–2020), pp.36–38. People's Publishing House, Beijing (2010)
3. China Central Radio and Television University National Open University Construction Plan, p. 23. China Central Radio and Television University Press, Beijing (2011)
4. Zhong, Z., Lin, A.: Interpersonal management: improving the learning ability of distance learners. J. Dist. Educ. **5**, 44–50 (2008)
5. Shen, X.: Tang Mingfei, President of Pudong South School of Shanghai Open University: a good leader of lifelong education for all. Online Learn. **1**, 56–59 (2021)
6. Brief Introduction of Changzhou Senior Vocational and Technical School. Reference of Politics Teaching in Middle School **4**, 106 (2021)
7. Duan, R., Tian, L.: Research on the evaluation of open education scholarship based on Delphi method and analytic hierarchy process – Taking the scholarship of National Open University as an example. Sci. Technol. Indust. **21**(1), 147–152 (2021)
8. Xin, Y.: One belt, one road, academic book development and going out. Media Forum **4**(2), 95–96 (2021)
9. A series of academic reports on "theory and practice innovation of open and distance education in the new era" held by modern distance education research. Modern Dist. Educ. Res. **33**(1), 25 (2021)

10. Tu, Z., et al.: Application of smooth anti adhesion master batch in BOPET film. Synth. Resins Plast. **38**(1), 52-54+60 (2021)
11. Tong, M.L., Zeng, D.X.: Research on the construction and application of Digital Forestry Platform: a review of technical basis of digital forestry platform. Forest. Econ. **43**(1), 100 (2021)
12. Qinghai radio and TV University renamed as Qinghai Open University. Qinghai Educ. **z1**, 7 (2021)
13. He, Y., Lv, L., Wei, J.: Problems and counter measures of multi media teaching materials in Open University of China. Knowl. Base **2**, 179–181 (2021)
14. Cao, T., Xu, B.: The current situation and Enlightenment of learning support service in Japan's Broadcasting University. Contemp. Vocat. Educ. **1**, 106–112 (2021)
15. Wu, J.: Design and research of online examination question bank system based on co construction and sharing mode. Microcomput. Appl. **37**(1), 59-62+66 (2021)
16. Xu, L., Xie, Q., Zhang, C.: Value orientation and realization path of online course in Open University in intelligent era. Adult Educ. **41**(1), 14–19 (2021)
17. Chen, D., Xu, Z.: Innovation in technology enabled Education: a learner centered approach to teaching: an interpretation of the report on Innovative Teaching (2020 Edition) of the Open University of the UK. Adult Educ. **41**(1), 20–24 (2021)
18. Jie, Y.: The current situation, Dilemma and development strategy of prefecture level radio and TV Universities – Based on the thinking of Henan Anyang radio and TV University. J. Anyang Inst. Technol. **20**(1), 66–69 (2021)
19. Jiang, Y., Xia, H.: Research on the development of China's elderly education since the reform and opening up: context analysis, theoretical forms and academic support. Adult Educ. **41**(1), 30–39 (2021)
20. Guangyu, J.: Reflection on the practice of prior learning accreditation in open universities in China based on experiential learning theory. Adult Educ. **41**(1), 82–87 (2021)
21. Wang, X., Jiang, Y.: A comparative study on the policies and regulations of Open Universities: a case study of Britain, the United States, Japan and China. Adult Educ. **41**(1), 88–93 (2021)
22. Zhao, Z.: Achievements, problems and suggestions on the policy of lifelong education system construction in China – Based on the ten-year implementation of the outline of national medium and long term education reform and development plan (2010–2020). Adult Educ. **41**(1), 8–13 (2021)
23. Wang, X., Liu, X.: Interpretation and Enlightenment of internal quality assurance system of Catalonia Open University. Foreign Educ. Res. **48**(1), 61–76 (2021)
24. Official Reply of Guizhou Provincial People's Government on approving Guizhou Radio and TV University (Guizhou Vocational and Technical College) to change its name to Guizhou Open University (Guizhou Vocational and Technical College). Bull. Guizhou Provin. People Gov. **1**, 40 (2021)

Research on Teaching Reform of New Energy Technology Based on Virtual Reality System

Yan Huo[✉]

College of Information Engineering, Shenyang University, Shenyang, China

Abstract. New energy technology is developing rapidly with the development of industry. In recent years, the demand for talents with new energy technology is increasing. In this paper, an innovative training mode of new energy technology based on virtual reality system is established to constitute an oriented training system from aspects of teaching and studying. The model formulation should cover the learned knowledge points. The research can provide a reference for building an education structure of advanced technology application and strengthening practical ability.

Keywords: New energy technology · Teaching reform · Virtual reality · Modern industry

1 Introduction

Energy is a key factor constraining the sustainable development of economy and society. With the gradual consumption of non-renewable resources (e.g. petroleum, ore, *etc.*) and the deterioration of ecological environment, it is urgent that a green and sustainable energy conversion and storage technology should be developed. Nowadays, with the rapid development of economy and science, too much energy consumption has led to energy depletion and environmental pollution seriously. Most people are advocating energy conservation and emission reduction. We should encourage automobile industry to reduce energy consumption and pollution, but it is more important that new technologies are developed to replace traditional energy. Therefore, new energy technology [1, 2] has become an effective way for overcoming the problem of energy shortage and environmental pollution. Under the general trend of the international energy transition, how to maintain the sustainable development of new energy is a major strategic issue. New energy technology is one of modern high technologies, including nuclear energy technology, solar energy technology, geothermal energy technology, ocean energy technology and so on.

The development and utilization of nuclear energy and solar energy have broken the traditional energy concepts that take oil and coal as the main body, and created a new era of energy. It is also one of the important directions for the cultivation of college students. However, there are some disadvantages for the education of new energy technology. For

M. A. Jan and F. Khan (Eds.): BigIoT-EDU 2021, LNICST 392, pp. 139–147, 2021.
https://doi.org/10.1007/978-3-030-87903-7_20

example, college students always face the abstract and conceptual cognition in paper textbooks. Then it causes the lacks of learning enthusiasm and practical ability.

Virtual Reality (VR) [3] was a useful technology for simulation field. The virtual environment established by virtual reality is made up of digital modeling and combination through real data, to establish the virtual scene in line with the design standards and requirements of engineering projects, and to truly reproduce the planning project. By using VR technology, teachers and students can expand space and compress time according to actual needs. VR can present virtual practical experiments that should obtain the results taking days and months in the real environment. This can give people an intuitive and visual display, which is the biggest benefit of VR education. VR can produce a full sense of experience and immersion. Therefore, we should introduce VR into the education of new energy technology which is designed to combine theory with practice. The research subject is to make positive contributions to the improvement of the talent training program and upgrade the quality of teaching in colleges.

2   Problems of New Energy Technology Teaching

Due to the short time of establishment for new energy technology, we lack the managing and education experience. Therefore, there are some problems in the teaching of new energy technology and its application. The problems are listed in Table 1.

2.1   Practical Teaching Problem

Teachers are unfamiliar with new energy technology, which leads to the focus of professional teachers' teaching work mainly on related theoretical knowledge points. For this teaching way, the practical teaching work will be neglected, resulting in students with more theoretical knowledge and lack of practical ability. Practical ability is an important aspect of training talents in higher colleges. It is difficult for higher vocational graduates without strong practical ability to meet the needs of enterprises [4, 5].

2.2   Assessment Method Problem

In the processes of the past teaching, the teaching assessment of new energy technology relied too much on the final exam. We can see the assessment method was too simplistic and single, which can not objectively and comprehensively evaluate the learning results of students. The single assessment method could not evaluate the real level of mastering the basic knowledge for students. It can not also measure the students' of ability learned for teachers' understanding of the learning effect of students. Another disadvantage is that the feedback of teaching effect is very late. Teachers cannot know the weak of teaching until the wrong answers are found in final exam. It is too late to remedy the teaching problems [6].

2.3 Experimental Safety Problem

In the process of experiment training, the safety of teachers and students is a key problem. For example, energy technology experiments always face inevitably high-pressure, high-temperature, toxic environments. For new energy technology, how to ensure the safety of the energy equipment in the process of new energy training. Meanwhile, some energy products can be harmful for the health of students without full training in the first time. So the safety methods should be improved as soon as possible. The three teaching problems of new energy technology are shown in Table 1.

Table 1. Three teaching problems for new energy technology.

No	Term	Description
1	Practical teaching	More theoretical knowledge teaching and less practical ability training
2	Assessment method	The assessment method was too simplistic and single
3	Experimental safety	High-pressure, high-temperature, toxic environments for experiment training

3 The Characteristics of VR Teaching Model

Virtual reality teaching is a new teaching model that allows students to have more autonomy and participate in a wider range of practical teaching and learning processes [7].

3.1 Realistic Practical Operation

For new energy technology courses, it is reasonable that the teaching of theoretical knowledge is relatively less and the practical application knowledge is more [8]. Cultivating practical ability in class is the key subjective to learn this course well. The teaching mode by VR system can simulate engineering cases according to the technical problems in actual engineering, so that students can get in touch with real projects and information. Students can not only observe the structure and shape of equipment through VR system, but also realize the energy production. At the same time, the software integration interface can be used to adjust the operating parameters of equipment in real time to realize the operation results.

3.2 Process Assessment with Interaction

A highly interactive teaching and learning process is important for process assessment. The interaction between teachers and students happens with ease during the teaching process with VR system, because communication is the main way in teaching, replacing the traditional teaching focusing on knowledge points. In each teaching step, the progresses of students are recorded in VR system. Teachers can evaluate the learning results of students in real time and they can revise the Teaching content and progress, according to the actual situation [9].

3.3 Safe Operation Environment

The practical operations of students are realized on VR system, and they can not be exposed to the high-pressure, high-temperature, toxic environments of the production site [10]. Then the safety is improved by using VR system. At the same time, the functions and structures of devices provided by VR system are similar to that of real devices, but the VR cost is much lower than the actual devices. We can effectively reduce the cost of teaching. With the application of VR technology, the damages of experimental and practical training are reduced [11].

4   Design of VR System

The overall system architecture is designed based on the combination of B/S mode and C/S mode to realize the teaching training of new energy technology (see Fig. 1). The system includes the five main modules, virtual experiment, on-line communication, quality evaluation, virtual classroom, and material database etc.

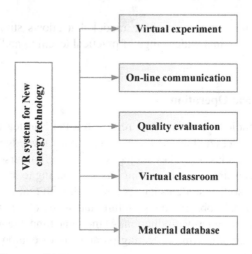

Fig. 1. The overall VR system architecture for new energy technology

4.1 Virtual Experiment Module

Virtual experiment module is an Internet or local virtual experiment platform, which is developed for experimental courses of new energy technology. The system can simulate the equipment used in the real experiment, provide an experiment environment similar to the real experiment, and provide the functions of experiment information management and progress monitoring.

4.2 On-line Communication Module

This function of on-line communication module is clear at a glance. It can provide a communication channel for students and teachers. The module is formed in the Internet era. How to let students and teachers more convenient feedback and receive information is the main design basis of the module [12].

4.3 Quality Evaluation Module

The design of the quality evaluation module is mainly divided into two roles: a manager role and a candidate role. The manager can add, delete, modify and view candidate information, can query statistical results and manage test questions; Candidates can view my information and upload profile information. The module can also evaluate the results of experiment operation in every steps for practical ability training. Some factors of the performance assessment for students are used in the module. The general evaluation of Gaussian is used as

$$f(x) = \frac{1}{\sqrt{2\pi}\sigma} e^{-\frac{(x-\mu)^2}{2\sigma^2}} \tag{1}$$

where μ is the mean of performance set, and σ is the variance of performance set.

4.4 Virtual Classroom Module

Classroom teaching based on VR is combined with engineering application, practical innovation and the latest research progress [13]. Students are encouraged to participate in the classroom according to their own interests. For some contents, which cannot be understood by traditional teaching methods, they can be helped to understand them by combining practice view. As shown in Fig. 2. The module can provide real-time interactive classroom, local courseware production, online-demand knowledge, students' learning behavior analysis.

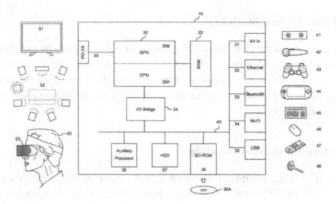

Fig. 2. VR system

4.5 Material Database Module

Database can consult and learn new energy technology papers, drawings and technical documents. It becomes the essential content during the training of students. The knowledge base contains new energy technology documents and multimedia materials. The whole knowledge is classified and sorted with graph method. Students can gain in-depth knowledge of new energy technologies by viewing the material database.

5 Overview of Virtual Reality

5.1 The Definition of Virtual Reality Technology

Virtual reality technology is the crystallization of scientific and technological progress since the 20th century, which embodies the latest achievements in computer technology, computer graphics, multimedia technology, sensor technology, reality technology, ergonomics, human-computer interaction theory, artificial intelligence and other fields. It is based on computer technology and exists in the computer. Some special input/output devices create a multi perception three-dimensional virtual world. In this virtual environment, users interact with things in the virtual environment through vision, hearing, touch and so on. All changes in the real world are vividly reflected in the virtual world, making people and things in the virtual world into a whole. People can directly experience the changes of the surrounding environment as if they were on the scene, and effectively realize human-computer interaction, including perceiving things and contacting the environment. Virtual reality has become an art in our life [14].

The definition of virtual reality technology can be summarized as follows: Virtual Reality (VR) is an advanced computer technology, which uses modern high-tech means to create a virtual environment, and enables users to "invest" in the environment through a variety of peripheral devices. This technology makes use of the natural way to interact with things in the virtual environment to achieve the purpose of human-computer interaction. VR technology allows users to use human basic skills to interact or operate objects in the virtual reality world, while providing visual, listening, touching and other intuitive and natural real-time perception [15].

5.2 Basic Characteristics of Virtual Reality

Virtual reality and users are interactive, which can make users feel immersive. Through the real-time interaction of vision, smell, hearing and other senses with nature, it greatly facilitates the user's operation and improves the user's work efficiency. It fully embodies the significant characteristics of immersion and interaction in the virtual world. In addition, the third feature of virtual reality system is imagination. In this way, immersion, interaction and imagination constitute the three characteristics of virtual reality technology [16]. American scientists burdeag and coiffet call it "the triangle of virtual reality technology", which concisely represents the three most prominent features of virtual reality technology, as shown in Fig. 3.

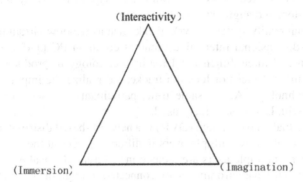

Fig. 3. Three "I" characteristic diagrams

Interactivity, there are people's participation and feedback in the virtual reality environment, people are an important factor in the virtual environment, people lead the change of things. At the same time, the reality in human-computer interaction is the premise and foundation, and the reality is completed through the effectiveness of other people's interaction. At the same time, human-computer interaction must be real-time. Real time refers to the virtual system can quickly respond to the various operations of users, human beings can use virtual reality technology, through natural skills, interact with the objects in the virtual environment [17].

Immersion, also known as temporary sensibility or immersion, is that users, as the main body, place themselves in the virtual world, let users change from the initiative of things to the participants of things, transcend the reality, and integrate themselves into the changes of the virtual world. It is considered to be the most important feature of VR, which is that the user's consciousness and illusion of real objects are mapped to the virtual environment. It has the characteristics of multi perception and autonomy. People have a variety of perception functions in the real world, and virtual reality should have these functions in the virtual environment. Multi perception means that virtual reality system should have all kinds of human perception functions. Autonomy means that all objects in the virtual environment have their own characteristics. The object can move independently in its own way and interact with each other. At present, our research on virtual reality is relatively late, and the major of technology is in its infancy. At present, the research on the system is still limited to visual immersion, auditory immersion, tactile immersion, olfactory immersion, body sensation immersion, taste immersion and other aspects, and the development is not mature. In addition to the above, immersion is also used to collect the influence of the three-dimensional image field of view, depth information (whether it is appropriate to the user's life experience), whether the user is suitable for interactive devices, and whether the tracking time and spatial response are accurate.

5.3 Classification of Virtual Reality

According to different standards, virtual reality systems can be divided into four categories: desktop virtual reality system, immersive virtual reality system, distributed virtual reality system and augmented reality system.

Desktop virtual reality system (pcvr), also known as window virtual reality system, is a set of three-dimensional interactive scene of common PC platform. It uses low-level workstation and three-dimensional reality technology to produce virtual scene. Participants use input devices or location trackers to realize the important features of virtual reality technology. At the same time, participants can set up various virtual environments at will. Low cost, widely used.

Distributed virtual reality system (DVR) is a network-based distributed virtual environment that can be used by multiple users in different places at the same time. In this virtual environment, multiple users are located in different physical environment locations, and multiple virtual environments are connected through the network, or multiple users participate in a virtual reality environment, and the computer interacts with other users through sharing the virtual space. In the distributed virtual reality system, through the network, many users operate on the virtual world and communicate with each other in many ways. Distributed virtual technology has wide application prospects in the fields of distance education and telemedicine.

Distributed virtual reality system (DVR) is a network-based distributed virtual environment that can be used by multiple users in different places at the same time. In this virtual environment, multiple users are located in different physical environment locations, and multiple virtual environments are connected through the network, or multiple users participate in a virtual reality environment, and the computer interacts with other users through sharing the virtual space. In the distributed virtual reality system, through the network, many users operate on the virtual world and communicate with each other in many ways. Distributed virtual technology has wide application prospects in the fields of distance education and telemedicine.

Augmented reality (or hybrid reality) is a kind of system that combines real environment with virtual environment. Users can enhance their understanding of the real world with the information provided by virtual objects. It has the characteristics of real-time interaction, virtual reality combination and three-dimensional registration. It can also add virtual objects in the real environment, reduce the cost of complex environment in real things, and process things. At present, the system is widely used in medical visualization, equipment maintenance and processing, leisure and entertainment.

6 Summary

New energy technology is an interdisciplinary major involving natural science, technical science and social science. VR has been explored and discussed in many application-oriented colleges. The talent training method should be adjusted according to the development situation of economy. The VR sysem should be used for knowledge points of new energy technology. It is important that the training construction with VR will meet the need to further improve the teaching efficiency.

References

1. Fan, W.Y., Wang, G.P., Liu, H.D.: Study in the application of PSIM software in the teaching of curriculums related to the new energy power generation. Adv. Mater. Res. **986–987**, 551–555 (2014)
2. Byun, Y., Park, N.: Analysis of international joint research into new and renewable energy technology. New Renew. Energy **14**, 4–11 (2018)
3. Xu, P., Si, H., Wang, Y., et al.: Application of virtual reality technology in the English teaching of the University. Adv. Mater. Res. **926–930**, 4469–4472 (2014)
4. Bai, Z., Lu, J.: Discussion based on 3dma X in urban virtual reality. Coal Technol. **28**(5), 175–177 (2009)
5. Liu, X., Xiong, Z., Cao, Q.: Discussion on the establishment process of virtual reality based on 3dma X. Western Explor. Project 2 (2009)
6. Wang, W.: Research on maximum power point tracking of solar photovoltaic power generation system. Xi'an Normal University (2009)
7. Fu, W.: Design and implementation of solar photovoltaic power generation monitoring system. University of Electronic Science and Technology, Xi'an (2007)
8. Zhang, Z., Ma, Q., Cheng, D.: Research on Control Technology in solar photovoltaic power generation system. Low Volt. Elect. Mod. Elect. Technol. **12**, 130–137 (2008)
9. Du, H.: Research on solar photovoltaic power generation control system. North China Electric Power University, University of Electronic Science and Technology, Beijing
10. Zhou, N., Yan, L., Wang, Q.: Research on the connection and dynamic characteristics of photovoltaic power generation in microgrid. Protect. Control Power Syst. 38(14): 119–127 (2010)
11. Guo, H., Su, J., Zhang, G.: Research status of microgrid technology. Sichuan Elect. Power Technol. **32**(2), 1–6 (2009)
12. Li, P., Zhang, L., Sheng, Y.: Microgrid Technology: an effective way for large-scale application of new and renewable energy grid connected power generation. J. North China Electr. Power Univ. **36**(1), 10–14 (2009)
13. Dwells, C.H.: Solar micro grids to accommodate renewable intermittency. In: 2010 IEEE PES Transmission and distribution Conference andExposition: Smart Solutions for a Changing World, 2010. IEEEPES Transmission and Distribution (2010)
14. Tian, S.: Research on Photovoltaic Power Generation Control Technology in Microgrid. Xi'an University of Technology, Xi'an (2010)
15. Yan, S., Yin, M., Li, Q., et al.: Research on related technologies of solar photovoltaic grid connected system. Technol. Front. 65–73 (2009)
16. Zhou, N.-N., Deng, Y.-L.: Virtual reality: a state-of-the-art Survey. Int. J. Autom. Comput. **6**(4), 319–325 (2009)
17. Zuo, W., Li, P., Cheng, J., et al.: Overview of microgrid technology and development. China Electric Power **42**(7), 26–30 (2009)

Research on the Image Tracking of Tennis Rotation Features Under Big Data Analysis

Yanlou Sun[✉]

Sanya Aviation and Tourism College, Sanya 572000, China

Abstract. Aiming at the problem of large error in tracking image data analysis in traditional image tracking system of tennis rotation feature, an image tracking analysis system of tennis rotation feature is proposed. The data are processed by holographic projection, the average constant of motion angle is optimized, the gate tracking algorithm is introduced to track the national standard of the image, and the threshold is calculated to realize the image tracking of tennis rotation feature. The experimental data show that the designed method can effectively track and analyze the tennis image, and solve the problem of large error in image data analysis.

Keywords: Tennis rotation feature · Image tracking · Gate tracking algorithm · Holographic projection · Data processing

1 Introduction

In modern tennis games, when the players are in the game, they have to judge the position of the ball each time they hit the ball. It can be said that whether they accurately judge the position will affect the quality of the return stroke to a certain extent. Only with accurate judgment, he can use the most reasonable and effective return action to hit a return ball of the highest quality, he can make his level play incisively and vividly and win the game. Therefore, the accuracy of the player's judgment of the direction of the ball hit by the opponent directly affects how to catch the ball in the game.

In tennis match, the angle of hitting, the coefficient of the ball in high speed movement and the contact with the beat speed make the tennis rotation in the sport. The tennis rotation will show a unique arc parabola of jumping, which gives people the beautiful track of tennis hitting technology. Tennis can be divided into four categories according to rotation and trajectory: the different rotating track types of tennis, up spin ball, bottom line ball and cutting ball 6 will show different flight curve, beating frequency and rebound angle. Therefore, to play the backkick which the opponent can not take precautions, it is necessary to judge and use different hitting techniques.

In order to effectively solve the problem of large error in tracking image data analysis, 3DLAN holographic projection technology is used and verified by comparative simulation experiment [1]. The accuracy of the designed tennis motion prediction trajectory direction automatic detection software is proved by experimental verification.

© ICST Institute for Computer Sciences, Social Informatics and Telecommunications Engineering 2021
Published by Springer Nature Switzerland AG 2021. All Rights Reserved
M. A. Jan and F. Khan (Eds.): BigIoT-EDU 2021, LNICST 392, pp. 148–154, 2021.
https://doi.org/10.1007/978-3-030-87903-7_21

2 Design of Image Tracking and Analysis System

2.1 Data Processing and Analysis

In order to improve the data analysis ability of the image tracking analysis system of tennis rotation characteristics designed in this paper, we can accurately measure the radius, mass, rotation angle, speed and other constants and variables of tennis in the process of movement through holographic projection, and calculate the friction, wind speed and other influencing factors. The essence of this technology is to transform the imported measured data into a nonlinear function equation, and substitute the momentum factors such as threshold, difference and ratio into the formula, so as to simulate the trend of tennis and let the players play high-quality return strokes according to the trend. The index forms of specific change parameters are as follows:

$$
z \begin{bmatrix} u \\ v \\ l \end{bmatrix} = \begin{bmatrix} \frac{1}{d} & 0 & u \\ 0 & \frac{1}{d} & v \\ 0 & 0 & l \end{bmatrix} \begin{bmatrix} R & t \\ 0 & 1 \end{bmatrix} \begin{bmatrix} x \\ y \\ z \\ l \end{bmatrix}
\tag{1}
$$

Where: Z is the number of pixels of the target tennis; u is the tennis characteristic coefficient; V is the speed of the tennis movement; it is the horizontal distance of tennis; D and D1 are the tennis radius and the deformation radius respectively. If there are n such pixels, you can get:

$$
S = \frac{D}{T} + \frac{D}{\frac{D}{B} + n\Delta t}
\tag{2}
$$

By measuring the repetition rate, the image tracking analysis system of tennis rotation features can keep stable running state, and the clarity of image pixels has increased. In this way, the future direction of tennis can be determined from the data source. The calculation formula of error friction is:

$$
f = \eta \frac{s}{h} V
\tag{3}
$$

2.2 Analysis of the Disadvantages of Fingertip Dynamic Process

In the process of fingertip tracking, the most important thing is the dynamic feature detection of multi frame image of fingertip. In the process of fingertip tracking, the role of fingertip is to identify the direction of finger movement. Because of the shape of the fingertip, in the detection process, the fingertip shape is used as the feature to obtain the coordinate position of fingertip, complete the tracking, and convert the color space of finger image into HSV mode, which is more conducive to the effective feature extraction, The new H component is analyzed to obtain the corresponding chromaticity map of the image.

$$
q_u = \sum_{i=1}^{n} \delta[c(x_i) - u]
\tag{4}
$$

3 Experimental Verification

In order to verify the accuracy of image tracking analysis of tennis rotation feature, a comparative simulation experiment is designed. In this paper, the image tracking analysis of tennis rotation characteristics in the process of hitting in a training ground is carried out [2]. In order to ensure the effectiveness of the experiment, the design method and the traditional method are used simultaneously.

3.1 Parameter Setting

In order to ensure the accuracy of image tracking analysis of tennis rotation characteristics, the fluctuation limit h is set to 65.32; the peripheral kinetic energy $D \times 1$ of tennis is set to 9.3; the saturation value ab of the captured motion area is set to 10 and t to 50 in the range of [0.66130]. According to the parameters set by the simulation, experiments are carried out, and the results are as follows.

3.2 Acquisition and Analysis of Experimental Data

As shown in Fig. 1, the percentage of data frequency hopping is very high in traditional methods, and some data are more than 2.0%, which seriously affects the ability of data analysis.

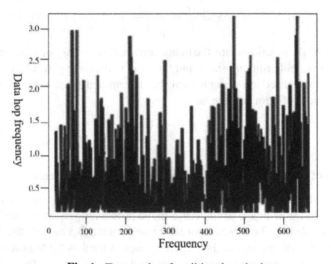

Fig. 1. Test results of traditional method

As shown in Fig. 1, the image tracking and analysis system of tennis rotation features designed in this paper can keep a low frequency hopping percentage of data, and the average data is below 0%, which indicates that the data is very stable and the analysis results are very stable, so the error will be very small.

4 Simulation of Fingertip Tracking Process Based on Image Processing

4.1 Introduction of Bayesian Classification Decision

In the current tracking process, multiple constraints are introduced, and there is a lag problem. Bayesian filtering method is used to filter the collected image features, that is, the known data information is used to construct the relevant posterior probability density of the state variables. According to Bayes' ability to simplify the multiple constraints, the optimal tracking process is completed. When the state of the fingers has different values, The confidence can be realized by observing data, so the optimal estimation of state is completed. It is assumed that the feature vector sequence of fingers is represented, K is used to describe the time series label, and state vector is used to describe the time series. Then the spatial model of finger motion state can be described in the following formula. Corner is defined as the point with sharp brightness change or the point with maximum curvature on the edge curve of two-dimensional image. The detection algorithms include: corner detection method based on scale space, corner detection method based on multi-scale filtering, adaptive corner detection method, corner detection method based on template, corner detection method based on geometric features, and corner detection method based on minimum brightness change, There are four criteria to measure corner detection algorithm: accuracy, localization, stability and complexity.

4.2 Feature Extraction of Fingertip Detection and Tracking Process

Due to the consistency of human skin color, the probability of skin color can be calculated by constructing a color statistical model. Given that the likelihood distribution of fingertip skin color is represented by P (R, G, B skin), and the distribution of non skin color pixels is represented by P (R, G, B nonskin), then the probability of skin color of any pixel is calculated according to Bayesian law. The shape feature of an image is formed because the physical and geometric characteristics of the object make the gray level of the local area in the image change significantly [3]. The commonly used shape based feature extraction and matching methods can be divided into two categories: boundary feature, region feature and point feature. The boundary feature mainly aims at the outer boundary of the object, that is, the image edge, which refers to the set of pixels with step change or roof change in the gray level of the surrounding pixels; while the region feature is related to the whole shape region.

5 System Feature Model in the Field of Motion Tracking

5.1 Principle Explanation of Motion Characteristic Model

In software design of domain engineering, feature and feature model are used to capture the commonness and difference of systems in a domain in domain analysis. Feature refers to an eye-catching aspect, quality or feature of a software system or system that is visible to the end user. Feature model consists of four parts: feature graph, feature definition, combination rules for features, and feature principle. Feature model in domain engineering is obtained by feature modeling, Feature modeling is the activity of modeling the common and variable attributes of concepts and their interdependencies, and organizing them into a consistent model, namely feature model. One of the characteristics of software engineering modeling is its representativeness in the field of software engineering modeling. The result of feature modeling is to generate a domain feature graph. The idea of feature graph is to explicitly represent the configurable aspect, and leave other aspects, such as structural modeling, to more appropriate symbols, such as entity relationship graph or object graph.

5.2 Domain Architecture Approach

In domain architecture, a full description of the software architecture requires multiple views. For example, the software architecture "4 + 1 view model", which was promoted by rational methodology Philippe kruchten, contains a logical view (class diagram, sequence diagram, collaboration diagram, activity diagram, and state transformation diagram), a process view (process diagram), a physical view (package diagram), a deployment view (deployment diagram), and a use case model. When designing a software architecture, not only the functional requirements of the domain, but also the robustness, adaptability, scalability, reusability, etc. The practical purpose of domain oriented software architecture is to quickly answer how feature nodes in feature models are reflected as components, and clearly reflect the relationship between components, including the differences (configurable) covered by feature models. This paper will design the domain software architecture based on the domain model diagram, mainly including the component diagram, the domain configuration generator, the running process view of the system architecture, and the class diagram designed with MVC mode. Finally, through the above design aspects, a hierarchical structure diagram showing the whole software hierarchy is obtained, It provides a clear and accurate description for the reuse of the whole architecture.

6   Multi Point Fusion Correlation Tracking Algorithm

6.1   Research Status of Correlation Tracking Algorithm

In the field of motion tracking algorithm, correlation tracking algorithm is the focus of the research, and it is also a tracking algorithm with a wide range of application. The image sequence in motion tracking has temporal correlation and spatial correlation. The adjacent two frames have little change. The motion of the target is reflected as the change of the image. The relative position change of the target can be obtained from the change of the image. Therefore, it is an ideal method to track the moving target in the image by using correlation tracking.

Correlation tracking takes the region image containing the tracking target as the template, and then searches the sub image region specified by the wave gate to judge whether to find the matching region with the template according to some similarity criteria. Because correlation tracking algorithm is a tracking method to directly find the closest region of the target template image in the image sequence, it directly performs the operation in the original image sequence, It can be used without the process of image segmentation and feature extraction, and is not limited by the shape, size and brightness of the target [4]. As long as the target has features, that is, the image has obvious gray changes in the region, it can be used, especially for large targets with obvious features and complex background.

6.2 Control Processing Explanation

In the control layer, the user generates the component configuration and sends it to the component function configuration generator of the model layer. In the control layer, the user sends the control command to the message manager through the tracking control command tool. If the message manager finds that it is a component function message, it sends the message to the component function executor of the model layer. In the model layer, the component configured by the user is called by the component function executor. When the component function executor calls the function of the corresponding component, it will call the component function configurator to determine whether the component function can be executed legally. If it can, it will call the specific component in the component library to respond to the specific message. In the model layer, tracking the component group will generate the moving object image data, The data needs to be transmitted to the tracking process display component in the view layer for tracking process display. The parameter solution component will generate the result data of the moving target parameter solution [5]. The data needs to be transmitted to the data table display, parameter curve display, data printing display and other components in the view layer for display or printing, Then new components can be added to the component library, so that the component library can be extended and reused.

7 Conclusion

In the process of fingertip tracking, it is necessary to set a large number of constraints on the background and foreground. The accuracy and efficiency of fingertip tracking using traditional algorithms are low. Therefore, a fingertip tracking method based on image processing is proposed. According to the Bayesian principle, the collected image is filtered, the posterior probability is calculated according to the Bayesian rules, and the constant is normalized, And through the establishment of color statistical model to calculate the skin color probability information, the fingertip tracking in the image region is completed. According to the relevant theory, the fingertip dynamic equation is constructed, so as to complete the fingertip tracking process. The experimental results show that the improved algorithm can improve the tracking accuracy and robustness, and provide strong technical support for the realization of human-computer interaction, It has great advantages.

References

1. Hu, J., Zhang, L., Ren, R., Li, H.: Image tracking technology in complex background of flight test. J. Nav. Aeronaut. Eng. Coll. **25**(01), 11–14+18 (2010)

2. Zhang, S.: Research on domain oriented image tracking software architecture and tracking algorithm. Xi'an University of Electronic Science and Technology (2008)
3. Zhou, F., Wang, H., Yu, X., Yan, DI.: Analysis of respiratory motion based on image tracking. Chin. J. Biomed. Eng. **05**, 652–657 (2007)
4. Yu, X., Zhou, F.: Respiratory motion analysis based on image tracking. Chin. Stereology Image Anal. **01**, 22–26 (2006)
5. Luo, S., Zhang, Y., Luo, F., Wang, Y.: Intelligent decision method for image segmentation based on rough set theory. Chin. J. Image Graph. **01**, 66–73 (2006)

"5 G" Research on System Design of Data Science and Big Data Technology Talent Training in Colleges and Universities

Xiaoying Zhang and Xin Lu[✉]

School of Science, Changchun University, Changchun 130000, China

Abstract. This paper focuses on the construction of data department and big data technology major in local colleges and Universities under the background of 5G. According to the characteristics and advantages of disciplines, it pays attention to the interdisciplinary and integration, closely follows the major needs of the country and the local, and connects the talent demand of enterprises and industries, with the guiding ideology of "through", "cross cutting", "collaboration" and "joint", It focuses on the comprehensive reform in the important links of professional development, such as training program, teaching staff construction, curriculum system construction, teaching reform and practical teaching. A set of new engineering talents training mode covering big data research and development and industrial application, reflecting the school running characteristics of local colleges and universities and the characteristics of data science and big data technology, and playing a demonstration and promotion role in local universities nationwide.

Keywords: 5G · Data science and big data technology · Local universities · Mathematics related disciplines · Talent training mode

1 Introduction

The traditional data analysis technology has been unable to meet the actual needs, and the explosive growth of data has put forward higher requirements for all walks of life. In the era background of 5G network technology, the ideological and political courses in Colleges and universities must be reasonably innovated, so as to better train the talents of the diversion. Although on the surface, the specific relationship between Ideological and political education and talent training in private colleges and Universities under the background of 5G is not obvious, and there is a suspicion that students are hard to piece together when studying on these factors [1]. However, for the relevant personnel with strong research ability, they are bound to be interested in and have the ability to complete the overall task, so as to promote the innovation of curriculum ideological and political education in private colleges and Universities under the background of 5G, To improve the quality of Ideological and political courses to implement and improve the talent training mode.

M. A. Jan and F. Khan (Eds.): BigIoT-EDU 2021, LNICST 392, pp. 155–164, 2021.
https://doi.org/10.1007/978-3-030-87903-7_22

Under the background of the rapid development of network technology in China, new media technology has also been constantly innovated. College students do not choose to rely on traditional television, newspapers and other media to obtain information, but use the network to clarify the relevant video and audio information. Thus, the development of the network is closely related to the implementation of modern education, As the upgraded version of 4G network technology, 5G network technology can make more devices access to the network, thus forming innovative data, which has a profound impact on college education. The above is basically the change brought about by 5G network era.

2 Related Work

The downlink transmission of 5G system adopts OFDM technology, and the time-domain transmission signal of the transmitter can be expressed as:

$$s(k) = \frac{1}{N} \sum_{n=0}^{N-1} X_n e^{\frac{j2\pi n}{N}}, k = 0, 1, \ldots, N-1 \tag{1}$$

Where n is the number of FFT points and X_n is the data modulated to the nth subcarrier in frequency domain.

When the time domain signal s(k) of the transmitter passes through the multipath fading channel and the Gaussian white noise is added, the time domain complex baseband signal of the receiver can be expressed as:

$$r(k) = [\sum_{d=0}^{D-1} h_d(k)s(k - \tau d - \delta)]e^{\frac{j2\pi n}{N}} + z(k) \tag{2}$$

Where $r(k)$ is the received signal, $h_d(k)$ is the channel coefficient of the D-th path, which obeys Rayleigh distribution, D is the maximum channel delay extension, τd is the multipath delay, δ is the time offset between the sender and receiver, and $z(k)$ is the additive white Gaussian noise.

This paper analyzes the characteristics of the main synchronization sequence in 5G system. Aiming at the problem that the traditional timing synchronization algorithm can not achieve fast synchronization in 5G system with large frequency offset, an improved timing synchronization algorithm based on segment correlation is proposed [2]. The algorithm pre stores the anti frequency and frequency domain sequences locally, decomposes the long correlation into the short correlation, and transforms the data to the frequency domain to achieve fast correlation, thus effectively reducing the computational complexity.

3 Relying on the Construction of Data Science and Big Data Technology Specialty in Local Colleges and Universities

3.1 Making Training Plan and Exploring Talent Training Mode

Learning from the successful experience of the reform of the talent training program of mathematics and computing science and the establishment of "Shaofeng School of mathematics", according to the guiding ideology of stratification + diversion and individualized development, and according to the talent training objectives, the traditional professional training mode is changed to separate undergraduate students, and the new training mode of "specialized courses + whole school supplementary courses, innovation laboratory, and school enterprise joint training" is implemented, Thus, it constructs a diversified and three-dimensional talent training mode of data department and big data technology specialty in local colleges and universities [3]. In 2018, relying on the undergraduate major of "data science and big data technology", our university applied for the construction of data science teaching experimental platform, including the purchase of software and hardware equipment and the construction of experimental environment through the central financial support for the reform and development of local colleges and universities.

3.2 Construction of Teaching Staff

In terms of optimizing the teaching staff in the University, we should break through the barriers of the University and deeply integrate the teaching staff of mathematics, statistics, computational science and other related applied disciplines. Promote the combination of big data mathematics with statistical basis, computer foundation and Application module. Build four core teams for big data collection, storage, analysis and visualization. In view of the problem that it is difficult for local universities to introduce high-level teachers in such disciplines as statistics and computer science, the university intends to give appropriate preference to policies and introduction efforts to strengthen the weak links of teachers. In terms of the construction of double qualified teachers team, centering on the "ability building as the core and system innovation as the driving force" of the construction of teaching staff, through the two-way docking of "University Research Institute (Institute) joint" and "school enterprise alliance", the construction of professional "double qualified" teachers team is carried out, and the young and middle-aged teachers with professional characteristics are guided and formed.

4   Teaching Research and Reform Simulation Analysis

Because data science and big data technology is a new major, considering its strong interdisciplinary and strong practicability, we should carry out teaching research and teaching reform in time. In terms of teaching and research, we should consider the characteristics of data science and big data technology, break the barriers between colleges and departments, build incentive measures for documents, and study the cross integration of mathematics, statistics, computer science and related application disciplines, and constantly optimize the knowledge structure of the major. Further build a complete talent training system of data science and big data technology from undergraduate, master, doctor to postdoctoral, and establish a benign interaction mechanism between discipline construction and data science and big data technology teaching. Professional teachers are encouraged to apply for teaching reform projects at provincial and ministerial level and publish high-quality teaching research papers. In the aspect of teaching reform, we should reform the teaching organization form, based on the in class teaching link, take

the extracurricular science and technology interest group, subject competition and innovative experimental plan project as the platform to carry out "on campus + off campus,
in class + extra-curricular" interdisciplinary training and school enterprise joint training
[4]. In September 2017, the main leader of big data undergraduate major of Xiangtan
University applied for the Ministry of education's new engineering research and practice
project "construction and practice of data science and big data technology in local universities" by the Ministry of education, and successfully obtained the project. Teaching
research and Reform data sample is shown in Fig. 1. From Fig. 2, we can see that under
the 5G background, the bonus of talent training is mainly concentrated on boys, while
girls are less. Therefore, we can see that under the 5G background, both boys and girls
can get more profits [5].

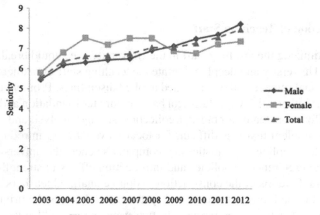

Fig. 1. Teaching research and reform

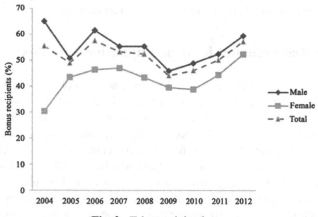

Fig. 2. Talent training bonus

5 Strengthening Practical Teaching

Because data science and big data technology are practical majors, it is necessary to have a good practical base as training support [6]. Take Xiangtan University as an example, Make full use of the Key Laboratory of Intelligent Computing and information processing of the Ministry of education, Hunan Key Laboratory of scientific engineering calculation and numerical simulation, Key Laboratory of engineering structure dynamics and reliability analysis of Hunan Key Laboratory of national defense science and technology and Hunan Provincial Patent Analysis and evaluation center Taiwan. Based on the industrial development of mathematics, statistics and data industry, we will explore the training mode of compound applied science professionals who master modern mathematics and statistics ideas and methods, have deep science foundation and strong engineering application ability, Xiangtan University is a pilot unit to study and formulate the talent training program and curriculum system of "data computing and application" of Applied Science, plan and carry out the construction of series of teaching materials, and promote and lead the "emerging engineering education" derived from science.

6 Mining Subject Words of Data Science Talent Demand

6.1 Overview of LDA Model

In most cases, LDA has two meanings. The LDA topic model in this paper refers to the latent Dirichlet allocation (LDA) model, which was constructed by David BLEI, Andrew ng and Michael I. Jordan in 2003 on the basis of PLSA. It will show the topic of each text in the text set in the form of probability distribution [7]. This is the core definition of generative model, which holds that all words in any text should conform to this rule. Bag of words (BW) is the most common method of LDA model. Each text is regarded as a word frequency vector, which makes it easier to model and analyze after text information is transformed into digital information. LDA model is usually composed of document (d), topic (z) and word (W), so it is also called three-layer Bayesian probability model, assuming that any document is a mixture of various topics, The model is generated from Dirichlet distribution by sampling, and the specific model flow is shown in Fig. 3. Taking the probability information into account in the original traditional space vector model can not only mine the topics in the data set, but also help to extract the hot concerns and related feature words in the data set for in-depth analysis.

6.2 Topic Number Selection

Firstly, input the data processed by feature, and then use M algorithm to solve the topic model to get the corresponding distribution of document, topic and word. The selection of topic number k is the most critical step in setting model parameters. If the value of K is too small, it will make many concepts can not be classified under the corresponding related topics, so they are wrongly summarized to the topics that are not suitable for them, and there is no way to accurately express the concepts that the document needs to convey, which is not conducive to extending the model to other talent demand mining. In

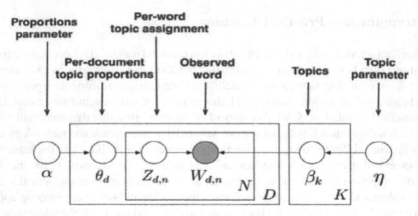

Fig. 3. Flow chart of LDA model

order to help more talent demand data to match with the theme model, the coarse-grained topics such as personal ability and educational background are selected as the theme of the theme model [8].

When the number of topics is set to 3, it means that only three different topics are extracted for analysis. In Fig. 4, we can see that topic 1 has a large number of words, including "data analysis", "algorithm", "machine learning", "data mining", "experience" and other words about professional knowledge, skills and work experience requirements of data science [9]. Topic 2 is mainly about the ability and quality of job seekers, such as "communication", "cooperation", "teamwork" and so on. Topic 3 is

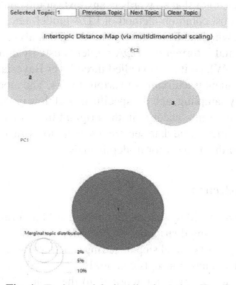

Fig. 4. Topic model visualization when K = 3

about the distribution of academic background words, such as "Statistics", "Mathematics", "postgraduate", etc. [10]. So when the model selects three topics, the result is that the three topics are professional knowledge and skills, personal ability and quality, and education background. From the above analysis of the distribution of words, the result is not ideal when the number of topics is selected as 3. The three topics are divided into multiple topics, so that it is difficult to determine the center of the topic.

6.3 Subject Term Extraction of Data Science Talent Demand

Because the number of topics selected in the best model is four topics, we need to manually delete some words that are not related to the topic content or have less information [11]. Finally, we extract ten most representative topic words from the original topic words. The number of topics selected in this paper is 4, and the corresponding topics are education background, work experience, personal ability and professional knowledge and skills of data science talents. Through the analysis of the keywords of the four categories under the demand of all data science talents, it can be seen that the requirements of professional knowledge and skills are rich and diverse, the work experience and education background are slightly different, and the personal ability and quality are basically the same [12].

In order to highlight the diversity needs of professional knowledge and skills, four types of representative talents are selected for comparative analysis [13]. For the analysis of professional knowledge and skills of data science talents: data analysis talents are mainly required to have good statistical foundation and analysis ability in the application of big data; the professional knowledge and skills of database R & D talents lie in database related knowledge, including database development and operation and maintenance; Algorithm engineering talents mainly need to master the knowledge of machine learning and natural language processing; artificial intelligence development talents need to be familiar with the content of computer and software development and other fields [14].

According to the comprehensive analysis of the other three topics of data science talent demand, the personal ability and quality requirements of the four types of talents are related to teamwork, sense of responsibility, logical thinking and understanding ability; for education background, the subject words of the four types of talents are distributed in the same sentence, It is mainly reflected in the "Statistics", "Mathematics" and "computer" professional "undergraduate" and "graduate" crowd; work experience is similar, which requires us to use word2vec algorithm model to adjust the results of LDA model, and further analyze the demand for data science talents. Through the simple analysis of the demand for four types of representative talents, the topic of LDA topic model is still relatively clear, but the setting of quantity and precision is too rough, which can not completely present the main content of the demand. Therefore, we need to use the word vector model to expand the in-depth mining of topic words [15].

6.4 Overview of word2vec Model

Word2vec is an open source tool for computing word vector in natural language processing. It is a word vector model based on neural network. It can be trained efficiently on millions of dictionaries and hundreds of millions of datasets, and its algorithm mainly includes cbow model and skip gram model. Cbow is the abbreviation of continua bag of words. Mathematically, it is equivalent to multiplying the vector of a bag of words model by an embedding matrix to get a continuous embedding vector. The model learns the expression of word vector from context's prediction of targe word. The essence of skip gram model is to calculate the cosine similarity between the input vector of input word and the output vector of target word, and to normalize it with SoftMan κ. In this paper, we calculate the cosine value between the numerical morpheme vectors which contain a lot of semantic information through the model, and reflect the degree of association between words in the topic through it, so that the words in the bag of words model will not appear large differences in dimension and semantics, which will affect the result analysis. At the same time, the running speed of the model also has certain advantages compared with other models [16].

6.5 Overview of Visualization Tools

Gepi is a JM based open source and cross platform data visualization software in the field of complex network analysis. It simplifies the network transformation into the form of nodes and edges, and uses them to express the internal structure of data and the relationship between various parts. The visualization model analysis provided by Gepi can be divided into two types: one is to present the location of nodes as a graph in a certain way by selecting different layout algorithms, and interpret the network relationship on this basis. The other is to analyze and explain the network relationship by selecting different statistical algorithms to calculate the overall characteristics, modularity and node centrality of the network.

Gepi provides 12 layout modes, including 6 main layout tools and 6 auxiliary layout tools. The most commonly used are: two force oriented algorithms, circular layout and Hu Yifan layout. Force oriented layout is divided into force atlas and force atlas 2. By imitating the gravitational and repulsive forces of the physical world, force oriented layout automatically generates beautiful network layout graphics until the forces are balanced, and fully shows the overall structure and Automorphism characteristics of the network. It has strong readability, so it plays a leading role in the selection of visual model layout.

Circular layout, also known as fruchterman Reingold layout, is a fr algorithm proposed after many improvements on elastic model. The nodes with edge connection should be close to each other, and the distance between nodes should not be too small, which are two indispensable principles of the algorithm. The algorithm assumes that the nodes in the graph are atoms in particle physics, and calculates the position relationship between the nodes in the form of simulating the position law of atoms until they enter the dynamic equilibrium state. The interaction of gravity and repulsion is considered in the calculation, and the theoretical law of particle physics must be followed. Yifan Hu, Yi Fan Hu proportion and VI I fan Hu multi-level layout are collectively referred to as Hu Yifan layout, which is suitable for large graphics processing, characterized by rough graphics, reducing the amount of calculation and improving the running speed.

Based on LDA topic model and word2wvec word vector model, this section extracts the expanded topic word set, and uses Gepi software to analyze the topic words of data science talent demand through network relationship analysis method. There are four main themes of data science talent demand: education background, work experience, professional knowledge and skills and personal ability and quality. Based on these four different demand topics, the demand of data science talents is visualized and analyzed, and the core demand of data science talents, the correlation between the internal keywords of each demand topic and the community relationship between each demand topic are mined.

7   Conclusions

On the basis of optimizing the teachers' team in the school, we should build a double qualified team, formulate appropriate teaching resource construction plan, and provide rich teaching resources for big data major through school enterprise cooperation based on the concept of "construction, application and sharing"; we should break down the barriers between colleges and departments and conduct research in the way of interdisciplinary integration; actively carry out multi-channel and multi-mode practice teaching relying on practice base, The experimental teaching mode of "foundation, synthesis and innovation" is constructed. Finally, a complete set of talent training mode is formed, which covers the research and development of big data and industrial application, reflects the school running characteristics of our university and the characteristics of our data department and big data technology specialty, and plays a demonstration and promotion role in the national local colleges and universities.

Acknowledgements.   The key Research topics for research on the reform of higher education of Jilin Province Education Department (SJZD19–04) .

References

1. Lu, L., Tian, Z., Zhou, M.: Fast frequency domain synchronization algorithm of main synchronous signal in TD-LTE system. Sci. Technol. Eng. **16**(10), 174–177 (2016)
2. Zhang, P.: Analysis on the cultivation platform of innovative talents in Colleges and Universities Based on 5g technology--Taking the construction of Applied Innovation Laboratory in Guangzhou as an example, human resources development, (9) (2020)
3. Chen, X., Zhou, L., Cao, Y.: Exploring the construction of talent training program for application-oriented Undergraduate data science and big data technology. Mod. Ind. Econ. Inf. **7**(23), 40–42 (2017)
4. Xinyou, L., Ge, L.: Research on big data talent cultivation in higher vocational colleges. J. Hebei Tourism Vocat. Coll. **22**(01), 88–90 (2017)
5. Ji, X.: Research and application of multi label text classification algorithm. Shandong University (2019)
6. Zhu, X.: Microblog recommendation based on wrd2vec topic extraction. Beijing University of Technology (2014)
7. Bonhard, P., Sasse, M.A.: Knowing me, knowing you' — Using profiles and social networking to improve recommender systems. BT Technol. J. **24**, 84–98 (2006). https://doi.org/10.1007/s10550-006-0080-3

8. Chen, X.: Research on some key technologies of text mining. Fudan University (2005)
9. Li, R.: Research on text classification and related technologies. Fudan University (2005)
10. Chen, Z., He, T.: Demand and cultivation of data science talents. Big Data, **2**(05), 95 (2016)
11. Cao, G., Hu, Z., Guo, J., Wang, Y.: Research on professional training requirements and curriculum of master of data science in the United States. Digit. Libr. Forum, (05), 38–45 (2018)
12. Yueliang, Z.: Characteristics and Enlightenment of talent training mode of foreign I schools data science project . Libr. Inf. Knowl. **04**, 109–118 (2018)
13. Qiangshen, W.: Domain keyword extraction: combining LDA with wor d2vec. Guizhou Normal University (2016)
14. Wang, J., Zhang, J.: Application of statistical model in Chinese text mining. Math. Stat. Manage. **36**(04), 609–619 (2017)
15. Shuyuan, N.: Research on the development of data science and personnel training. Stat. Inf. Forum **34**, 117–122 (2019)
16. Zhang, K., Shi, T., Li, W., Qian, R.: Research on wrd2vec optimization strategy based on statistical language model. Chin. J. Inf. **33**(07), 11–19 (2019)

Big Data Analysis of the Formative Factors of Text Meaning in Ancient Chinese Literature of Higher Education

Jie Hang[✉]

Xian FanYi University, Xian 710105, Shaanxi, China

Abstract. The research on the meaning of ancient literary texts has always been a hot topic in literary criticism. Traditional criticism holds that the meaning of the text should be traced back to the original meaning of the author when he created the text, and the author's meaning is the meaning of the text. Under the background of big data, it is of great significance to analyze the generative factors of the text meaning of ancient literature in higher education.

Keywords: Ancient literature · Text meaning · Particularity · Generative factors

1 Introduction

What kind of thoughts and ideas can be realized only with the help of a certain text form, so this form has become an important means and carrier of human ideological and cultural communication after word of mouth. The formation of this form depends on language and characters. Since with the help of character symbols, both western alphabetic characters and Chinese characters have direct and obscure language features. Any text is not entirely meaningful, which leaves room for different readers to understand it differently. In this way, since the birth of the literary text, it has been a long and continuous process theory of literary text accepted, understood and interpreted by people. Its own existence is not the key, but the important thing is to explore the language intention hidden behind the form of the text. In western literary theory, the research on this aspect belongs to the special field of hermeneutics [1]. Of course, this paper does not intend to follow the footsteps of many western theoretical schools, but only hopes to use some theoretical achievements of Western hermeneutics to provide some help for exploring the meaning generation of ancient Chinese literary texts.

2 Common Data Mining Algorithms

The so-called frequent pattern mining refers to the commodity sets found in the commodity transaction database records. The frequency of these commodity sets is higher than a threshold. These frequent commodity sets are called frequent patterns. The idea

M. A. Jan and F. Khan (Eds.): BigIoT-EDU 2021, LNICST 392, pp. 165–174, 2021.
https://doi.org/10.1007/978-3-030-87903-7_23

of frequent patterns is very simple. First of all, the number of occurrences of each single commodity is counted, which constitutes a dimension table. Then, according to the one-dimensional table, a two-dimensional table is generated by pairwise combination of commodities.

The so-called relationship mining, is worth mining out the causal relationship between the various projects [2]. The basis of relation mining is frequent pattern mining. It's easy to get relations through complex pattern mining. For example, we get a frequent set. The typical data mining algorithm is shown in Fig. 1.

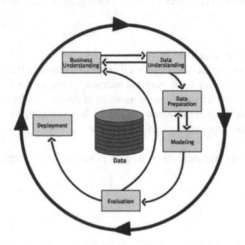

Fig. 1. The typical data mining algorithm

3   The Particularity of Chinese Ancient Literature Text

If we regard the generation of text meaning as a process of understanding, then for us, people's understanding process starts from the text symbol system, and then gradually integrates its associative factors on this basis. As far as ancient Chinese literary texts are concerned, the function of the Chinese character expression system to readers' understanding mainly lies in two aspects.

3.1   The Imagery of the Structure of Chinese Characters

Character is the most basic factor of the existence of text, which is "the writing symbol of language, the conventional visual signal system of information exchange between people". These symbols should be able to flexibly write the language composed of sound, so that the letter can be sent to distant places and to future generations. "The universal characters can be divided into phonetic characters and ideographic characters, both of which can record the language in the form of symbols, so that the information can be preserved and the limitation of time and space can be broken [3–5]. Compared with the western alphabetic characters, Chinese characters have very special ideographic characteristics, especially the vast majority of them are the combination of sound, form and meaning. The ancients paid attention to the description of the physical characteristics at the beginning of the word making, so that people can know the meaning of the hieroglyphs such as "water".

From the perspective of information acquisition, human beings get more information from the outside world through the visual symbol system. Compared with the western alphabetic characters, the Chinese characters with rich physical characteristics have a large amount of written Chinese information. Therefore, they are more suitable for the implicit and rich style of font shape, and even can associate with the principles and objectives of life, as sung in the first popular song.

$$l = \{I1, I2, I5\} \tag{1}$$

Then the subset set can be obtained by permutation and combination:

$$\{I1, I2\}, \{I1, I5\}, \{I2, I5\}, \{I1\}, \{I2\}, and \{I2\} \tag{2}$$

In relation mining, a very useful relation pattern mining: the relation patterns of miassoc:

$$A_{quan1} \wedge A_{quan2} \Rightarrow A_{cat} \tag{3}$$

3.2 The Duality of Written Language

Chinese people pay attention to "harmony between man and nature", which is the highest realm they pursue all their lives. Even if they can't climb this height, they also strive for natural harmony. They have the philosophy of unity of opposites and the aesthetic principle of "unity of opposites" and attach importance to the "symmetry and harmony" of things in pairs. Some scholars attribute it to thinking. This characteristic has a very important impact on the formation of the whole Chinese traditional culture. As for the literary creation we are going to talk about, just as the Chinese people pay attention to symmetry and harmony in art design, they are also striving for a kind of balance and harmony in the complicated strokes and radicals of Chinese characters [6–8]. For example, the knots of many Chinese characters in "Tanaka" are axisymmetric, There are also some slightly different but generally symmetrical left and right sides - the structure of these Chinese characters is symmetrical.

With the Chinese people's symmetrical paintings and pianbian, they are striving for balance and harmony. For example, most of them are axisymmetric, and some of them are slightly different on the left and right sides, We can feel a special way of thinking and aesthetic of Chinese people from it. This symmetry and balance is fully reflected in the creation of characters.

4 Research Results

4.1 Statistics of Topics

Through the vectorization of the original document and the analysis of the structural topic model, 15 research topics of AI education are obtained. Each topic is made up of high ratio keywords of AI education, which are arranged from big to small, and they represent the main meaning of this topic. Each root will also be distributed in different topics with different ratios, which represents the degree of their thematic meaning in different topics. The topic of AI education is expressed by the matrix of topic word "and" word topic". Through the synthesis of the meaning of words with a high proportion of topics and the review of relevant literature, two peer experts are invited to assign a topic to each topic after many consultations.

4.2 Topic Model Estimation

Taking the literature on "artificial intelligence + education" as the analysis data source, the country and time of the literature were set as the influencing factors of the change of topic ratio. The regression equation constructed by 15 topics was estimated by using the estimated effect function of structural topic model, and the uncertainty parameter in the model was set as "Goba". Table 2 shows the p value of each topic under the influence of country and time ($P < 0.001$ is very significant, $P < 0.05$ is significant, $P < 0.1$ is not significant). From the perspective of topic preference, there are significant differences between China and the United States except topic 7, topic 8 and topic 15. From the time development of the topic, there is no significant change in the intensity of all topics in the process of time development.

4.3 Topic Preference Estimation

Estimating the relationship between metadata and topic is the core advantage of structural topic model. The mapping function of estimating effect can deal with the results of estimating effect of artificial intelligence education topic model. For the differences between China and the United States, researchers can use the difference option to map the changes of topics from one specific value to another. Using the structural topic model package (STM), taking the country as the covariate, the two values of the covariance parameters are set to China and the United States respectively, and the specific names of 15 topics are specified. Finally, the topic preference diagram of "artificial intelligence + education" research in China and the United States is drawn, as shown in Fig. 1. Chinese researchers mainly focus on four topics: Games and intelligent agents, intelligent teaching system, intelligent medical and nursing, and educational intelligent software [9, 10]. American researchers mainly focus on knowledge management system, educational robot, intelligent research field, educational intelligent technology, and machine learning, The number of "Ai + education" topics preferred by American researchers is significantly more than that in China.

4.4 Topic Content Comparison

The analysis of preference differences between Chinese and American researchers for a certain topic content can reveal the specific content of research preference in detail. The content preference difference analysis of the same topic can be realized by using the content (CO η ten) parameter covariates of the structural topic model. First, we estimate the structural topic model with country covariates, and then use the plot function to draw a comparative diagram of each topic between China and the United States. Researchers can analyze 15 topics one by one and conclude the preference differences between China and the United States in the same topic.

5 The Theory of Text Meaning in Ancient Chinese Literary Theory

5.1 "The Distinction Between Words and Meanings"

In the philosophy of language in this historical period, "speech" refers to the meaning of expressing speech, while "meaning" refers to the origin or fundamental law of all things in the universe (equivalent to the purpose, destination and realm of showing life in Laozi and Zhuangzi's Philosophy), More often, people are used to analyze it from the perspective of philosophy; but as for the ancient Chinese literary texts, it has become a deep analysis of the relationship between words and meanings in literary discourse [11–13]. The proposition of "Confucius is not complete, not full of meaning" in Zhouyi · Xici is the first of its kind, and a series of discussions on "words" and "meaning" are launched.

5.2 "Against the Will with the Will"

The understanding of "Zhi" has always been controversial. Some people think that "Zhi" is a verb, which means "record", that is, the record of historical facts. However, the author thinks that although these three views are different, they are not as incoherent as the understanding of "meaning". They can be integrated together, and they do not have essential differences. No matter how it is interpreted, "Zhi" can be attributed to the original intention of the author or the work. Of course, it is not enough to have "intention" and "ambition", but also to be able to "reverse". The literal understanding of "inverse" is to trace and explore what "Zhi" is; in fact, it is also equivalent to the process of "fusion of Horizons" in western theories. Readers need to eliminate the distance between themselves and the text and the author in time and space, and organically integrate their own "realistic horizon" and the author's "initial horizon", so as to grasp the exploration of "Zhi".

5.3 The Interaction Effect of Topic

It is estimated that the topic intensity of AI + education research is closely related not only to country or time, but also to the interaction effect of country and time. The conclusion of the topic model provides an explanation for the influence of country on topic intensity. In the process of structural topic model fitting, the product of country and time is taken as the independent variable of topic intensity, and then the estimated effect function is used to estimate and draw the fitting trend chart of each topic intensity one by one [14]. As shown in Fig. 2. Taking topic 13 "intelligent medical and nursing" as an example, this paper uses the effect estimation function to describe the adjustment effect of countries on the temporal change intensity of topic.

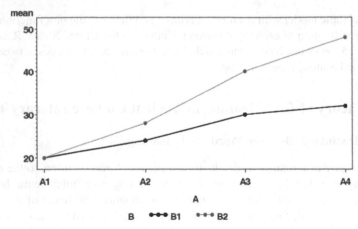

Fig. 2. Interactive effect of topic

6 Analysis and Discussion

By analyzing the structural theme model of "artificial intelligence + education" literature in China and the United States, this paper reveals that researchers in China and the United States mainly focus on 15 topics, including educational robots, educational intelligence software and intelligent teaching system. According to the model of "artificial intelligence + education" and the topic preference model of Chinese and American researchers, the differences between Chinese and American topics are mainly reflected in the number and content of topics. The differences of topic contents are mainly reflected in the comprehensiveness of the research on "artificial intelligence + education" and the differences of their interests, involving educational artificial intelligence software, artificial intelligence teaching and learning, intelligent learning evaluation, educational robot and other topics.

6.1 Topics of Common Concern

The development and application of artificial intelligence education technology is a topic of common concern of researchers in China and the United States. There is no significant difference in the three topics of knowledge management system, intelligent processing system and machine learning. The topic of knowledge management system mainly focuses on the knowledge representation algorithm and model construction of artificial intelligence system, The purpose of this paper is to explore how agents can solve complex problems, improve reasoning ability or help students understand complex behaviors through knowledge management, knowledge representation algorithm and complex system understanding, such as knowledge representation and extensible computing object network design of knowledge-based system, structural behavior function model for complex system understanding, intelligent complex problem solving based on metaphor Algorithm, knowledge understanding based on golden ratio and genetic algorithm, interactive multimedia representation development system, cognitive activities and reasoning based on purposive topological reasoning theory. The topic of intelligent processing system mainly focuses on manufacturing, semantic understanding, sharing decision and effective learning. Intelligent manufacturing is the key to improve the quality of processing process [15, 16]. Intelligent manufacturing is realized through automatic video monitoring, knowledge reasoning to process safety, human-machine physical system and optimization, It can also be used as a framework or ruler to understand the impact of semantic technology on individuals and society, design educational courseware, design the process and content attributes of decision-making system, and measure the effect of human use of cognitive resources. Machine learning topics mainly focus on machine learning process, data mining, student learning performance and learning evaluation supported by artificial intelligence. As far as machine learning is concerned, intelligent machine can read and understand unstructured text through problem feature oriented attention mechanism.

6.2 Estimation of Country Moderating Effect

6.2.1 Overall Differences in Topic Content

First of all, there are significant differences in the number of topics and research scope of AI + education between China and the United States. The research scope of the topic "artificial intelligence education" in the United States is more comprehensive, more specific and more profound than that in China, covering 11 topics such as educational artificial intelligence technology, machine learning, educational robot, knowledge management, machine learning and learning evaluation, However, Chinese researchers only focus on educational intelligence software, educational games, intelligent medical and nursing, intelligent teaching system and other topics. There are also great differences in the research topics of "artificial intelligence + education" between China and the United States [17]. The United States is inclined to the innovation of artificial intelligence education technology, especially the research of machine learning, educational robot and intelligent knowledge management system; in contrast, Chinese researchers pay more attention to the teaching and learning application of artificial intelligence education technology, such as educational games, teaching agents, intelligent medical treatment and intelligent teaching system.

6.2.2 Content Comparison of Single Topic

The content covariates in the structural topic model can realize the difference preference comparison between China and the United States for each topic content, and the high proportion words of China and the United States for each topic are drawn by using the plot function. As follows, we will analyze the difference words of China and the United States in each topic one by one, The review of educational AI robots is a topic of common concern of Chinese and American researchers, covering travel, medical care, intelligent education and super AI. The United States pays more attention to the development of artificial intelligence and risk prevention strategies, while it is difficult for China to identify the high proportion of research preferred words. From the perspective of Topic 2 "educational AI", American researchers pay more attention to the topic content than Chinese researchers. They mainly focus on the purpose and value of AI education from a macro perspective. It is difficult to distinguish the topic content that Chinese researchers pay attention to. From the content of Topic 3 "learning and education", researchers in China and the United States pay almost the same attention to topic content. Chinese researchers tend to pay attention to the influence of AI on learners, online learning, resource recommendation and learning performance, while American researchers tend to focus on teaching design, learning tasks, interactive conversation and learning experience supported by AI. The comparison of high-frequency words in topic 4 learning evaluation shows that Chinese researchers mainly focus on neural network, intelligent algorithm and evaluation accuracy, while American researchers mainly focus on intelligent evaluation, event prediction and measurement model. There are significant differences between China and the United States in the topic of educational intelligence technology [18–20]. The United States pays particular attention to learner learning supported by artificial intelligence technology, while Chinese researchers focus on specific research fields such as intelligent education, development and innovation, and intelligent platform.

7   The Generative Factors of the Text Meaning of Ancient Chinese Literature and Simulation Analysis

In the process of generating the meaning of ancient literature text, some factors play a fixed role in the whole process of readers' reading, which mainly refers to the text itself. The text has its own inherent stipulation. "Once it gets rid of the speaker's immediacy, the text can go beyond the historical, psychological and sociological limitations of the speaker listener context." The text is a kind of permanent existence, which is independent of the perception of the acceptor. Its existence does not depend on the aesthetic experience of the acceptor, and its structure will not change because of people's events [21–24]. "In other words, once the text is formed, it will be out of the author's" control", all the text symbols that make up it have been determined, and the meaning of the text on the level of these text symbols will also change At the same time, it is fixed. Regardless of the author's original intention and the reader's participation in understanding activities, the structure, rhetorical devices, stylistic style and writing skills of the text itself are fixed.

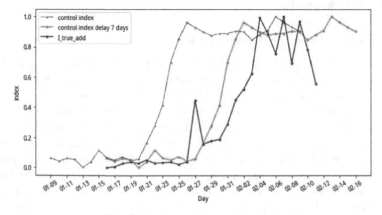

Fig. 3. Text generation with bigdata

We do the simulation as shown in Fig. 3. From Fig. 3, we can see that with the gradual enhancement of the algorithm, that is, with the passage of time, the generated words and words also show an upward trend, which shows that the algorithm proposed in this paper has played a certain effect.

8 Conclusions

In this paper, the author always emphasizes that the generation of text meaning is a dynamic and complex "generation" process, and tries to analyze the influence of various factors on the readers' understanding of the text meaning item by item through the systematic framework of "decoding the text from the readers" and combining the comprehensive factors such as reading situation, etc., The meaning of ancient Chinese literature text is relatively ignored in the generation. The system they set up is generally aimed at all texts. There is no special research on my unique text. We can't take the ancient Chinese text by its number in many of their theoretical frameworks. There are many unique text nouns and symbols between the lines. Therefore, the significance of this paper is to try to re-establish a system to analyze the generation process of text meaning, which can be used to study the meaning generation of ancient Chinese literary texts in the current environment on the premise of ensuring that the generalized texts are established.

Acknowledgements. Supported by Xi'an Translation Research Team (XFU18KYTDC01).

References

1. [song] Zhu, X.: Notes on the Chapters and Sentences of the Four Books Notes on the Analects of Confucius. Beijing: Zhonghua Book Company (1995)
2. Xiuyan, F.: Textology: a Systematic Study of Textualism [M], p. 12. Peking University Press, Beijing (2004)
3. Gadamer, H.: Shanghai: Shanghai Far East publishing house (2003)
4. Pan, D.: Character, interpretation and tradition: the modern transformation of Chinese interpretation tradition. Shanghai: Shanghai Translation Publishing House, October 2003
5. Brief introduction to the war narrative of Murakami's works in the war memory crisis of contemporary Japanese literature. J. Ningbo Univ. (Hum. Ed.) **33**(05), 133 (2020)
6. Chen, Y.: Practice of digging the theme meaning of text from the perspective of deep learning. Basic Foreign Lang. Educ. **22**(04), 76–82+108–109 (2020)
7. The intertextuality construction and significance expression of Xi Li Pumi funeral ceremony "giving sheep Zi" and the text -- Taking nansongyuan, Lanping County, Yunnan Province as an example. J. Kunming Univ. **42**(04), 102–107 (2020)
8. Guikui, L.: From the perspective of intertextuality the jidacheng of the dream of Red Mansions. Chin. Cult. Res. (03), 84–93 (2020)
9. Ling, Y.: Story, ability and Metaphor: a brief discussion on the mutual text structure and meaning interpretation of Lu Xun's the moon running. J. Taiyuan Univ. (Soc. Sci. Ed.) **21**(04), 60–65 (2020)
10. Dong, Y., Guo, S.: Exploring visual text meaning in English teaching in the internet environment. J. Higher Educ. **25**, 48–51 (2020)

11. Liyazhi, G.: Re mention of feelings: experience aesthetics and text interpretation. J. Xi'an Univ. Archi. Technol. (Soc. Sci. Ed.) **39**(04), 88–93 (2020)
12. Lu, J., Chen, K.: Context parameters, textual interpretation and meaning confirmation – on the constraints of context on interpretation. Philos. Res. **08**, 90–97 (2020)
13. Pang, H.: Boundary of interpretation and hermeneutics of literature and art of hesch. Foreign Lit. (03), 1–10+156 (2020)
14. Qiao, Z.: The teaching model of college English reading in higher vocational colleges based on the theory of personal cognition. Overseas Engl. **15**, 157–158 (2020)
15. Xu, Y.: The double construction of compound borrowing, time form and text internal and external meaning -- narrative research of the mystery of the beloved Van Gogh and the stars. New film works, (04), 72–75 (2020)
16. Hongju, Z., Shi, J.: The analysis of the cultural communication path of Hongsao in the perspective of coding decoding theory. J. Party Sch. Jinan Municipal Committee Communist Party China **04**, 96–99 (2020)
17. Yuan, Z.: Innovative design of primary school Chinese literacy teaching with Chinese. Sch. Educ. **20**, 38 (2020)
18. Zhang, W.: Exploring the significance of multi text reading in the teaching of Chinese reading in senior high school. New Curriculum **33**, 67 (2020)
19. Zhao, L.: Barry brumet's communication thought research. Shandong University (2020)
20. Shiqun, S.: Acceptance of classical poetry and literature: an hermeneutics observation of literature text teaching in normal university. J. Guangxi Norm. Univ. Sci. Technol. **35**(04), 94–97 (2020)
21. Li, Y.: The penetration of Chinese traditional culture in the lower grade Chinese texts and its educational significance . Prose Hundred (New Chin. loose Page) **08**, 190 (2020)
22. Fu, F.: Analysis of the guiding strategies for the deep reading teaching reform in junior high school Chinese. Read. Writ. Calculation **23**, 115 (2020)
23. Wang, J., Yuchunxia, H., Wufengyun, Chen, L., Suli, H.: The significance of the same book heterogeneity of teachers is significant -- the text characteristics and Class Innovation Discussion in the whole book teaching perspective (8). Ref. Chin. Teach. Mid. Sch. (23), 4–8+20 (2020)
24. Baiwu, J.: Textual interpretation and contemporary significance of historical records in Jiangnan under KangQian. Knowl. Libr. (16), 9+193 (2020)

Construction of Practical Teaching Platform for Mobile Communication System Under Big Data

JingMei Zhao[✉], Hairong Wang, and Min Li

College of Optoelectronic Engineering,
Yunnan Open University, Kunming 650223, Yunnan, China

Abstract. Aiming at the problems existing in the practical teaching of mobile communication system, a practical teaching platform of mobile communication system is established, which combines the computer LAN under the background of big data as the transmission carrier, supplemented by virtual signaling and business link interface. The platform fully simulates all the functions and signaling process of the actual communication system, so that students can not only master the structure framework of the whole mobile communication system, but also master the workflow and signaling process of the communication system. At the same time, the platform can also provide secondary development and other functions to enhance students' design and development ability.

Keywords: Big data background · Mobile communication · Teaching platform

1 Introduction

At present, many colleges and universities have set up the major of communication engineering [1]. Under the current situation, the Ministry of communications attaches more and more importance to the practice of teaching. Experiment is a very important part of practice, which can arouse students' learning enthusiasm, give full play to students' subjective initiative, and cultivate students' rigorous and realistic scientific attitude.

Higher vocational education must take the improvement of quality as the core, reform and innovation as the driving force, continuously deepen the reform and accelerate the development, so as to make the higher education have a new improvement in the quality of personnel training, a new breakthrough in the school running system and mechanism, a more reasonable structure layout, and an increasing social service ability, so as to realize a historic leap forward in the connotation development.

Undoubtedly, the conference pointed out the core issue of the construction of educational connotation the deep integration of school and enterprise. The deep integration of school and enterprise is an important means for higher vocational colleges to deepen teaching reform, improve teaching quality and serve local economy. The key to the success of the deep integration of school and enterprise is to form a unique, post oriented and systematic practical teaching system based on working process.

M. A. Jan and F. Khan (Eds.): BigIoT-EDU 2021, LNICST 392, pp. 175–184, 2021.
https://doi.org/10.1007/978-3-030-87903-7_24

2 Construction of Practical Teaching Platform for Mobile Communication System

In view of the problems existing in the existing communication experiment and the rapid development of computer technology, it is necessary to reconstruct the communication experiment teaching platform through software simulation with the help of computer software virtual technology. Inspired by this, we simulate various functional entities including MS (mobile station), visitor location register (VLR), home location register (HLR), media gateway MGW (media gate way) and MSc server (mobile switching center server) through software virtual technology, With the help of virtual signaling and service link interface, a complete mobile communication system practice teaching platform is constructed by using computer LAN as transmission carrier. Relying on this platform, students can intuitively "touch" the structure and working process of mobile communication system, and let students establish the overall concept of mobile communication system [2–4]. In addition, communication students establish the overall concept of mobile communication system. In addition, through the development of relevant experiments on the platform, students can transfer the experiments of independent knowledge points done in the experimental box to this platform. By integrating the experiment of independent knowledge points into this platform, students can understand the influence or role of each independent knowledge point in the whole mobile communication system, deepen the understanding of distributed knowledge points, and consolidate the overall concept of mobile communication system.

From Fig. 1, it can be seen that from 2014 to 20120, the construction of big data platform of national universities is on the rise, so we know that this scale must be very large.

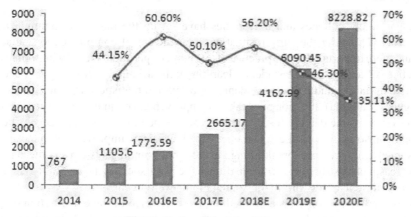

Fig. 1. Scale of big data platform

2.1 Software Design of Mobile Communication System Practice Teaching Platform

Consider a set of ergodic band limited functions with the highest frequency of W cycles/s. Let the distribution density function of amplitude of n continuous sampling points be as follows:

$$p(x_1, \cdots, x_n) \tag{1}$$

The entropy (per degree of freedom) of the system set is defined as:

$$H' = - \lim_{n \to \infty} \frac{1}{n} \int \cdots \int p(x_1, \cdots, x_n) \log(x_1, \cdots, x_n) dx_1 \cdots dx_n \tag{2}$$

We also define an entropy H (per second), which is divided not by N, but by the duration t (in seconds) of n samples.

When the noise is Gauss white thermal noise, there are:

$$H' = \log \sqrt{2\pi e N} \tag{3}$$

$$H' = \log 2\pi e N \tag{4}$$

For a given average power n, the entropy of white noise is maximum. This is derived from the maximization property of the Gaussian distribution given above.

2.2 Open Compound Talents Training

1) The teachers of the course group publicize some independent knowledge points in the scientific research to the students and allow them to carry out their own research. Students can also directly participate in the scientific research of cooperative projects and obtain the training of scientific research ability and scientific research management.

2) The curriculum team of international teacher team training mode sends teachers and students to Hong Kong, Europe, the United States and other universities to listen to teachers' lectures, exchange with overseas students, invite famous overseas professors to teach, offer lectures and offer new courses [5–7]. These measures not only improve the quality of teachers, but also open up students' knowledge vision, and play an important role in students' all-round development. Through the construction of multi-layer and multi-mode open "mobile communication system" practice teaching new mode, the course group has completed the seamless, smooth and compound talent training for students, and achieved a series of teaching research results. Over the past five years, the course team has published hundreds of papers, compiled 8 works and applied for more than 100 patents, of which 31 have been authorized. Through the cultivation of multi-level practice teaching, students improve their practical ability [8].The construction of curriculum and teaching materials has been fruitful. The course of "mobile communication system" in our school was rated as the excellent course of Sichuan Province in 2010. In 2006, the textbook

of modern wireless and mobile communication technology was completed, and the upgraded textbook of mobile communication system and network was completed in 2012. The course group has also compiled textbooks on related topics, mainly including spread spectrum communication in 2002, differential frequency hopping communication principle and application and modern coding theory and application in 2007, and Iterative Equalization Technology in wireless communication in 2011. In addition, the course group has compiled three editions of the guidance book of mobile communication system experiment for the developing practical teaching system of mobile communication system.

3   Realization of Software Platform and Simulation Analysis

Realization of Software Platform

1) Low cost, high reliability
   The platform is implemented by software, its new cost and update cost are far lower than that of hardware platform, and its reliability and stability are better than that of hardware experimental platform [9, 10]. After the technology is updated, only the software module of the corresponding network element needs to be updated, which greatly reduces the update cost of experimental equipment.
2) Normative
   The program design adopts modular and layered design. Different modules such as network transmission, signaling processing, voice processing and interface display are designed independently with clear and standard levels.
3) Support multi-user and multi task
   In this simulation software platform, it can support multiple students to participate in the operation and observe the operation effect independently.
4) Expansibility
   After the virtual simulation platform is built, many mobile communication system related experiments can be developed based on this platform. For example: SMS codec experiment, channel codec experiment (such as convolutional coding, Viterbi decoding experiment), interleaving and deinterleaving experiment, at instruction experiment and SMS Gateway experiment.
5) Support multi format
   Simulation of mobile communication systems (csmcdma, LTE, etc.) supported by the platform in parallel.
6) Consistency
   The configuration method or parameters of the platform should be consistent with the actual mobile communication system; the operation interface should be beautiful, and the operation methods and habits should be close to the reality [11–13]. For example, when the mobile station function simulation entity is implemented, the operation interface of the mobile station is consistent with the actual mobile phone, and the interface is beautiful and generous.

7) Integrity (macro)

Whether it is communication process demonstration or system control, or development experiment, students can see the operation process of "overall" platform, so as to better and more clearly understand the overall structure, working process and principle of mobile communication system, and organically connect the independent and scattered knowledge points in the theoretical teaching in class.

8) Simple maintenance

The simulation experiment software platform is realized by using mature computer LAN technology and computer software. The structure is simple and clear, and it is easy to maintain. Compared with B/S (Browser/server) mode, RCP provides more powerful functions for users through plug-in development [14, 15]. By using SWT (standard widget Toolkit) to design U form and extending GEF (graphical editing framework) plug-in to process graphics, the interaction between users and applications is greatly improved. Oracle database is used for data storage. Oracle 9 is a powerful object-oriented database oriented to Internet computing environment, which changes the way of information management and access. Oracle complies with the industrial standards of data access language, operating system, user interface and network communication protocol; supports high-performance transaction processing of large database and multi-user; implements security control and integrity control, which provides technical guarantee for data security.

3.1 Simulation Analysis

We selected the data of communication data platform from February 1 to March 4, as shown in Fig. 2. From Fig. 2, we can see that in these platform cycles, with the increase of time, the platform construction data shows an upward trend, but there is a fluctuating trend. This shows that the platform data is unstable. But from another aspect, we can know that in the construction of this platform, the external data we use is inconsistent, which leads to this situation. In the past five years, the curriculum team has established contact with well-known universities at home and abroad, exchanging teachers and students and

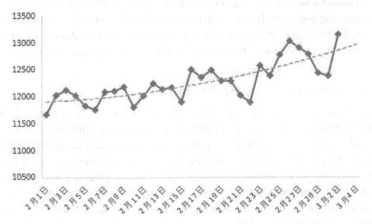

Fig. 2. Communication data platform data

academic exchange [16]. There are 12 teachers in the course group, accounting for 80% of the total number. The course group has received many foreign professors. And sent many teachers and students to attend the international conference. The course team has carried out a number of project cooperation with well-known enterprises at home and abroad, including Intel, DoCoMo, Samsung, Huawei and ZTE, to provide students with all-round opportunities to participate in scientific research.

4   The Design Idea of Practical Teaching System

4.1   Construction of Practical Teaching System

4.1.1   Construction of Ideas

In the construction of practical teaching system, the hierarchical and progressive system structure is designed, following the practical teaching training idea of "basic knowledge learning ability training competition strengthening", that is: first, solid basic knowledge learning, on this basis to strengthen the ability training, and finally through participating in various subject competitions to improve students' practical and innovative ability. For the learning of basic knowledge, the basic experiment module is designed, including verification experiment based on "communication principle comprehensive experiment box", design experiment using DSP, arm and FPGA respectively [17]. Through the communication principle experiment box, students can verify the basic modulation and demodulation principle, encoding and decoding technology, system performance analysis and other theoretical knowledge in the point-to-point communication system, It is realized by the development system of ν SP, FPGA and arm [18]. Aiming at the cultivation of strengthening ability, the system simulation experiment module is designed, including the communication system simulation experiment based on ssystemview 5 and the communication system simulation experiment based on MATLAB 6. As shown in Fig. 3. The communication system from simple to complex is designed through the software simulation platform, In the application-oriented experiment module, including software radio experiment and CDMA mobile communication system experiment, more complex modulation and demodulation and encoding and decoding technology are realized in software radio system, It can touch the structure and working process of the actual system, exercise hands-on ability, and finally select excellent students to participate in various competitions to improve students' scientific and technological innovation ability.

4.1.2   Design of Practical Teaching System

The following points are highlighted: (1) each module is not only connected with each other, but also has its own system, supporting the corresponding theoretical teaching content, combining basic experiments with comprehensive and designed experiments, combining modern teaching methods and means such as multimedia technology, virtual technology and network technology, and making full use of advanced and colorful experimental teaching resources, Combined with guiding students to compete in electronic design and self-made instruments and equipment, teaching, discussion, independent experiments, in class and out of class experiments are combined to develop students'

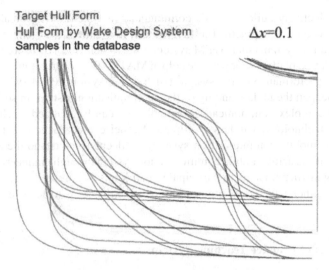

Target Hull Form
Hull Form by Wake Design System $\Delta x = 0.1$
Samples in the database

Fig. 3. Target system of practice teaching.

ideas and cultivate their innovative spirit; (3) in the course architecture design, each module is designed step by step, from simple to complex, from basic to comprehensive, from design to innovation, and at the same time, it takes into account and radiates other related majors, For example, students of non electrical majors are recommended to do the confirmatory experiments in the basic experiment module and system simulation module, students of electrical non communication majors are recommended to do the basic and system simulation experiment module, students of communication majors can choose the three modules in turn, and students with good foreign language level can choose bilingual teaching MATLAB and communication simulation experiment [19]. It can also be used as an open experiment and postgraduate experiment project. (4) The design of experiment content is closely related to the training goal of application-oriented talents, highlighting the cultivation of students' engineering practice ability and professional ability.

4.2 Practical Teaching Target System

Firstly, through the hierarchical and progressive architecture, it helps students to establish the overall concept of the communication system; secondly, through the verification, design and comprehensive experiments, it helps students to establish a comprehensive and multi-level understanding of the communication system from the overall architecture to the specific modules and the corresponding key technologies, It plays a positive role in promoting the cultivation of students' engineering practice ability and professional ability; finally, it cultivates students' ability to comprehensively use basic communication knowledge and experimental technology, so as to meet the needs of cultivating high-quality engineering application-oriented talents of communication and information [20]. In the experiment of DSP and FPGA development system, students can design the core functional modules (such as fr filter, 2fs κ module, etc.) in the communication system, and selectively build the basic communication system to organically combine theoretical knowledge with practical ability. In the system simulation, students can design the communication system (such as FM system, PS κ system, etc.) by using Systemview software simulation platform, Or with the help of MATLAB software simulation platform to analyze the performance of the system (such as am system, PCM system, etc.), so as to further deepen the understanding of the communication system. In software radio system, more complex communication technologies can be realized, such as GMSK demodulation technology, wireless multipath channel characteristic experiment, etc., On the CDMA mobile communication system, we deeply understand the architecture and workflow of the actual mobile communication system through engineering practice. Finally, by organizing students to participate in all kinds of competitions, students can apply and strengthen their knowledge and practical skills, so as to achieve the teaching goal.

4.3   The Effect of Practice Teaching System

(1) It can stimulate students' innovative consciousness, improve their comprehensive practical ability and scientific research quality, and become the enlightenment education of scientific research [21, 22]. After entering the graduation thesis and graduate stage, the students' scientific research ability is highly praised by the tutors. (2) stimulate the students' interest in communication basic experiment and system experiment. In recent years, more and more non communication majors have taken this course, and some students have taken advantage of their spare time or summer vacation to further expand their experiments in the laboratory. (3) many students have won national and provincial awards in the National Undergraduate Electronic Design Competition and Challenge Cup competition: (1) in 2007, they participated in the National Undergraduate Electronic Design Competition, He won 1 first prize and 4 third prizes in Shaanxi competition area; ② he won 1 first prize in Shaanxi competition area in 2008; he won 1 s prize in national embedded competition of national Bo Chuang cup; ③ in 2009, he won 1 s prize in national challenge cup, 1 first prize in Shaanxi and 1 s Prize in Shaanxi; he took part in NoC college students' science and technology innovation competition, He won one national first prize and the highest trophy; he took part in the National Undergraduate Electronic Design Competition and won one second prize in Shaanxi competition area.

5   Conclusion

The mobile communication practical teaching platform introduced in this paper can realize the connection between communication basic courses and communication professional courses, and directly apply the coding and decoding technology in the basic courses to the overall system of mobile communication. In the aspect of communication technology, the system control from terminal, wireless air interface and signaling, base station, VLR, HLR and network exchange is completed. Students can not only master a specific technology, but also grasp the whole network architecture and communication system. The design and development of mobile communication engineering training simulation system make full use of modern information technology to develop a virtual training environment covering the whole process of mobile communication engineering, which provides an effective way for students to combine theory with practice before practical operation. It can enable students to have a comprehensive understanding and understanding of the basic principles, key technologies and typical systems of mobile communication, strengthen the engineering practice ability, cultivate students' systematic thinking, analyze and solve the relevant problems in the mobile communication system, and lay the necessary foundation for the practical mobile communication system engineering construction, maintenance, management, learning of new knowledge and research of new problems. At the same time, the system can also carry out pre job training of mobile communication basic skills for the marketing, service, management and equipment maintenance personnel of mobile communication business department.

References

1. Wang, R., Zhang, Y., Ming, Y., et al.: Research and practice of innovation and practice ability training of electronic information college students. Nanjing J. Electron. Electr. Educ. **28**, 283 (2006)
2. BA, S., Zhou, S.: Exploration of undergraduate practice teaching reform, Chengdu: Exp. Sci. Technol. **4**, 53–56 (2006)
3. Wang, X., Wang, Y., Zhao, Y., et al.: Construction of characteristic practice teaching base . Chengdu: Exp. Sci. Technol. **8**(6), 136–137 (2010)
4. Zhang, Y., Xiang, W.S., et al.: Modern Exchange Principle . Science Press, Beijing (2012)
5. Xiayi, G., Bohu, L., Chai, X., et al.: Overview of big data platform technology. J. Syst. Simul. **26**(3), 489–496 (2014)
6. Chaoyang, Z., Jiye, W., Deng, C.: Research and design of power big data platform. Power Inf. Comm. Technol. **13**(006), 1–7 (2015)
7. Zhu, C., Wang, J., Deng, C.: Research and design of power big data platform. Power Inf. Comm. Technol. (2015)
8. Wang, Z.: Analysis of power marketing informatization construction based on big data platform. Inner Mongolia power technology
9. Shiyi, F.: Collaborative research on traditional data warehouse of Hadoop big data platform. Donghua University (2014)
10. Hao, R., AI, Q., Xiao, F.: Research on power consumption behavior analysis framework based on multi big data platform. Power Autom. Equipment, **37**(008), 20–27 (2017)
11. Zhang, H., Wenli, L., Wang, Y.: Design of fault detection system for spacecraft communication signal equipment based on CPCI bus. Comput. Measure. Control, **29**(02), 1–4+9 (2021)
12. She, S., Zhao, Y., Yuee, H., Cao, Y., Yuan, Z.: Design of LNG cylinder level monitoring system based on wireless communication. Comput. Measure. Control, **29**(02), 14–19 (2021)
13. Li, L., Wang, S., Wang, L.: Initial synchronization technology of LEO satellite communication based on 5g standard . Comput. Measure. Control **29**(02), 150–154 (2021)
14. Ke, C., Li, M.: Overview of wide beam antenna technology . Technol. Innov. **04**, 161–163 (2021)
15. Liu, Y., Zhang, B.: Design of automatic identification and control system for abnormal signal of power robot carrier communication. Autom. Instrum. (02), 179–181 (2021)
16. Zhao, B., Chen, Z., Zhao, L.: Design of frequency stabilization communication simulation system for quantum lidar based on MATLAB . Laser J. **42**(02), 161–165 (2021)
17. Lei, Y., Xiong, H., Gou, Z., He, S.: Research on operation control strategy of 5g communication base station energy saving power supply system. Hydropower Energ. Sci. **39**(02), 150–155 (2021)

18. Tian, S., Zhai, S.: Influence of communication protocol on stability of networked control system. Autom. Technol. Appl. **40**(02), 66–68+93 (2021)
19. Lin, X., Bai, S., Shi, J., Duan, Y., Wei, Y.: Precise load con-trol system based on 5g communication network. Zhejiang Electric Power **40** (02), 61–67 (2021)
20. Tian, Z.: Analysis of intelligent safety production management system in Ruilong coal mine. Energ. Energg. Conserv. **02**, 219–220 (2021)
21. Xu, L.: Design of communication liquid level monitoring system between PLC and LabVIEW based on Modbus RTU. Inform. Technol. Inf. (02), 111–113 (2021)
22. Mo, X., Wangjing, J., Dongzhen, G.: Application of wireless communication technology in police intelligent physical training system. J. Chifeng Univ. (Natl. Sci. Ed.) **37**(02), 13–17 (2021)

Construction of University Students Innovation and Entrepreneurship Resource Database Based on Collaborative Big Data Analysis

Lili Zhang(✉)

Changchun Institute of Technology, Changchun 130012, Jilin, China
zhangll@ccit.edu.cn

Abstract. In recent years, with the continuous promotion of teaching reform in our country, the society has higher and higher requirements for talents. Combined with collaborative big data, this paper analyzes the construction of university students' innovation and entrepreneurship resource database. According to the requirements of teaching reform, colleges and universities should strengthen their own work construction, attach great importance to the construction of university students' innovation and entrepreneurship resource database, and provide a large number of innovative talents for social development.

Keywords: Collaborative big data · College students · Innovation and entrepreneurship · Database construction

1 Introduction

Innovation and entrepreneurship education is to adapt to the current stage of economic transformation and upgrading, to cultivate talents with innovative spirit, innovative consciousness and creative ability as the goal, set knowledge, professional and innovation and entrepreneurship as one of the quality education. Since 1997, some domestic colleges and universities have carried out preliminary exploration on Entrepreneurship Education. In 2002, Tsinghua University, Renmin University of China and other nine universities were established as the first batch of Pilot Universities of entrepreneurship education by the Ministry of education. In 2010, the Ministry of Education formally adopted the concept of innovation and entrepreneurship education in "opinions on vigorously promoting innovation and entrepreneurship education in Colleges and universities and college students' independent entrepreneurship work", defined it clearly, established its development direction, and gradually entered the stage of diversified discussion. In 2015, the general office of the State Council issued the implementation opinions on deepening the reform of innovation and entrepreneurship education in Colleges and universities, which stressed that the comprehensive development of innovation and entrepreneurship education is a strong support for the construction of an innovative country and the realization of two centenary goals [1]. Deepening innovation and entrepreneurship education in

M. A. Jan and F. Khan (Eds.): BigIoT-EDU 2021, LNICST 392, pp. 185–193, 2021.
https://doi.org/10.1007/978-3-030-87903-7_25

colleges and universities has become an important topic to promote the transformation and upgrading of national quality and teaching mode, the construction of database is shown in Fig. 1.

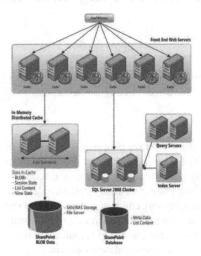

Fig. 1. Construction of database.

2   Collaborative Filtering Algorithm

As an effective technical means to deal with "information overload" in the context of big data, it has become one of the most widely used recommendation technologies. However, with the deepening of research, it is found that the traditional collaborative filtering algorithm still has many problems to be improved. data sparsity is one of the most significant problems. Therefore, many experts and scholars have put forward many methods to improve the algorithm, The correlation weight coefficient is introduced to improve the effectiveness of the algorithm. In literature 5, the Jaccard similarity coefficient is combined with the scoring correction formula to reduce the impact of item non correlation. it is proposed to mine the potential relationship between scoring users [2–5]. To make the recommendation results more reliable, literature introduces the item correlation coefficient to calculate the user similarity according to the item type and score. Uses the confidence function to map the user's implicit feedback to the confidence probability, and proposes a heterogeneous confidence optimization algorithm based on random gradient descent.

Definition1: Let u be a universe, and given a mapping relation $\mu_A : U \rightarrow [0, 1], x \rightarrow \mu_A \in [0, 1][0, 1]$, where the mapping μ A is called the membership function of fuzzy set a, represents the membership degree of κ to fuzzy set a.Although the above algorithms have optimized the effectiveness of the algorithm, there are still some problems, such as the accuracy of the algorithm still needs to be improved in the case of data sparsity, the measurement factor of user similarity is relatively simple, and the mining of potential relationship between users is not deep enough. In view of the above problems, this paper proposes a collaborative overcomputing method which integrates multi-layer similarity and trust mechanism. According to the result of neighbor recommendation, the algorithm is as follows:

$$\mu_{good}(x) = (x - 1)/4; 1 \leq x \leq 5 \tag{1}$$

$$\mu_{good}(x) = (5 - x)/4; 1 \leq x \leq 5 \tag{2}$$

3 Sample Sources and Research Methods of Database Construction

3.1 Basic Principle of Ant Colony Algorithm

To improve the innovation and entrepreneurship resource platform, colleges and universities should improve the existing network platform for college students' innovation and entrepreneurship, introduce corresponding policies for college students' innovation and entrepreneurship, break professional barriers, realize interdisciplinary integration, expand the scope of innovation and entrepreneurship, integrate innovation and entrepreneurship education into all aspects of Cultivating College students, and strive to provide a platform for college students' innovation and entrepreneurship [6, 7]. Stimulate college students' interest in innovation and entrepreneurship, promote college students to cultivate enough ability and confidence, and make college students change from passive employment mode to "independent entrepreneurship" mode. Innovation and entrepreneurship platform can provide relevant policy guidance for college students, provide business registration and start-up fund support for innovative projects, and promote projects with good market prospects, so as to promote the successful incubation of excellent project achievements.

3.2 Reform of Talent Training Mode

Colleges and universities should further carry out the reform of College Students' teaching and student status system, implement flexible education system, support students' innovation and entrepreneurship, further improve the professional curriculum, increase innovation and entrepreneurship practice credits, and realize the conversion of professional curriculum credits, stimulate college students' innovation and entrepreneurship power, and cultivate students' innovation and entrepreneurship ability [8]. The establishment of a relatively complete network course, teachers will put the course on the Internet, students can be flexible elective, get the corresponding credit. Professional curriculum should be combined with the needs of social development, adding new industry related courses such as Internet and Internet of things, and increasing entrepreneurship quality education and entrepreneurship training courses for college students. In addition, we encourage college students to take an active part in Internet plus college students' innovation and entrepreneurship related competition, so that their ability in "double innovation" can be improved, and the reform of talent training mode in universities and colleges will also be promoted.

4   Optimize the Teaching Staff of Innovation and Entrepreneurship

The construction of the teaching staff of innovation and entrepreneurship education needs time, and colleges and universities can optimize it based on their own training and introduction. On the one hand, colleges and universities should strengthen the construction of innovation and entrepreneurship discipline, grasp the mainstream development direction of social economy, seize the opportunity, actively guide their teachers to carry out frontier research, promote the transformation of outstanding innovation and entrepreneurship project achievements, and provide talent reserves for innovation and entrepreneurship education [9–11]. On the other hand, with the help of industry university research platform, colleges and universities play the role of school enterprise joint academic tutor, cooperate with outside institutions, undertake various innovation and entrepreneurship activities, and actively invite experts in various fields, entrepreneurial elites, entrepreneurs and other outstanding talents to serve as part-time instructors of innovation and Entrepreneurship education, so as to create a diversified structure of teaching staff [12]. Furthermore, colleges and universities should actively invite elites with rich entrepreneurial experience at home and abroad to give lectures on innovation and entrepreneurship for college students by using their own experience. In the process of close contact with college students, colleges and universities should stimulate college students' innovation and entrepreneurship inspiration and find partners for innovation and entrepreneurship projects (see the Fig. 2).

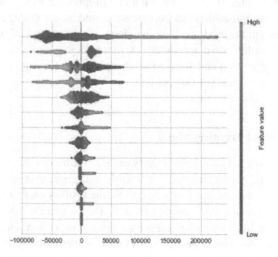

Fig. 2. Optimize the teaching staff of innovation and Entrepreneurship

5   Optimize the Teaching Staff of Innovation and Entrepreneurship

The construction of the teaching staff of innovation and entrepreneurship education needs time, and colleges and universities can optimize it based on their own training and introduction. On the one hand, colleges and universities should strengthen the construction of innovation and entrepreneurship discipline, grasp the mainstream development direction of social economy, seize the opportunity, actively guide their teachers to carry out frontier research, promote the transformation of outstanding innovation and entrepreneurship project achievements, and provide talent reserves for innovation and entrepreneurship education. On the other hand, with the help of industry university research platform, colleges and universities play the role of school enterprise joint academic tutor, cooperate with off campus institutions, promote.undertake various innovation and entrepreneurship activities, and actively invite experts in various fields, entrepreneurial elites, entrepreneurs and other outstanding talents to serve as part-time instructors of innovation and entrepreneurship education, so as to create a diversified structure of teaching staff [13]. Moreover, colleges and universities should actively invite elites with rich entrepreneurial experience at home and abroad to give lectures on innovation and entrepreneurship for college students by using their own experience. In the process of close contact with college students, colleges and universities should inspire college students' innovation and entrepreneurship inspiration and find partners for innovation and entrepreneurship projects.

6   Demand Analysis of Teaching Resource Database Platform

In the development of a project system, the primary task is to master the target function of the project system, that is to clarify the user's requirements for the function of the project system. That is to say, we should make clear what the purpose of project system development is, what problems to solve for users, and what the project system should do. Before the development of the project system, we need to do a very good demand analysis of the project system [14]. Only by doing a good demand analysis and making clear the functional objectives of the project system, can we lay a good foundation for the later design and development work, and the later work can be orderly progress.

6.1   Feasibility Analysis

Feasibility of system development and implementation system related developers have certain software system development ability, and have obtained professional and technical qualifications (titles) of computer software level qualification examination series: information system project manager, software designer, etc., which provides human resource guarantee for the development of the system [15]. The technical feasibility platform is based on J2EE architecture and is developed with B/S mode, At present, it is rich in mainstream development technology, mature web application systems based on J2EE architecture, and various references provide effective technical support for the platform. The construction of economically feasible teaching resource library platform is one of the sub projects of "backbone colleges" construction project of Guangxi Vocational and technical college, The core function of the platform is to effectively manage and integrate teaching resources [16]. The platform includes portal function, identity verification module, shared resource library sub module, excellent course release management sub module, network teaching management sub module and system management sub module.

6.2 System Non Functional Requirement Analysis

The non functional requirements of the system, that is, the requirements other than the functions realized by the software system, including the operating environment, reliability and security of the system. The teaching resource library platform is a teaching information system supported by computer network technology and modern education technology [17]. It should also have the following characteristics: reliability. For any information system, reliability is the most important, Teaching resource database platform is no exception. If there are frequent hardware or software failures in the process of using, the teaching resource database platform will lose the significance of serving teaching.

The advanced technology platform adopts the current mainstream technology framework and applies advanced technology solutions to ensure the application cycle of the platform.

Normative, according to the international and national standards of educational resources construction, realize the sharing and mutual use of resources. [18–19] Security, we should fully consider the security of the system, so as to ensure the security mechanism and data security of the system, including reliable software and software development technology architecture, developing and writing code to avoid common loopholes and defect functions; hardware, using redundant data storage technology and backup disaster recovery technology.

Portability, using J2EE development architecture and three-tier B / s system structure, the system program and database can run on Linux platform, can also run on Windows platform, at the same time in the event of conventional or catastrophic failure can be quickly transplanted and enabled.

Scalability: in terms of hardware, the storage device supports the flexible configuration and expansion of the system to ensure the user's demand for data storage in the use process; in terms of software, As shown in Fig. 3 the use process should be able to find and repair bugs in time, and provide secondary development interface to facilitate docking with other application systems.

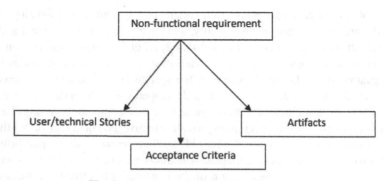

Fig. 3. System non functional requirements.

6.3 User Role Analysis

Select courses. After logging in, students can choose the courses of their department. For courses of other departments, as long as you submit an application and the instructor who created the course sets the permission to access the course, you can access it. For example, users of computer network technology major in computer department can choose all courses offered by computer network technology major by default. For courses offered by culture and communication department, students of computer network technology major can also visit courses offered by culture and communication department as long as the teacher who created the course allows students of computer network technology major to visit [20]. This function can well adapt to the situation of setting up public elective courses in the whole hospital. After entering the course, students can browse the course introduction, syllabus, teaching plan, etc., and download courseware, training guide, etc. Course learning and discussion is the process of learning the course, and students can learn the course content online. After studying and discussing, students should complete their homework according to the relevant requirements of the course and submit the electronic documents and other attachment materials. Interaction and discussion: after students submit their homework, they can use chat room online group discussion, or use the forum created by teachers to discuss related problems offline, and ask teachers questions about difficult learning points. Check the teacher's comments. Students can check the teacher's comments on the homework in time after the teacher corrects the homework. Course evaluation, the teacher's course content, teaching plan, courseware, training guide, flash animation, etc. are scored and evaluated.

7 The Fourth Chapter is the Overall Design of Teaching Resource Database Platform

7.1 Design of Teaching Resource Database Platform

Based on the idea of teaching centered design, the platform of ideological teaching resource library is developed with full consideration of the actual teaching situation of students and teachers. Compared with the traditional shared teaching resource database, it integrates the network teaching function and the excellent course release and evaluation function, fully arouses the enthusiasm of students and teachers, and has positive significance to change the traditional teaching methods and promote the construction of education informatization. The teaching resource database project system follows the development standards in the field of software engineering [21]. UML not only supports object-oriented analysis and design, but also supports the whole process from requirement analysis to software development. In the design phase, UML is used to draw the sequence diagram and interaction diagram, which makes the development and design process more simple and clear.

7.2 General Design Principles of Teaching Resource Database Platform

The teaching resource base platform is a teaching service system for teachers and students in the whole school. In order to better serve teaching, improve teaching quality and the quality of talent training in the school, in the overall design of teaching resource base, we should grasp the following principles and practical principles, which is the basic principle of the design and development of any information system. The main users of teaching resource database platform are teachers and students. How to make users find resources conveniently and quickly and use teaching platform is a problem that must be considered in system design. The function of the platform must be close to the actual needs of teachers and students [22, 23]. At the same time to meet the needs of personalized. User participation principle the information system of user participation management is regarded as man-machine interface, and the management information system that users need to use is directly involved in the work, and becomes a part of the construction system work in each stage. The use case in the teaching resource database platform describes the process of teachers and students' online teaching and uploading resources. Therefore, in the design process, we must take the user as the principle and take the use case of user role as the object to design.

8 Concluding Remarks

Colleges and universities should actively establish long-term cooperative relations with state-owned enterprises, joint ventures and other large companies to build bridges for college students' innovation and entrepreneurship. Colleges and universities can make full use of educational resources to form a multi campus school enterprise cooperative education mechanism by combining with advantageous colleges and universities in the same region or other regions. According to the needs of enterprise talents, colleges and universities invite enterprise management and technical talents to participate in entrepreneurship curriculum, such as entrepreneurship preparation, enterprise operation management and other courses, and formulate corresponding talent training programs. By using the school enterprise cooperation mechanism, we can provide college students with internship opportunities in winter, summer and weekend, so that more college students can increase their practical experience, stimulate their innovation potential, and use the school enterprise platform to realize the transformation of achievements and put into entrepreneurial action.

Acknowledgements. Jilin Province Department of Education "13th five-year plan" Social Science Research Project, Jilin Province College Students Innovation and Entrepreneurship Resources Database System Research and design (employment special) (JJKH070533JY).

References

1. Deng, Y., Wu, M., Pan, J.: Improved collaborative filtering algorithm based on articles and its application. Comput. Syst. Appl. **28**(1), 182–187 (2019)
2. Zhang, C., Zhu, Y.: Item based collaborative filtering recommendation algorithm based on user age perception. Comput. Technol. Dev. **29**(6), 95–99 (2019)
3. Li, K., Wan, P., Zhang, D.: Collaborative filtering recommendation algorithm based on improved user similarity measure and score prediction. Mini Comput. Syst. **39**(3), 567–571 (2018)

4. Jing, T., Zichun, C., Xiaoqiang, Z.: Hesitant fuzzy multiple attribute decision making method for triangular fuzzy numbers. J. Yibin Univ. **19**(6), 81–85 (2019)
5. Si, L., Liu, Y., Zhou, J., One belt, one road, the national economic management database resource classification system construction. Libr. Forum, 1–8 (2021). http://kns.cnki.net/kcms/detail/44.1306.G2.20210402.0854.002.html
6. Chen, H., Qian, Q., Wan, X.: Research on the development strategy and path of Anti Japanese War archives resources from the perspective of digital humanities. Shanxi Arch. 1–10 (2021). http://kns.cnki.net/kcms/detail/14.1162.G2.20210309.1027.002.html
7. Han, C.: Human resource management system and its active database construction. Mod. Electron. Technol. **44**(05), 161–165 (2021)
8. Liu, N.: Research on the construction of Tianjin traditional music resources database. Fujian provincial business association. Collection of papers of South China education informatization research experience exchange meeting 2021. Fujian provincial business association: Fujian provincial business association, pp. 851–854 (2021)
9. Li, X., Zhu, P., Kong, W., Liu, L., Liu, W.: Construction and service release of multiple information database of uranium resources in Erlian Basin. World Nucl. Geol. Sci. 37(04), 257–262 (2020)
10. Li, X.: Strategies for building ceramic cultural resources database. Dig. World **12**, 83–84 (2020)
11. Wang, R., Zheng, Y.: Research on the construction mode of cultural heritage information network of Yongding River Cultural belt in Xishan. Beijing Cultural Museum **01**, 115–122 (2020)
12. Wang, Z., Chen, L., Cheng, A., Jiang, H., Bao, J.: Analysis on the construction of forest resources and infrastructure investigation technology in state owned forest farms of Zhejiang Province. East China Forest Manager **34**(04), 58–61 (2020)
13. Zhang, Z., Zhu, P.: The construction of Library Digital Resource Service Mode under the background of "Internet plus". Anal. Mod. Contemp. Lit. (39), 90–91 (2020)
14. Huang, Y., Zhang, X., Liu, C., Yang, Z.: Construction of standardization of agricultural resources information database in Guizhou Province. Guizhou Agric. Sci. **48**(08), 132–139 (2020)
15. Ma, J., Sun, X., Nie, J., Yang, T., He, W., Wang, S.: Construction and application of spatial database technology application teaching resource database. J. Kunming Metallurgy Coll. **36**(04), 45–49 (2020)
16. Liu, L.: Database construction. Application of Arc GIS in the construction of mineral resources planning database. China Metal Bull. (07), 67–68 (2020)
17. Feng, X., Li, X., Yan, Y., Li, J., Liu, M., Wu, Y.: Research on the construction of public health emergency data platform based on knowledge entity. Knowl. Manage. Forum, 5(03), 175–190 (2020)
18. Yang, T.: Database construction of science and technology information video resources based on streaming media technology. Softw. Guide **19**(05), 181–185 (2020)
19. Yu, Z., Wang, J.: Construction and implementation of Zhuhai geographic information public platform. Surv. Mapp. **42**(06), 253–259 (2019)
20. Shi, Y., Yang, L.: Research on construction and application of relational data model of Humanities and Social Sciences Database. Mod. Intell. **39**(12), 19–27 (2019)
21. Lin, F.: Exploring the construction of "aggregation, management and utilization" system of natural resources big data. Land Resour. Intell. **11**, 13–18 (2019)
22. Wang, Y., Liang, H., Cheng, W., Li, E.: Research on the construction of characteristic database of Confucian Literature – Taking the library of Qufu Normal University as an example. Res. Pract. Innov. entrepreneurship Theory **2**(21), 168–170 (2019)
23. Liu, Z.: Constructing the judicial regulation mode in the era of algorithm. Procuratorial Situation **21**, 18–19 (2019)

Cooperative Foreign Language Teaching Model with Data Mining Algorithm

Xuelian Su(✉)

Sichuan Vocational and Technical College, Suining 629000, Sichuan, China

Abstract. This paper constructs the practice teaching mode of curriculum cooperation mechanism, and expounds its application in the cooperative foreign language teaching practice teaching under the data mining algorithm. This teaching mode will quickly penetrate the latest development and the latest methods of artificial intelligence into the computer practice. Starting from the construction of artificial intelligence curriculum system and multi-disciplinary cooperation mechanism, it will cultivate students' practical ability and innovation ability.

Keywords: Data mining algorithm · Cooperative foreign language teaching mode · Ant colony algorithm

1 Introduction

With the rapid development of global information and industrialization, talents in the field of big data and artificial intelligence are in a state of shortage for a long time. It is particularly important for local colleges and universities to cultivate students with practical operation and innovation ability required by employers, This paper constructs a new teaching mode, curriculum collaborative mechanism practice teaching mode, and expounds its application in artificial intelligence practice teaching [1]. The curriculum collaborative mechanism practice teaching mode expounds the talent training methods to comprehensively improve the project practice ability and application innovation ability from the aspects of talent training objectives, computer project practice training, daily classroom teaching arrangement, teaching performance evaluation, etc., It is of great significance to broaden the learning vision of big data majors and exercise the ability of integrating theory with practice.

1.1 The Connotation of the Practice Teaching Mode of Curriculum Coordination Mechanism

The framework of the practical teaching mode of course collaboration mechanism is shown in Fig. 1. The practical operation of big data technology includes the establishment of Acer basic big data platform and the management of software and hardware environment, data storage and encryption processing, the setting of corresponding data

M. A. Jan and F. Khan (Eds.): BigIoT-EDU 2021, LNICST 392, pp. 194–203, 2021.
https://doi.org/10.1007/978-3-030-87903-7_26

field mode, the adoption of batch processing algorithm, the Research on the application direction of results, etc., The purpose is to comprehensively improve the practical ability, guide students to seek programming and algorithm design independently, and understand the theory and method of artificial intelligence from the program through the actual Multi Programming and multi computer. The framework of practice teaching mode of curriculum cooperation mechanism is shown in Fig. 1.

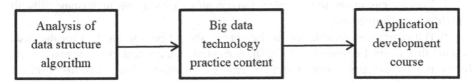

Fig. 1. The framework of the practice teaching mode of curriculum collaborative mechanism

1.2 Curriculum System Design

Big data mining and analysis are realized by means of computer programming, and basic programming courses should be set up in the course system. Practice one is to collect data, crawl experimental data, clean and encrypt data. Practice two is basic experiment, database design, construction of data cloud computing development platform and corresponding model modeling, The purpose is to make students learn to find the target data, combine with mathematical modeling cases, and solve the process of target task pertinently [2]. Practice 3, through data visualization technology, data analysis and batch processing technology, shows the mining results of the experimental project in a variety of ways, and improves the ability of analyzing and solving practical problems. Practice 4, cultivate students' comprehensive practice and business ability: Practice 5, practice 3, practice 3, practice 3, practice 3, practice 3, practice 4, practice 4, practice 4, practice 5, practice 4, practice 5, practice 5, practice 5, practice 5, practice 5, Set up a comprehensive experiment course, as the basis of follow-up teaching.

2 Introduction of Case Embedded Collaborative Teaching Mode

This teaching mode includes two aspects: one is the collaborative planning of course group, which integrates the knowledge of each course with cases as a link, so as to help students build a complete knowledge system and effectively enhance their overall grasp of knowledge; the other is the exploration of research-based teaching mode around cases, Taking engineering cases as the main line to cultivate students' ability to analyze and solve problems.

2.1 Collaborative Planning of Curriculum Group

The selection of engineering cases is not more important than refinement. It is mainly one or two cases. It does not need to cover all the knowledge points of all courses. It is more about teaching students to understand engineering concepts through cases and master the methods to solve problems. There are several principles in the selection of Cheng cases.

① Engineering cases should be common in engineering practice, with real engineering background.

② The knowledge covered by the case should be able to run through the teaching of the course group.

③ The case can not be too complex, too complex case students understand difficult and difficult to implement.

④ The equipment suitable for the laboratory should be selected first for the case, and the final control scheme should be implemented on the actual experimental device. Integrating the above principles, the engineering case of this teaching mode firstly selects the liquid tank control system in the laboratory, and uses this case to run through the teaching of each course. Secondly, the evaporator simulation device is selected as an advanced case to carry out comprehensive training in the industrial network control system.

2.2 Each Course is Based on Case Study

Research teaching is a "student-centered" teaching mode, which can arouse students' thinking through problems or projects, realize deep learning of knowledge, stimulate students' interest and potential in learning, and improve classroom atmosphere and teacher-student relationship, which is also the trend of teaching development in the future [3]. Different courses have different emphasis on foundation and application, and different research teaching modes and methods are adopted. This teaching mode adopts research-based teaching method based on problem, task and project. "There are many basic knowledge of sensors and detection, and the problem-based discussion classroom teaching mode is mainly used in the teaching process." the process control instrument and device course has basic knowledge and engineering application, and the problem + "task" teaching mode is adopted in the teaching process. "Industrial network control system" is a project-based teaching mode in the application teaching process. An improved data mining structure is shown in Fig. 2.

2.3 Similarity Model

The basis of classical rough set model is indistinguishable relation. When there are missing attribute values in the data (which is very common in database), the inseparable relation or equivalent relation can not cope with this situation. In order to improve the ability of rough set, many authors have proposed using similarity relation instead of indistinguishable relation as the basis of rough set. Practice shows that the similar model has better performance than the classical rough set model in practice. When solving the problem of missing values in database, a simple similarity relation can be defined as (where "origin" represents not knowing or not caring):

$$\tau_c(x, y) = \{x \in \cup, y \in \cup\} \tag{1}$$

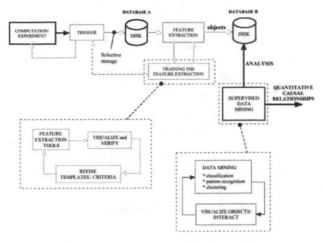

Fig. 2. Improved structure for data mining

The principle class of genetic algorithm provided in this paper is different from other kinds of genetic algorithms in other fields, mainly in the representation and fitness function. The representation method defined by the author is that each bit string represents one item of the discernibility matrix, that is, the discernibility attribute set of two objects. When a bit is 1, the attribute exists, otherwise it does not exist. Thus, each bit string is a candidate for reduction. The fitness function is defined as follows:

$$F(v) = (N - Lv)/n + 2Cv/(m^2 - m) \tag{2}$$

3   Practical Effect Simulation

3.1   The Teaching Effect is Widely Recognized by Students

The case embedded collaborative teaching mode has been applied in 2013–2015 automation class of our university, and achieved good results, More than 360 students in 15 classes benefited. In July 2018, through the statistical analysis of 124 valid questionnaires collected from the two courses, 955% of the students said that they like the discussion based classroom teaching, 982% of the students thought that the given thinking questions can help students understand the course content, and 853% of the students thought that the discussion based classroom teaching is conducive to students' concentration and improvement of learning effect. 100% of the students agree that the project-based teaching method is adopted in the course of industrial network control system. The simulation of teaching effect is shown in Fig. 3.

We have carried on the simulation analysis from five dimensions, from Fig. 3, we know that in the algorithm proposed in this paper, our proposed teaching effect is the best.

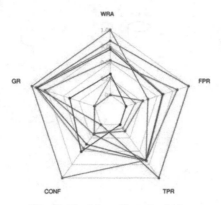

Fig. 3. Teaching effect simulation

3.2 The Students' Comprehensive Ability to Use Knowledge has Been Improved

In the 2016–2018 "Siemens Automation challenge", the degree of undergraduate students' participation has increased year by year. Because bisai is an undergraduate and research subject, the subject of bisai is to investigate the design and implementation of control scheme for complex production process, such as heating furnace, boiler, reactor and other production devices [4]. In previous years, the undergraduate students of bisai middle school were subordinate to di Ren, who really participated in the design and debugging of the project. In the past two years, 80% of the participating teams are undergraduate students, who undertake the main design and debugging work, and even many teams of undergraduate students form teams to participate in the project.

4   Research Design

Based on the investigation of the development of College English Teaching under the computer network environment in China, the essential task of this paper is to explore the ecological teaching mode. In this chapter, we will focus on the overall design process of this empirical study. The specific structural arrangement is as follows: firstly, establish research problems, select research subjects and variables, and clarify research methods and means, then list research design and process steps in detail, finally collect research data and text information, and carry out classified discussion and ecological analysis [5].

4.1   Research Questions and Hypotheses

Based on the current situation of College English Teaching in the computer network environment, this paper summarizes the maladjustment in the integration of traditional English teaching and modern education methods from the micro perspective, analyzes the maladjustment by using the principles and laws of educational ecology, and tries to explore the strategies and suggestions for reasonable optimization of teaching, It aims to promote the healthy development of College English Teaching in China towards the ecological direction of "compatibility, dynamic and benign". The author intends to focus on the following four aspects of research issues, to make a detailed investigation of College English Teaching under the computer network environment, and to clarify whether there is imbalance in the process of integration of computer network and College English Teaching in the University [6]. 2. Summarize the performance and influence of imbalance objectively and in detail. 3. Use the principles and mechanisms of educational ecology, From the perspective of individual ecology, group ecology and system ecology, this paper analyzes the maladjustment and its causes; 4. According to the ecological education concept of harmony, connection and development, starting from the objective reality of micro niche (teachers, students and teaching environment), we have the courage to innovate and look for Countermeasures to alleviate the maladjustment, In order to build and optimize the ecological foreign language teaching mode of "teacher led student-centered" under the computer network environment [7].

4.2   The Maladjustment of the Combination of College English Teaching Elements and Computer Network Environment

From the perspective of educational ecology, any healthy development of subject teaching has its own relatively unique teaching elements. These elements run in and coordinate with each other in the long-term teaching practice, and gradually form an external balanced and stable language teaching environment and an internal coordinated development of teaching ecosystem. Traditional foreign language teaching is no exception [8]. However, when the computer network technology enters the foreign language curriculum, the traditional teaching elements are impacted by new teaching elements (such as multi-media, network content, technical methods, etc.), which will lead to the reorganization of teaching structure and the change of teaching arrangement (such as the design of teaching activities, the positioning of teachers and students' roles, the choice of teaching mode, etc.) [9]. The running in of the new and old elements requires a long-term process of compatibility and imbalance. The imbalance will inevitably break the original ecological balance of traditional foreign language teaching, and the unbalanced teaching system will immediately lead to more new imbalances. From a macro point of view, the maladjustment of curriculum integration at this stage mainly includes: the maladjustment of teaching concept and teaching practice, the maladjustment of national education policy and school specific situation, the maladjustment of technology and application, the maladjustment of new teaching mode and traditional teaching system, the maladjustment of teaching process and teaching management, etc. With the development of curriculum integration in depth, various new forms of maladjustment will emerge one after another and exist for a long time in the macro and micro fields of teaching, which directly affects the normal operation of the whole foreign language teaching system and seriously restricts the healthy and orderly development of modern foreign language teaching in China. The author believes that it is necessary for us to carry out all kinds of research to effectively intervene and actively regulate these disorders in time [10].

4.3 The Construction of College English Ecological Teaching Mode Under the Computer Network Environment

If we want to analyze the causes of maladjustment and explore the countermeasures to solve the maladjustment, it is difficult to achieve only relying on the traditional teaching theory [11]. Although traditional teaching theories can well explain the essence of human learning activities, the imbalance in teaching has posed a challenge to these traditional theories, mainly because the elements of the integrated teaching system have changed. The existing teaching system involves not only teaching and learning, but also more elements such as technology and resources. Therefore, in order to find the causes and Countermeasures of the imbalance and restore the dynamic harmony of the teaching system, we should re-examine our foreign language teaching from the perspective of ecology on the basis of the advantages of traditional theories, that is, to investigate the relationship between the internal elements of the foreign language teaching system and the external development environment according to the principles of ecology and the theory of educational ecology, This paper discusses the characteristics and functions of foreign language teaching ecology, analyzes the basic laws of its evolution and development, and explores the teaching mode of students, such as the setting of curriculum objectives, the design of teaching activities, and the construction of teaching mode. The domestic research on Ecological foreign language teaching mode has made some achievements in both macro and micro fields. In this paper, the author aims to put forward some specific strategies for mode optimization based on the empirical research results, hoping to contribute to the domestic micro research in this field [12].

5 Theoretical Framework

The theoretical research of computer network technology applied to foreign language teaching is the result of the continuous optimization of combination and integration based on the theoretical research of various disciplines. With the development of College English teaching reform, the research of educational ecology has attracted close attention of the educational circles. It provides a new way of thinking for foreign language teaching research and enriches the theory of foreign language teaching. The integration of computer network and College English curriculum by using educational ecology is helpful to analyze and solve the system imbalance and ecological imbalance in College English teaching. The construction and optimization of ecological teaching mode and the realization of the ecological teaching of College English are the development and perfection of the theory in practice, and also a new perspective and experience of the reform of College English teaching. In this chapter, the author will introduce the theory of educational ecology and its basic principles and ecological laws, analyze the relationship between ecological theory and foreign language teaching in the computer network environment, and try to define the ecological teaching mode under the consideration of this theory [13].

5.1 The Theory of Educational Ecology

It is generally accepted that the term "educational ecology" was formally put forward by American scholar Lawrence Cremin (LA.) in his book "public education" in 1976.

The research on Educational Ecology in China started later than that in the west, and the research in Taiwan was earlier than that in the mainland. Since the beginning of the 21st century, due to the rapidly changing natural and social environment, economic globalization and other factors, people pay more and more attention to ecological problems, and the expectation of high-quality and efficient ecological education is gradually increasing, which promotes the research of educational ecology to be prosperous. For many years, the basic theory of educational ecology has been widely discussed by scholars at home and abroad. Many influential research results have been obtained by applying the theory to the practice of micro education [14].

Educational ecology is an interdisciplinary subject, which sprang up in the mid-1970s and spans the two fields of pedagogy and ecology. It is the product of the penetration and application of ecological principles and research methods in pedagogy. Specifically, ecology studies the law and mechanism of interaction between life system and environmental system; pedagogy studies the law of educational development, the influence of society on education, and the status and role of education in social development; while educational ecology takes pedagogy and ecology as the theoretical basis, draws on the research methods of these two disciplines and develops, According to the basic principles of ecology, especially the principles and mechanisms of ecosystem, ecological balance and coevolution, this paper studies various educational phenomena and their causes, and then grasps the law of educational development [15].

5.2 Basic Concepts of Ecology

The understanding of the basic concepts of ecology will help us to clarify the research objectives and standardize the research contents. There are many kinds of concepts in the subject. Here, the author mainly introduces the related projects of educational ecology involved in this study, and illustrates with examples of foreign language teaching in the computer network environment. 1. Species, key species, population and community species refer to the biological groups with certain morphological and physiological characteristics and existing in a certain natural distribution area. In different levels of ecosystem, due to the different basic units, the species in the ecosystem will be different. In the foreign language teaching under the computer network environment, we can regard students, teachers, teaching materials, computer hardware and software facilities, teaching environment, teaching methods, learning methods, school management and other teaching elements as the basic species in the ecosystem. Key species refer to the species whose disappearance or weakening will cause fundamental changes in the whole community and ecosystem. The number of individuals of key species may be large or rare, and their functions may be single or diverse. In the foreign language teaching ecosystem under the computer network environment, students, teachers and teaching methods are relatively important. These key species may also change with the development of teaching process. For example, in the multimedia classroom teaching dominated by teacher explanation, the leading role of teacher is more important, while in the process of student-centered network autonomous learning, the dominant position of student will be highlighted. Moreover, only teachers and students teaching groups without appropriate teaching methods can not form a healthy education ecosystem. Population refers to the collection of individuals of the same species living in a certain space, which is composed of multiple individuals of the same species in the space. There are differences among individuals. In the foreign language teaching ecosystem under the computer network environment, as a large fixed population, students' individual thinking and activities often have an effect on the development of the population. At the same time, students are also affected by the activities among the populations, It is an important issue to explore the interest cultivation and ability development of individual students and student groups in order to improve the teaching quality in this teaching environment. Community refers to the aggregation of all biological populations gathered in a certain area or ecological environment at a specific time. As mentioned above, students, teachers, teaching materials, computer hardware and software facilities, teaching environment, teaching methods, learning methods, school management and other teaching elements interact and correlate with each other in the foreign language teaching under the computer network environment, forming a teaching ecosystem community.

6 Conclusions

In the teaching of the course group, the embedded engineering cases are used to plan the teaching tasks of each course in the course group, and then the research teaching is carried out around the cases in the teaching of each course. The embedded engineering cases connect the knowledge points of each course, help students build a complete knowledge system, strengthen the concept of engineering application, and cultivate students' ability to analyze and solve problems through research teaching. The combination of the two parts improves students' ability to use comprehensive knowledge to solve complex problems.

References

1. Zhang, R., Wang, Z., Wang, Z.: Exploration of research teaching in Engineering Physics Teaching based on cases. Beijing: China Univ. Teach. (9), 63–64 (2013)
2. Zhang, J.: Exploration of College English curriculum group construction. Harbin: Heilongjiang High. Educ. Res. (2), 152–155 (2018)
3. Long, S., Jiang, J., Li, Q.: Computer basic curriculum group based on the curriculum system of "integration of liberal and professional education". Beijing: Comput. Educ. (5), 28–41 (2019)
4. Yang, Y., Wang, W., Meng, B.: Research on the improvement of engineering education training mode aiming at improving the ability to solve "complex engineering problems". Wuhan: Res. High. Eng. Educ. (4), 63–67 (2017)
5. Cao, C.: Introduction to ecology. Higher Education Press, (5), 64 (2002)
6. Chen, J.: More perfect requirements and clearer direction -- a new interpretation of the 0 τ version of College English teaching requirements. Audio Vis. Foreign Lang. Teach. (1), 43 (2008)
7. Jianlin, C.: Integration of computer network and foreign language curriculum. Shanghai Foreign Lang. Educ. Press 6, 16193–16197 (2010)
8. Junxia, D.: Review of humanistic teaching method. Foreign Lang. Teach. 5, 195 (2001)
9. Guorui, F.: Educational ecology. People's Educ. Press 7, 35 (2000)
10. Ge, F.: Modern Ecology. Science Press, (3), 59 (2008)

11. Kekang, H.: Design and development of modern educational technology and high quality network courses. China Audio Vis. Educ. **6**, 9 (2004)
12. Jia, H.: Application of mind map in Network College English Teaching. J. Changchun Normal Univ. (Humanit. Soc. Sci. Ed.) (2) (2014)
13. Department of higher education, Ministry of education. College English teaching requirements. Shanghai Foreign Language Education Press, (10), 37 (2004)
14. Shang, Y.: General Ecology (2nd Edition). Peking University Press, (2), 189236 (2002)
15. Dingfu, W., Wenwei, Z.: Educational ecology. Jiangsu Education Press **1**, 156183194 (2000)

Data Mining Algorithm in Volleyball Match Technical and Tactical Analysis

Yongfen Yang(✉)

Kunming University, Kunming 650217, China

Abstract. This paper analyzes the difficulties in the application of data mining algorithm in the technical and tactical analysis of sports competition, and puts forward a scheme for mining the key factors of winning volleyball matches. By applying the data mining algorithm based on Markov process and calculating the system reliability difference, the problem of finding the key action conversion process in volleyball competition is solved. In order to solve the problem of data acquisition speed, this paper proposes a method to solve the problem of data acquisition in real-time by setting the threshold of data acquisition. The design experiment shows the correctness and feasibility of the scheme.

Keywords: Data mining · Markov process · Data preprocessing · Technical and tactical analysis

1 Introduction

In the volleyball match, the result of the analysis of on-the-spot technical and tactical data is an important basis for coaches to carry out strategic deployment. Therefore, the collection, statistics and analysis of on-the-spot technical and tactical data have become the key to affect the coach's decision-making. However, it is difficult to complete the statistical work of data by hand. Due to the speed restriction, it is impossible to record a large number of, comprehensive and accurate data, and it is impossible to conduct scientific and systematic analysis of the collected data in a limited time, which can not meet the data requirements for scientific guidance. Coaches can only rely on experience and extraordinary experience.

1) Data collection. Because volleyball is to change the score by turns, and in the process of each round, there are 12 players in two teams at the same time. Volleyball requires high speed, which makes the recorder have to record the situation on the field in a very short period of time, including a series of data such as player number, ball landing area, technology and tactics, which requires high real-time performance.

2) Data analysis. Because there are many tactical changes and contingency in volleyball, it is necessary to prevent the possibility of making wrong analysis results based on a small amount of data. There is a high demand for the ability of systematic analysis of data. It is more and more necessary to apply the data mining algorithm to the technical and tactical analysis of volleyball matches.

M. A. Jan and F. Khan (Eds.): BigIoT-EDU 2021, LNICST 392, pp. 204–213, 2021.
https://doi.org/10.1007/978-3-030-87903-7_27

In this paper, we propose a method to analyze the key factors of winning volleyball match by using data mining algorithm based on Markov process. In view of the above two application difficulties, the solutions to meet the real-time requirements in the data acquisition process and the solutions to eliminate ambiguity in the data preprocessing process are given, and the correctness and feasibility of the solutions are demonstrated through experiments.

2   Principle of Data Mining Algorithm Based on Markov Process

The idea of data mining algorithm based on Markov process is: the mining object is regarded as a system composed of multiple states, and the transition process between states conforms to the semi Markov process. The state transition probability matrix is obtained by statistics [1]. The system reliability is calculated by using the matrix, and then the increment is set to calculate the system reliability difference. Thus, the sensitivity of reliability to transition rate is analyzed. The method to calculate the system reliability is: let Q_{ij} be the transition probability from state I to state state, and let $C1n$ be the system reliability from initial state to state state, then we can calculate it by the following equation.

$$\begin{bmatrix} C1n \\ C2n \\ \vdots \\ C3n \end{bmatrix} = \begin{bmatrix} Q11 & Q12 & \cdots & Q1n \\ Q21 & Q22 & \cdots & Q2n \\ \vdots & \vdots & \vdots & \vdots \\ Qn1 & QN & \cdots & Qnn \end{bmatrix} \begin{bmatrix} C1n \\ C2n \\ \vdots \\ Cnn \end{bmatrix} \tag{1}$$

The system reliability from the initial state to the successful state is obtained. The calculation method of system reliability difference is: after calculating the system reliability, each item in the state transition probability matrix is added with a small increment, and then the system reliability is recalculated by using the added state transition probability matrix. The difference between the former and the latter is the system reliability difference, and the larger the difference is, It indicates that the value changed in the state transition probability has greater influence on the system reliability.

3   Application Research

When the algorithm is applied in volleyball match, the series of actions from serve to score and their conversion relationship can be regarded as a system, and each action is only related to the previous action. Each action is regarded as a state, and the transition between actions is regarded as a state transition. According to the principle of the algorithm, the change of the state transition rate can affect the change of the system reliability. Therefore, the sensitivity analysis of the system reliability to the state transition rate between each pair of actions can be used to determine whether the state transition rate of each group of actions has a significant impact on the final result, According to the data mining algorithm based on Markov process and the characteristics of volleyball match, the scheme of applying the algorithm in volleyball match is made.

3.1 Data Acquisition

In order to collect the data needed to realize the data mining algorithm, it is necessary to record the execution process of each technical action in volleyball match. Because there are many changes in the athletes' technical and tactical movements in the competition, and all kinds of actions are completed in an instant, so it is challenging to record the process of the competition. In order to solve this problem, a script description language based on process is designed. This language uses mnemonic method to code the basic technical movements in volleyball match [2]. However, the design of this script description language requires the recorder to record a series of information including team member number, technical action, technical type, start area and end area, which makes the recorder work a lot.

In order to collect competition data more quickly and accurately, it is necessary to design a pattern based script description language. The so-called script description language based on pattern refers to: since most of the tactics used by both sides of the game have certain patterns in high-level volleyball matches, these patterns can be used as some formulas in the process of script description language design, and only a small amount of information about this mode is needed in the user's recording of the game process, Other information can be inferred from the tactical rules of volleyball matches.

In terms of acquisition mode, the main advantage of keyboard acquisition mode is faster acquisition speed, and the disadvantage is that the accuracy is slightly lower than that of mouse acquisition. Mouse acquisition has advantages in acquisition accuracy, but due to the limited collection interface, it can not cover all scripts, so it has certain limitations. In this scheme, the combination of mouse acquisition and keyboard acquisition is used to improve the collection efficiency.

1) Search the script data table to count the most frequently used scripts and their usage frequency.
2) Script sorting.
3) Output frequent scripts to acquisition interface.
4) For frequent scripts, the mouse collection method is used; for non frequent scripts, keyboard acquisition method is used.

3.2 Data Preprocessing

Data preprocessing can improve the quality of data and make the process of data mining more effective and easier. In this scheme, the necessary data preprocessing includes data classification and data ambiguity elimination.

4 Data Classification

Because the script description language based on pattern is designed for technical and tactical information, it is necessary to classify and integrate the collected data and integrate the data in different data tables to build a new small data warehouse. The separation of data warehouse and operational database is due to the different structure, content and usage of data in these two systems [3]. Decision support needs historical data, but the operational database generally does not maintain historical data. In this case, although the data in the operation database is rich, it is often insufficient for decision-making. For different data types, such as offensive and defensive processes, different data warehouses will be generated. The data classification process is shown in Fig. 1.

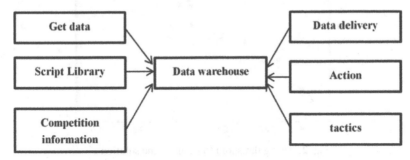

Fig. 1. Data classification process.

5 Simulation for Solutions

When applying the data mining algorithm based on Markov process to volleyball match, we must solve the problem of ambiguity in the process of action conversion. The origin of this problem is decided by the rules of volleyball match. In volleyball match, at most three consecutive actions can be carried out in each half court, which leads to the ambiguity in describing the process of action conversion.

In the process of preprocessing the collected data, the solution is as follows:

1) Set a threshold. The method of threshold setting is: due to the characteristics of volleyball match, the frequency of some action conversion process is very large, and some action conversion process only appears several times in the whole game. According to the collected state transition rate of action combination, take such a value, which can divide the numerical combination of state transition rate into two intervals [4]. The interval of state transition rate containing larger value is interval a, and the interval containing smaller value is interval B, which is close to the maximum critical value of B interval [5]. Figure 2 shows the solution simulated by threshold method.

2) For the action combination (a interval element) whose state transition rate is greater than the threshold value, the original practical significance is maintained and the reasonable action conversion is carried out. Figure 3 shows the use of different components to simulate the solution.

3) The action combination (b-interval element) whose state transition rate is less than or equal to the threshold value is ignored, or its value is incorporated into the action combination with similar actual meaning.

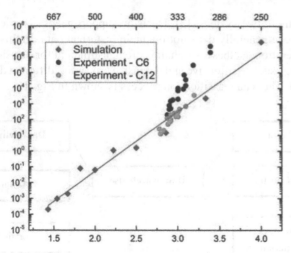

Fig. 2. Using threshold to simulate the solutions.

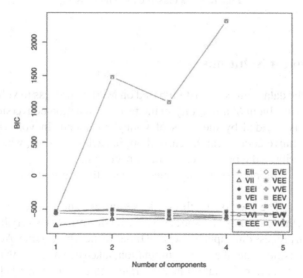

Fig. 3. Using the different components to simulate solutions.

6 The Design and Implementation of the Statistical Analysis System for Volleyball Match

6.1 System Design Objectives

The goal of the system design is to realize a volleyball match on-the-spot technical and tactical statistical analysis system which integrates the functions of data acquisition, data statistics, data analysis, data management and video management. Through this software, coaches can record and analyze the game data by computer, so as to provide technical support for coaches' strategy making [6].

The detailed design objectives are as follows: to build a basic platform with the functions of collection, statistics, analysis and management of on-the-spot technical and tactical information of volleyball matches. It can realize the collection, statistics and analysis of technical and tactical data, as well as the management of team, player and game information. At the same time, video detection and video location technology are applied to realize the function of playing video according to script. Based on the in-depth analysis of the technical and tactical composition of volleyball match, the script description language suitable for describing the technical and tactical process of volleyball match is designed, and the parser of the language is written. Applying data mining algorithm to deeply mine a large number of data, this paper proposes a data mining algorithm which is more suitable for sports computing. The main algorithms to be applied are: applying Markov process principle to data mining, looking for the state transition process which has the greatest impact on the system reliability in the state from serving to scoring. The association rules algorithm is used to find the association information between two or more actions to help coaches make strategic deployment. The application of classification and prediction technology to athletes in the game of different technical movements and technical and tactical combination of classification, while establishing a prediction model, given a player or the whole team in the past game for a specific tactic of technical and tactical information, predict their future face a certain tactic of countermeasures [7, 8].

6.2 System Function Structure Design

The function of the system mainly includes four parts: data management, data acquisition, data analysis and processing, system management. In addition, it also involves the application of video positioning technology. Data management is mainly the management of team information and player information, data collection is mainly through the mouse collection and keyboard collection of two collection schemes to collect data, data analysis and processing is through the massive data mining, so as to get some coaches interested in information, including the function of data statistics and analysis. Video management mainly involves some content of video positioning [9].

7 Research on Data Acquisition Solution

7.1 Design Idea of Script Description Language

The basic idea of script description language is: most of the techniques and tactics used by both sides in volleyball matches have the characteristics of fixed patterns, which are represented by non ambiguous codes. At the same time, in the process of recording the game, a series of description methods and reasoning rules are customized. You need to input as few codes as possible, and use the logic reasoning realized by computer to supplement the remaining codes and infer Hide the information, so that users can get as much information as possible [10].

According to the basic design idea of script description language, the basic design principles of script description language are summarized as follows: intelligent reasoning - through the analysis of volleyball skills and tactics, summed up as many patterns and rules as possible, users use script description language to describe the game data, input as little data as possible, get as much information as possible, Effectively shorten the length of a large number of code. Simplified memory - in order to make users with different professional backgrounds, different age groups and different cultural levels be proficient in the use of script description language as soon as possible, Pinyin abbreviations are used to record competitions. Pinyin, as one of the bases of Chinese language knowledge, is more suitable for Chinese memory habits and can meet the needs of users at different levels. Key record - in the process of script description language recording, due to the fast pace of volleyball match and the ever-changing situation, it is unrealistic to require users to record all data in real time and accurately in a short period of time, which must be recorded selectively. The script description language designed in this paper supports users to record the most interesting and final data first, Select the key process with the most reference value and analysis value to record [11].

7.2 Code Design of Script Description Language

The script description language uses Chinese Pinyin as the description carrier, which is convenient for users of different ages and cultural backgrounds [12]. In the process of designing the code, we mainly follow the following principles:

The first is convenient memory. For the choice of characters, the first letter of Pinyin is generally chosen, such as "F" for serve, "g" for attack and "1" for block. The advantage of this choice is that as long as people can say the word, they can immediately associate with the code that represents the phrase, which provides great convenience for memory. There is also a problem in doing so, that is, the ambiguity of letters. For example, the first letter of two words is the same character. In the process of code design, in this case, two different solutions will be adopted according to the specific needs: one is to use another character to represent one of the words, which is not the first letter of the word, but has other connections with the word, which can also facilitate the memory. The second method is that words with different meanings can be described by the same symbols if they are not in the same category. This ambiguity can be overcome in the process of script parsing through the design of script parser. In addition, the design of script parser supports the case discrimination of the same character, so it expands the availability of characters [13].

The second is convenient collection. Because the process of collecting data is mainly completed through the keyboard, so in the process of selecting characters, this topic tries to choose those keys that users can easily touch, or only need to press a key to record characters. The code design of script description language is shown in Fig. 4.

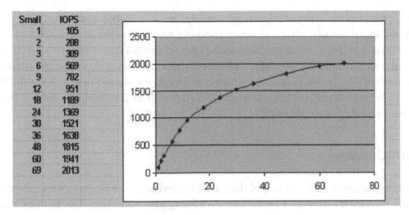

Small	IOPS
1	105
2	208
3	309
6	569
9	782
12	951
18	1189
24	1369
30	1521
36	1638
48	1815
60	1941
69	2013

Fig. 4. Code design of script description language

7.3 Working Steps of Script Parser

1) Field split, split based on double spaces.

2) Judge whether it is valid one by one (single code, mode code or invalid code), add the valid code and invalid code into the code table on the right, and the invalid code is highlighted to prompt modification.

3) For the valid code, analyze it one by one. The correct code before parsing is entered into the code table, and the parsed code is entered into the parsing table.

4) If the field to be parsed is a single code (it is already a valid code at this time): judge whether there is an action. If the action is empty, the default action is serve. In the single code, only the action must exist, and the default area information is supplemented according to the action. Inference of default information. Data storage.

5) If the field to be parsed is a pattern code (it is already a valid code at this time), for the pattern field split into several segments, the analysis starts from the last segment. First, the mode number and action field are put into the back section of the analysis mode. If there is a mode number, the technology type is put into the database. If there is a mode effect, the action effect of several records is put into the database. In the front part of the analysis mode, if there is no action, the default is spike action. Analyze the non pattern fields from the back to the front in turn. If there are repeated actions, the first one analyzed (that is, the one close to the pattern field) shall prevail. That is, if the content storage unit to be entered is not empty, the current content to be entered will be abandoned [14].

Script translation: when the currently used scripts are no longer suitable for the needs of users and need to change the script format or the characters contained in the script, it is inconvenient for users to analyze the data described by the old scripts because a large number of previous scripts still exist in the database [15]. In order to unify the old and new scripts, the system provides a script editor, which can make the old scripts in the database be translated into new scripts. In the process of translation, the script structure design is involved. In order to meet the different needs of users, the script editor provides a new script syntax customization module. The grammar customization module provides six fields for editing, namely: team member number, team member action, technical details, effect, start area and end area. In addition, at the end of the six fields, you can add any separator character. These elements can be freely defined by the user. After the new grammar rules are successfully customized, the script translation can be carried out. In the process of script translation, we should first find out the list of scripts to be translated, and then start the translation. The new script is written into the database instead of the old script.

8   Conclusions

This paper studies the difficulties of data mining algorithm applied to sports calculation, and puts forward a scheme of applying data mining algorithm based on Markov process to analyze volleyball match techniques and tactics. In the process of data collection, the scheme improves the collection efficiency by searching for frequent scripts. In the process of data preprocessing, a method of setting threshold is proposed to make the data meet the application conditions of the algorithm, The experimental results show that the scheme can dig out the data results that can not be obtained by visual observation and simple statistics, but have important significance for coaches to guide the game, and can mine the key action conversion process. The scheme has a certain role in mining the technical and tactical information of volleyball matches. The next step is to apply the algorithm to the analysis of key area transfer process and key tactics transfer process in volleyball match, as well as the application of association rules and classification prediction algorithm to the analysis of volleyball match technical and tactical information.

References

1. Yujia, Z., Zhao, H., Jiewei, W.: Research on the application of data mining algorithm in volleyball match technical and tactical analysis. Comput. Appl. **12**, 3027–3029 (2006)
2. Hanm, K.M.: Data Mining Concepts and Technologies 1. China Machine Press, Beijing (2005)
3. Editorial board of Encyclopedia of safety science and technology. Encyclopedia of safety science and technology, China Labor and social security press, Beijing (2003)
4. Rong, J., Ben, H.: Application of association rules in the analysis of volleyball skills and tactics. Glamour China **28**, 133–134 (2008)
5. dataproject[EB/OL. http://www.dataproject.com
6. Zhang, J.: The application of data in sports news report. Chin. Journalist (2003)
7. Jiawei, H., Micheline, K.: Data Mining Concept and Technology. Mechanical Industry Press (2005)
8. Olivia Parr RUD, data mining practice, China Machine Press (2003)
9. Orlando, F.: Data mining and knowledge discovery: theory, tools, and technology. Proc. SPIE. **11**, 259-264 (2000)
10. Agrawal, R., Imielinski, T., Swami, A.: Mining association rules between sets of items inlarge databases. Proc of Very Large Data Bases (1993)
11. ZPawlak Rough Sets: Theoretical Aspects of Reasoning about Data. KluwerAcademic Publishers, Boston (1991)
12. Lianjie, S.: Encyclopedia of Safety Science and Technology. China Labor and Social Security Press, Beijing (2003)

13. Zhao, H., Wang, G., Gao, Y.: Abstract model of software architecture. J. Comput. Sci. (2002)
14. Zhao, H., Sun, J., Wang, G., Gao, Y.: Component based software reliability model. Minicomputer Syst. 730–736 (2002)
15. Langley, P.: Elements of Machine Learning. Morgan Kaufman, San Franciso (1996)

Data Mining Technology in Higher Vocational "Double Helix" Hybrid Teaching

Mingyang Li[✉]

Shandong Vocational College of Science and Technology, Weifang 261053, China

Abstract. In this paper, a hybrid teaching personalized learning system model based on data mining technology is proposed. The data mining technology is used to discover each student's personality, learning behavior, learning feedback information and teaching information that the teacher is interested in, adjust the teaching strategy in time, and make the teaching content and teaching activities suitable for the student's personality. The personalized learning system based on this model truly embodies the educational concept of teaching students according to their aptitude. Hybrid teaching is the development trend of college teaching, but there are not many colleges and universities that push forward the reform of hybrid teaching mode as a whole at the school level. On the basis of the literature research on the strategies for promoting the reform of hybrid teaching mode at home and abroad, this paper constructs the strategies for promoting the reform of Hybrid Teaching Mode in Colleges and universities in China from five aspects: top-level design, training system, quality monitoring and information feedback mechanism, incentive system and service system. Practice has proved that the promotion strategy creates a mixed teaching mode reform atmosphere in Colleges and universities, improves teachers' teaching ability, and improves students' enthusiasm and ability to learn.

Keywords: Personalized learning · Data mining · Teaching strategy · Hybrid teaching · Reform · Security system · Research

1 Introduction

Web based learning system with the help of Internet However, most of the existing learning system models are static, lack of interaction, lack of personality, and students can not learn on demand. In addition, a large number of useful teaching information accumulated on the site has not been used, such as user's access log, registration information, question answering information, test results, assignments, exchange information, learning progress Degree and so on, which causes a great waste of resources. These shortcomings limit its further development. Therefore, it is necessary to introduce intelligent and personalized services, which is also a trend of web-based learning system development. With the rapid development of the Internet, modern online education is developing rapidly in the United States, Britain, Australia, Japan and other countries with

© ICST Institute for Computer Sciences, Social Informatics and Telecommunications Engineering 2021
Published by Springer Nature Switzerland AG 2021. All Rights Reserved
M. A. Jan and F. Khan (Eds.): BigIoT-EDU 2021, LNICST 392, pp. 214–222, 2021.
https://doi.org/10.1007/978-3-030-87903-7_28

high degree of information technology. The hybrid teaching reform has been success-fully implemented in many foreign universities and formed a systematic hybrid teaching reform management and practice mode. Its excellent experience is worth learning from domestic universities [1].

Although blended teaching is a hot research topic, but from the analysis of the results of literature review at home and abroad, there are few research results on the implementation of institutional level blended teaching. In foreign countries, Siemens pointed out in the Research Report "welcome Digital University: on distance, hybrid and online learning", that "in our literature search, we only found an article that systematically combed the implementation of Hybrid Teaching at the institutional level". At the same time, some scholars pointed out the gap in the current research: "first of all, the research on the role of teachers in blended learning is not enough. Due to the lack of information on the policies and implementation of teacher training and University blended learning, the research on the views of teachers and institutions is also insufficient." There are not many colleges and universities that carry out the reform of mixed teaching mode in domestic colleges and universities, and there are few successful experiences. Therefore, how to build an effective strategy to promote the reform of mixed teaching mode in Colleges and universities is an urgent problem to be solved.

2   Basic Design Idea

The personalized learning system based on Web can guarantee students' learning in two aspects: first, adopt the teaching design based on learning, students should be able to choose the knowledge learning they need according to their own needs, and teachers can choose the learning strategy suitable for students according to their knowledge background, that is, "personalized teaching"; second, Take effective learning navigation mechanism to ensure that students can learn smoothly, and will not be delayed or even give up learning due to difficulties without timely help (see the Fig. 1).

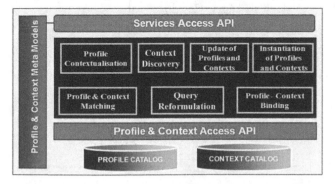

Fig. 1. Services assess API

3   The Realization of Personalized Learning System Based on Web

3.1   Formal Definition of Course Related Association Rules

This paper takes the students' scores in the course selection database of a college as an example to formally describe the relevant association rules of courses [2].

Definition

$$Let I = \{i_1, i_2, \cdots i_n\} \quad D = \{t_1, t_2, \cdots t_n\} \tag{1}$$

is the set of student's achievement database records, where $t_i \subseteq I (1 \leq i \leq N)$ only contains the course names that meet the conditions, for example, courses that meet the requirements of excellent or higher than 80 scores. The association rules are implicit in the following form:

$$p_1 \wedge p_2 \wedge \cdots p_n \rightarrow q_1 \wedge q_2 \wedge \cdots \wedge q_m \tag{2}$$

The definitions of confidence and support are given below.

3.2   Framework of Personalized Learning System

This paper constructs an intelligent and personalized network teaching system. Figure 2 is the functional framework of the system.

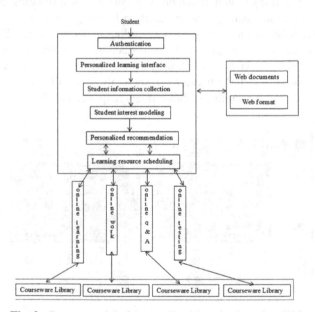

Fig. 2. System model of personalized learning based on Web

The web-based personalized learning system is based on Browser / server architecture, which realizes the management of students, teachers, educational administration and the system under the synchronous interactive teaching mode and asynchronous interactive teaching mode. The main difference between it and the general distance learning system is that learners will not see the same page after logging in the system [3], but a personalized learning environment closely related to their own. Including: completion and correction of online assignments, test scores, frequently erroneous knowledge points, learning progress of courseware, participation in Q & A, BBS questions, course selection, student records and teacher comments, etc. These information are based on the characteristics of the learners, different learners will see the system interface will be very different. The system can make full use of the rich information accumulated on the education platform and allocate teaching resources reasonably according to each student's different personality. Using the modules of Web Mining and data mining to analyze and accumulate various information on the site regularly, such as various server logs, interactive information between students and the site, etc., to mine some interesting patterns and rules.

4   The Effect of Promoting the Reform of the Mixed Teaching Mode in the Third Three Universities

In 2015–2016 academic year, the school carried out a one-year practice of hybrid teaching mode reform. In order to analyze the effectiveness of the promotion strategy of hybrid teaching mode reform, this study conducted a questionnaire survey, interview and performance comparison study.

From December 29, 2015 to January 4, 2016, this study conducted a student satisfaction survey on the first batch of students who participated in the reform of hybrid teaching mode, and collected 2354 papers and 2336 effective papers. From July 1 to July 15, 2016, a student satisfaction survey was conducted on the second batch of teachers and students who participated in the reform of hybrid teaching mode. 78 scores were collected from teachers, 76 valid papers, 2364 from students and 2349 valid papers. From March to June 2016, a questionnaire survey was conducted on the training satisfaction sent to the college. On July 20, 2016, the final examination results of the compulsory course of the reform of mixed teaching mode were carried out analysis. Through the analysis of the above data, we can conclude the achievements of the reform and guarantee system of Hybrid Teaching Mode in one year.

4.1   The Reform of Hybrid Teaching Mode Promotes the Transformation from "Teaching" Centered to "Learning" Centered

Through one year's construction, the school has constructed 103 courses of mixed teaching mode reform. The first batch of 33 courses participated in the reform of hybrid teaching mode, 32 of which passed the acceptance; the second batch of 70 courses participated in the reform of hybrid teaching mode, 69 of which passed the acceptance. It can be seen that at present, there are 101 courses of mixed teaching mode reform in the school that have passed the acceptance. These courses have completed the input of basic information of courses, the construction of teaching resources such as courseware and question bank, and carried out teaching activities such as homework, discussion and test by using the network at different levels. In addition, learning has also approved the construction of resources of network integrated teaching platform 192 courses. As of July 21, the online integrated teaching platform has attracted more than 870000 visitors, 412 teachers and 8914 students have logged in to the platform. The course construction of mixed teaching mode reform plays an important role in deepening the reform of classroom teaching mode, improving the comprehensive quality of students and promoting the transformation of teaching mode from "teaching" centered to "learning" centered.

4.2   The Reform of Mixed Teaching Mode Improves Teachers' Teaching Ability

Since the reform of the mixed teaching mode, 19 trainings have been organized by the joint teacher development center of the Modern Educational Technology Department of the academic affairs office and the Institute of educational technology of Tsinghua University, with 1265 teachers trained. Through a series of training, teachers' understanding of the necessity of teaching reform in our school has been improved, and teachers' teaching ability has been improved.

In the second half of 2016, the Modern Education Technology Department of the academic affairs office, together with the teacher development center, promoted the mixed teaching mode in 11 colleges (departments) including the school of media, the school of music and the foreign language teaching department of the University. The results show that 93.9% of the teachers think it is necessary for the school to reform the classroom teaching mode, and 89.2% of the teachers are willing to try the mixed teaching mode in the future classroom.

A series of training activities in the reform of mixed teaching mode provide a platform for teachers of different colleges and disciplines to exchange teaching methods. Teachers who did not participate in the reform actively applied for the reform of hybrid teaching mode. Teachers who participated in the reform began to carry out the construction of online courses and hybrid teaching design, and their teaching design ability was significantly enhanced. Data shows that in the first batch of excellent teaching plan selection activities in our school, 24 of the 39 teachers who won the prize participated in the reform of hybrid teaching mode, accounting for 61.5% of the total number of winners [4].

4.3   The Reform of Mixed Teaching Mode Improves Students' Enthusiasm for Learning

Because the hybrid teaching mode emphasizes the organic combination of online learning and offline learning, it effectively mobilizes students' enthusiasm for learning through different forms of teaching activities such as pre class preview, pre class self-test, on class discussion and off class homework. Through one year of practice, students gradually adapt to and recognize the hybrid teaching mode. According to the survey results of the second semester of 2015–2016 academic year, 82.89% of teachers think hybrid teaching improves students' learning enthusiasm; 95.24% of students think hybrid teaching improves their learning enthusiasm, 2.12% more than last semester; 93.24% of students like hybrid teaching mode, increased compared with last semester 2.1%; 92.8% of the students are very satisfied with the online submission of homework, teaching evaluation and discussion, an increase of 6.16% over the previous semester; 94.1% of the students think that part of the course learning on the Internet is very valuable, an increase of 3.9% over the previous semester [5].

The simulation results and improved results are shown in Fig. 3 and Fig. 4.

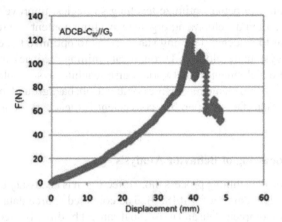

Fig. 3. The simulation results for hiberd helix teaching

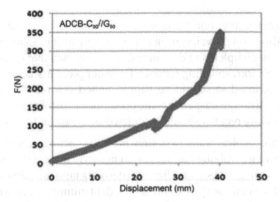

Fig. 4. Improved simulation results

5   Application of Data Mining Technology in Network Teaching

In front of this paper, through the analysis of the characteristics and problems of network teaching, expounds the necessity of the application of data mining technology in network teaching. The network teaching platform and teaching management system have relatively systematic teaching record data and log files, which makes the application of data mining feasible [6].

(1) For students. Through the data mining of online learning behavior, student performance and other information, we can recommend to students, help to improve their

learning behavior, learning resources and learning tasks, and recommend good learning experience and learning strategies to students, so as to improve the effectiveness of online learning [7]. (2) For teachers. Through the analysis of students' interest in learning, learning trajectory, and students' evaluation of course schedule, content arrangement and teaching methods, we can provide more objective feedback information to teachers, so that teachers can better optimize teaching strategies, improve teaching process and improve curriculum development; organize teaching content and reconstruct teaching plan according to students' learning status, so as to optimize the effect of network teaching. (3) For system developers. Through data mining feedback information, gradually improve the development of network course teaching system platform functions, make the teaching system elements organic combination, background management more intelligent, more scientific and operable, convenient for teachers, course managers and students.

5.1 Data Preprocessing of Behavior Analysis

In order to make the data mining process more effective, it is necessary to provide a clean, accurate and complete data set for it [8, 9]. The collected source data is often prone to data redundancy, incomplete data and noise, so it can not be directly used for data mining. In order to effectively carry out data mining, the following data preprocessing can be carried out on the source data set:

(1) Missing data processing. For the missing attribute values in the data set, you can choose manual filling, single value filling, class mean filling, mathematical inference and other methods to complete the data items; for some data that cannot be completed, you should delete the corresponding records. For example, for students with incomplete examination results, the data should be supplemented completely after checking the results; for students who are suspended, dropped out or transferred from school in the table, the corresponding records should be deleted by ignoring tuples.

(2) Data conversion. Typical transformation methods include: reorganizing classified variables, transforming symbolic variables into numerical variables, and mathematical transformation (including data standardization, deriving new variables, etc.). The specific conversion method depends on the specific task of data mining, the data mining algorithm and technology used. For example, K-means clustering algorithm is not suitable for dealing with discrete attributes; for association mining algorithm, interval quantization is needed for the attribute value corresponding to each tuple of the original data set, and finally transformed into discrete attributes. After the completion of data conversion, according to the purpose and task of mining and the selection of mining algorithm, according to the method of attribute selection, the original data table is transformed into the target data table [10, 11].

5.2 Establishment of Behavior Analysis Model and Application of Results

In the analysis of network learning behavior, the analysis goal determines the selection of data mining algorithm and the establishment and application of model [12]. For example, to analyze the relationship between students' online learning behavior (such as test times, pass points, etc.) and online learning effect, we can choose mining algorithms such as association rules and decision tree, to understand the association between various learning behaviors and their impact on learning performance through association or decision rules, and to predict the learning effect of different learning behaviors; Cluster analysis can be used to analyze the learning characteristics of different learning behavior groups. Through cluster analysis, we can understand the learning commonness of students in the same learning behavior group and the difference of performance among different group [13].

The purpose of data mining is to discover the knowledge and rules hidden in the data, and to apply the mining results to solve or improve the related problems. For example, the results of association analysis and cluster analysis in the above example can be used to improve the following shortcomings of online teaching:

1. Relevance analysis is helpful for teachers to recommend learning behaviors, learning resources and learning tasks that can help students improve their learning effect; by recommending successful learning experiences to students, it can guide students to formulate effective learning strategies, so as to improve their learning ability and improve the efficiency of online learning [14, 15]. 2. Through cluster analysis, the classification model of students' characteristics is established, which is helpful for teachers to find students' potential learning problems and help students correct their learning behaviors in time [16]. it is also helpful for students to check their own shortcomings and determine the direction of their efforts by comparing with the classification model. In addition, we can also do a good job in the early teaching design through clustering analysis of students' characteristics, so as to provide reference for personalized teaching. According to students' different mastery of knowledge points and different demands for knowledge and teaching resources, the teaching structure, teaching links and teaching resource design of online courses should be adjusted in time to provide personalized teaching contents and learning resources recommendation for different types of learning groups, so as to realize hierarchical teaching; To provide learners with what they want quickly and effectively, so as to teach students in accordance with their aptitude [17].

6   Conclusions

Because the mixed teaching mode has changed from passive listening to class to active learning, students' reading ability, teaching information acquisition, analysis and processing ability, and oral expression ability have been enhanced. The results show that 93.7% of the students think that the hybrid teaching mode improves their learning ability; 76.2% of the teachers think that the hybrid teaching mode improves the teaching effect compared with the traditional teaching mode.

In order to understand the promoting effect of the mixed teaching mode reform on the teaching effect, this study makes a horizontal and vertical comparative analysis on the written examination results of the second batch of mixed teaching mode reform courses. The final examination of the school uses the question bank to produce questions and the flow to judge papers, so the final examination results can reflect the learning effect of the students to a certain extent. The data shows that in the 30 courses with final examination, the average score and pass rate are higher than that of the control class, 22 courses, accounting for 73%.

References

1. Drysdale, J.S., Graham, C., Spring, K.J., et al.: An analysis of research trends in dissertations and theses studying blendedlearning . Internet High. Educ. **17**, 90–100 (2013)
2. Graham, C.R., Woodfield, W., Harrison, J.B.: A framework for institutional adoption and implementation of blended learning in higher education. Internet High. Educ. **18**, 4–14 (2013)
3. Porter, W.W., Graham, C.R., Spring, K.A., et al.: Blended learning in higher education: institutional adoption and implementation. Comput. Educ. **75**, 185–195 (2014)
4. Zhao, J., Xu, L., Situ, H.: Data mining based intelligent learning system DMBILS. Microcomput. Inf. 7–3, 212–214 (2006)
5. Bishui, Z., Hongbiao, X.: Design and implementation of web-based intelligent learning system. Comput. Eng. Des. **26**(11), 3130–3132 (2005)
6. Guodong, Z.: Documentary research on University Digital Campus and digital learning, p. 19. Peking University Press, Beijing (2012)
7. Daokai, G., Shaogang, Z., Shunping, W.: Methods and Applications of Educational Data Mining. Educational Science and Technology Press, Beijing (2012)
8. Zhang, M., Ding, X.: Online education: from opportunity growth to integration into the mainstream and steady development -- Enlightenment of online education series survey and evaluation in the United States on the development of online education in China. Open Educ. Res. **4**, 10–17 (2006)
9. He, Y.: Research on Effective Teaching of Online Courses. Xi'an: Shaanxi Normal University Press, 10 (2011)
10. Li, T., Fu, G.: Research status and trend analysis of educational data mining at home and abroad. Mod. Educ. Technol. (10) (2010)
11. Jin, L.: Big data and informatization teaching reform. China Audio Vis. Educ. (10), 8–13 (2013)
12. Bing, L.: Web Data Mining. Tsinghua University Press, Beijing, 6 (2009)
13. Zhibo, C.: Data Warehouse and Data Mining, p. 93. Tsinghua University Press, Beijing (2009)
14. Shiping, L.: Data Mining Technology and Application, pp. 154–155. Higher Education Press, Beijing (2010)
15. Zhibo, C.: Data Warehouse and Data Mining, p. 94. Tsinghua University Press, Beijing (2009)
16. Ji, X.: Application Examples of Data Mining Technology. China Machine Press, Beijing 200:53
17. Zhibo, C.: Data Warehouse and Data Mining, p. 137. Tsinghua University Press, Beijing (2009)

Development of APP Visual Design Under the Background of Big Data

Jiao Zhong[✉]

Jiangxi Normal University Science and Technology College, Gongqingcheng 330077, Jiangxi, China
zhongjiao0312@sina.com

Abstract. With the wide application of big data analysis technology in all walks of life, this paper proposes the development of APP visual design based on the background of big data, focuses on the application of big data in visual design, studies the application of big data visual technology in information transmission, transforms system data into common icons, describes the current situation of big data storage technology, and studies the big data visual design. This paper analyzes the application of big data storage technology in the monitoring system, and puts forward the corresponding data security measures.

Keywords: Big data · Visual design · Teaching logic · Divergent thinking

1 Introduction

The arrival of the era of big data is impacting all walks of life, to a certain extent, it puts forward new requirements for the talent training of visual communication design major, but there are some serious problems in the traditional creative course of visual communication design. Domestic colleges and universities need to change the teaching and training mode, innovate teaching logic and teaching practice in order to cultivate talents to adapt to the development of the times and social needs, especially some Newly Upgraded Undergraduate Colleges and universities need to achieve positive transformation in teaching.

Visual design is a subjective form of expression and result of eye function. Compared with visual communication design, visual communication design is a part of visual design, which is mainly for the conveyed object, namely the audience, and lacks the demand for the designer's own visual demand factors. Visual communication is not only conveyed to visual audience but also to designer. Therefore, in-depth visual communication research has paid attention to all aspects of visual feeling, which is called visual design more appropriate.

M. A. Jan and F. Khan (Eds.): BigIoT-EDU 2021, LNICST 392, pp. 223–232, 2021.
https://doi.org/10.1007/978-3-030-87903-7_29

In the end, what is visual design? Nieh, who is involved in computer professional research, will answer your doubts [1]. Visual design is the expression means and results of the subjective form of eye function. Similarities and differences between visual communication design and visual communication design: visual communication design is a part of visual design, which is mainly aimed at the audience, and lacks the appeal of visual demand factors of designers. Visual communication is not only conveyed to the visual audience, but also to the designer himself. Therefore, in-depth research on visual communication has focused on all aspects of visual experience, which is called visual design more appropriate.

Based on the analysis of the design status of at the present stage, the main content of its visual communication design is still graphic design, which is commonly called "graphic design" by professionals. "Visual communication design" and "graphic design" contain no big difference in the design category at this stage, "visual communication design" and "graphic design" in the conceptual category of distinction and unity, there is no contradiction and opposition (see the Fig. 1).

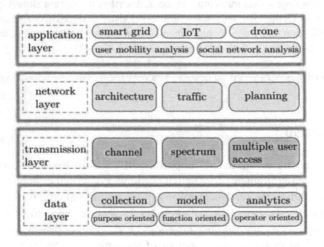

Fig. 1. Frame for big data

2   Big Data Analysis

2.1   Definition of Big Data Analysis

Big data analysis is the product of rethinking data science and exploring new models in a data intensive environment. Its core idea is to use a more effective way to manage massive data and extract value from massive data [2, 3]. It is the core of the concept and method of big data. It refers to the process of analyzing massive data with various types, rapid growth and real content, and finding information that is conducive to decision-making from the mass data.

2.2 Data Visualization

Data visualization is a technology originated from computer graphics. Its function is to directly display the implicit information in the data. According to the types of displayed data, we can divide the displayed data into composite data and non combined data. Therefore, the visualization of display data can also be divided into combined data visualization and non combined data visualization. According to the different data sources, we can divide the modern data visualization into two parts. The first part is generated by computer graphics processing, the other parts are generated by image processing technology, and then displayed in the form of interaction [4].

The purpose of data visualization is to convey and exchange information to the outside world, which is generally realized by means of graphics. However, this is not to say that data visualization will become tedious because of its function, or extremely complicated because of its colorful appearance. How to effectively convey ideas and concepts, this is the difficulty of technology. In order to solve this problem, we need to develop aesthetic form and function simultaneously, so as to realize effective communication. Complex data sets are difficult to observe. This problem can be solved by data visualization. The method to realize this problem is to intuitively convey the key aspects and characteristics of data.

2.3 Big Data Storage Technology

With the continuous development of centralized monitoring system, the types and quantity of signal equipment monitored by the monitoring system are more and more. Coupled with the characteristics of real-time monitoring of the monitoring system, the amount of data collected is very large, which poses a great challenge to the system storage. Therefore, the system puts forward higher requirements for data storage, It is necessary to strengthen the system data storage function to meet the storage demand of the rapid growth of data (see the Fig. 2).

At present, the transportation equipment management system stores a lot of information related to the technical status of the equipment [5–7]. They store a lot of relevant data through the traditional data storage method. However, due to the independent construction and application of each professional system, there is a lack of unified data storage method between them, so there are great difficulties in data sharing and comprehensive application. With the vigorous development of big data technology, the data collected by professional transportation equipment management system is becoming more and more diversified and massive. At present, the integrated transportation equipment management system has been built and put into use, accumulating the amount of data up to Pb level. Analyzing the data types, most of the data types are semi-structured or unstructured, and are growing rapidly. It can be seen that in the face of the explosive growth of massive data, the traditional storage management technology has been unable to meet the growth needs of massive data processing, calculation and analysis as well as comprehensive application.

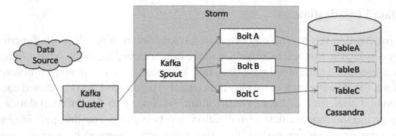

Fig. 2. Big data storage technology

3   System Function Realization Based on Big Data Mining

3.1   Time Series Mining Algorithm

There are many factors to determine the sample length, which are affected by many factors, among which the energy leakage effect of field signal and the resolution of different frequency harmonics play a decisive role. Because we can't analyze the infinite length sample, we can only intercept the limited part of the sample to analyze. According to the general rule, the sample length of the sampling value with complete operation time stage is the whole operation time stage of the equipment being monitored [8].

The calculation method of distribution density is to take V continuous segments from I to j, and then calculate the distribution density of effective data distributed on them. See Formula 1.

$$\rho_{ij} = \frac{\sum_{p=1}^{j} N_P}{\sum_{p=i}^{p=j} 1_{mp}} (i = j - v, 0 < j \leq m - v) \tag{1}$$

Where l_{mp} the length of the line m_p;

Calculation of estimated normal value:

The estimated normal value is obtained by calculating the effective sampling values in the interval with the largest linear density distribution.

The estimated normal value R_0 is shown in Eq. 2

$$R_0 = \frac{1}{m}(\sum_{i=i_0}^{i_0+v} \sum_i rs) \tag{2}$$

In the formula, m refers to the value of effective sampling points in the interval corresponding to the maximum linear density. It refers to not only the values of individual points, but also the values of all effective sampling points.

This paper briefly introduces the above-mentioned processing process, including the following three steps: the first is grouping, grouping the collected data by equipment, the second is preprocessing, and the third is comparative analysis. When the equipment is in normal state, the equal interval sampling curve is basically similar, and the collected data are compared and analyzed, Only the data distributed within the error range are valid values.

Because of equal interval sampling, so the sampling time is equal, then the change trend can be determined by the increment of R. That is, $K_{ij-ij+1}$ and K_{0j-j+1} are the same sign

$$K_{ij-ij+1} \times K_{0j-j+1} \geq 0 \tag{3}$$

It can be transformed into the following formula, as shown in formula 4:

$$(R_{ij} - R_{ij+1}) \times (R_{0j} - R_{0j+1}) \geq 0 \tag{4}$$

When one of them is 0, further determination is made according to the relative increment of amplitude.

4   Construction of Narrative Strategy Model in Visual Communication Design and Simulation

Visual communication design narrative is a narrative type different from the traditional language and literature narrative. It is very different from the traditional language and literature narrative in terms of historical development, narrative media, narrative perspective and space-time structure. After a thorough study of narrative theory, it is the focus of this paper to propose a complete and feasible narrative strategy [9]. Starting from the current development and problems of visual communication design and narratology, this chapter analyzes the internal and external factors involved in narrative text, and attempts to establish a feasible narrative strategy model is the focus of this chapter. Firstly, it analyzes the theme, media and discourse of narration from the inside of the narrative text. Secondly, it systematically studies the narrative of visual communication design from the external factors of the text, such as the influencing factors, evaluation indexes, objectives and tasks, and communication methods, so as to establish a set of universal and operable narrative design strategies, In order to provide some reference and guidance for the design narrative practice.

The construction of strategic model is a systematic project, and its influencing factors are an important research level that can not be ignored [10]. It is more conducive to the complete establishment of the strategic model to systematically sort out the relevant influencing factors before constructing the strategy model. The narrative of visual communication design is a qualitative research. The research on its influencing factors is different from the quantitative research, so it needs specific analysis. First of all, we need to analyze the relevant characteristics of visual communication design narrative in the communication process with the help of communication theory, and then summarize its influencing factors. (see Fig. 3) Secondly, on the basis of communication theory, combined with the specific text construction content of design narrative, this paper studies the relevant influencing factors of its existence, so as to provide support for the establishment of narrative strategy model.

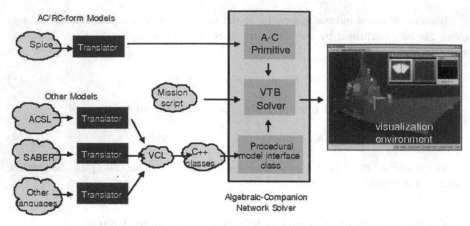

Fig. 3. Visual simulation model.

5   The Double Attribute of Character Symbol

As a process of expressing people's cognition and thinking, it has the function of expressing semantics and is a very important visual symbol system. With thousands of years of history, it presents two important functions of ideography and phonology, and finally integrates form, image and meaning. The final discourse relationship is highly abstract. Combining the phonetic function of language, it can achieve the function of language expression well [11–14]. At the same time, due to the expression of the form of words, it also has the aesthetic function of expressing the meaning through graphics, which is the problem of its graphic meaning and discourse system mode.

5.1   Linguistic Symbolic Content

It is self-evident that what kind of speech happens in today's media, and the importance of words in it is to achieve the transmission of the meaning of words in a way of carrying. Saussure, a Swiss linguist, once said that words and speech are the two main components of language, which have to go through historical evolution [15]. The language system used by people from generation to generation is what people say, which involves many factors such as words and grammar, among which some standards and norms have been established. Language is an important form for individuals to express what they feel and think in society. Language itself needs to be presented through words and words. The essential difference between auspicious language and characters lies in the visual form of reading. However, due to the influence and function of pronunciation and meaning, the function of speech is an important way to express vision and convey the core connotation of design. At the same time, it also has the function of assisting and supplementing the image system. Even if this form of expression of graphic meaning has the image and decoration, it also needs to be carried out on the basis of the combination of language and words.

5.2 Visual Symbol Form

Because characters have their special visual symbol form, they can be expressed by visual carrier, which requires people to understand the rules of characters and technology. In the visual design, the symbolic text expresses its content perfectly through its presentation form, and the text also has a perfect interpretation of the graphics. Man is a kind of visual animal. With the help of its unique form of expression, words can make people produce specific emotions [16]. The image of words and the expression of its meaning are two inseparable parts. From the perspective of semiotics, that is to say, the meaning of symbols and the expression of symbols are two interdependent attributes. If a symbol or character can't express its emotion and meaning perfectly, it has no soul, no culture and can't influence people's positive energy, and the character is just an empty body; similarly, if it only has soul, no external form, the soul has no place to place, and can't attract people's attention in form, Only with boring content is not perfect text form.

6 "Language" Art of Visual Communication

Text has its unique way of expression, so the value of visual communication design can not be replaced. Text can mark the graphics through its language expression ability, and also can complement the visual graphics. This complementary creative way provides more innovation space and more possibilities for the creators.

6.1 Graphics of Characters

Through the structure of the whole system of characters and the embodiment of various shapes, the graphitization of characters can be reflected, which includes seal cutting, printing, writing, technology and other forms. There are two aspects to the graphic trend of Chinese characters. The first is that the rules of creating Chinese characters in our country can reflect the image meaning of Chinese characters [17, 18]. From the Chinese character symbol system, it can be concluded that for some European and American countries, their early development tends to phonetic notation, which is difficult to contact with the relevant historical origins, The value of Chinese character form can be traced back to the source of historical development. Through the simulation of various natural objects, it can reflect the establishment mode and logic of the overall structure of Chinese characters. Nowadays, Chinese characters present a relatively simple way of expression, with a certain degree of cognition. With the transformation from anti lock to simple mode, the typical characteristics of Chinese characters have been preserved, and even the expressions related to logic and pictophonetic escape have appeared. This is also an important form of visual creative expression of characters. The decoration and visualization design of characters eventually promote the development of characters in the direction of graphics, and the constituent elements mainly include image elements, decorative patterns and characters. With the help of decorative images, the image has stronger appeal and embodiment. In addition, through the expression of traditional Chinese characters, it is not confined by decorative pictures and images. In the category of "diagram", it also includes the overall arrangement of diagrams, charts and pictures, sometimes even completely composed of words [19]. This is not only in line with the concept of today's schema, but also reflects the distinction between graphic recognition, graphic theory and graphics. The final visual effect is more prickly, and also has the expressive power of the superposition of graphic information.

6.2 Creating the Schema of "Meaning"

Creativity, as we often say, does not only refer to the innovation on the surface, but also the expression of new concepts in visual design. It is a visual expression method with new significance. In general, the meaning of text design is often reflected in the combination of text and pictures, which also shows the complementary situation of text and pictures. Words and pictures are inseparable, and also an important part of the combination of noumenon and metaphor [20–23]. As shown in Fig. 4. The combination of words and pictures is based on pictures and assisted by words, which has the largest proportion in the design. The text only plays a certain auxiliary role in the design of the picture, including the content of the title and accompanying text. The text of the title is often innovative, and it will also express the meaning of the title to a certain extent.

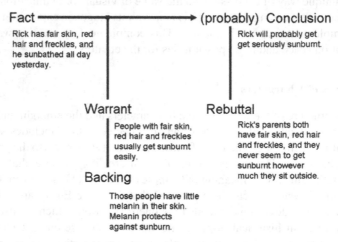

Fig. 4. The schema of creating "meaning".

7 Conclusion

Visual communication design narrative is an important part of contemporary visual narrative. It is different from classical narratology in terms of narrative media, choice of narrative perspective, and temporal and spatial structure of narrative. Its narrative research is also constantly absorbing the research results of other disciplines, gradually expanding the expression form and communication mode of narrative, and with the progress of social science and technology, it is still in the process of continuous development. As a member of modern visual narrative, visual communication design narrative has almost all the characteristics of modern visual narrative, and is an open narrative category. It always meets new technology and new ideological trend with an open attitude, and constantly changes and develops the narrative form and expression form.

The evolution of human history is closely related to the evolution of human history. However, in the process of its development, with the continuous change of narrative media, the narrative form has undergone great changes, which has played a different impact on people's lives in different periods. In the primitive period, visual narrative is an important means of communication for human life [24]. With the emergence of characters, the shortcoming of insufficient linear time of visual narrative is magnified and gradually replaced by language narrative with linear characteristics. Language narrative began to become the mainstream of narrative, and visual narrative gradually changed from the mass narrative behavior in the primitive period to the artistic behavior of the minority. With the development of mechanical reproduction technology and the maturity of photography and film technology, visual narrative began to revive in an all-round way, and became the mainstream of the times, setting off a new wave of visual culture.

References

1. Jianyi, C.: Research and application of intelligent operation and maintenance platform for big data of Telecommunications. Railway Commun. Signal **55**(11), 54 (2019)
2. Zheng, X., Feng, Q.: Real time fault diagnosis expert system based on fuzzy neural network inference. Comput. Eng. Appl. **42**(3), 226–229 (2006)
3. Keqiang, L.: Research and application of turnout monitoring system for high speed passenger dedicated line. China Railway **4**, 38–41 (2009)
4. Liang, G.: Construction and application of big data in safety production command center. Railway Commun. Signal **53**(7), 32–42 (2017)
5. Di, W.: Enlightenment of pop art on strengthening youth ideological guidance through visual design – Based on the introduction of pop art development experience in the East and the West. Chin. Communist Youth League **07**, 59–60 (2017)
6. Li, G., AI, H., Cheng, J.F., Yang, X.: Font form design. People's Posts and Telecommunications Press, 201705.137
7. Xiaochen, D.: On the development of visual design style in UI design. Drama House **07**, 171 (2017)
8. Huang, J.: Ways to improve user viscosity of fitness app. Nanjing Academy of Arts (2017)
9. Yizhiyuan, emotional design for children [d.2017]
10. Zichao, X.: On the development of Chinese characters and modern visual design. Popular Lit. Art **23**, 111–112 (2016)
11. Yang, F.: Development of Fangcun visual design art. Hunan Normal University (2016)
12. Tao, Y.: Research on precise design of life service app interface based on user micro experience. East China Univ. Sci. Technol. (2016)
13. Hua, G.: Application of "development of traditional patterns" in modern visual design. Comput. Knowl. Technol. **12**(15), 210–211 (2016)
14. Ying, W.: Visual design of spatial cognitive educational games based on situational development. Jilin Educ. **21**, 4–5 (2016)
15. Zhong, X.: Research on the application of emotional design in mobile app interface. Southwest Jiaotong Univ. (2016)

16. Liu, Z.: Research on application rules of mobile UI visual design. Xi'an Acad. Fine Arts (2016)
17. Qingfeng, L., Wang, Y.: Research on perceptual cognition in visual design based on Mobile Music App. Grand View Fine Arts **01**, 122–123 (2016)
18. Xinghai, C., Huan, Y., Haijin, L.: Research on the development of mobile interface visual design aesthetics based on efficiency. Packag. Eng. **36**(16), 107–110 (2015)
19. Wang, J.: Research on the development and application of mobile terminal icon design. Northwestern Univ. (2015)
20. Jinbiao, Z., Xiaomei, Z., Ge, J.: Newspaper visual design under the development of media convergence. Design **05**, 125–126 (2015)
21. Xia, Y.: Development, evolution and reconstruction of visual design foundation course. Tianjin University (2014)
22. Yuan, Z., Hua, F.: Current situation and development trend of visual design of teaching website. Manage. Obs. **18**, 76–77 (2013)
23. Guan, M., Zhou, F.: On the development trend of mobile phone interface visual design multi dimensional design. Beauty Times **01**(02), 106–108 (2013)
24. High quality sustainable development concept and international exchange of visual design -- quoted from Helmert Langer's personal homepage. J. Suzhou Inst. Arts Crafts, (01), 15–16 (2012)

Digital Combination Prediction Model for Online English Teaching at the University of Iowa in the Context of Big Data

Ying Huang^(✉)

Wuhan Institute of Design and Sciences, Hubei 430200, China

Abstract. For many years, effectiveness English teaching, a large number of educators and front-line College English teachers have been working hard, and have made achievements, but also exposed some problems. The ecological teaching mode based on cloud computing platform combines cloud computing and ecological teaching concepts, which can help teachers abandon the traditional "filling" teaching mode, support autonomous learning, achieve the balance of educational resources, and inject new vitality into the traditional audio-visual teaching methods.

Keywords: Cloud computing · College English · Ecology

1 Introduction

Traditional teaching methods are. On one hand, the classroom capacity is limited, on the othe hand, the utilization of teaching resources is seriously insufficient. These deficiencies are particularly obvious in the teaching practice and experiment. In recent years, cloud computing has developed rapidly because of its service-oriented resource delivery mode and dynamic and easy access to expand and integrate online resources. The combination of cloud computing and online teaching, especially practical teaching, will bring opportunities [1].

2 Design of Online Teaching Practice Platform

2.1 Object Model Design of Online Teaching Practice Platform

The object model of online teaching practice platform provides a flexible and extensible object model for online teaching practice scene, which is the basis of subsequent online teaching practice platform design. The design of the object model of online teaching practice should be based on two considerations. First, the model can achieve rapid response and complete efficient interaction between users and platforms. Second, the model can customize different component capabilities to meet the diverse needs of users.

M. A. Jan and F. Khan (Eds.): BigIoT-EDU 2021, LNICST 392, pp. 233–237, 2021.
https://doi.org/10.1007/978-3-030-87903-7_30

2.2   Recommended Function Design

The learning resource recommendation module will first split the professional training requirements, and cooperate with the user management module to obtain three key information of users, namely, school, major and grade. Through the three keywords to refine the user profile, through the text analysis of professional training requirements to extract the current user keyword list, the keyword list will also be used as the classification navigation of Learning Resource Recommendation page. Users can click different keywords to filter out the recommended content list under the corresponding keywords (Fig. 1).

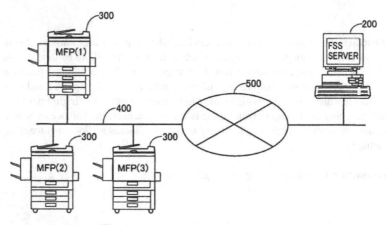

Fig. 1. Recommended function design

Keyword extraction here is based on the recommended content list. Text is similar to page rank algorithm. It is a graph based sorting algorithm [2].

Page links between web pages. Through a voting mechanism, links between web pages are interpreted as votes between pages to determine the importance of pages. Therefore, the implementation formula of page rank algorithm is as follows:

$$S(V_i) = (1 - d) + d \times \sum\nolimits_{j \in In(V_i)} \frac{1}{|Out(V_j)|} S(V_j) \tag{1}$$

coefficient, which is generally taken as 0.85. Since the page rank value of the web page is not zero, it is necessary to introduce the damping coefficient. The text rank algorithm takes words as nodes, establishes the connection by establishing co-occurrence relationship between word nodes, and introduces the concept of weight for edges. When the algorithm is used to extract keywords, we regard words as nodes, so there is no similarity between two words. Generally, the weight between words is set to 1 by default, so the algorithm is basically similar to page rank algorithm. The principle of text rank algorithm can be explained as that the importance of word I is determined by the sum of the weight of word I and the word J before word I and the weight between word J and other words. The implementation formula of text rank:

$$WS(V_i) = (1 - d) + d \times \sum\nolimits_{V_j \in In(V_i)} \frac{\omega_{ji}}{\sum_{V_k \in Out(V_j)} \omega_{jk}} WS \tag{2}$$

The text rank algorithm first splits the given text into whole sentences. After the split sentences are segmented, the stop words are filtered out. composed of the keyword candidates generated in the previous step. Then, the connection between nodes is constructed by co-occurrence relationship. The weight of each node is initialized by the formula to measure the importance in page rank, and the iterative calculation is continued until convergence. The weights of the converged nodes are arranged in descending order to get the T words with the highest weight and generate the initial candidate keyword pool. Finally, the words are marked in the original text, and added to the keyword sequence.

There is no significant difference between the experimental class and the control class. In the case of no significant difference in the success rate of serve, the students in the experimental class mastered the details of serve better than the students in the control class, and the deduction points in the technical score were significantly less than those in the control class: in the batting action, most students in the control group lacked the action of pressing wrist and pushing at the moment of batting, which made the ball spin; only with the help of the strength of the arm, the students in the control group did not have the action of pressing wrist and pushing at the moment of batting, The students in the experimental class mostly use the power brought by kicking and turning to close the abdomen, and then cooperate with the coordinated force of the whole body, so that the ball is easier to cross the net, but also increases the power and aggressiveness of the service.

Implementation of online teaching practice platform.

2.3   Overall Implementation Scheme of the Platform

The overall architecture scheme and functional module division of online teaching practice platform have been introduced in detail [3]. Based on the previous introduction, this section will first introduce the overall implementation scheme of online teaching practice platform. The platform code implementation structure is shown in Fig. 2.

2.4   The Concrete Level Division of Code Realization of Online Teaching Practice Platform

1) Web page presentation layer
Through the web page and the user direct interaction, based on JSP, and using strust2 tag to complete the page data loading and rendering. The data comes from the action layer, and at the same time, the data input by the user will be transmitted to the action layer for subsequent business logic processing.

2) Dao layer
It is mainly responsible for the specific operation of entity objects, which can also be understood as the direct data interaction with the underlying database, mainly including data search, addition, editing and deletion.

3) Model layer
It is mainly responsible for data persistence, mapping data entity and database data by mapping relationship, and realizing interaction with database through object-oriented

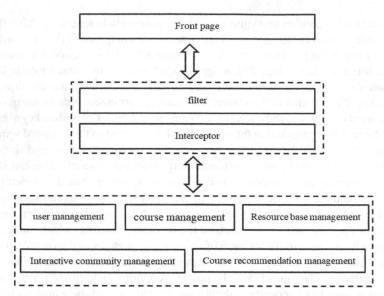

Fig. 2. Code implementation structure of online teaching practice platform

idea [4]. JPA is used as O/R mapping framework, and Hibernate is used to implement. The following sections will introduce the specific function module implementation.

The user management module provides three main functions: user login, user management and group management, including user login page, user management page (user list, new user, search user) and group management page (group list, group member list and group edit). This section will focus on these three main page implementations in detail.

As an important function module of online teaching practice platform, learning resource recommendation module selects the most suitable content for current users from the massive content through the aggregation of learning content, so as to realize module include three aspects: keyword extraction based on professional training requirements, content aggregation based on keywords and text similarity calculation of keyword content.

Professional training requirements come from the training programs of different majors. The are generally set according to the major and grade. Each major is required by different academic objectives, and there are differences in curriculum requirements in different grades of each major. The platform first carries on the text analysis to the university training plan, obtains the preliminary training plan keyword list. In the process of user registration, the platform will require users to fill in three information: school, major and grade, which can be modified at any time. Through these three key information platforms, the relevant content of professional training requirements will be positioned as the source of keyword extraction. Based on the textrank algorithm, the keyword list obtained from the keyword extraction of the training requirements will be used as the classification navigation of the Learning Resource Recommendation page. The navigation will change dynamically according to the modification of the school, major and grade recommended content under different keywords by clicking the navigation keywords.

In the teaching experiment, because the students in the experimental class learned MOOCS before class, they can quickly establish the initial movement representation in the initial stage of each technology. Teachers only need to explain the movements a little, students can quickly understand the essentials of the movements and practice, no longer need teachers to repeat in the classroom, for students to save learning time in the classroom, leaving more time to carefully polish their own movement technology, at the same time, it also increases the opportunity and time for teachers to guide students one-to-one. In the classroom practice of students, correct each student's movement, and provide guidance for students to further improve their movement skills, and flexibly master various sports skills in volleyball. Compared with the students in the experimental class, the students in the control class need teachers to repeatedly emphasize the basic essentials of the action in the classroom. Teachers also need to spend more time explaining the details of the action and demonstrating. At the same time, students need to spend more time in the classroom to practice in order to grasp the action essentials of various techniques. Class time is limited. Facing all students, teachers spend enough time on explanation and demonstration, and the time left for students to practice independently is correspondingly reduced. Teachers can not take into account the learning situation of each student.

3 Conclusion

The ecological English teaching based on cloud computing interaction, and sharing. It can break the dependence of traditional E-learning on computer technology, break through the limitation of time and space, bring students a new learning experience, stimulate students' learning autonomy, and form a teaching ecological chain of "teaching, learning, evaluation and reform". College English teachers should always bear in mind that students are the most important ecological subject. College English teaching should not only pay attention to students' language knowledge input, but also regard students as an organic life body. Therefore, more their emotions and complete life education should be given to students, so that the ecological English teaching can develop in a positive and healthy way.

References

1. Dai, W.: Building a high-level teaching reform exchange platform to promote foreign language teacher education and development. Foreign Lang. (5) (2010)
2. Wang, D.: A study on the validity of college English assessment. Chin. Foreign Lang. (2) (2010)
3. Wang, C., et al.: Technical Manual of Waste Gas Treatment Engineering. Chemical Industry Press, Beijing (2012)
4. Zhang, Z.: Photocatalytic oxidation/absorption process for treatment of malodorous waste gas from waste incineration power plant. Energy Environ. Protection (5) (2012)

Ecological Teaching Mode of College English Based on Cloud Computing Platform

Rui Zhang(✉)

Xi'an Fanyi University, Chang'an District, Xi'an 710105, Shaanxi, China

Abstract. Over the years, in order to improve the effectiveness of College English teaching, a large number of educators and front-line College English teachers have worked hard on the road of College English teaching reform, and have made some achievements, but also exposed some problems. The ecological teaching mode based on cloud computing platform combines the teaching mode of cloud computing and ecological teaching, which can help teachers abandon the traditional "cramming" teaching mode, support autonomous learning, meet the needs of students' personality development, realize the balance of educational resources, and inject new vitality into the traditional audio-visual teaching method.

Keywords: Cloud computing · College English · Ecological teaching mode

1 Introduction

In July 2007, the Ministry of Education officially promulgated the requirements for College English teaching, which has become an important guiding document for College English Teaching in China. Since then, colleges and universities in China have carried out the reform of College English Teaching in full swing.

Traditional teaching methods are limited by time and space. On the one hand, the classroom capacity is limited, on the other hand, the utilization of teaching resources is seriously insufficient. These deficiencies are particularly obvious in the teaching practice and experiment. In recent years, cloud computing has developed rapidly because of its service-oriented resource delivery mode and dynamic and easy to expand resource scheduling ability [1]. The combination of cloud computing and online teaching, especially practical teaching, will bring opportunities to solve the problems encountered in the current teaching practice.

2 Design of Online Teaching Practice Platform

2.1 Object Model Design of Online Teaching Practice Platform

The object model of online teaching practice platform provides a flexible and extensible object model for online teaching practice scene, which is the basis of subsequent online

M. A. Jan and F. Khan (Eds.): BigIoT-EDU 2021, LNICST 392, pp. 238–247, 2021.
https://doi.org/10.1007/978-3-030-87903-7_31

teaching practice platform design. The design of the object model of online teaching practice should be based on two considerations. First, the model can achieve rapid response and complete efficient interaction between users and platforms. Second, the model can customize different component capabilities to meet the diverse needs of users.

2.2 Recommended Function Design

The learning resource recommendation module will first split the professional training requirements, and cooperate with the user management module to obtain three key information of users, namely, school, major and grade. Through the three keywords to refine the user profile, through the text analysis of professional training requirements to extract the current user keyword list, the keyword list will also be used as the classification navigation of Learning Resource Recommendation page. Users can click different keywords to filter out the recommended content list under the corresponding keywords.

Keyword extraction here is based on text rank algorithm. Text rank algorithm is similar to page rank algorithm. It is a graph based sorting algorithm, which is based on page rank algorithm [2].

Page rank is an algorithm proposed by goge to calculate the importance of web pages by the number of links between web pages. Through a voting mechanism, links between web pages are interpreted as votes between pages to determine the importance of pages. Therefore, the implementation formula of page rank algorithm is as follows:

$$S(V_i) = (1 - d) + d \times \sum\nolimits_{j \in In(V_i)} \frac{1}{|Out(V_j)|} S(V_j) \tag{1}$$

d is the damping coefficient, which is generally taken as 0.85. Since the page rank value of the web page is not zero, it is necessary to introduce the damping coefficient d.

The text rank algorithm takes words as nodes, establishes the connection by establishing co-occurrence relationship between word nodes, and introduces the concept of weight for edges. When the algorithm is used to extract keywords, we regard words as nodes, so there is no similarity between two words. Generally, the weight between words is set to 1 by default, so the algorithm is basically similar to page rank algorithm. The principle of text rank algorithm can be explained as that the importance of word I is determined by the sum of the weight of word I and the word J before word I and the weight between word J and other words. The implementation formula of text rank algorithm for keyword extraction is as follows:

$$WS(V_i) = (1 - d) + d \times \sum\nolimits_{V_j \in In(V_i)} \frac{\omega_{ji}}{\sum_{V_k \in Out(V_j)} \omega_{jk}} WS \tag{2}$$

The text rank algorithm first splits the given text into whole sentences. After the split sentences are segmented, the stop words are filtered out, and only the words with the specified part of speech, such as nouns and adjectives, are retained to form the candidate word bank of keywords. Then the candidate keyword graph G = (V, E) is constructed, where V is the node set, which is composed of the keyword candidates generated in the previous step. Then, the connection between nodes is constructed by co-occurrence relationship. Only when the corresponding words exist in the window of length K. The weight of each node is initialized by the formula to measure the importance in page rank, and the iterative calculation is continued until convergence. The weights of the converged nodes are arranged in descending order to get the T words with the highest weight and generate the initial candidate keyword pool. Finally, the words are marked in the original text, and the adjacent words are combined into multi word keywords and added to the keyword sequence.

3   Implementation of Online Teaching Practice Platform

3.1   Overall Implementation Scheme of the Platform

The overall architecture scheme and functional module division of online teaching practice platform have been introduced in detail [3]. Based on the previous introduction, this section will first introduce the overall implementation scheme of online teaching practice platform. The platform code implementation structure is shown in Fig. 1.

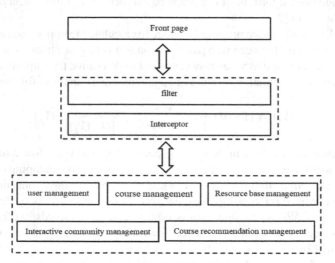

Fig. 1. Code implementation structure of online teaching practice platform

3.2   The Concrete Level Division of Code Realization of Online Teaching Practice Platform with Simulation

1) Web page presentation layer

Through the web page and the user direct interaction, based on JSP, and using strust tag to complete the page data loading and rendering. The data comes from the action layer, and at the same time, the data input by the user will be transmitted to the action layer for subsequent business logic processing.

2) Action layer

The action management layer is mainly responsible for the interaction between the front-end web page and the business logic layer. The action layer calls the relevant methods to process the requests from the front-end page by calling the interface provided by the business logic layer. The business logic layer will return the processed data to the action layer, and the action layer will return the result data to the front-end page to display the results.

3) Business logic layer
The business logic layer is responsible for the processing of the manager of data objects and the management of methods. Sometimes it can be understood as the encapsulation of Dao layer methods.

4) Dao layer
It is mainly responsible for the specific operation of entity objects, which can also be understood as the direct data interaction with the underlying database, mainly including data search, addition, editing and deletion.

5) Model layer
It is mainly responsible for data persistence, mapping data entity and database data by mapping relationship, and realizing interaction with database through object-oriented idea [4–7]. JPA is used as O/R mapping framework, and Hibernate is used to implement. The following sections will introduce the specific function module implementation.

The user management module provides three main functions: user login, user management and group management, including user login page, user management page (user list, new user, search user) and group management page (group list, group member list and group edit). This section will focus on these three main page implementations in detail (see Fig. 2).

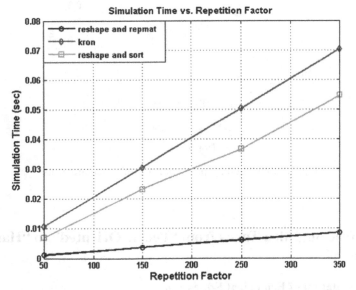

Fig. 2. Repetition factor and model layer for teaching platform

As an important function module of online teaching practice platform, learning resource recommendation module selects the most suitable content for current users from the massive content through the aggregation of learning content, so as to realize personalized learning resource recommendation. The main functions of learning resource recommendation module include three aspects: keyword extraction based on professional training requirements, content aggregation based on keywords and text similarity calculation of keyword content.

Professional training requirements (see Fig. 3) come from the training programs of different majors in Colleges and universities [8, 9]. The training programs are generally set according to the major and grade. Each major is required by different academic objectives, and there are differences in curriculum requirements in different grades of each major. The platform first carries on the text analysis to the university training plan, obtains the preliminary training plan keyword list. In the process of user registration, the platform will require users to fill in three information: school, major and grade, which can be modified at any time. Through these three key information platforms, the relevant content of professional training requirements will be positioned as the source of keyword extraction. Based on the textrank algorithm, the keyword list obtained from the keyword extraction of the training requirements will. The navigation will change dynamically according to the modification of the school, major and grade to meet the learning needs of students at different stages. Users can filter out the recommended content under different keywords by clicking the navigation keywords.

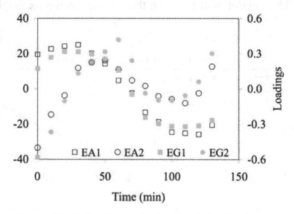

Fig. 3. Simulation for professional training requirements

4   The Teaching Idea Turns from "People Oriented" to "Harmony Oriented"

4.1   Interpretation of Ecological Education

The term "ecology" first appeared in ecology, which refers to the relationship between all animals and plants living in a certain region (or space) and between animals and plants and their environment. It emphasizes the mutual connection, interaction and functional unity of various factors in the system, and contains the meaning of system, overall connection, harmony, symbiosis and dynamic balance [10]. There are mainly three understandings of Ecological Teaching: (1) teaching and its theory developed under the guidance of ecological theory; (2) teaching about ecological theory, that is, the content of teaching is related to ecological problems, such as environment (environmental protection) course; (3) teaching developed for the purpose of practicing ecological theory, which mainly refers to teaching under the guidance of ecological theory, At present, these specific teaching forms are still in the construction of a text [11]. Therefore, it can be said that there is still a lot of room to explore the ecological teaching mode in university education. In such a social and historical context, how to cultivate the ecological talents of graphic design major, So that they can become all-round development people with modern wisdom, cultural people, people who seek insight through and understand the whole, and people who often get "sesame open discovery"? We refer to and draw lessons from the current ecological teaching theory, combined with the teaching experience of graphic design major, summarize and explore the following paradigms and methods.

4.2 People Oriented

The direct consequence of the proposition of "people-oriented" is the expansion of human selfish desire, the depletion of natural resources and the rampant prejudice of his life. Therefore, we must start from ourselves, give up the superiority of being human and turn our values to harmony [12–14]. As far as graphic design major is concerned, the changes brought about by the development of the current environmental situation will certainly bring strength to the creative work of graphic design. Although people may not believe that designers and creative people can change our climate and environment, they can really arouse people's awareness of environmental protection, which is also the embodiment of their own sense of social responsibility. On the one hand, they need to bring good suggestions to enterprises and find a balance between ecological and social impact. On the other hand, their task is to change the embarrassing situation of ecological concept, and integrate the real ecological concept into design practice and commodity publicity. In view of this, the teaching content of graphic design courses should keep pace with the times, Under the premise of making students learn to recognize, do things, cooperate and survive as the core of the education goal, the ecological design idea is penetrated into each core course. In particular, it should be emphasized that ecological design is also the embodiment of ethical values. Rather, it is the practical expression of the core concept and professional ethics that designers should uphold [15]. Therefore, it is necessary to create a new course – "graphic design ecology", which is based on the original teaching system. In this course, we mainly teach the graphic design related to the ecological environment, teach the concept of ecological design, as well as the effective ways and relevant cases to achieve the big design with "small cost", so as to establish the students' permanent, harmonious and ecological design concept.

5 Developing "Dynamic Three-Dimensional" Self-organizing Teaching with Ecological Teaching Form

5.1 The Importance of Dynamic Team Teaching

The so-called dynamic is to break the static and single teaching clue of the teacher, follow the principle of self-organization of the ecosystem, and show the situation of lively teaching content and interactive teaching links. To a certain extent, the self-organization, automatic and spontaneous organization of students is better than the interference and mandatory command of authoritative organizations [16]. As shown in Fig. 4. Therefore, the development of "self-help proposition" teaching form can maximize the promotion of students' creative initiative and the independence and expansibility of ideas. Since then, the author has been trying to apply this principle in teaching, among which the most prominent effect of "self-help proposition" teaching is the "folk art course" of graphic design major. In this course, the teaching leaders release more space for students. On the premise of grasping the teaching context and outline, students can combine their own interests to choose the subject content, such as "doll, paper-cut, batik calligraphy". After receiving the topic, the students immediately put in great enthusiasm to explore, from the static text and picture collation, to the acquisition of dynamic multimedia materials, are integrated with a unique perspective. In the process of learning and playing, students have already realized the connotation and true meaning of folk art, which is finally reflected in their major assignment of "folk element development and design".

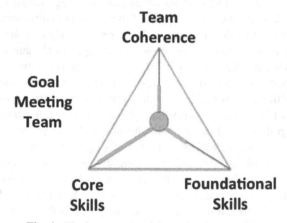

Fig. 4. The importance of dynamic team teaching

5.2 Ecological Teaching Subject "Students, Teachers and Students"

Two way dialogue means equal status and mutual respect, which is the most balanced and ecological attitude and accomplishment between the two objects. Under this premise, the teacher is no longer the instigator of knowledge, not the absolute speaker of discourse, but only the objective aspect of the teaching subject. It is true that the guidance of teachers is important, but the ecological teaching subject pays more attention to the relationship between teachers and students and the relationship between students and students [17–19]. That is to say, as the constituent factors in the system, they should be equal, interrelated, and interact to form an organic whole. In such a dialogue atmosphere, through the establishment of students' classroom, or 3° classroom, to give students more space and freedom as the main body of teaching, and continue the excellent "classroom in class" into "lecture after class", forming a systematic, integrated and personalized education mechanism of graphic design specialty. In this link, the graphic design major of Dalian University of technology is still in the stage of seeking a breakthrough in teaching practice, but the group teaching and discussion teaching methods in graphic design courses can still provide a realistic basis for this idea.

5.3 Ecological Teaching Platform, Attaching Equal Importance to "Humanity" and "Technology"

Graphic design, as a major closely related to social environment, natural environment and humanistic environment, is no longer a simple solution to design problems in the general sense. We should pay more attention to social humanities and natural disciplines in the adjustment of knowledge structure and curriculum system, and combine specialty setting and modular curriculum to meet social needs. In the past, most of the practical courses we emphasized were limited to professional classrooms, which is certainly a necessary link in teaching and practice. However, the closed classroom teaching is more about producing idealistic works, which not only has poor operability and neglects the problems of material technology and construction technology of design works, but also is empty and unrealistic, I love the self-expression of the design like twigs [20–24]. Out of the classroom, on the one hand, directly participate in the process of market practice and project implementation; on the other hand, deeply understand the social groups, experience the life of others, not only catalyze the sublimation of creative thinking, but also the emotional experience personally. At present, the teaching syllabus of graphic design major in Dalian University of technology has been adapting to this trend and constantly adjusting, with the course of "printing technology and process" as the representative, which focuses on the purpose of multi-dimensional teaching. For example, the teachers of graphic design use social resources to guide students to visit the printing factory, so that they can directly learn and communicate with the printing technicians, and strive to give students the most rare social experience. And the introduction of social service design projects into the classroom, or even the establishment of "Internship cooperation" and "manager lecture" relationship with some fixed design companies and printing plants, are beneficial to expand the teaching platform and produce good teaching results.

6 Conclusion

The ecological English teaching based on cloud computing has the characteristics of interaction, real-time and sharing. It can break the dependence of traditional E-learning on computer technology, break through the limitation of time and space, bring students a new learning experience, stimulate students' learning autonomy, and form a teaching ecological chain of "teaching, learning, evaluation and reform". College English teachers should always bear in mind that students are the most important ecological subject [25]. College English teaching should not only pay attention to students' language knowledge input, but also regard students as an organic life body. Therefore, more attention should be paid to the development of their emotions and complete life education should be given to students, so that the ecological English teaching can develop in a positive and healthy way.

References

1. Jigang, C.: Research on College English Teaching in China from the Perspective of Applied Linguistics. Fudan University Press, Shanghai (2012)
2. Dai, W.: Building a high-level teaching reform exchange platform to promote foreign language teacher education and development. Foreign Lang. (5) (2010)
3. Wang, D.: A study on the validity of college English assessment. Chin. Foreign Lang. (2) (2010)
4. Wang, S. Wang, T.: A survey of the current situation of college English teaching in China and the reform and development direction of college English teaching. Chin. Foreign Lang. (9) (2011)
5. Zhao, Y.: Research on ecological teaching mode of business English translation in the context of Smart Education. Sci. Technol. Horizon **29**, 73–75 (2020)
6. Lu, L.: Research on the construction of English ecological teaching based on immersion teaching mode. Educ. Teach. Forum **41**, 177–178 (2020)
7. Shen, P.: Research on the construction of college foreign language ecological classroom under the online live teaching mode. Sci. Technol. Commun. **12**(12), 124–125 (2020)
8. Yang, X.: On the construction strategy of primary school English ecological classroom. Campus English **17**, 204 (2020)
9. Zhao, Y.: Exploration on the practice teaching mode of children's infiltrating outdoor activities from the perspective of ecology. Theor. Res. Pract. Innov. Entrepreneurship **3**(4), 142–143 (2020)
10. He, M.: An empirical study on ecological college English blended teaching mode. Campus English **6**, 1–3 (2020)
11. Zhang, Q., Zhou, X.: Analysis on the construction of ecological teaching mode of marketing subject under maker education. Market. Circles **3**, 108–109 (2020)
12. Xiao, X.: Construction of ecological teaching mode of tea culture from the perspective of educational ecology. J. Sichuan Univ. Arts Sci. **29**(6), 140–144 (2019)
13. Yang, J.: Research on the construction of ecological English teaching mode in higher vocational colleges under the concept of blended learning. Overseas English **9**, 72–73 (2019)
14. Li, X., Yan, X.: Research on ecological teaching of Chinese language under internet environment. J. Shanxi Radio TV Univ. **24**(01), 18–22 (2019)
15. Zhang, Y.: The ecological construction of mixed mode of college English teaching under the background of "Internet plus". Think Tank Era (37), 218–224 (2018)
16. Luo, Y.: Approaches to the integration of information technology and education and teaching from the perspective of educational ecology. Modern Vocational Educ. **12**, 78 (2018)
17. Shi, Y.: Construction of ecological teaching mode of college English – taking Sanya university as an example. English Teacher **17**(23), 11–14 + 45 (2017)
18. Song, J.: Exploration and practice of ecological teaching of college English embedded in flipped classroom teaching. J. Heihe Univ. **8**(8), 135–136 (2017)

19. Ma, L.: Exploration on Ecological Model of High School Biology Classroom. Shandong Normal University (2017)
20. Deng, X.: Research on the reform of Ideological and political theory course in application-oriented colleges from the perspective of ecology. J. Chongqing Vocational College Electron. Eng. **26**(1), 147–151 (2017)
21. Zhao, R., Jin, Z.: Ecological research on interpretation teaching mode for business English majors in higher vocational colleges. Hebei Vocational Educ. **1**(1), 96–99 (2017)
22. Zeng, X., Liu, Z.: A preliminary study on the ecological teaching mode of college English under the network environment. Inform. Record. Mater. **17**(6), 123–124 (2016)
23. Jiang, H.: Maker education and college English teaching reform and innovation. J. Ningbo Polytech. **20**(4), 35–38 (2016)
24. Ji, H.: Application of ecological classroom teaching mode in the teaching of secondary vocational clothing specialty. Exam. Wkly. **26**, 167 (2016)

On the Application of Data Mining Algorithm in College Student Management

Xiaofei Sun[1](✉) and Yunan Zeng[2]

[1] Nanjing University of Finance & Economics, Nanjing 210023, China
[2] Nanjing Niuding Technology Co., Ltd, Nanjing 210023, China
9220080008@nufe.edu.cn

Abstract. With the rapid development of computer technology and Internet technology, information technology has been widely used in all walks of life. Big data, Internet of things, artificial intelligence and intelligent manufacturing have become the pronouns of the times. This paper mainly expounds the use of data mining for scientific and convenient management in college student management, which is more conducive to the healthy growth of students.

Keywords: Data mining · College student management · Application analysis

1 Introduction

With the development of information technology, information technology has shown its advantages in the management of colleges and universities. Under the background of information technology, the management of colleges and universities will be more standardized and scientific, Therefore, education authorities at all levels have regarded information technology as an important manifestation of the level of running a university. However, a new management model will always face many opportunities and challenges instead of the old one. The main problems are as follows: some older managers often have rich management experience and always think that they have experience [1]. The new management model is not easy to use, They will find all sorts of reasons to stop new management models. At present, most of the students in Colleges and universities are post-95 students. They are active in thinking, do not accept too restrictive management, and have a strong ability to accept new things and new technologies. In particular, they are good at computer and mobile phone operation, and the information management mode is relatively easy for them to accept. Although the school attaches great importance to information construction, it needs to invest a lot of human, material and financial resources, which affects the speed of information process.

M. A. Jan and F. Khan (Eds.): BigIoT-EDU 2021, LNICST 392, pp. 248–257, 2021.
https://doi.org/10.1007/978-3-030-87903-7_32

2   The General Process of Data Mining

2.1   The Role of Data Mining in University Management

The subjective factors in the management of students are still too heavy and lack of scientific analysis and basis. On the one hand, the managers manage the students according to the provisions of various documents, but the documents only stipulate various conditions, procedures or quota limits. If the students who meet the conditions exceed the quota limit, How to select the truly excellent students objectively and reasonably depends on the subjective judgment of the managers because there is no specific operation method in the management documents.

2.2   The General Process of Data Mining in University Management

Data cleaning and integration eliminate noise or inconsistent data, combine multiple data sources together, select and transform data, retrieve data related to analysis from database, and transform data into a suitable form for mining. For example, data mining is the basic step by summarizing or aggregating operations, and data patterns are extracted by intelligent methods. According to a certain interest measure, pattern evaluation can identify the really interesting patterns that represent knowledge [2]. Knowledge representation uses visualization and knowledge representation technology to provide users with mining knowledge. The general process of data mining can be shown in Fig. 1.

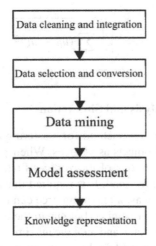

Fig. 1. Basic process of data mining

3   Fuzzy Mining Algorithm

In the process of data mining, data mining algorithm is the most important. By using fuzzy mining algorithm and data extracted from data warehouse, we can find the individual types existing in the current organization. In addition, we can judge which of these types an individual belongs to.

3.1   Pattern Discovery

The sample set to be classified is established on all data records of data warehouse. The object to be classified is called sample, such as sample set. The specific attributes should be quantified. The quantified attributes are called sample indicators, and there are m indicators. This is the case:

$$u_i = \{u_{i1}, u_{i2}, \cdots u_{im}\} \ i = 1, 2, \cdots, m \tag{1}$$

Therefore, these original data should be standardized, and the standardized value of each data can be calculated according to the following formula.

$$u_{ik} = \frac{w_{ik} + u_{ik}}{S_k} \tag{2}$$

Then the standard deviation of the original data is calculated according to the following formula:

$$S_k = \sqrt{\frac{1}{n} \sum_{i=1}^{n} (u_{ik} - u_k)^2} \tag{3}$$

3.2   Cluster Algorithm Analysis and Simulation

This paper uses the maximum tree method for clustering analysis, that is to construct a graph with all the classified objects as vertices. When $\neq 0$, the vertices can connect an edge. The method is to draw one of the vertices first, and then connect the edges in the order of row from large to small. It is required that there is no loop until all the vertices are connected. In this way, a maximum tree can be obtained. Each edge of the tree can be weighted by a certain number, but due to the different connection methods, the maximum tree can not be unique, and then the maximum number is taken into the cut set, that is, those weights are removed. In this way, a tree is cut into several disconnected subtrees [3]. Although the largest tree is not unique, the subtree is the same after taking the cut set. These subtrees are the patterns of inductive discovery in data warehouse. We use the cluster algorithm to simulate the manager system which is shown in Fig. 2. We can get thus results, with the increase time, the synchronous rate is increase. This also shows that our management system is safe and stable.

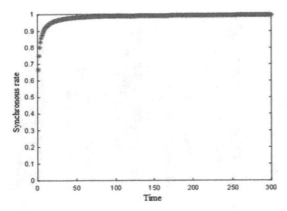

Fig. 2. Cluster algorithm and simulation for the manager system

3.3 Forecast

The first mock exam is used to get the average index of each mode and the average index is obtained by pressing the formula:

$$M_{ij} = \sum u_{kj}/P(i = 1, 2, \cdots, s, j = 1, 2, \cdots, m) \tag{4}$$

The total number of patterns. K is the total number of records that the pattern (i.e. the second pattern) is pushed out by in the warehouse. The first mock exam sample X and the first mock exam set the sample close to which model, and predict the whole situation from the overall situation of the model.

4 The Inevitability of Applying Big Data Technology to Student Management

In all colleges and universities, teaching is the center, and student management is the basis to ensure the work of teaching center. If the student management can not keep up with the teaching work, it can not be guaranteed, so the student management is related to the healthy and safe growth of students. The traditional management of students is often based on the experience of the management staff. The top-down meeting is used to manage the students. This way is often a long cycle and inefficient. As time goes on, it is not suitable for the management of the post-95 and post-2000 students. In the Internet age, contemporary college students acquire knowledge faster, and the way to acquire knowledge is also more. The way in the past is certainly not good. Through big data technology, students' life and learning situation in school are recorded, and each student's life and learning habits are analyzed through data mining, so as to better understand students' learning situation, mental health status and living habits, and achieve more accurate help, so that each student can live and study healthily and happily in school. When we add the noise to the system, we find the ARI has some changes which is shown in Fig. 3.

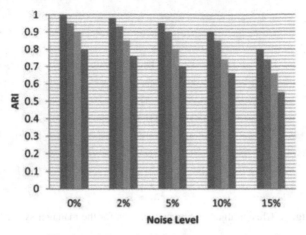

Fig. 3. Adding noise for the manage system

5   Application of Big Data Technology in Student Management

Colleges and universities are places where elites are concentrated, and all kinds of intellectuals are gathered. There are many kinds of data collected by the data center of colleges and universities, including not only relational data, but also non relational and semi relational data. These data must be stored in the data warehouse, The data platform is established to record the activity track of each student from getting up in the morning, brushing teeth, washing face, eating breakfast, reading early, doing morning exercises until the light off in the evening. The data stored in the data warehouse for a period of time is analyzed and modeled, Cleaning, integration, standardization, mining, finally mining out the students' learning situation every day, predicting when the students get up, when they go to the classroom, when they have three meals, when they turn off the lights to have a rest and so on [4]. No link is accurately recorded [5]. These information are used to classify the students, which students have good learning initiative and which students have good learning habits, Which students are fond of playing, which students are serious in class, which students have special skills, and so on. Through this information, we can help and guide students with unhealthy psychology through data mining, so as to make students healthy, Happy learning in school, the application of data mining technology in student management can accurately predict each student's learning situation, and the students' life performance from getting up to resting [6].

6   Establishment and Management of Data Warehouse

6.1   Data Extraction, Transformation and Loading

Data cleaning, transformation and loading is a very important stage in the process of data warehouse generation. In this paper, we extract the information of campus activities and academic achievements of grade 2005 students from the card information system, library information system and educational administration system of Beijing Jiaotong University [7].

The processing of missing values and irregular values of data, such as the basic information of students, is imported by various early systems. Because of the non-standard among various systems, null values and irregular values appear in this section. Therefore, only the filtering function can be used to filter out null values and irregular values for manual correction. Data impurities and inconsistent data should be treated differently according to the situation, and can not be deleted. For example, the average score of the course should be queried for each student of the class or grade or major. If the student's score in the course is −0.1 or −0.2, it will be considered as disqualification or not attending the course, and the course will be calculated as zero; if all the courses of the student are empty, If it is determined that the student's studies have changed and cannot be compared with other students as valid data, all data of the student will be deleted. Repeat the results, the same person has more than one score in the same course, if there are make-up examination and re examination results, take the results of the first examination [8].

Because the object of our analysis is the students' scores of the courses in each college, and the courses with the course type of public courses are the courses with the unified proposition and examination of the whole school, the scoring standards of these public courses are consistent. If the course type is the achievement of professional basic courses and professional courses, and there are differences in the scores of various majors in each college, if we want to make a comparative analysis of the students' scores of the whole school, We must first normalize some grades, such as introducing Z value in statistics to compare students' course grades, and then increase the grade of field grades. The definition of Z value is to surpass different grades, colleges and inconsistent scoring standards, and replace students' course performance with a standard value, so that students' course performance can be judged by the same standard. Because this paper analyzes the course scores of the students of the same grade in the same college, there is no need for Z-value conversion, that is, the student scores in the original educational administration system are available [9].

6.2 Building a Cube

After loading the business data to be analyzed into the data warehouse, it lays the foundation for meeting the needs of students' management and decision-making. The future operations are based on the data warehouse with business data. However, the multidimensional analysis of data is not mainly for data warehouse, but for the subset extracted from data warehouse, such as data mart and multidimensional data set (also known as data cube). Before analyzing the data, you need to create a cube based on the analysis topic. There are two ways to organize multidimensional data sets: star and snowflake. In this paper, the organization of multidimensional data sets is mainly star. In the star model, business data are concentrated in the fact table, so as long as you scan the fact table, you can query, and there is no need to associate multiple huge tables. At the same time, complex queries can be completed through the comparison of various dimensions, drill up and drill down operations. So using star structure can improve the performance of query [10].

7 Application System Analysis and Design

7.1 Design Goal

The system can analyze and process a large number of data in the campus information system, find the information that needs attention, and realize the conversion of original data to valuable knowledge. The number of records in campus information system is usually large and complex [11]. Therefore, it is a key problem to extract, filter, transform and integrate a large number of data in order to discover knowledge. At the same time, to enable users to participate in the process of analysis and mining, the system should have a good interactive function and friendly interface.

7.2 The Architecture of the System

The system architecture mainly includes the following five layers: data presentation layer: providing human-computer interaction interface, accepting the corresponding requests of users, and returning the corresponding query results to users [12].

Data analysis layer: process the corresponding data request of the upper layer. Multi-dimensional analysis module provides multidimensional data management environment and realizes OLAP function. Data mining module is used to discover the relationship between data and make model-based prediction. Data storage layer: realize data management, accept relevant data submitted by data processing layer, and save them to relevant data warehouse and multidimensional data set according to the logic set by the system. Other models support data mining. Data processing layer: it can extract, transform, clean and load data from different data sources. Data source: the data source of the system, mainly including the record data of students' dining, access control and Internet access in the all-in-one card system; the basic information of students in the educational administration system, course selection results and students' Graduation destination information; the borrowing record data of students in the library system.

7.3 Evaluation of Data Mining Results

A data mining system often obtains thousands of patterns or rules after running one (Group) mining algorithm. Association rule mining is a typical example. The execution result of association rule algorithm can get thousands of association rules even if it processes a small record set, such as tens of thousands of records. However, only a small part of these thousands of rules have practical application value. So how to evaluate the results of data mining effectively in order to get a practical model is very important. Usually, we can judge and screen from the subjective level of users and the objective level of the system. The subjective level mainly includes the following four criteria: easy for users to understand; potential value; novelty; and being able to determine the effectiveness of new data or test data. The objective criteria are mainly based on the structure and statistical characteristics of the mined patterns. For example, an objective evaluation criterion for association rules is confidence (probability), which indicates how trustworthy the rule is, that is, confidence $(a = > b) = P(BA)$. It is usually necessary to combine the two to find the really valuable and interesting knowledge. The establishment of data mining model generally includes two stages: the first stage is to establish the basic mining structure, in which the mining structure includes the mining algorithm and the input and output attribute set; the second stage is to optimize the parameters of the mining algorithm in the mining structure. The two phases use different data sets. Firstly, the training data set is used to try a variety of mining models, and then the test data set is used to evaluate these mining models to find the best mining model [13].

At present, the vendors of BI solutions include IBM, Oracle, Sybase, Microsoft and other database products. In addition, they also include Cognos, business objects and Brio, SAS, a company with data mining and advanced statistical tools, and NCR, a manufacturer focusing on large-scale data warehouse. Here is a brief introduction to the BI solutions of IBM, Oracle and Microsoft IBM:IBM The company provides a set of business intelligence (BI) solutions based on visual data warehouse, including isual warehouse (VW), Essbase/DB2 OLAP server, IBM db2udb, as well as front-end data presentation tools (such as Bo) and data mining tools (such as SAS) from third parties, VW is a powerful integrated environment, which can be used not only for data warehouse modeling and metadata management, but also for data extraction, transformation, loading and scheduling. Essbase/DB2 OLAP server supports dimension definition and data loading. Essbas/DB2 OLAP server is a hybrid of ROLAP and MOLAP. Strictly speaking, IBM itself does not provide a complete data warehouse solution, the company adopts a partner strategy [14]. For example, its front-end data presentation tools can be Bo of business objects, Roach of lotus, impromptu of Cognos or query management facility of IBM; Essbase supporting arborsoftware and DB2 OLAP server of IBM (jointly developed with arbor) in multidimensional analysis; SAS System in statistical analysis. Oracle: the architecture of oraclebi solution is divided into three layers: data acquisition layer. All ETL processes can be stored in Oracle10g database by the ETL script generated by Oracle Warehouse builder, which is a tool provided by Oracle data warehouse. According to the requirements of the data warehouse system, the data can be extracted and loaded into the data warehouse system regularly, Oracle 10g database realizes the centralized storage and management of various types of data in data warehouse system, and supports the storage of massive data by using partition technology [15]. In the data presentation layer, Oracle provides a new business intelligence solution, Oracle Bi EE, OLAP analysis and development tools (JDeveloper+bi beans) and data mining tools (Oracle data miner), which show the results of statistical analysis in various ways.

8 Epilogue

This paper focuses on the research and implementation of university student management decision support system based on data mining. In this paper, firstly, the necessity and feasibility of applying data mining technology to students' management decision-making are expounded. Then, it introduces the concepts of data mining, data warehouse, online processing analysis and their relationship, as well as the related mining algorithm and its key parameters applied in this paper. Then, the architecture of decision support system based on data mining, the source of data, the data preprocessing process of data extraction, transformation and loading (ETL), and the construction, processing and access of multidimensional cube and mining structure are described.

Finally, the analysis results of multidimensional cube and data mining model are introduced in detail, and the practical significance of these results in college student management decision-making is analyzed. In this paper, data warehouse, online analytical processing and data mining technology are applied to the decision support system of student management, which has a certain reference significance for the construction and development of teaching management decision support system. The research of this paper is based on the data of students in the computer college, and the conclusions and rules are universal and special compared with the whole school. Among them, students' dining, students' borrowing and the correlation of students' scores belong to the common characteristics of various colleges; students' Internet access, students' courses, students' clustering and the correlation of students' Graduation destination have their special characteristics.

Due to the limitation of time and ability, there are still some deficiencies in this study, which need to be further studied and improved, mainly in the following aspects: 1. This study only takes the students of a certain term in the computer college as the research object, so it is necessary to further expand the scope of students, such as the horizontal expansion in the scope of colleges, and the vertical expansion in the scope of students' grades. 2. The selection of related attributes. In the attribute selection of data collection, there are many other factors that do not affect students' performance, such as the analysis of students' professional interests, the analysis of students' learning objectives, and so on. The corresponding questionnaire is needed to count the information in this aspect. Therefore, the data information used for data mining may not be the best data set. 3. The mining algorithm is improved. This study uses the mining algorithm of the mining tool, and does not improve the corresponding algorithm according to the actual mining needs.

References

1. Chen, W.: Data Warehouse and Data Mining Tutorial. Beijing Tsinghua University Press (2006)
2. Yu, L., et al.: Principle and Practice of Data Warehouse. Beijing People's Posts and Telecommunications Press (2003)
3. Han, J., Kamber, M.: Data Mining Concepts and Techniques. Morgan Kanfmann Publishing (2000)
4. Wei, Z.: SQL Server Development Guide OLAP Beijing Electronic Industry Press 2001 Shen Zhaoyang SQL sever20ap solution. Beijing Tsinghua University Press (2001)
5. Xu, H.: Data Warehouse and Decision Support System. Beijing Science Press (2005)
6. Jian, P., Wei, C., Chang, Z.: J. Data Cube Algebra Softw. OLAP **10**(6), 561–569 (1999)
7. Li, J., Gao, H.: A multidimensional data model for data warehouse. Acta Sin. Sin. **11**(7), 200908–200917
8. Jia, J., Kamber, M.: Translated by Fan, M., Meng, X., et al. Data Mining Concepts And Technologies. Beijing Machine Press (2003)
9. Ian H. Witten EIBE Frank, New Zealand, translated by Dong, L., Qiu, Q., Ding, X., Wu, Y., Sun, L., Practical Machine Learning Technology for Data Mining. China Machine Press, Beijing (2006)
10. Hand, D., et al.: Principles of Data Mining. Beijing Machinery Industry Press (2003)

11. Oliviaparrrud: Data Mining Practice, translated by Zhu, Y. Beijing Machinery Industry Association (2003)
12. Li, X., Jin, F., Yu, H.: The architecture of decision support system based on data warehouse. J. Hefei Univ. Technol. (Nat. Sci. Ed.) (2003)
13. Data Mining Tutorial HTP/mdn2 Microsoft. COM/zh – Cnllibrary
14. Zhang, J.: Data Mining Extension Language DMX.Windowsitpromagazine International Chinese Edition (3) (2007)
15. Lv, W., Huo, Y., Lv, B.: Introduction and Improvement of c-2005. Beijing Tsinghua University Press (2006)

Research on Dynamic Evaluation of University Collaborative Education System Based on Global Entropy Method Under Big Data

Huang Muqian[(✉)]

Software Engineering Insitute of Guangzhou, Guangzhou 510990, Guangdong, China

Abstract. Based on the analysis of the essence of global entropy in collaborative education under big data, this paper constructs the evaluation model of university collaborative education from the overall perspective. Through the analysis of the regional innovation system from the perspective of "structure function", it is found that the dynamic evaluation of university collaborative education system. Cooperative education in Colleges and universities has a certain direction. Through this study. It expands the breadth and depth of the application of collaborative education in Colleges and universities.

Keywords: Global entropy · Cooperation · Education system

1 Introduction

With the increasingly fierce competition in the international market, innovation has become an effective way and means for all countries and regions to establish competitive advantages, enhance their comprehensive strength and enhance their international status. China has also promoted innovation drive to the height of national strategy [1]. In the report of the 19th National Congress of the Communist Party of China, the great decision of "improving the ability of independent innovation and accelerating the construction of an innovative country" was put forward. As an important part of the national innovation system, regional innovation system is an effective tool to enhance regional innovation ability and promote the sustainable growth of regional economy. It is also the foundation and foothold of building an innovative country. Comprehensive and scientific evaluation of regional innovation system and understanding of the operation of regional innovation system have important practical significance for clarifying regional development advantages, formulating scientific and effective regional science and technology innovation policies, and promoting the improvement of regional innovation ability.

2 The Formation of the Concept of Collaborative Innovation

In his speech on the 100th anniversary of the founding of Tsinghua University, General Secretary Hu Jintao (2011) stressed that colleges and universities, especially research-oriented universities, are not only an important base for training high-level innovative

M. A. Jan and F. Khan (Eds.): BigIoT-EDU 2021, LNICST 392, pp. 258–267, 2021.
https://doi.org/10.1007/978-3-030-87903-7_33

talents, but also an important source of innovative achievements in basic research and high-tech fields [2–4]. They should actively promote collaborative innovation through institutional innovation and policy project guidance, We should encourage colleges and universities to carry out in-depth cooperation with scientific research institutions and enterprises, establish strategic alliances for collaborative innovation, promote resource sharing, jointly carry out major scientific research projects, achieve substantive results in key areas, and strive to make positive contributions to the construction of an innovative country. Collaborative innovation is the first concept put forward by Peter gloor, a researcher of Sloan Center of Massachusetts Institute of technology in the United States. He believes that collaborative innovation is a network group composed of self-motivated personnel to form a collective vision and achieve a common goal through network communication and cooperation. Domestic scholars have also elaborated on the concept of collaborative innovation. Yan Xiong (2007) proposed that when collaborative innovation is enlarged to the macro level, the main operation form is industry university research collaborative innovation. Lin Tao (2013) analyzed from the perspective of synergetics that the university collaborative innovation system is composed of subsystems (universities, scientific research institutions, enterprises, etc.) and elements (talents, knowledge, technology, information, etc.) in the system, as well as the relationship flow between them. Different innovation subjects achieve synergy through resource sharing. The essence of collaborative innovation system in Colleges and universities is a grand innovation organization mode for the realization of major knowledge innovation and scientific and technological innovation. It is a multi collaborative network innovation mode with universities, enterprises and scientific research institutes as the main body and supplemented by government, intermediary organizations and service platforms (Li zuchao, 2012) [5]. The Collaborative Innovation led by colleges and universities clearly focuses on the major national needs, industry common key technologies and regional strategic industry development. Through resource integration, under the cooperation of the government, service institutions, finance and other relevant departments, it has made great progress and breakthroughs in theoretical innovation and technological research.

3   The Manifestation of Collaborative Innovation Dilemma

3.1   The Policy Environment of Collaborative Innovation is not Perfect

Collaborative innovation is a new scientific research management mode in China in recent years, and it takes colleges and universities to take the lead in applying for the "2011 plan" as the specific implementation mode. As it is an exploration process and requires colleges and universities to occupy a dominant position in collaborative innovation, it is difficult to carry out substantive collaborative innovation under the current situation of imperfect domestic policy environment. For this reason, Li zuchao (2012) pointed out that the policy environment suitable for collaborative innovation needs to be improved, which is embodied in the following aspects: at present, the domestic laws and regulations on collaborative innovation have not been issued or detailed, so it is difficult to have rules to follow; the intellectual property system is not clear enough, so it is difficult to quantify and implement the definition and distribution of achievements and interests; the intellectual property system is not clear enough; The government's policy guidance in personnel employment and assessment, tax loans and other aspects is still difficult to mobilize the enthusiasm of all parties involved; the construction of common service platform and resource sharing system needs to be strengthened [6]. He Haiyan (2012) believes that the current traditional mode of scientific research organizations in Colleges and universities in China is limited to cooperation within the University, and there is a lack of scientific research platforms, channels and means for resource sharing among colleges and universities and between colleges and research institutes; there is a lack of strategic cooperation between colleges and enterprises, and it is difficult to clearly agree and implement responsibilities and rights.

3.2 It is Difficult to Cooperate in the Process of Collaborative Innovation

The members of collaborative innovation come from universities, scientific research institutes, government (industry) departments and local (industry) enterprises. There is a lot of room for improvement in the aspects of subject confirmation, leadership role positioning and task division. Li zhongyun (2011) believes that the crux of the difficulty lies in the differences in the purposes of various units. Colleges and universities attach importance to scientific research achievements, enterprises emphasize technological breakthrough, local governments need economic benefits, and participants also have great differences in innovative ideas and values. Ma Zhiqiang (2012) proposed that school enterprise collaborative innovation is a sincere cooperation process in which both sides complement each other to achieve greater innovation benefits. However, due to the influence of moral hazard factors such as asymmetric information and opportunistic behavior, school enterprise collaborative innovation has the typical characteristics of dynamic game [7, 8]. Considering the evolutionary game analysis of school enterprise collaborative innovation, this paper explores the conditions for realizing the optimal game state of both sides, Construction of university service value model based on the background of University Enterprise Collaborative Innovation.

4 Path Selection of Collaborative Innovation

4.1 Define the Orientation of All Parties and Establish a New Flexible Organizational Structure

In the collaborative innovation system, we should first make clear the positioning of all parties: colleges and universities are the subject of collaborative innovation responsibility for basic research oriented to the frontier of science and technology; industry departments (enterprises) are the subject of collaborative innovation responsibility for application research oriented to industry common key technologies; local governments are the subject of collaborative innovation responsibility for achievement transformation oriented to regional industrial transformation and upgrading. In order to carry out the above three types of collaborative innovation, universities need to establish flexible boundaryless organizations: first, break the barriers of cooperation within the University, realize the cooperation within the University, build interdisciplinary research platform, and realize data sharing; second, communicate with scientific research institutes, build laboratories and R & D bases, share software and hardware resources and information network, and jointly declare major scientific research projects; The third is to build industry university research alliance and scientific and technological achievements Incubation Park with off campus units (Li zhongyun, 2011) [9–12]. In the flexible organization, involving personnel from different units and systems, the information superhighway can be used as the connection way and hub to strengthen the communication between the personnel of all parties and break through the differences.

4.2 Pay Equal Attention to Basic Application and Promote the Connection Between Disciplines and Industrial Clusters

Discipline specialty industry chain is an effective carrier of collaborative innovation in Colleges and universities. Colleges and universities should plan and allocate professional talents and educational resources around the needs of industrial clusters, and develop and expand the disciplines and subject clusters urgently needed by industrial clusters. In this regard, Luo Weidong (2012) proposed that it is an inevitable trend for colleges and universities to promote the new mode of combination of industry, University and research, which is mainly characterized by collaborative innovation of discipline groups and industrial clusters. In the cooperation of industry, University and Research Institute, the subject cluster and industrial cluster form an alliance in mutual cooperation, so that the production factors such as labor force, capital, information and technology can flow reasonably among enterprises, universities, scientific research institutions and other organizations, cross coordination and complementary advantages, so as to form an organization with expansion power: emphasizing the combination of disciplines, majors and social needs for expansion; Take the platform management as the specific governance mode, play the intermediary role of the platform, and enhance the expansion ability through the integration of disciplines, majors, funds, production factors and other resources (Hu CHIDI, 2012). Tang Anbao (2012) analyzed the current situation and existing problems of subject clusters in Jiangsu Province, and proposed to make full use of the advantages of existing subject clusters in southern Jiangsu and Nanjing, connect with industrial clusters, build a collaborative development platform of subject industry with universities as the main body, and consolidate the manufacturing subject cluster, which is the strongest and largest subject cluster in Jiangsu Province, At the same time, we should increase the cultivation and investment of relatively weak subject clusters in agriculture, forestry, animal husbandry and service industry.

5 Entropy and Performance of Regional Innovation System

Entropy is the first concept in physics, and then it is gradually applied to the field of social science through the continuous research and development of scholars. Therefore, to clarify the meaning of entropy in the performance evaluation of regional innovation system, it is necessary to sort out and analyze the development process of entropy theory.

5.1 The Concept and Development of Entropy

The concept of entropy originated from the classical thermodynamic theory. In 1865, Clausius first proposed the concept of entropy when he studied the Carnot theorem [13]. Then, in 1877, Austrian physicist Boltzmann further clarified the statistical properties of the second law of thermodynamics on the basis of Clausius entropy. In 1948, American mathematician Shannon, the founder of information theory, introduced the concept of Boltzmann entropy into information theory in his article "mathematical principles of communication", and put forward the concept of information entropy, pointing out that information entropy is a measure of the uncertainty or disorder degree of random events, that is, information entropy:

$$H(X) = - \sum_{i=1}^{n} p_i \log_c p_i \tag{1}$$

At present, information entropy is widely used by scholars. Information entropy theory extends the concept of entropy from physics research to random event set of any level and category, greatly expanding the meaning and application scope of entropy. Therefore, information entropy, also known as pan entropy or generalized entropy, is widely used in sociology, economics, informatics, life science, chemistry and other fields, In order to measure the uncertainty or disorder degree of any system operation, this paper will also use information entropy to build the structural performance evaluation model and functional performance evaluation model of regional innovation system.

5.2 The Significance of Entropy in RIS Performance Evaluation

From Clausius entropy to Boltzmann entropy and then to information entropy, the research object of entropy gradually develops from closed system to open system, from equilibrium state to non-equilibrium state, from thermodynamic process involving energy transformation to non thermodynamic process, and from equal probability event to non equal probability event [14]. The Jordan condition and application scope of entropy are constantly released and expanded. Although there are some differences among Clausius entropy, Boltzmann entropy and information entropy in concept, their connotation comes down in one continuous line, that is, entropy is a measure of the degree of "uselessness, disorder, confusion or uncertainty" of the system. At present, the most common method used by scholars is information entropy, which measures the degree of dispersion of index data by calculating the information entropy value of the index, and determines the index weight according to the amount of information carried by the index data. It can be said that entropy has great advantages in determining the index weight, but few scholars pay attention to the "system structure information" carried by entropy, such as the system structure order degree information reflected by entropy (Fig. 1), and no scholars clarify the essential meaning of entropy (rather than entropy weight) in the performance evaluation of regional innovation system.

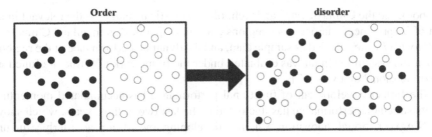

Fig. 1. Diagram of ordered structure and disordered structure

6 Performance Evaluation Model of Regional Innovation System

6.1 Construction of Evaluation Index System

Regional innovation system is a complex self-organization system. Considering the collectability and consistency of index data, this paper deconstructs regional innovation system into knowledge innovation subsystem, technology innovation subsystem and innovation support subsystem. Among them, the main body of the knowledge innovation subsystem is universities and various scientific research institutes, which are the leading force of knowledge innovation, while the main body of the technological innovation subsystem is enterprises. They apply the new knowledge formed by the knowledge innovation subsystem to the production practice of enterprises, form new technologies and new processes, and realize the market value of innovation, Innovation support subsystem provides environment guarantee and policy support for knowledge innovation subsystem and technology innovation subsystem.

6.2 Structural Performance Evaluation Model

If there are n evaluation indexes and M evaluation objects, there is a cross section data table x every year

$$X = (X^1, X^2, X^3, \cdots, X^T), X^T = x_{ij}(t) \tag{2}$$

Where: $x_{ij}(t)$ represents the j-th index value of the regional innovation system in year t.

7 Empirical Analysis on Performance Evaluation of Regional Innovation System and Simulation Analysis

7.1 Empirical Analysis

In the aspect of data collection, there is a big difference in the statistical indicators between the science and technology statistical yearbook before 2009 and the science and technology statistical yearbook after 2009. Considering the consistency and comprehensiveness of the evaluation indicators, according to the constructed regional innovation system performance evaluation index system [15, 16]. From the China Statistical

Yearbook and the China Science and technology statistical yearbook, the relevant index data of 31 provinces (autonomous regions and municipalities) in mainland China from 2009 to 2017 were selected as sample data, and the dynamic performance of the regional innovation system in China was evaluated under the framework of "structure function" two dimensional analysis [17].

The further development of functional performance has accumulated momentum, and its structural performance has fluctuated in the last two or three years, which shows that Ningxia regional innovation system is developing towards a new orderly structure. Tianjin regional innovation system represents another kind of evolution track, that is, with the improvement of system function performance, its structural performance is also rising slowly, which is a very unique evolution mode. This shows that Tianjin regional innovation system has great development prospects. Under the background of the great historical opportunity of coordinated development of Beijing, Tianjin and Hebei, Tianjin regional innovation system continues to deepen regional collaborative innovation of science and technology, By optimizing the ecological environment for scientific and technological innovation, striving to improve the transformation ability of scientific and technological achievements, and giving full play to the first role of talents, Tianjin regional innovation system is being built into an important source of independent innovation and an important source of original innovation in China.

7.2 Simulation Analysis

In this paper, we use entropy as the main research tool, and use entropy in teaching evaluation, so we can get the distribution of teaching evaluation model. we can get the following conclusions: In teaching evaluation, the whole evaluation presents normal distribution, so the simulation results shown in Fig. 2 meet this requirement In teaching evaluation, we hope to make an evaluation around a certain center, and Fig. 3 is also in line with the expectation. In Fig. 3, the blue part we see is the one with high evaluation [18–23]. This shows that the method we use is correct and positive.

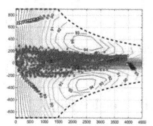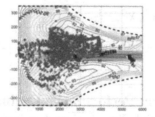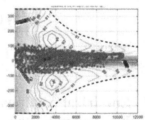

Fig. 2. Entropy distribution

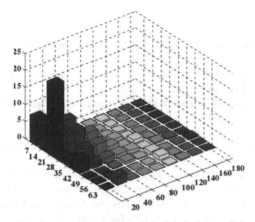

Fig. 3. Algorithm evaluation simulation model

8 Conclusions

Entropy represents "structural potential energy" of regional innovation system, and measures "prospect", "space" and "possibility" of system structural evolution. The higher the entropy value is, the smaller the dispersion degree of the index performance level of each subsystem of the regional innovation system is, the closer the structure of the system is to the equilibrium state, and the less space and possibility for the system to evolve. On the contrary, the smaller the entropy value is, the more discrete the index performance level of each subsystem of the regional innovation system is, the more the structure of the system deviates from the equilibrium state, and the system will further develop and evolve in a certain direction under the influence of "structural potential energy", until it approaches the equilibrium state again and forms a new system structure.

Restricted by various historical factors, the functional performance of China's regional innovation system is still "high in the East and low in the west". The functional performance of the regional innovation system in the eastern provinces is generally high, followed by the central region, and the western region is the lowest. However, the functional performance of the regional innovation system in the northeast region is similar to that in the central region. From the perspective of development trend, the average functional performance of the eastern region, the central region and the western region shows a rising trend year by year, especially the central region, which has the fastest growth rate, rapid development and obvious catching up momentum, while the average functional performance of the regional innovation system in the Northeast region shows a declining trend year by year, Northeast China is changing from a region with relatively high innovation performance to a region with relatively low innovation performance.

Acknowledgements. This work is supported by 2017 Project of Specialty Construction of "Logistics Management" in Teaching Quality and Teaching Reform Project of Undergraduate Universities in Guangdong Province ([2017] No. 214 Document of Division of Higher Education, Department of Education, Guangdong Province).

References

1. Wen, J., Feng, G.: Heterogeneous institutions, enterprise nature and independent innovation. Econ. Res. **47**(3), 53–64 (2012)
2. Bai, J., Jiang, F.: Collaborative innovation, spatial correlation and regional innovation performance. Econ. Res. **50**(7), 174–187 (2015)
3. Liu, S., Guan, J.: Evaluation of innovation performance of regional innovation system. China Manag. Sci. (1), 76–79 (2002)
4. Schrodinger: What is Life. Peking University Press, Beijing (2018)
5. Lin, J.: Construction of future institute of technology: cultivation of future technology leading talents. Educ. Res. Tsinghua Univ. **42**(01), 40–50 (2021)
6. Zhang, Q., Lei, L., Zhang, Z.: Research on the cooperative education mechanism of the two major classes of innovation and entrepreneurship education in colleges and universities – taking the "three innovation" space of Nanjing Agricultural University as an example. Heilongjiang Animal Husbandry Vet. **4**, 145–149 (2021)
7. Wang, Y.: A study on the education mode of English majors in local undergraduate universities through the cooperation of government, university and enterprise. J. Anshan Normal Univ. **23**(1), 71–74 (2021)
8. Liao, B.: Characteristic analysis and development path of demonstrative vocational education group – based on the analysis of the first batch of cultivation units of demonstrative vocational education group (alliance). Vocational Tech. Educ. **42**(6), 45–49 (2021)
9. Hou, X.: Practice characteristics and 2035 strategy of collaborative development of vocational education in Beijing Tianjin Hebei. Vocational Tech. Educ. **42**(6), 57–61 (2021)
10. Li, J.: Curriculum system construction of innovation and entrepreneurship education in colleges and universities based on the perspective of collaborative education. Polit. Teach. Ref. Middle School **7**, 98 (2021)
11. Yao, H.: Research on collaborative education mechanism of youth patriotism education in the new era. Knowledge Seeking Guide **8**, 79–80 (2021)
12. Min, J.: Research on the collaborative education of music aesthetic education and ideological and political education in colleges and universities. Exploration of Decision Making (middle) **2**, 46–47 (2021)
13. Guo, D., Liu, J.: Parents volunteers enter the campus to cooperate in education and development. Jiangxi Educ. (6), 15 (2021)
14. Chen, L.: Cultivating class teacher's collaborative leadership in home school cooperation. Jiangsu Educ. **15**, 19–20 (2021)
15. Zhang, S., Chen, J.: Ways to improve the efficiency of smart media content dissemination. China Publish. **4**, 11–14 (2021)
16. Zhao, F.: Exerting the cooperative education effect of mental health education and ideological and political education. J. Huaibei Vocational Tech. College **20**(1), 42–44 (2021)
17. Xie, J.: Research on the construction of "curriculum ideology and politics" in higher vocational colleges based on collaborative education. Ind. Technol. Forum **20**(04), 123–124 (2021)
18. Chen, X., Wang, Y., Li, H., Han, C.: Analysis on the construction of practical training platform from the perspective of industry education integration. China .Inform. **24**(4), 227–228 (2021)
19. Chen, S.: A study on the training of business English majors based on service study. J. Hubei Open Vocational College **34**(3), 168–169 (2021)
20. Li, J., Chen, W., Dong, L.: Research on cultivation path of postgraduates in artificial intelligence field under the background of "new engineering" construction. Degree Postgrad. Educ. **2**, 29–35 (2021)
21. Rao, Z.: Implementation path of petrochemical talent quality improvement under the background of higher vocational enrollment expansion. J. Hubei Open Vocational College **34**(3), 35–36 + 39 (2021)

22. Su, X., Wang, L.: Research on the current situation and path optimization of undergraduate professional tutors and counselors' collaborative education under the concept of "three complete education". Journal of Hubei Open Vocational College **34**(3), 29–31 + 34 (2021)
23. Zhang, D., Gao, P., Huang, J., Wang, Y., Wei, J.: Exploration and practice of interdisciplinary talent training mode of construction and environmental protection specialty under the background of new engineering. Higher Architectural Educ. **30**(1), 1–9 (2021)

Research on Evaluation Mechanism of Innovation and Entrepreneurship Team Management Based on Data Mining Classification

Shuang Qiu[✉]

Hubei University of Medicine, Shiyan 442000, China

Abstract. In order to improve the quantitative performance appraisal mechanism in the existing innovation and Entrepreneurship Talent Management System, a research scheme based on data mining technology is proposed. The combination of decision tree algorithm and cluster analysis is applied to the quantitative performance appraisal system, so as to explore the relationship between the appraisal results and various factors. Kmeans clustering algorithm is used to evaluate and analyze the team members, which is roughly divided into four levels in the form of classification rules. According to the evaluation level and the core attributes of entrepreneurial team, the detailed final individual quantitative assessment score table is generated by using the decision tree algorithm. Taking the actual data of an entrepreneurial team as the sample to test, analyze and verify, the test results show that the proposed scheme has better accuracy, and provides strong decision support for talent team management.

Keywords: Data mining · Evaluation index · Performance evaluation · Quantitative performance · K-means clustering · Decision tree algorithm

1 Introduction

With the rapid development and large-scale popularization of computer technology, information collection and analysis has become a key problem in the development process of major enterprises and institutions. The 21st century has entered the era of big data. With the application of various computer-aided technologies such as office automation, information equipment and database software, massive data information has been produced. However, how to efficiently analyze and process these rapidly expanding data, and provide decision-making services and technical support for the business development of the Department, has become a difficult problem to be solved by the process supervision and control system, especially the innovation and entrepreneurship team management [1].

M. A. Jan and F. Khan (Eds.): BigIoT-EDU 2021, LNICST 392, pp. 268–277, 2021.
https://doi.org/10.1007/978-3-030-87903-7_34

Data mining is an interdisciplinary subject that appeared in the 1990s, involving research results from database technology, knowledge engineering, probability and statistics, pattern recognition, neural network, visualization technology and other fields. The essential goal of data mining is to extract the hidden and valuable information and relationships from a large number of noisy, incomplete, fuzzy and random data. At present, the application of data mining in quantitative performance appraisal management system has become a hot research direction. Literature 5 proposes a human resource assessment system based on data mining. Literature 6] using the Apriori algorithm of data mining association rules to comprehensively analyze the students' scores, not only can we know the students' mastery of knowledge, but also can explore the internal relationship between courses. (7) data mining technology is applied to mine and integrate the information with potential value of enterprises and relevant information, so as to obtain more valuable information for evaluating enterprises and use project assessment to improve efficiency. Through the above research and analysis, it is found that the existing performance appraisal methods based on data mining all adopt single decision tree or association rule analysis, and the selection of member attributes involved in performance appraisal is not accurate [2].

Therefore, this paper proposes to apply decision tree algorithm and cluster analysis to quantitative performance appraisal system, in order to reveal the valuable information hidden behind the performance appraisal. Firstly, K-means clustering algorithm is used to evaluate and analyze the team members, which are roughly divided into four levels in the form of classification rules. Then, I3 decision tree algorithm is used to generate the final individual quantitative assessment score table according to the evaluation level and the core attributes of entrepreneurial team. Taking the actual data of an entrepreneurial team as the sample to test, analyze and verify, the test results show that the proposed scheme has good clustering accuracy and evaluation accuracy, which provides strong technical support for decision-making management and improves the work efficiency of innovation and entrepreneurship team management.

2 Data Mining Definition

Data mining brings together research results from machine learning, pattern recognition, database statistics, artificial intelligence and other fields. The large-scale popularization of computer produces massive data. Data mining processes and analyzes massive data by integrating the technical achievements of the above disciplines. Data mining is the key step of knowledge discovery process, as shown in Fig. 1.

A large amount of business information is digitized and key information is collected, preprocessed and transformed, as well as reasonable model selection, from which valuable hidden associated information can be extracted to assist management decision-making. Data mining can effectively improve business competitiveness and team operation efficiency. Through data mining technology, we can find two unrelated data, but at the same time, it is related to other third-party data, so as to indirectly establish a hidden connection through the network, so as to facilitate the transmission and analysis of information. The research goal of this paper is to build a quantitative performance evaluation mechanism of innovation and entrepreneurship team based on data mining

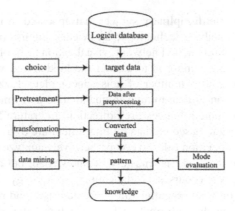

Fig. 1. Data mining knowledge discovery diagram

technology, so as to explore the relationship between the evaluation results and members' work-related factors.

3 Research on Quantitative Performance Appraisal Method Based on Data Mining

3.1 Analysis of Assessment Index

For the innovation and Entrepreneurship Talent Management System, the performance evaluation indicators are shown in Fig. 2, including achievement indicators, daily evaluation indicators and individual evaluation indicators [3].

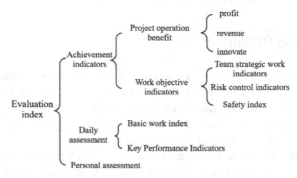

Fig. 2. Performance appraisal index system

3.2 Performance Appraisal Grade Evaluation Based on K-means Clustering

As a distance based partition clustering algorithm, K-means clustering algorithm has the advantages of simple structure, high efficiency and wide application range. K-means clustering algorithm is generally optimized by the objective function shown in Eq. (1):

$$E = \sum_{j=1}^{K} \sum_{x \in C_j} \|x - m_j\|^2 \tag{1}$$

Where e is the clustering criterion function and K is the total number of clusters.

3.3 Quantitative Performance Evaluation Based on ID3 Decision Tree Algorithm

The key of ID3 decision tree algorithm is to calculate the information gain and entropy according to the idea of recursion. The initial entropy is calculated as follows:

$$S(I) = \sum_{i=1}^{c} \left(\frac{N_i}{N}\right) \log_2 \left(\frac{N_i}{N}\right) \tag{2}$$

In order to get more accurate evaluation results, seven core attributes are set in the performance appraisal database to build ID3 decision tree.

4 Test Results and Simulation Analysis

4.1 Test Configuration

The experimental hardware environment parameters are: Windows 7 operating system, CPU is i7 processor, 4GB memory. The test data comes from the actual historical data of an entrepreneurial team in recent two years. The team is divided into four project groups, with a total of 38 people.

4.2 Result Analysis

K-means clustering algorithm and I3 decision tree algorithm are used to calculate the performance evaluation scores of all members in a group [4].

The results show that the personal performance evaluation score is consistent with the actual personal performance evaluation results, and the accuracy of the data reaches 92%, which can meet the actual application needs in accuracy. In addition, through the data mining method, the efficiency of quantitative performance appraisal has been greatly improved, which has verified the advanced nature and effectiveness of the method [5].

We use 5000 students to simulate our algorithm, and the results show that with the increase of time, the quality and effect of teaching is also increasing, which also shows the effectiveness of our algorithm which is shown in Fig. 3.

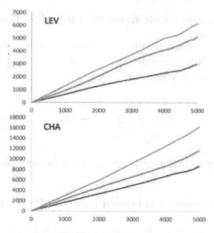

Fig. 3. Algorithm verification

5   Theoretical Basis and Concept Definition

5.1   The Theory of Team Building

Team building theory is generally divided into: cooperative competition theory, team member participation theory and constructive conflict theory [6]. In college students' innovation and entrepreneurship team, team members come from different family backgrounds and discipline backgrounds, and are good at different professional fields. According to the theory of cooperative competition, a team should have the team goals that the team members jointly identify. If a team does not have a common goal, the team members only consider their personal performance, which is easy to lead to the phenomenon of fighting separately, and they will ignore other team members. The team as a whole has no cohesion. When the team is in a competitive state, the team members try to maximize their own interests, Knowledge and resources will be blocked or even attacked and destroyed, resulting in poor team performance. Therefore, the team should first establish a common goal, and guide team members to work hard to achieve the common goal. Give full play to their own professional advantages, share the information and resources of the team, team members share their areas of expertise, cooperate with each other, and strive to get a higher level of performance. Team member participation theory can further stimulate the initiative and enthusiasm of team members by guiding team members to participate in and interact with decisions concerning their own interests. When team members participate in team decision-making and management, they will have a sense of team ownership. They are more likely to identify with the team goals set by their own participation, and they are more active in implementing decisions. The theory of construction conflict shows that team building should focus on the formation of cooperative relationship among team members. In the process of team project, team members may have different views on the same problem due to different backgrounds. Reasonable team conflict will make good preparation for the formation of high-quality decision. When the cooperative relationship is really formed, the team members will take the common goal of the team as the core, exchange views frankly, discuss valuable elements, fully communicate and reach consensus. Therefore, through the treatment of constructive conflict, team members have more recognition of team goals, and the team relationship is more consolidated [7].

5.2 The Theory of Team Management

Team management theory is the concentrated embodiment of human thought in western management theory [8]. This theory is mainly based on the understanding of human nature. Since the beginning of the 20th century, there have been four major changes in the western organization management theory based on human nature: the hypothesis of "economic man", "social man", "self actualized man" and "complex man". The essence of team management is the process that team managers gather their team members together through management methods, establish common team goals, guide team members to cooperate with each other, make the original loose sand members form team cohesion and become an integral team, make full use of limited human resources, and produce much higher than individual work performance. The main viewpoints of team management theory are as follows: first, team members should have different professional backgrounds and play different roles in team activities to ensure the effective operation of team projects. Secondly, team leaders are influenced by many factors when they are in team management. Third, there are different types of team leadership. Fourth, the effective operation of the team needs to have interrelated conditions to maintain. Fifthly, the success of team work depends on the degree of social identity and social performance. Sixth, in the process of team management, the team leader should authorize the members to a certain extent. When members have the right to participate in decision-making, their sense of identity with team goals will be strengthened, and their decision-making execution will also be improved. Seventh, the respect between team leaders and members is the key to the success of team management. Eighth, cultivate the team's innovative spirit [9].

5.3 Group Psychology and Group Behavior Theory

People have the basic psychological needs of socialization. The members of College Students' innovation and entrepreneurship team form a group for various reasons, and have a common team goal. This goal enables members to gather together [10]. In the process of team cooperation, the overall performance is far higher than the performance level obtained by individual struggle, which not only realizes the team goal, but also realizes the goal pursued by individual. The team integrates all members, determines the common goal of the team through consultation, and guides the members to work hard for it, so as to generate team cohesion in the process of the team project, enhance the team awareness of the members, and make the team members work together and help each other to achieve the common goal; under the role of team cohesion, the team members fully communicate and exchange in the team activities, Enhance the harmonious relationship between team members, and create a good team atmosphere. Team atmosphere has a reverse effect on team cohesion, which helps to improve the overall work efficiency of the team, and further stimulates the enthusiasm and initiative of team members in the process of activities, so as to promote the production of high team performance [11].

6   Research Conclusions and Suggestions

6.1   Members of Innovation and Entrepreneurship Team Lack Practical Experience, But They Have High Interest in Innovation and Entrepreneurship

When organizing the basic information of team members, it was found that more than half of the respondents in the college students' innovation and entrepreneurship competition had only one experience, while the number of participants three or more times was relatively rare. The statistical analysis of the reasons of the members shows that most of the participants participate in the competition with strong interest in innovation and entrepreneurship, and the situation of receiving the encouragement of the team members is relatively small. It can be seen that although college students lack practical experience, they are still enthusiastic about innovation and entrepreneurship [12].

We can find that colleges and universities provide support for college students' innovation and entrepreneurship team from different aspects, such as school publicity, instructors and team internal characteristics. In the connection of instructors, it is found that most of the teams have fixed instructors or innovation and entrepreneurship trainers; and most of these instructors have practical entrepreneurial experience, which can make up for the lack of team cooperation and actual entrepreneurial experience of college students to a large extent, So as to better improve the team performance of innovation and entrepreneurship team [13].

The effective improvement of College Students' innovation and entrepreneurship team performance can stimulate college students' entrepreneurial passion and improve their entrepreneurial willingness. The purpose of the university students' innovation and entrepreneurship competition is to stimulate the university students' innovation and entrepreneurship intention, so the team project audit is relatively loose in the early stage of the competition, and gradually strict in the later stage. According to the final score distribution of the team, 11.6%, 15.6%, 548%, 13.9% and 4.1% of the total team did not receive awards, department level, school level, provincial level and national level awards, respectively. Through the analysis of team performance and entrepreneurial intention, we can find that college students' innovation and entrepreneurship competition has a positive impact on College Students' entrepreneurial intention [14].

6.2   Countermeasures and Suggestions

This study analyzes the main factors affecting the performance of College Students' innovation and entrepreneurship team. In the actual management and construction tasks of the team, how to improve the performance of College Students' innovation and entrepreneurship team, so as to stimulate college students' entrepreneurial intention, is an important issue that university leaders, scientific research managers and team leaders must face. This section concludes the previous research work, and gives feasible and operational suggestions for the reference of the builders and managers of College Students' innovation and entrepreneurship team [15].

First of all, the team leader should make clear the motivation of the members to participate in the college students' innovation and entrepreneurship team. College Students' innovation and entrepreneurship team members are generally divided into three categories: the first type is interest oriented, with clear goals and strong execution, and will actively enrich the team's capital reserves from various aspects. The second is the competition utilitarian type, the participation of the team is only to obtain the corresponding credits, competition experience or awards. Although such members may not be interested in the direction of the team project, their strong purpose will bring efficient execution. The third type is the human persuasion type, which is generally because the team leader finds out and persuades the team leader to participate in the team without the intention of taking the initiative. Secondly, through the analysis of member information, the project tasks are arranged. Through communication with members, the team leader can collect as much information as possible, analyze the personal goals of members and the corresponding abilities required by the project. After mastering the above information, according to the principle of "differential treatment", targeted arrangements should be made to promote the integration of individual goals and team goals, so as to maximize the realization of project goals while ensuring the stability of members' participation in the project. Taking the interest oriented members as an example, the team should empower them to make their own team work plans and arrange their work, so that they can feel greater personal responsibility in the process of team work, stimulate their autonomy to participate in team work, encourage them to persist in research interest, and further cultivate the subjective consciousness and initiative consciousness of the interest oriented members. Finally, implement the evaluation and set the goal. In team work, specific and clear goals make team members have a stronger sense of commitment. Team members' commitment to goals will affect team performance. Therefore, the goal setting should be in line with the requirements of all parties involved in the process of achieving the team goal [16]. The clear division of the team goal is that the team members can timely understand the stages of the project, as well as the tasks of each stage and the standards to be achieved. This combination of specific team goals and project-oriented team norms can enhance the effectiveness of team members to complete team tasks and achieve team goals.

7 Conclusion

In this paper, the decision tree algorithm and cluster analysis are applied to the quantitative performance appraisal system. Firstly, K-means clustering algorithm is used to evaluate and analyze the team members, which are roughly divided into four levels in the form of classification rules. Then, ID3 decision tree algorithm is used to generate the final individual quantitative assessment score table according to the evaluation level and the core attributes of entrepreneurial team. The actual test results show that the proposed scheme has good clustering accuracy and evaluation accuracy, which has a certain reference significance for quantitative performance appraisal system.

Limitations of research tools. This study was conducted from different levels. In addition to the basic information, the addition of the three scales results in a long questionnaire, which makes the subjects feel tired and affects the quality of the test. In order to reduce the number of invalid questionnaires, I use the paper version of all the questionnaires, and send them to each team one by one, and explain the reasons with the subjects, and ask them to fill in carefully. This is of great help to the quality of questionnaire filling, but it still can not avoid the phenomenon that individual subjects do not want to do or give up halfway, which has a certain impact on the recovery rate and efficiency of the questionnaire. In the future questionnaire design, we should pay more attention to the time and patience of the subjects, and reduce the number of questions without affecting the test results. In terms of research methods, due to objective reasons, some members of the team are not convenient to interview, and privacy issues may also cause resistance. The lack of guidance in the interview and the lack of in-depth analysis lead to defects in research methods.

References

1. Pei, X.: Analysis of human resource assessment system based on data mining. Enterprise Reform Manag. **18**, 57–58 (2016)
2. Wang, M., Zang, C.: Realization of information system in engineering construction field. Inform. Technol. **5**, 206–209 (2014)
3. Yang, J., et al.: Data forwarding control strategy for opportunistic networks with fuzzy control. Syst. Eng. Electron. Technol. **38**(2), 392–399 (2016)
4. Li, Z., et al.: Clustering network data collection method based on hybrid compressed sensing. Comput. Res. Dev. **54**(3), 493–501 (2017)
5. Xu, B.: An empirical study on the interaction between formal and informal organizations based on group dynamics. Nankai Econ. Res. **4**, 21–27 (2005)
6. Wang, L., Ge, J.: The formal limit of organization and informal organization. Zhejiang Soc. Sci. (4), 56–61 + 127 (2009)
7. Guo, X., Shi, S., Wen, L.: Incubation and cultivation of College Students' innovative team: a study of Chinese youth (11), 106–108 (2008)
8. Song, F.: The current situation, problems and suggestions of college students' participation in "innovation and entrepreneurship." Macroecon. Manag. **1**, 67–71 (2018)
9. Yuan, J., Tian, X.: Development status and path selection of innovation and entrepreneurship education in colleges and universities under the background of new normal – based on the investigation and analysis of eight colleges and universities in the "Yangtze River Delta region." Modern Educ. Manag. **6**, 35–41 (2018)
10. Wen, Z., Chen, Y.: The learning process of the entrepreneurial competition team: organizational ideas. In: Proceedings of the Symposium on the Practice of Creativity. Chengchi University, Taipei, pp. 608–631 (2003)
11. Yu, Y., Huang, Z., Yu, M.: Thinking and policy suggestions on the construction of scientific and technological innovation team in colleges and universities. Ding R & D Manag. **26**(002), 129–132 (2014)
12. Qi, X., Qi, E., Shi, Z.: Cross level influence of organizational structure characteristics on product innovation team performance: an empirical study based on Chinese manufacturing enterprises. Sci. Sci. Technol. Manag. **34**(3), 162–169 (2013)
13. Lu, P., Zeng, W.: Management and countermeasures of university innovation team construction. Heilongjiang Higher Educ Res. **8**, 86–88 (2010)
14. Yao, K., Cui, X.: Study on the interaction between formal organizations and informal organizations from the perspective of externality. J. Fudan (Soc. Sci. Ed.), **55**(6),143–150 + 180 (2013)

15. Yang, F.: The influence mechanism of top management team leadership behavior on team performance: a case study. J. Manag. **8**(4), 504–516 (2001)
16. Zhao, J., Shao, D.: Target deviation and correction strategy of quantitative design of administrative performance appraisal index. Soc. Sci. Front **08**, 260–264 (2018)

Research on Hybrid Teaching Pattern Design in Applied Colleges and Universities Under the Background of Big Data

Lili Qi[✉]

Heihe College, Heihe 164300, Heilongjiang, China

Abstract. This study aims at the problems in the theoretical research and teaching practice of Hybrid Teaching Based on BP neural network. In the aspect of theoretical research, it combs the research status of blended teaching at home and abroad. Combined with practical teaching experience, this paper constructs a hybrid teaching mode based on intelligent teaching platform, and optimizes classroom teaching design from five aspects: learning task list, teaching resources, teaching activities, teaching evaluation and teaching strategies. Teaching experiment research in teaching practice. And the teaching experiment was carried out.

Keywords: BP neural network · Applied University · Mixed teaching mode

1 Introduction

With the reform of education and teaching mode and the network of modern university education, a new education and learning mode MOOC has emerged. Using new technology and platform, MOOCS puts forward a new learning method and using system. The system includes a variety of learning resources, such as exercises, discussion topics, supplementary assignments, short videos and after-school tests. At the same time, it adopts a new teaching strategy and method, which combines "online learning", "flipped learning" and "mutual learning". 2. Students do not need to study at a fixed time and place, but can use the little time in their life to learn through online counseling and interaction, It strengthens the discussion and exchange between teachers and students, and organizes the teaching resources more reasonably, so that students can learn step by step according to the plan. However, teachers and students in MOOCS can not communicate face to face like in traditional class, which is not conducive to effective communication between teachers and students [1]. Therefore, at present, the traditional classroom teaching can not be completely replaced by MOOCS classroom. We can adopt a hybrid teaching mode, combining MOOCS and traditional classroom teaching methods, so that they can show their strengths in teaching.

M. A. Jan and F. Khan (Eds.): BigIoT-EDU 2021, LNICST 392, pp. 278–284, 2021.
https://doi.org/10.1007/978-3-030-87903-7_35

2 Related Concepts and Theoretical Basis

2.1 Definition of Related Concepts

2.1.1 Blended Teaching

Blended teaching combines the advantages of online teaching and face-to-face teaching, and has become one of the effective carriers for the deep integration of information technology and education. According to the literature review, the difference between blended teaching and blended learning is very small, so many studies do not distinguish the two, and the two can be mixed. Due to the continuous innovation of the supporting environment of blended teaching and the deepening of the theoretical research of blended teaching, scholars at home and abroad have given the concept of blended teaching from different angles, but there is no unified definition.

$$Y = Y_{f-1} modn \tag{1}$$

$$key_1 = Y^a = g^{ba} modn \tag{2}$$

$$r_a = (Y_2) aQ^{-1} modn \tag{3}$$

He Kekang, Li Kedong and Li Jiahou, the famous scholars in China, have given the concept of blended learning. Professor he Kekang believes that blended learning inherits the dual advantages of traditional learning and online learning, which not only plays the leading role of teachers' guidance and supervision, but also plays the main role of students' autonomous learning. Professor Li Kedong believes that blended learning is a teaching method that integrates face-to-face teaching and online learning to reduce costs and improve efficiency. Li Jiahou believes that blended learning is to select the appropriate part from many teaching factors to complete the teaching objectives. Foreign scholars Singh & reed believe that blended learning is a learning way to achieve the optimal learning effect through the combination of "appropriate" learners, time, learning technology and learning style.

2.1.2 Flipped Classroom

"Flipped classroom" is translated from "the FD classroom", which is a new teaching mode and a "reversal" of the traditional classroom. In the traditional classroom, the mastery of knowledge depends on Teachers' teaching in the classroom, while in the flipped classroom, the mastery of knowledge depends on students' autonomous learning before class. In the traditional classroom, the internalization of knowledge relies on homework exercises after class, while in the flipped classroom, the internalization of knowledge relies on the teaching activities such as question answering and discussion, group exploration and achievement display organized by teachers in the classroom. Professor Zhang Jinlei believes that flipped classroom is the reverse arrangement of knowledge teaching and knowledge internalization, changing the role of teachers and students in traditional teaching, and re planning the classroom time. Flipped classroom teaching mode emphasizes students' autonomous learning and inquiry learning. Learners can choose their favorite learning methods to complete the learning of new knowledge. Classroom time is reserved for teaching activities such as question answering and inquiry.

Flipped classroom is a new paradigm of teaching reform [2]. Flipped classroom emphasizes mastering basic knowledge through preview before class and completing knowledge internalization in class. Hybrid teaching emphasizes the combination of online and offline, and the hybrid online and offline teaching platform can be used before, during and after class.

2.1.3 Intelligent Teaching Platform

Professor Zhu zhiting pointed out that by building an ecological learning environment with technology integration, smart education, based on the principle of "accuracy, optimization, thinking, sharing and creation", enables teachers to carry out effective teaching and learners to obtain personalized learning services with good experience, In order to cultivate talents with good thinking quality and strong innovation ability, intelligent teaching platform makes teaching activity monitoring, data acquisition and analysis convenient and efficient, provides decision support for the implementation of precision teaching, optimizes the teaching process, realizes personalized teaching, improves students' thinking quality and cultivates innovation ability. The "precision" here mainly involves the accurate determination of whether the learning occurs and whether the learning task is completed on schedule, and providing accurate assistance for learners through data analysis. Intelligent teaching platform can realize individual learning, efficient learning, immersion learning and continuous learning, which provides material basis for the realization of intelligent education.

2.2 Relevant Theoretical Basis

Constructivism theory proposes that the main goal of learning is "meaning construction", and the basis of learning is "situation", "communication" and "cooperation". The situation establishes the link between the new and old knowledge and experience, and the learning of "meaning construction" takes place in the specific problem situation. Therefore, the design teaching should first design the problem situation suitable for students' learning. "Meaning construction" is accomplished by students' autonomous learning, "cooperation" and "communication." cooperation "means group cooperation and exploration," communication "means teaching interaction, including teacher-student and student student interaction.

The Enlightenment of constructivism theory to this study: constructivism emphasizes the autonomy of learning and the self construction of knowledge. The design of learning task list in this study fully reflects the dominant position of students' learning. The intelligent teaching platform has rich teaching resources and creates a good learning "situation". Group assignments/tasks need to be completed through group "cooperation", Discussion and question answering, and achievement presentation embody "communication".

3 The Construction of Teaching Mode

3.1 Analysis of Typical Teaching Mode

In teaching practice, many researchers put forward different hybrid teaching models, such as the hybrid teaching model based on umu interactive learning platform, and the hybrid teaching model based on wechat platform, superstar learning, two classroom and other intelligent teaching platforms. These hybrid teaching models are relatively simple, and the theoretical height is not enough.

Some well-known scholars at home and abroad also put forward the mixed teaching mode, which is of universal significance. The following is a list of three representative scholars' mixed teaching mode.

Cohen, an American scholar, analyzes the system as a whole and puts forward the octagonal framework to explain the system well. As shown in Fig. 1, in the octagonal framework, hybrid teaching can be divided into eight dimensions: teaching, organization, technology, interface design, evaluation, management, resources and ethics. These eight dimensions are interdependent and work together to improve the teaching effect by optimizing the online and offline teaching structure. Among them, teaching and evaluation mainly refers to teaching objectives, contents, platforms, methods and strategies; organization and management mainly refers to the management of digital learning and the maintenance of learning environment; technology and interface design mainly refers to the provision of software and hardware technology and good user experience; resources and ethics mainly refer to open teaching resources and equal and friendly learning atmosphere.

3.2 Improvement After Class

Students are required to complete exercises online to consolidate their knowledge in time; on the other hand, to carry out professional experience. Around the key and difficult points of this chapter, the teacher designed a social service activity of "going into the community, health guidance – paying attention to diabetes". The whole activity takes students as the main body. Students are required to prepare propaganda posters and PPT in advance, and carry out blood glucose and blood pressure detection on site [3]. The setting of this link, the combination of work and study, and the unity of knowledge and practice, has greatly stimulated the students' learning enthusiasm, cultivated the students' ability to analyze and solve the problems, and this kind of feeling obtained through personal practice is easier to be deeply rooted in the depth of the students' hearts, never forget, and promote the spirit of China, The comprehensive evaluation method which combines the usual performance, practical training performance and examination results is implemented. While examining the systematicness, integrity and scientificity of students' knowledge, we should focus on examining students' knowledge application ability and operation skills, so as to avoid the phenomenon of "high score but low ability". Diversified curriculum evaluation methods can timely reflect the learning situation of students, provide useful feedback to students, help students build self-confidence, stimulate and cultivate their interest in learning.

4 Sakai Based E-learning Space

4.1 Analysis of Saka Concept

Saka is a free, open source online collaboration and learning environment, developed and maintained by Saka members. It was initiated by Indiana University, University of Michigan, Stanford University and Massachusetts Institute of technology in 2004 as an open source curriculum and teaching management system (CMS) development program, mainly to replace the system or related business software system independently developed by each university. Sakai's collaboration and learning environment is a free educational software platform with shared source code, which is mainly used for teaching, research and collaboration. Network learning space is the basic learning space for teachers and students based on the network environment. It is the product of the deep integration of information technology and education and teaching. As shown in Fig. 1. It breaks through the limitations of I space, teaching methods and teaching resources of traditional teaching mode, so that students can make full use of their spare time to learn, It promotes the learning and communication between teachers and students, teachers and teachers, students and students. It can not only improve teachers' teaching level, but also improve students' self-learning ability.

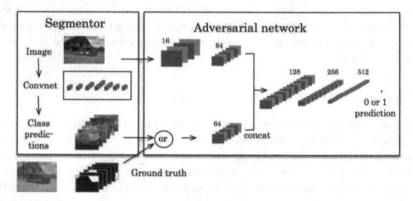

Fig. 1. Analysis of Saka concept

4.2 The Current Situation of Usage and Popularization in China

At present, Sakai has been introduced into many universities in China, such as Shanghai Fudan University and Beijing University of Posts and telecommunications. Under the guidance of the Provincial Department of education, Sakai's research and development has achieved rapid development. Now more than 30 universities have used Saka based online learning space. The Saka based e-learning space platform is designed with Java as the development language, mysq as the database, Tomcat as the background server, and VMware virtualization as the deployment environment. Four virtualization servers are deployed on two physical servers, two web application servers for network teaching, one mysq database server for network teaching data storage, and one od middleware server for data storage and load balancing.

5   The Design of Three Mixed Teaching Mode

Sakai based online teaching process learning platform includes five roles: system administrator, educational administrator, College (Department) supervisor, teacher and student.. The system administrator mainly includes four modules: system management, course statistics, course evaluation management, course construction and application statistics management. Among them, the system management module mainly includes four sub modules: user management, role management, module management and log management; the course statistics module mainly includes four sub modules: School macro data statistics, current semester data statistics, teacher data statistics and course basic item statistics; the course evaluation management module mainly includes two sub modules: evaluation rule setting and evaluation result inquiry; The module of curriculum construction and application mainly includes two sub modules: Statistics of overall curriculum construction and statistics of teachers' curriculum construction [4]. College (Department) supervisor mainly includes three modules: course statistics inquiry, course evaluation inquiry, course construction and Application Statistics. Among them, the course statistics module mainly includes three sub modules: the current semester data statistics of the Department, the data statistics of the beginning teachers and the statistics of the basic items of the course; the course evaluation module mainly includes the result inquiry sub module of the course evaluation of the Department; The course construction and Application Statistics module mainly includes two sub modules: the overall course construction statistics of the Department and the teachers' course construction statistics. Teacher users mainly include curriculum construction, curriculum management, curriculum resource construction, online interactive communication, homework and test release and correction, student performance management, curriculum construction and application query, student learning statistics query and other modules.

6   Concluding Remarks

As an application-oriented university, it is our duty to cultivate high-quality application-oriented talents and serve the development of economy and society. Curriculum teaching is the foundation to achieve the goal of cultivating high-quality applied talents. Therefore, in the final analysis, the realization of talent training goal is the reform and improvement of curriculum and curriculum system. The construction of network learning space provides an open and free network learning environment for teachers and students, makes full use of information technology teaching means and network teaching platform, and improves the teaching effect and the quality of personnel training. The hybrid teaching mode, which combines classroom teaching and network teaching platform, will become the mainstream direction of teaching mode reform in the future.

Acknowledgements. Heilongjiang Province higher education teaching reform research project name: mobile Internet era applied colleges and universities mixed teaching model reform research, number: SJGY20190457.

References

1. Zhang, C.: Research on the application of o2o Teaching Mode in Senior High School Geography Teaching. Liaocheng University (2019)
2. Ding, X.: Research on Blended Teaching Effect of Neuroscience Based on Pan Ya Platform. Northwest University for Nationalities (2019)
3. Yu, X.: Research on the Application of Blended Teaching Based on Cloud Class Platform in Senior High school English Listening Course. Northwest Normal University (2019)
4. Xu, L.: Research on the Application of Blended Teaching in Senior High School Comprehensive Practical Activity Curriculum. Northwest Normal University (2019)

Research on the Application of Big Data Analysis Auxiliary System in Swimming Training

Qingtian Zeng[✉]

School of Public Administration, Chongqing Vocational College of Transportation, Chongqing 402247, China

Abstract. In the field of sports, the success of computing technology in various industries has been paid more and more attention by the majority of sports teachers and coaches. We know that the intervention of high technology in the field of sports teaching and training not only improves the scientific training and teaching to a new stage, but also brings about the rapid improvement of the level of competitive sports. It is the development direction of modern physical education and training to apply high-tech achievements to sports training and physical education. The emergence of computer-aided system technology provides a powerful impetus for the development of sports to a faster and better level. This paper attempts to expound the composition and characteristics of the computer-aided system and the particularity of swimming teaching and training, applies the computer-aided system technology to swimming teaching and training, and probes into and analyzes the feasibility of the application of the computer-aided system in swimming teaching and training.

Keywords: Computer aided system · Swimming teaching · Swimming training · Application · Feasibility

1 Introduction

With the rapid development of computer technology, its application has penetrated into all areas of society, effectively promoting the development of social information. Mastering and using computer technology has become an essential skill for people. Especially in recent 10 years, computer technology is widely used in aerospace, aircraft design, mechanical design, environmental simulation, three-dimensional animation and so on. With the increasingly fierce sports competition and the continuous improvement of sports training level, high-tech means are more and more widely used in sports competition and training teaching. However, the computer aided system (CAS) has not been developed in the field of sports due to the difficulty in designing funds for technical talents. We know that the intervention of high technology in the field of sports teaching and training not only improves the scientific training and teaching to a new stage, but also brings about the rapid improvement of the level of competitive sports. It is the development direction of modern physical education and training to apply high-tech achievements to sports

M. A. Jan and F. Khan (Eds.): BigIoT-EDU 2021, LNICST 392, pp. 285–290, 2021.
https://doi.org/10.1007/978-3-030-87903-7_36

training and physical education [1]. With the development of physical education, the improvement of sports training level and the deepening of specialization, athletes and coaches are forced to improve their sports technology, training methods and high-tech equipment to adapt to the fierce competition of high-level competition. It also makes PE teachers use high-tech means to improve the quality of PE teaching and provides new methods and ideas. Therefore, the application of computer-aided system in swimming teaching and training has a very broad prospect and great significance.

2 Application of Computer Aided System in Swimming Teaching and Training

2.1 Composition and Concept of Computer Aided System

Computer aided system includes: Computer Aided Design (CAD), computer aided manufacturing (CAM), computer base education (CBE), etc.

CAD is to help all kinds of designers to design. Because of the computer's fast numerical calculation, strong data processing and simulation ability, CAD technology has been widely used. For example: innovative design of swimming technique and quantitative analysis of world excellent swimmers' technology. CAD not only improves the speed of technological innovation, but also improves the innovation quality and speed of movement.

Computer aided education CB includes: Computer Aided Instruction (CAI), computer aided test (CAT) and computer management instruction (CMI). Computer aided instruction (CAI) is a kind of teaching method that uses computer as teaching medium to teach learners. It makes great changes in the teaching mode, teaching content, teaching method of educators. Teachers in class (one person, one pen, one brush) has gradually become history. With the popularity of computer, CA has gradually become an important means of teaching. At present, there are three kinds of CAI: Network CAI, multimedia CAI, intelligent caicat and CMI, which is to use the development of computer and network technology to carry out teaching management (exchange and transmission of teaching documents, test papers, etc.) and examination.

CBE includes: CAI (Computer Assisted Instruction), cat (computer aided test) and CMI (computer management instruction). Computer aided instruction (CA) is a kind of teaching method that uses computer as teaching media to teach learners. It makes great changes in the teaching mode, teaching content, teaching methods of educators. Teachers in class (one person, one pen, one brush) has gradually become history. With the popularity of computer, CA has gradually become an important means of teaching. Now there are three kinds of CAI: Network CAI, multimedia CAI and intelligent CAI. With the development of computer and network technology, cat and cm are used for teaching management (exchange and transmission of teaching documents, test papers, etc.) and examination [2].

2.2 Hardware Equipment Required by Computer Aided System

The hardware environment of CAD and CAA mainly includes: host computer (PC, PC workstation), graphics and image processing system (input device and output device).

With the improvement of the level of competition, the athletes and coaches put forward higher requirements in the design innovation of technical movements and learning advanced technology. If athletes can learn from foreign advanced technology, and then according to their own characteristics to form a set of nearly perfect technology, is the key factor to win the competition. In swimming training, the training and innovative design of movement technology still stay in the stage of artificial design based on experience, only relying on Coaches' language explanation and demonstration, athletes try carefully.

3 The Basic Theory of the Auxiliary Software for Monitoring the Load of Sports Training

3.1 The Basic Theory of the Auxiliary Software for Monitoring the Load of Sports Training

Concept refers to "the practice process in which people summarize the common essential characteristics of things in the process of repeated practice and cognition, and leap from perceptual knowledge to rational knowledge". Concept is a knot that people use to understand and master the net of natural phenomena, a stage in the process of cognition, and the beginning of problem solving. It is the connection and interpretation of the connotation and extension of things, and the scientific logic of logical judgment of the development of things. Therefore, this research software will start from the concept of "sports training load", analyze the essence of things, seek truth from facts, and lay a scientific and rigorous theoretical foundation for the whole design and research work [3].

Sports training is a process in which people try to change some aspects of the body's ability through some training means and methods. Sports training is a process in which people take the initiative to transform the body's shape, structure and function. It can be concluded that swimming training refers to the process of actively changing the quality and ability of the human body through the means and methods of swimming professional training, so that the human body can adapt to the development of swimming events in shape, structure and function.

3.2 Establishment of Swimming Training Load Monitoring Auxiliary Software Monitoring System

The monitoring system is the main body of this research design. In order to make this research design more practical and scientific, the indicators contained in the system must have the following principles: the established indicators must have been published in the current specialized works of swimming training theory in China, and the research results based on the indicators have proved that the indicators have practical monitoring significance, The principle of utility and evaluation: the selected indicators must be at least those that can be monitored by sports teams at or above the provincial level, and have public research records of continuous human body monitoring. The monitoring means, monitoring purposes and simple evaluation methods have been relatively mature in practical application, and can not be limited to laboratory conditions, Moreover, the principle of quantitative statistics and analysis relevance is for the purpose of more objective comprehensive evaluation, and the selected index can at least be correlated with another index in the system determined by this study, which can be used for comprehensive evaluation. Except for single load function index.

The monitoring of swimming training load is divided into two parts: training intensity monitoring and training course load monitoring. Training intensity monitoring is mainly aimed at the body reaction of athletes with different energy supply systems under specific training intensity in the process of training course; training course load refers to the monitoring of athletes' functional status with the training stage as the measurement unit after the training course (Fig. 1).

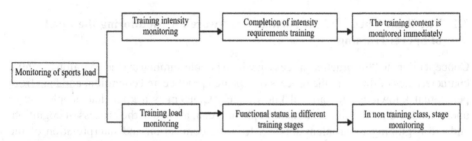

Fig. 1. Monitoring content of sports load monitoring system

4   Design and Application of Swimming Training Load Monitoring Assistant Software

The main function of the software is to input the "basic information" through the coach, and to classify, form, record and analyze the input information through the program written by the database language in advance. All the formulas and classification methods of background calculation are carried out according to the monitoring points and rules of each index in the above index system.

The working principle of this software is an auxiliary tool for coaches to calculate, classify and judge the existing and needed requirements such as index meaning, trend change and concept interpretation. Therefore, at the beginning of analysis, coaches need to input the information to be analyzed according to the format requirements of the software. The basic information to be input includes the personal basic information of the team members: including: name, gender, main event swimming style, main event distance, training years, date of birth, basic information of exercise monitoring (each class): main event amount, minor event amount, rowing amount, kicking amount, coordinated swimming amount, mixed swimming amount, technical swimming amount, intensity swimming amount; The results showed that: 1) the amount of leisure time, low intensity oxygen, aerobic anaerobic threshold, maximum oxygen uptake, lactate tolerance, peak lactate, peak lactate speed/burst strength; Basic information of exercise load monitoring) the intensity (heart rate, blood lactic acid index) of each energy system training project completed in each training class: random swimming, low intensity aerobic, aerobic anaerobic threshold, maximum oxygen uptake, lactic acid tolerance, lactic acid peak, lactic acid peak speed/explosive power; 2) According to the test requirements of the indicators, the morning pulse, blood pressure, sonogram, hemoglobin, serum testosterone, cortisol, serum creatine kinase, blood urea and urine protein were tested before, during and after the training stage [4].

The process of training is a process of coexistence of individuality and generality. We should not only adopt the training means suitable for a certain training purpose, but also combine the individual characteristics of athletes in the application process. Therefore, personal characteristics is the most critical comparative reference evaluation standard in this study. Only when there is comparison can there be difference. Facing the same training purpose and the same training method, different athletes will have different reactions. Therefore, it is necessary to carry out similar reference based on the same event and gender. Any theory will develop to the refinement stage in the development process, Through successful examples, the index system can not meet the index reference of athletes with different characteristics, but to a certain extent, the standard system based on excellent athletes' test indexes can provide reference for the indexes collected in the same test environment.

5 Conclusion

From the perspective of instrumental innovation, the "swimming training load monitoring auxiliary software" provides coaches with instructions for observing and supervising the changes of relevant indexes such as exercise volume, exercise intensity and exercise load in the training process of swimmers, and the rules of changes through programmed and systematic means; compares with reference standards; and forms the work of purposeful sorting, recording and analysis, It is a special assistant software for the design of motion computer data platform technology. The target users of the auxiliary software are coaches and the target monitoring objects are athletes. According to the rules of swimming training, the main monitoring scope of this auxiliary software is: (1) monitoring of exercise volume; (2) monitoring of exercise load; thus four core operation templates are designed: (1) basic information record; (2) monitoring index record; (3) monitoring index monitoring evaluation; (4) monitoring means and method instructions. According to different monitoring requirements, the monitoring contents are divided into: (1) according to the monitoring time limit, It can be divided into real-time monitoring after class, daily monitoring and training stage monitoring; (2) according to the monitoring content, it can be divided into training amount monitoring, training intensity monitoring, training load and energy status evaluation; (3) according to the different ways of monitoring index calculation and analysis, it can be divided into: total training amount monitoring, seven energy supply system intensity plan prediction and completion monitoring, vascular system monitoring, oxygen operation ability system monitoring, and energy management system monitoring; Skeletal muscle and tissue cell damage monitoring; material and energy metabolism monitoring.

References

1. Lu, Y., et al.: Scientific Training and Monitoring of Swimming. Beijing Sport University Press (2007)
2. Hong, P.: Research on training monitoring of elite swimmers. China Sports Sci. Technol. (7), 42–44 (2003)
3. Cui, W.: Research and implementation of computer aided analysis system for exercise load calculation. Com. Eng. Appl. **21**, 217–220 (2003)
4. Tian, M.: Sports Training. People's Sports Press (2000)

Research on the Development Model System of "Integration of Industry and Education" in Colleges and Universities Based on Cloud Computing

Yuan Wang[✉]

Party Committee Office, Principal's office, Changshu Institute of Technology, 99 South Third Ring Road, Changshu 215500, China
wangy@cslg.edu.cn

Abstract. With the development of "industry and education integration" in Colleges and universities, the development mode system of "industry and education integration" in Colleges and Universities under cloud computing is constructed. It has changed the traditional mode of education and training, and found a new way to train professional talents.

Keywords: Cloud computing · Industry education integration · Development model

1 Introduction

Cloud computing is a trend of interaction and evolution between cloud computing and traditional industries under the promotion of Internet technology innovation in knowledge society. It is a new form and new format of Internet development under the innovation of computer technology. Cloud computing platform has the characteristics of diversification, wide coverage, high efficiency and low cost, which will produce the reconstruction of learning mode and production mode in various fields of social economy, and create new organizational mode and organizational form.

2 The Impact of Cloud Computing on Higher Education

The development and wide application of cloud computing, especially the mobile Internet technology, provides a new technical means for the modernization of education. It is changing people's learning methods, methods and habits unprecedentedly, and will inevitably lead to profound changes in the mode of private higher education [1]. Through the impact of e-commerce on traditional formats, we can foresee the basic form of the development of private higher education in the future and the opportunities faced by higher education in the era of cloud computing (see Fig. 1).

M. A. Jan and F. Khan (Eds.): BigIoT-EDU 2021, LNICST 392, pp. 291–300, 2021.
https://doi.org/10.1007/978-3-030-87903-7_37

Fig. 1. Structure of cloud computing

2.1 The Internet Promotes the Rise and Development of Open Education Resource Movement

The Internet has the characteristics of wide area, cross-border and so on. It can make the dissemination and learning of human knowledge very convenient, and provide a broad platform for the aggregation and sharing of educational resources.

2.2 Internet is an Ideal Way to Realize the Integration of Production and Education in Higher Education

At present, although the traditional university enterprise cooperation in higher education has been greatly developed, through the cooperation platform built by enterprises and schools, enterprises can enter schools and students can enter enterprises, which provides an effective path for the cultivation of high-quality and professional talents in Colleges and universities. In addition to the large-scale private colleges and universities, many colleges and universities hardware scale and teaching staff, faced with uneven investment, trapped in the level of teachers, hardware shortage and policy and other objective conditions, making higher education as a whole of higher education, many high-quality education resources cannot play their due effectiveness. The traditional mode of university enterprise cooperation in higher education has been unable to meet the growing demand of university enterprise cooperation personnel training. Therefore, cloud computing platform based on Internet can become an ideal way for enterprises and schools to cooperate [2]. With the development and innovation of broadband network and mobile Internet technology, anyone who can access the Internet can receive higher education.

In addition, students can choose their own training units according to their own professional learning path and with the help of school enterprise cooperation platform, so as to independently control the learning rhythm and master the necessary vocational skills before entering social work. In this way, it is not limited by time and region, which improves the flexibility and convenience of cultivating talents by the integration of production and education in higher education, and provides the possibility for the professional, professional and personalized talent training of higher education.

2.3 Internet Promotes the Innovation of Teaching Mode in Colleges and Universities

The wide coverage, long distance and high sharing of Internet make knowledge production and communication break through the regional and spatial limitations, which will cause substantial and structural changes in personnel training mode and management mechanism of private colleges and universities, and inevitably lead to the reconstruction of the original curriculum system and teaching system. On the one hand, the new teaching mode only solves the problem that students enter colleges and universities to acquire knowledge in a variety of ways, but it involves the practical mastery of specific professional skills, which cannot be well solved, and the vast majority of professional knowledge remains in theory. In order to solve the problem of professional skills training in the process of personnel training, we can choose the appropriate off campus training base or cooperative units to enter the school "factory" form through the school enterprise cooperation platform built by cloud computing technology to carry out professional and professional talent training.

3 Based on Apriori Algorithm

In this paper, aiming at the massive data in cloud computing environment, the Apriori algorithm of data mining is analyzed and studied. By improving the algorithm, it can be used in MapReduce programming model. Apriori algorithm is a classic algorithm for mining frequent itemsets based on Boolean association rules [3]. It was first proposed by Agrawal et al. The basic principle of the algorithm is to scan the transaction database for many times to calculate the support of the item set, calculate all the frequent itemsets, and then get the association rules. The algorithm is based on two important properties.

Property 1: when k-dimensional data itemset m is a frequent itemset, the necessary condition is that all $k - 1$-dimensional subsets are frequent itemsets.

If $m - 2$ is not a frequent subset, then it is not frequent. Algorithm process:

(1) The frequent itemset L1 is obtained by scanning the transaction database. If L1 is not empty, L1 will join and prune to generate candidate set C2 with length of 2;
(2) Then all subsets CL of each transaction in the transaction database in C2 are calculated, and all candidate itemsets with length of 2 in CT are counted and added with 1.
(3) After scanning the transaction database, find all the itemsets whose count is not less than the minimum support in the candidate set C2, and obtain the frequent two item set L2.

(4) Loop the above steps to process the newly obtained frequent item set. When no frequent item set is generated, the loop is terminated.

The algorithm is described as follows:

Input: D (transaction database), min_Sup (minimum support)

Output: l (all frequent itemsets in transaction database d)

The specific algorithm is as follows:

L1 = find_frequent_I-itemsels (D); // find the L1 set of frequent 1 itemsets

For k = 2;Lk − 1 ≠ φ;k++

{G = apriori_gen (LK − 1); // connect k − 1 with itself to generate candidate k itemset Ck

For each transaction t ∈ D // scans D for counting

{Ct = subet (Ck,t,); // candidate set

For each candidate c ∈ Ct

c. count+ +; // counter plus 1

Lk = (c ∈ Ck|c.count ≥ min_sup) // all frequent itemsets with minimum support

Return L = {all LK} // find the set L of all frequent sets. Through the algorithm process, it is not difficult to see that Apriori algorithm needs to traverse the transaction database multiple times. For example, if the database with massive data is operated, it will cost a lot of computing time and memory space, and the I/O load is very large. Cloud computing has the characteristics of parallel distribution. By improving Apriori algorithm and applying it to cloud computing environment, the efficiency of data mining can be improved according to the parallelism of cloud computing.

4   Reform Strategy of Industry Education Integration Education Mode in Colleges and Universities Under the Condition of Cloud Computing

The trend of deep integration of cloud computing technology and education is irreversible. Colleges and universities should systematically plan, actively explore the development path of higher education industry education integration under the background of cloud computing, and vigorously promote the informatization development of school enterprise cooperation in private colleges and universities.

4.1   Actively Promote the Comprehensive Application of Information Technology in Teaching Practice

On the basis of teaching practice, we should build our own information platform for school enterprise cooperation. Through the construction of a network platform for school enterprise cooperation with rich content and convenient use, the school enterprise two-way selection of on-line and off-line is planned to be pilot, and the design of learning situation design and teaching resources is carried out, so as to gradually realize the integration of "teaching and doing".

4.2 We Should Choose the Best and Organize Network Resources Seriously

The university enterprise cooperation network platform based on cloud computing helps to show the characteristic specialties of private colleges and universities, increase the popularity of enterprises, expand the popularity of school enterprise cooperation platform, and guide more schools and enterprises to participate. Thus, in the teaching process, private colleges and universities can further cultivate talents suitable for the needs of the society, make students feel the learning atmosphere of "learning for practical use and specialized in technology", and further improve students' learning enthusiasm. Therefore, on the basis of teaching practice, the school should timely push out its strongest, most advantageous and characteristic mature teaching methods and methods on the network platform [4–6]. The success of school enterprise cooperation network platform lies in its efforts to provide students with the best courses and personalized learning services. This requires schools to choose carefully, teachers to prepare carefully, scientific design, need to revise the traditional curriculum teaching mode and system structure to adapt to the new requirements of production teaching integration.

4.3 Innovating Incentive Mechanism and Strengthening the Construction of Teaching Team

The teaching mode of integration of production and teaching is not the unilateral behavior of schools or enterprises, but requires the joint efforts of schools and enterprises to build a professional teaching service team to cooperate with each other, especially in the training of "double qualified" teachers. Theoretical knowledge can be explained thoroughly in class, and design operation and demonstration can be completed skillfully in the training field. Private colleges and enterprises should actively promote the construction of curriculum team, build incentive mechanism to encourage teachers to participate, support and promote teachers to enter the production practice line, constantly improve practical skills and improve their own knowledge system. We can create a teaching service system based on cloud computing platform to promote the specialization and integrated management of teachers, and transform teaching from individual labor to team cooperation (see Fig. 2).

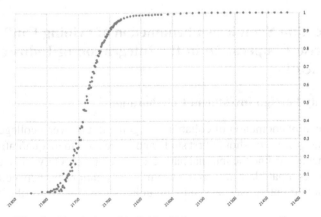

Fig. 2. Simulation of individual labor to team cooperation

4.4 Innovating the Management System of Private Higher Education

The development of the teaching mode based on cloud computing platform will bring impact on the current higher education teaching system, and will cause the deconstruction and reconstruction of the relationship between the school-centered management mode and the enterprise centered management mode. Under the current school enterprise cooperation management system, the "factory in school" between private colleges and enterprises in China makes it impossible for enterprises to effectively supervise and control, resulting in the students' practice process being basically isolated and closed from the outside world, resulting in the disconnection of learning content and external demand, making cars behind closed doors, and serious waste of teaching resources. In addition, the "factory out of school" school cannot be effectively managed, resulting in the students' practical teaching and theoretical teaching cannot be unified – resulting in the formation of university talent knowledge system [7–9]. Therefore, in the face of the historical development opportunities in the Internet era, enterprises and colleges and universities should fully understand and attach importance to it from a strategic height, and speed up the fundamental reform of university teaching mode and management system (see Fig. 3).

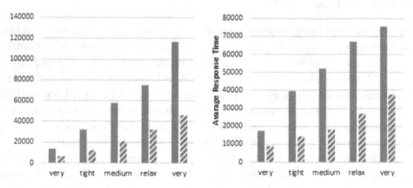

Fig. 3. Number of university teaching mode and management system

5 Judgment on Symbiosis Phenomenon of "Going Out" of School Enterprise Cooperation in the Integration of Industry and Education

5.1 University Foreign Investment Collaboration

As for the social phenomenon of collaborative going out between colleges and foreign investment enterprises, we should first study and judge their main qualitative parameters (the factors that determine the internal nature and its changes. Only when the qualitative parameters are compatible can there be a symbiotic relationship between colleges and foreign investment enterprises as symbiotic units, Since the start of the "double high" construction and quality improvement and excellence training plan of colleges and universities, especially vocational colleges, under a series of forward-looking, overall and systematic top-level designs of the government, the vitality of colleges and universities has been stimulated, and the improvement of internationalization level is undoubtedly one of the connotation construction of high-quality development, Through international exchanges and cooperation, foreign aid, study in China, standard education, overseas education and other forms, we can test the comprehensive strength of running a school from the perspective of internationalization, show and radiate the construction achievements in a wider world, and serve the social and economic development. As shown in Fig. 4, in this era, Chinese enterprises are required to start early and make great contributions in the process of going out, However, we should also be soberly aware that enterprises not only face external business environment risks, but also have problems such as cross-cultural identity conflicts, inconsistent professional standards, lack of overseas applied technical talents, etc. These objective realities restrict the cross-border investment of most enterprises to a certain extent and cannot achieve the expected results. Through the successful practice of some enterprises going out.

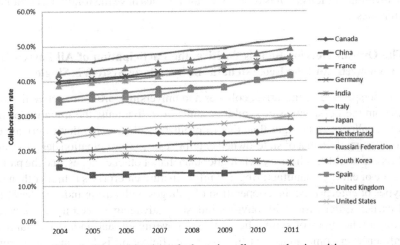

Fig. 4. The cultivation of talents in colleges and universities

5.2 The Cultivation of Talents in Colleges and Universities

Colleges and universities, especially vocational colleges, cultivate high-quality management and application-oriented talents that enterprises urgently need overseas. Overseas projects of enterprises can provide colleges and universities with practice places and guidance for students, and both sides can adapt to each other in the supply and demand of human resources and talent training [10]. The majors of effective schools basically cover all fields of national economic development, Both sides adapt to each other in terms of industrial structure and industry demand; colleges and universities gather high-quality educational resources and technological innovation capabilities, which not only provide intellectual support, technical support and innovation support for enterprises' overseas development, but also strengthen and highlight the international level and comprehensive strength of the school itself, so that both sides can achieve win-win in resource sharing. Under the specific conditions of "going out", it is imperative for colleges and universities to cooperate with foreign investment enterprises. The main quality parameters of colleges and universities can choose the international level, and the main quality parameters of foreign investment enterprises can choose the cross-border investment level [11–13]. The two are compatible and interact with each other. The integration of industry and education between colleges and foreign investment enterprises reflects a symbiotic phenomenon and has a symbiotic relationship.

6   Characteristics of Industry Education Cooperation Between Universities and Foreign Investment Enterprises

Based on the quantitative and qualitative analysis of the symbiosis degree, symbiosis coefficient, collaborative behavior and organizational model between universities and foreign investment industry, it is found that "universities have the following collaborative characteristics".

6.1   The Overall Development is Positive, the Participation of All Parties is Unbalanced, and the Ecosystem in the Symbiotic System is Single

From the perspective of initiative, colleges and universities are more active than enterprises. From the perspective of symbiotic units or individuals, the initiative of higher vocational colleges is stronger than that of ordinary colleges, and that of private enterprises is stronger than that of state-owned enterprises; "the participation degree of double first-class and double high colleges is higher than that of other colleges, and the participation degree of energy, manufacturing and infrastructure enterprises is higher than that of other types of enterprises; the cooperation of colleges with strong industry background started earlier, Enterprises with obvious industrial advantages need more cooperation. According to the types of symbiotic units, the government and industry organizations have influence in some areas, but the overall participation is low, and they have not played their due role in the symbiotic system.

6.2   The Modes are Diversified in Form and Degree, and the Profits of All Parties are Asymmetric, Tending to Integration

The cooperation of foreign aid training, overseas guidance or training should be based on the needs, and the main input of the school and the beneficiary is the enterprise; the cooperation between the school and the enterprise includes modern apprenticeship training, entrusted directional talent training, school enterprise joint education base, school enterprise joint development of professional or technical standards, joint scientific research and so on, However, the public welfare nature of most schools determines that their investment is the main one and the harvest is the auxiliary one, while the market nature of enterprises determines that their investment is less and efficiency is the most important one. The school enterprise joint construction of overseas school running entities requires schools and enterprises to build together for a long period of time, and joint investment and win-win results for each other [14–16]. The school's high enthusiasm for cooperation is faced with the hesitation of most enterprises, but the demands of overseas diversified operation and talent localization urge a small number of enterprises to test the water first, and take the evolutionary steps towards integration.

6.3 The Symbiotic Environment has Changed, the Reactions of All Parties are Inconsistent, and it is Evolving for a Period of Time in the Process of Adjustment

Since then, the international situation has been complex and changeable, with many uncertain factors. There are differences in risk prevention of "going out". Due to the sensitivity of the market to the market, the enterprises' ability to anticipate and guard against the risks is stronger than that of the schools. Most of the whole industry's overseas investment is in a country with good business environment or a very profitable industry area. Since, most enterprises in China have fully demonstrated the risk resisting ability of the enterprises in the host country. However, the international education field has been greatly impacted by the recent international epidemic. The offline operation of studying in China, running schools abroad and international exchange projects has basically pressed the pause button [17–20]. Although the online mode has been expanded, its effect needs to be comprehensively evaluated.

7 Conclusions

In short, the sustainable development of the teaching mode based on cloud computing platform needs not only the active participation of colleges and universities as educators themselves, but also the internal reform of the education system itself. With the rapid development of cloud computing technology, the education mode of integration of production and education will be more diversified, professional and personalized, so that learners can acquire high-level technical knowledge, and schools and enterprises as providers can obtain enough sustainable development power.

Acknowledgements. General topics of the planning and construction center of the ministry of education in 2017: exploration and practice of innovation and entrepreneurship education reform in application-oriented undergraduate colleges based on the concept of integration of industry and education, Number: 201702093031.

References

1. Strive to pursue the research innovation of combining theory with practice – Introduction to researcher Jiang Changyun, deputy director of Institute of industrial economy and technological economy of national development and Reform Commission. World Agriculture 05, 214–217 (2016)

2. Xia, Y.: Research on industrial integration model based on product perspective. Shanxi University of Finance and Economics (2016)
3. Wu, J.: Research on the development mode of aviation cold chain logistics from the perspective of industrial integration. Civil Aviation University of China (2016)
4. Zhao, X.: Government support research based on the Internet plus agriculture and animal husbandry platform. Inner Mongolia Agricultural University (2016)
5. Yuan, Y.: Foreign experience of China's cultural and creative industry development. Jilin University of Finance and economics (2016)
6. Huang, X., Yang, Y., Zhong, L.: Research on the integration mode of China's manufacturing industry and cultural and creative industry from the perspective of circular economy theory. Sci. Technol. Progr. Countermeas. **33**(06), 71–75 (2016)
7. Li, X.: Research on the innovation path of low carbon tourism formats based on the theory of "industrial integration." J. Southw. Univ. Nationalities (Humanit. Social Sci.) **37**(02), 126–130 (2016)
8. Dai, C.: Research on the construction of agricultural products logistics network in industrial integration development based on synergy theory. J. Linyi Univ. **38**(01), 86–90 (2016)
9. Yan, W.: Integration mechanism of tourism industry from the perspective of self-organization theory. Social Scient. **01**, 91–96 (2016)
10. Li, B., Liu, P., Dou, Y.: Study on tourism industry integration development of agricultural cultural heritage based on self-organization theory – taking Ziquejie terrace in Xinhua County of Hunan Province as an example. J. Central South Univ. Forest. Sci. Technol. (Social Science edition), **9**(06), 60–66 (2015)
11. Li, X.: On the theory and method of cultural creativity and its application. Res. Cult. Ind. Chin. Museums **00**, 24–29 (2015)
12. Liu, M.: Pragmatic solution to the integration of rural primary, secondary and tertiary industries. Heilongjiang Grain **12**, 26–27 (2015)
13. Zhou, J., Qiu, H.: Research on the integration and development of China's cultural industry based on grey theory. Yue Jiang Xue J. **7**(05), 72–79 (2015)
14. Zou, Y.: Theoretical and empirical research on the integration development of media industry and sports industry – taking Liaoshen region as an example. China Sport Science Society. Collection of abstracts of the 10th National Sports Science Conference 2015 (I). China Sport Science Society: China Sport Science Society, 1331–1332 (2015)
15. Wang, S.: Theoretical analysis on industrial integration of China's modern agricultural development path. Agric. Econ. **10**, 34–35 (2015)
16. Ruan, J.: Research on the development of pension hotels based on the theory of industrial integration. Southeast University (2015)
17. Qu, T., Li, Y.: Research on the transformation and upgrading mode of tourism industry based on the theory of industrial integration – taking Guizhou Province as an example. Mod. Bus. **24**, 40–42 (2015)
18. Liu, W., Tang, Y.: Literature review of industrial tourism development in Liaoning Province from the perspective of industrial integration. Econ. Res. Guide **23**, 111+120 (2015)
19. Guo, X.: An analysis of the development of Guangxi cocoon and silk industry from the perspective of industrial integration – reflections on the construction of the 21st century Maritime Silk Road. J. Guangxi Normal Univ. (Philos. Soc. Sci. Ed.) **51**(04), 15–19 (2015)
20. Ying, S.: Research on the integration mechanism of cultural industry and information industry at the enterprise level. J. Zhejiang Provin. Party Sch. **31**(04), 98–103 (2015)

Research on the Improvement of Learning Effect of Mental Health Education Curriculum Based on Big Data

Yan Zhang[✉]

Xi'an Vocational and Technical College, Xi'an 710071, Shaanxi, China

Abstract. This paper studies the concept, process, method and technology of big data technology, introduces in detail the research on the improvement of the learning effect of big data technology in the students' mental health education course, uses big data technology to establish the students' psychological problem model in the algorithm, and gives the classification rules, so as to provide some useful reference for the psychological consultation work in Colleges and universities. The experimental results show that this method has a certain practical value for the construction of preventive mental health education mode for college students.

Keywords: Data · Mental health · Education curriculum

1 Introduction

Many schools have set up special institutions for mental health education, counseling or consultation to survey the mental health of freshmen, conduct relevant psychological tests on students one by one, and establish students' personal mental health files on this basis. However, since its establishment, many psychological counseling centers in Colleges and universities have been in the plight of shortage of psychological counseling staff and funds. In addition, although they have conducted a survey on the mental health of freshmen and accumulated a large amount of data, the processing of these data is still in the stage of traditional analysis and statistics. How to establish a scientific and efficient early warning mechanism for students' mental health is a severe challenge for the current mental health education in Colleges and universities [1]. One of the effective ways to solve the above problems is to use data mining technology to find out the hidden information from the existing college students' personal mental health archives database and provide support for the planning and decision-making of mental health education in Colleges and universities.

2 Introduction and Application of Data Mining Algorithm

Data mining (DM) is to extract knowledge that people are interested in from a large number of data. These knowledge are implicit, unknown and potentially useful information, and the extracted knowledge is expressed as concepts, rules, rules and other forms.

M. A. Jan and F. Khan (Eds.): BigIoT-EDU 2021, LNICST 392, pp. 301–310, 2021.
https://doi.org/10.1007/978-3-030-87903-7_38

The process of data mining is a human-computer interaction process which is composed of multiple steps connected and repeated. A complete data mining process generally includes data preparation, data mining, expression and interpretation of results. There are many methods and technologies of data mining, including decision tree method, neural network method, association rule method, genetic algorithm, statistical analysis method, visualization method, rough set theory method and fuzzy mathematics method. Among them, decision tree method is the most popular data mining technology [2–4]. Compared with other data mining methods, decision tree method has less computation, can process continuous and discrete data, and can generate understandable rules. The research content of this topic is to build decision tree with C4.5 algorithm. This paper forecasts the mental health status of college freshmen, finds out the most likely attributes to affect their mental health, so that the school can scientifically and effectively warn, intervene as soon as possible, and focus on prevention of students' mental health problems, so as to make the school's psychological counseling work more targeted and purposeful, and improve the level and efficiency of psychological counseling work.

3 Principle of C4.5 Algorithm

3.1 Principle of C4.5 Algorithm

There are ID3 algorithm and C4.5 algorithm in common use. The core idea of ID3 decision tree algorithm is to use the information principle to select the attribute with the largest information gain as the classification attribute, recursively expand the branches of the decision tree, and complete the construction of the decision tree. However, in practical application, the following problems exist in the information gain function of ID3. The more branches of the test attribute, the greater the information gain value [5]. However, the more branches of the output does not mean that the test attribute has a better prediction effect on the unknown objects [6]. Therefore, in C4.5 algorithm, people propose to use the information gain rate as the basis for the selection of test attributes. C4.5 algorithm is an improved ID3 algorithm, which is based on the following principles.

Definition 1: let S be a set of S data samples. Assuming that the class label attribute has m different values, m different classes are defined $C_i : (i = 1, 2 \ldots, m)$. Let s be the sample number of class C. Where p is the probability that any sample belongs to C and is estimated by s, s. For a given sample classification, the expected information is as follows:

$$I(S_i, S_2, \ldots, S_m) = -\sum_{i=1}^{m} \frac{S_i}{S} \log_2 \frac{S_i}{S} \tag{1}$$

Definition 2: the information gain of a is the expected entropy compression caused by knowing the value of attribute a, and the formula is:

$$Gain(A) = I(S_1, S_2, \ldots, S_m) \tag{2}$$

The main idea of C4.5 algorithm is: assuming t is the training set, when constructing the decision tree for T, the attribute with the largest gainratio (x) value is selected as the splitting node, and t is divided into n subsets according to this standard. If the classes of tuples in the i-th subset T are consistent, the node becomes the leaf node of the decision tree and stops splitting [7, 8]. For other subsets of t that do not satisfy this condition, the spanning tree is recursively constructed according to the above method until the tuples of all subsets belong to one category.

3.2 Application of C4.5 Algorithm

When freshmen enter school every academic year, many colleges and universities will make a thorough investigation on the mental health status of students. Schools with a high degree of information have stored these data in the database of student management system. To find out the hidden information from the massive data of the existing college students' personal mental health archives database, so as to provide decision support for college mental health education [9]. For example, through the analysis of the students' mental health archives database system, we can find the main factors that affect students' psychological problems, provide the basis for the school to carry out the planning and decision-making of mental health education, and further explore the new methods of early prevention and intervention of students' psychological barriers. In this paper, C4.5 algorithm will be used to create a decision tree, the specific implementation plan is as follows.

4 Data Preprocessing

In general, the collected data is incomplete, harmless and noisy, so it is necessary to preprocess the data to improve the quality of data mining objects, which helps to improve the accuracy and performance of the mining process. Data preprocessing usually includes the following three steps.

4.1 Data Cleaning

In the data of students' psychological problems, some interesting attributes lack values, which can be filled by data cleaning. For a large number of vacant items, the method of ignoring tuples is used to delete them. For individual vacancies, the method of manual filling is used. The filling method uses most of the attribute values on the attribute to fill in the vacant attribute. After data cleaning, the total number of records is 960.

4.2 Data Extraction

Data extraction is sometimes referred to as data sampling or data simplification. It is based on the understanding of the discovery task and the content of the data itself, to find the characteristics of the expression data that depend on the discovery target, so as to reduce the size of the data, so as to minimize the amount of data on the premise of keeping the original data as much as possible (see Fig. 1).

In this paper, we extract the attributes that have a key impact on students' mental health from the database of College Students' psychological problems, and determine the four attributes of introversion, family harmony, family income and mental genetic disease to be mined respectively.

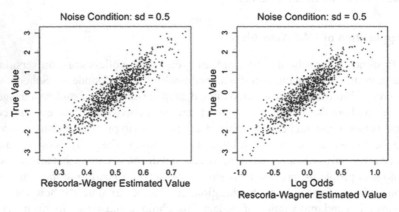

Fig. 1. Simulation with data extraction

4.3 Data Conversion

The purpose of data transformation is to transform the data information after data cleaning into a form suitable for mining, and establish a real analysis model suitable for mining algorithm. For example [10–15]. The "family income" field in mental health database was originally a continuous value. When data mining, this attribute must be transformed into a discrete value (the attribute name is changed to "economic difficulty", and the value is yes or no) to be suitable for classification mining task.

After data preprocessing, there are 960 student records. In order to evaluate and predict the decision tree model, 1/3 of the records are reserved as test data, and 2/3 of the record data are reserved as the training set of decision tree model, with a total of 640 records.

4.4 Building Decision Tree

The training set is taken as an example to illustrate the generation of decision tree model of students' mental illness. The steps of C4.5 algorithm to establish decision tree model are as follows: (1) calculate the information gain rate of each test attribute in Table 2; (2) select the attribute with the largest information gain rate as the root node, and divide the data set according to its value, if the attribute has only one value, stop dividing; (3) recursively execute (1)–(2) for each sub data set. According to the above formula, calculate the information gain rate of each test attribute, select the attribute with the largest information gain rate from the calculation results as the root node, divide the sample, lead to multiple branches, then calculate the division of each branch node, repeat the above steps, complete the division of each branch (see Fig. 2).

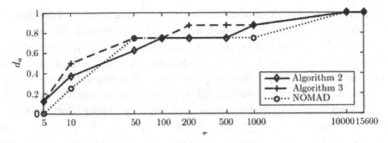

Fig. 2. Simulation of decision tree

5 Enhance the Understanding of College Students on the Course of Mental Health Education

Changing the traditional ideas is the key to let students accept new things independently. For college students' mental health education, it is a new teaching task with the development of society and economy, and it is a new teaching field. In order to promote the further development of this teaching interaction, the premise is to actively change the traditional teaching ideas of teachers and students. In this regard, we should do the following: first, we should fully understand the significance of College Students' mental health education [16]. For the teaching of College Students' mental health education, it is not only the transmission of students' knowledge, but also a new teaching concept. It is a teaching belief integrated with a variety of teaching tasks. It is a teaching content constructed by different types of teaching activities. Therefore, this kind of teaching activity is a new discipline different from the traditional teaching methods. The realization of College Students' mental health education course is not only aimed at the students with psychological problems or the students with psychological obstacles, but also in the face of the majority of teachers and students in the school. At the same time of obstacle prevention, it is to optimize students' psychology and use advanced teaching mode, so as to steadily improve students' overall psychological quality and make them face the challenges of life with a more positive attitude.

6 Building a Strong Team of Mental Health Education Teachers

For the study of this subject, its main purpose is not only to let students master a lot of theoretical knowledge, but also to cultivate students' comprehensive psychological quality. Mental health education course is a special course which is different from other professional disciplines. It is different from other courses. Specifically, its curriculum planning and activity organization are relatively special. For example, in the arrangement of psychological courses, the school generally chooses the compulsory course mode for the students to study the theory of psychology and discuss the psychological problems in the classroom. For a small number of students with special circumstances, it can adopt the mode of elective courses or organize psychological lectures to popularize psychological knowledge [17–19]. As far as the teaching content of psychological course is concerned, its main body is to focus on students' psychological needs and learning interest, integrate the psychological template with the characteristics of the times, further approach students' life practice, use the teaching method with scientific interest, set up interesting psychological education courses, and attract students to actively participate in psychological learning. For the mode of psychological classroom, it should establish diversified classroom teaching, effectively combine psychological theory with psychological cases, reappear psychological accident cases in the course teaching, and arouse students' thinking, so as to better mobilize students' independent participation. At the same time, the formation of a large team of psychological teaching teachers to help the orderly development of psychological teaching tasks, on the current situation of psychological teaching, the completion of the teaching task of psychological teachers put forward strict requirements, so do a good job in Teachers' psychological course guidance, the establishment of University Teachers' psychological qualification certification system, can better promote the development of their work.

7 Establishing an Integrated Curriculum System of College Students' Mental Health

7.1 Effective Coordination of the Relationship Between Mental Health Education Curriculum and Psychological Counseling

Science and health education course is the popularization of psychological knowledge for all teachers and students in Colleges and universities, while psychological consultation is the psychological adjustment for different students. Through the effective combination of the two, we can integrate their psychological theory into the specific practice and build a unified whole. In short, teachers should strengthen the psychological education of students in the course task, guide students to establish a new understanding of psychological counseling, change the old ideas, and understand that psychological counseling is not only aimed at people with psychological diseases [20]. Let students redefine psychological counseling and get out of their own misunderstanding of psychological counseling.

7.2 Optimizing the Relationship Between Mental Health Education and Moral Education Curriculum

Most of the objects of psychological education courses in Colleges and universities are school students, whose main goal is to educate talents, while the development of school moral education courses is to cultivate students' all-round development ability as the theme, and their goals are the same. On this basis, the Education Department of our country has made a distinction between mental health and moral education curriculum. As far as moral education is concerned, it is mainly to effectively distinguish right from wrong in the ideological field, so as to improve the ideological understanding to a certain extent. As far as mental health education is concerned, it is mainly to strengthen students' psychological construction, while excavating students' psychological potential, To help students develop in a positive and healthy way.

7.3 Perfect the Relationship Between Mental Health and Ideological and Political Education

College Students' mental health curriculum is not equal to ideological and political education. They have essential differences and cannot be replaced by each other. However, they can integrate with each other through their own common points. First, promote the complementary advantages of the two. Through daily study, strengthen the ideological and political education of students, promote students to establish a correct outlook on life, world outlook, realize their life value. On this basis, through a series of Ideological and political teaching, it will be integrated into the students' psychological education, and guide the students to go out of the psychological misunderstanding independently. Second, optimize the psychological feedback system. We should build a bridge between the psychological education department and other colleges and universities, and exchange the psychological problems of modern college students in their mutual communication. Third, increase the mutual complement of education team [21]. We should give full play to the key role of the student management team in mental health education, increase the dialogue between students and teachers, popularize psychological knowledge to them, and increase the accumulation of their mental health awareness, so as to help them shape good psychological quality.

8 Carrying Out Teaching Reform and Innovating Teaching Mode

8.1 Case Study Method

The development of mental health education curriculum can not only teach students a lot of psychological knowledge, but also help students adjust their mental state and optimize their psychological quality in life. Therefore, if teachers only teach psychological theory knowledge in mental health education, they should further enhance their psychological adjustment ability on this basis. In this regard, teachers should focus on grasping the students' psychological situation, collect the common psychological problems in students' daily life, and put them into a book for students' reference. The third line is shown in the Fig. 3. While attracting students' heated discussion, teachers should teach students' psychological adjustment function. It should be noted that when teachers choose teaching cases, they should pay attention to the actual control of the cases, arouse students' resonance, and use the most representative psychological cases to actively participate in the mental health education curriculum.

8.2 Action Training Method

With the development of mental health education curriculum, the behavior training method can promote the change of students' cognition to a certain extent, use specific behavior patterns, constantly optimize students' behavior mechanism, and effectively correct students' bad behavior in psychological teaching. For example, in real life, some students do not dare to express their opinions, can not control their emotional problems, and finally make it unable to adjust the interpersonal relationship around them. In this regard, teachers can enhance students' confidence in the psychological teaching course,

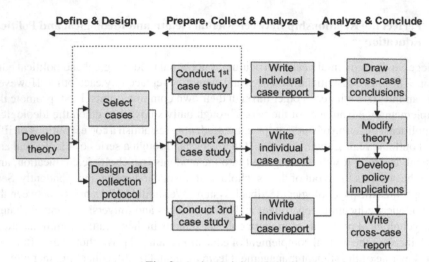

Fig. 3. Action training method

and optimize their behavior awareness through behavior training method. Specifically, organize the students to be divided into two groups, one is the opposite party, the other is the opposite party [22–24]. The teacher can encourage the other group of students who are not confident to learn to refuse, to express their own ideas, and to establish their own unique personality while letting others understand themselves.

9   Concluding Remarks

This paper mainly introduces the decision tree technology in data mining, and introduces the whole process of C4.5 algorithm mining in college students' psychological education. The data comes from the psychological test of 2010 freshmen in my college. Before data mining, data cleaning, extraction, conversion and other data preprocessing are carried out to lay a good foundation for further mining work. In the process of building the decision tree model, C4.5 algorithm is used. The algorithm recursively selects the attribute with the maximum information gain rate as the test attribute, and finally generates the decision tree model of whether the students have psychological diseases, and thus generates easy to understand classification rules, which provides some useful reference for the school's psychological counseling work. The follow-up work of this topic is to find out more attributes that are highly related to students' mental health. Using other mining algorithms (such as neural network, genetic algorithm) to further improve the prediction accuracy.

References

1. Wang, Y.: Practice of "teaching learning doing" mode in college students' mental health education. J. Beijing Print. Univ. **28**(12), 149–152 (2020)

2. Chang, E.: Research on teaching design of mental health education in secondary vocational schools. Mod. Vocat. Educ. **02**, 86–87 (2021)
3. Zhu, S.: The status quo and thinking of mental health education in primary and secondary schools – based on the investigation and analysis report of the status quo of mental health education in primary and secondary schools. Mental Health Educ. Prim. Second. Schools **36**, 4–10+14 (2020)
4. Ma, S.: Research on classroom teaching mode of college students' mental health education. J. Anhui Vocat. Coll. Electron. Inform. Technol. **19**(06), 73–75 (2020)
5. Ding, G.: Research on the Implementation Difficulties and Countermeasures of Mental Health Education Curriculum in Urban High Schools. Henan Normal University (2016)
6. Wang, J., Wang, X., Chen, D., Zhang, D.: Application of experiential teaching in college students' mental health education. China Med. Educ. Technol. **30**(02), 129–131 (2016)
7. Xu, R., Song, C.: Application of teaching mode under constructivism learning theory in college students' mental health education. West. Leather **38**(06), 284–285 (2016)
8. Qian, J.: Enlightenment of MOOC teaching on teaching reform of college students' mental health education. J Qiqihar Teachers Coll. **02**, 130–132 (2016)
9. Guo, Y.: Some ideas on the rational use of experiential learning method in university mental health curriculum. J. Jiamusi Vocat. Coll. **01**, 189–190 (2016)
10. Wan, R., Lan, W.: Classroom teaching experiment based on the learning needs of College Students – taking the course "mental health education for college students" as an example. J. Guizhou Normal Univ. (Nat. Sci. Ed.) **33**(06), 47–50 (2015)
11. Liang, P.: Exploration and practice of mental health education curriculum reform in secondary vocational schools. Curriculum Educ. Res. **35**, 237–238 (2015)
12. Qi, X.: Research on the development of mental health education curriculum in kindergartens in Shijiazhuang. Hebei Normal University (2015)
13. Pu, X., Zou, T.: Enlightenment of American social emotional learning on the construction of mental health education curriculum for college students in China. J. High. Educ. **20**, 21–23 (2015)
14. Wu, M.: Teaching research on mental health course in higher vocational colleges from the perspective of students' subjectivity. China Adult Educ. **16**, 147–149 (2015)
15. Hu, Y., Zhao, B.: Analysis of mental health status of higher vocational students and countermeasures. Acad. Theory **24**, 66–68 (2015)
16. Zhang, J.: Research and practice of social emotional learning curriculum in the United States. Development Psychology Committee of Chinese Psychological Society. Abstracts of the 13th Annual Conference of Development Psychology Committee of Chinese Psychological Society. Development Psychology Committee of Chinese Psychological Society: School of psychology, Department of education, Northeast Normal University 1 (2015)
17. Liao, X.: Development and practice of mental health education curriculum in junior middle school. Central China Normal University (2015)
18. Liu, K.: The current situation and reflection of mental health curriculum in Colleges and universities from the perspective of subjectivity education. Sci. Educ. Wenhui (Next Issue) **06**, 136–137 (2015)
19. Peng, Z.: Research on network curriculum design under the concept of blended learning. Shanghai Normal University (2015)
20. Chen, Y.: Exploration and practice of college students' mental health education curriculum reform. Xueyuan **17**, 65–66 (2015)
21. Lei, J.: Exploration of practicing "experiential learning" in "mental health education" course of secondary vocational school. Sci. Educ. Literat. Collect. (Next Issue) **05**, 101–102 (2015)

22. Zhang, X.: A road worth exploring: research learning practice of mental health education and color curriculum in secondary vocational schools. Jilin Educ. **10**, 149 (2015)
23. Liu, S.: Beauty education lays the foundation for a beautiful life. Future Educ. **z1**, 124 (2015)
24. Meng, R.: Exploring a new way to implement mental health education curriculum in Colleges and universities MOOC. Asia Pac. Educ. **07**, 126+124 (2015)

Research on the Integrated Development Model of Rural Tourism and E-commerce Under the Background of Big Data

Kang Liu[✉]

Sanya Aviation and Tourism College, Sanya Hainan 572000, China

Abstract. The combination of e-commerce and tourism has played a great role in promoting the overall development of tourism, improving the tourist experience and improving the tertiary industry of tourism destination. Based on the research background and basis of tourism e-commerce, this paper puts forward the development mode of "rural tourism + e-commerce". In addition, this paper also constructs the basic system of the comprehensive application of rural tourism e-commerce, which can provide reference for the decision-making of rural tourism related departments and the increase of farmers' income.

Keywords: Rural tourism · E-commerce · Application system

1 Introduction

Since the 1980s, tourism has gradually become an important pillar industry in China. In 2019, there will be 6.01 billion domestic tourists and 5725.1 billion yuan of domestic tourism revenue. While the overall tourism industry is booming, the number of rural tourists is also increasing. In the first half of 2019, the total number of rural tourists has exceeded 1.5 billion. In recent years, with the requirements of the policy and the continuous improvement of the infrastructure in rural areas, the share of rural tourism in China's tourism industry is gradually becoming heavy, and it has become one of the focuses of China's tourism development. Rural tourism is rich in content, products and highly inclusive. The integration of "rural tourism + e-commerce" is a new mode of integrated development. Through the integration of rural tourism resources in different regions, the accurate connection between destination resources and market can be realized, and the economic structure of rural tourism can be optimized by seizing the opportunities brought by e-commerce. The establishment of Huaxia tourism network marks the rise of tourism e-commerce in China, and then such websites in China have entered a stage of rapid development [1]. Yan Li and other scholars believe that the rapid development of rural tourism in China is due to the rapid development of urbanization, the increase of urban population density, and the increasingly serious environmental pollution. Zhang yugai and LAN Guiqiu believe that the development of rural tourism must build an information platform to meet the needs of different tourists and consumers.

M. A. Jan and F. Khan (Eds.): BigIoT-EDU 2021, LNICST 392, pp. 311–319, 2021.
https://doi.org/10.1007/978-3-030-87903-7_39

In addition, foreign scholar Williams thought in 1993 that the combination of information technology and tourism will become an inevitable trend. Now, this research result has been well proved that the e-commerce industry chain of rural tourism with farmhouse as the core has been quite mature [2]. The Fig. 1 shows the e-commerce platform architecture.

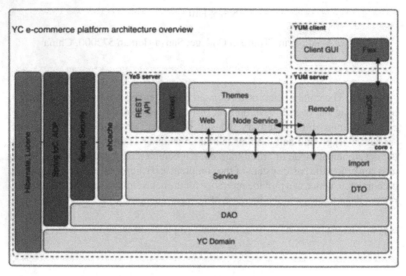

Fig. 1. The e-commerce platform architecture

2 The Importance of "Tourism E-commerce+" in the Integration of Rural Tourism and Targeted Poverty Alleviation

2.1 It is Helpful to Realize the Deep Promotion of Rural Tourism

In terms of the development of rural tourism, although its scale and scope are gradually increasing, its main tourism brand is not outstanding, and it does not have sufficient influence. People have many shortcomings in their understanding, and it is the main problem faced by the current rural tourism work that it does not give full play to the publicity effect. Rural tourism is a key project to promote local targeted poverty alleviation, but because of the weak promotion and publicity, it is still a gap with the target of targeted poverty alleviation. Although tourism resources are very high-quality, they are hidden in the areas where the traffic is not convenient [3]. These resources can bring sufficient impetus to the local economic development. If we want to promote the development of local targeted poverty alleviation through rural tourism, tourists are absolutely essential factors. Without tourists visiting the current tourism area, the economic development will inevitably be greatly impacted, and targeted poverty alleviation cannot be realized naturally. Only by means of Internet can tourism resources and brand products exist in rural areas be transmitted to the outside world and attract more people to travel. As a new mode of e-commerce work in the new era, tourism e-commerce+, which is not only limited to building a complete product line on the network, but also can promote the current rural tourism resources through the network, and effectively strengthen the visibility of the current rural tourism area. It is not only helpful to sell all kinds of brand products, but also can create a more rapid and complete platform with higher quality, which is an important content of promoting the marketing of ecotourism [4].

2.2 It Helps to Realize the Deep Marketing of Rural Tourism

In the process of building rural tourism system, the folk workshops such as oil workshop and tofu workshop opened by old crafts are not only the visiting places of traditional folk arts and crafts, but also the processing places of various products. In this process [5], it is necessary to recruit and accurately help poor people participate in the production work before, and help them get rid of poverty, Using the technology of "tourism e-commerce+" to promote these products and various agricultural and sideline products such as poultry and pork raised by the poor people to the outside world, effectively reduce the problems in the process of sales. This can not only promote local characteristic products, but also attract corresponding manufacturers to invest or purchase in the current region by means of network system. At the same time, they can also be sold to all parts of the country by means of network, and comprehensively strengthen the reputation of local products and effectively open the current product sales market. In the process of rural tourism management, all kinds of tourism services can be released to the corresponding e-commerce platform and sold. This can not only facilitate the corresponding tourists, but also comprehensively improve the current service level and effectively achieve the important goal of targeted poverty alleviation [6–8].

3 Suggestions on Integration of Rural Tourism and Targeted Poverty Alleviation Under the Background of "Tourism E-commerce+"

3.1 Change the Existing Misconception, and Promote the Deep Combination of Rural Tourism and Targeted Poverty Alleviation

Under the guidance of the information age, the basic state of rural tourism and targeted poverty alleviation work needs to be fundamentally changed. The role of e-commerce in rural tourism publicity is not underestimated, which is an important impetus to integrate rural tourism and targeted poverty alleviation. Under the important goal of targeted poverty alleviation, the importance of "tourism e-commerce+" in rural tourism must be changed by both the government and the individual, and their understanding should be improved in an all-round way, and the important role of correctly applying "tourism electronic commerce+" technology in the current situation must not be separated from rural tourism and targeted poverty alleviation, Instead, we should integrate the three together and promote the current rural tourism content with the help of "tourism e-commerce+" to help the local targeted poverty alleviation. We should take the targeted poverty alleviation as the ultimate goal of rural tourism work, and "tourism e-commerce+" is a new development means. In order to make rural areas have prosperous tourism industry, and then obtain high economic benefits and effectively get rid of poverty, at the same time, it is necessary to introduce the important technology of "tourism e-commerce+" to guide the participation of the masses of poor farmers and provide corresponding employment channels for them [9].

3.2 "Tourism E-commerce+" Should Play "Experience Card" Widely in Rural Tourism Work

In the context of "tourism e-commerce+", if we want to promote the handover of local tourism and targeted poverty alleviation in an all-round way, so that the two can be integrated, we must play "experience card" widely [10]. Specifically, in the process of basic development, we should correctly apply the characteristics of "Tourism e-commerce+", Through the new mode of online experience and offline experience integration, the influence of rural tourism work is expanded, more public will be attracted to travel, driving the sustainable development of local economy, and ensuring that rural areas can get rid of the poverty problems existing in the current place by the strength of "experience card" [11]. Online travel experience must fully show attractiveness, through the third party e-commerce platform or build its own marketing platform, and display local tourism specific products with the help of video, text and pictures. At the same time, WeChat official account and micro-blog can also be set up to display local characteristics for foreign tourists in the Internet. If the conditions permit, then you can add corresponding VR video to more vividly present the tourism color of local solitary residence, strengthen online experience activities in an all-round way, and expand the local influence. After online experience is completed, tourists will go to rural tourist sites to experience offline, so that the goal of targeted poverty alleviation will be achieved. Of course, in order to guarantee the long-term development of the current tourism economy, tourists must be given sufficient sense of access when performing offline services, let tourists feel the visit by themselves, and strengthen the tourists' trust in all aspects. In this process, online experience needs to focus on attracting tourists' attention and make them have a full sense of identity and exploration [12]. Then, it is to deepen the tourists' memory by offline experience, so that they can feel the actual tourism experience without violation, and then help to cultivate their love for rural tourism resources [13].

4 Data Mining Technology

4.1 Basic Concepts of Data Mining

Data mining is also known as knowledge discovery in database. It aims to build a descriptive or predictive model, and discover the knowledge and hidden patterns or trends by processing a large number of data. The popular understanding is that data mining can help users solve practical problems and make correct decisions and judgments by extracting useful information from a large number of data. The explosive growth of data volume is the fundamental reason for the emergence of data mining technology [5]. At the same time, data mining is widely used in a variety of fields, especially in finance, industry, risk management and analysis, and telecommunications. For example, in the data analysis process of IDS, data mining technology can be introduced to find valuable information in a large number of security audit data, and then use these valuable information to achieve higher accuracy and detection rate. The main steps of data mining are shown in Fig. 2 below:

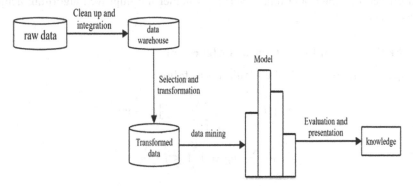

Fig. 2. Shows the main steps of data mining

5 Application of Data Mining Technology in Intrusion Detection

In the process of intrusion detection, data mining technology is very good at mining hidden and valuable information and knowledge from a large number of security audit data, and then build the acquired knowledge into an understandable rule set or pattern, Finally, the constructed rules or patterns are used to efficiently detect whether there are activities violating security policies or suspicious attacks in the network data [14].

Association rule is a kind of technology which aims to find out relevant data items from a large number of data. Association rules are usually divided into two steps to find the relationship between data items. The first step is to find all frequent itemsets that meet the minimum support threshold. The role of the second step is usually more critical than the first step, that is, to transform the discovered item sets into association rules that meet the confidence threshold requirements. However, it needs huge computing resources when facing large data sets. On the basis of unsupervised learning, association rules can be expressed by "if-then" model. At the same time, apron and RP growth algorithms in association rule technology are widely used in recommendation system and intrusion detection [15]. When only using the traditional association rule technology to detect whether there are suspicious activities or behaviors in large-scale network data, it will mistakenly regard the discovered item set as equally important performance and consume a lot of time and resources. Therefore, in view of the shortcomings of the traditional association rule technology, it needs to make effective improvements before it can be used to detect attacks in the network [16].

6 Research on Fuzzy C-means Algorithm

At present, clustering analysis plays an important role in intrusion detection, because it can not only detect new attacks, but also provide high-quality data sets as a preprocessing technology. Therefore, this chapter focuses on the principle of fuzzy c-means algorithm, proposes the weighted Euclidean distance and uses the density method to select a good initial clustering center, and finally verifies whether the improved algorithm achieves the desired effect [17].

6.1 The Principle of Fuzzy C-means Algorithm

The objective function of FCM algorithm is as follows:

$$\min J(U, V) = \sum_{i=1}^{c} \sum_{j=1}^{n} u_{ij}^{m} \| x_j - v_i \|^2 \tag{1}$$

$$s.t. \sum_{i=1}^{c} u_{ij} = 1, 1 \le j \le n \tag{2}$$

In Formula 1 and 2, m is called fuzzy weighted index.

6.2 Improvement of Fuzzy C-means Algorithm

The process of clustering analysis is divided according to the dissimilarity of each data, so choosing a reasonable measurement standard will have a great impact on the good or bad clustering results. Generally speaking, the traditional clustering algorithm uses the distance in the feature space as a metric to calculate the dissimilarity between different data. When distance is used as dissimilarity measurement, the smaller the distance between data, the more similar they are and the more likely they are to be divided into the same cluster; the larger the distance between data, the greater the difference between them and the less likely they are to be divided into the same cluster [18–20].

7 The Integration Development Strategy of Rural Tourism and E-commerce in China

7.1 Analysis of the Problems in the Integration of Rural Tourism and E-commerce in China

As a new model, "rural tourism + e-commerce" has a series of known and unknown problems, such as: difficult quality standard certification, lagging development of cold chain logistics system, imperfect logistics distribution system, lack of management in all aspects of production and distribution, inaccurate marketing channels, fierce competition of homogeneous products, etc. The development of "rural tourism + e-commerce" integration mode not only needs the support of government policies and enterprises' practical actions, but also needs to establish a perfect system, strictly manage every production and service link and attract capital investment. At the same time, we need to build e-commerce platform, strengthen personnel training, speed up infrastructure construction, and use the Internet to form integrated marketing publicity [21].

7.2 Simulation for Rural Tourism and E-commerce in China

The government can introduce a series of targeted policies and plans according to the local situation and referring to the relevant national, provincial and local policies and plans, as the policy guarantee for the development of "rural tourism + e-commerce" mode [6, 22].

Rural tourism has a vast territory. In order to integrate with e-commerce, professional promotion and operation teams are needed to settle in the countryside, collect merchant information (rural B & B, farmhouse [23], restaurants, entertainment places, etc.) and carry out later operation and maintenance. The mode of enterprises helping rural areas can reduce the financial pressure of local government, quickly open up local marketing channels, form multi advantages and mutual resource sharing, and form a virtuous circle of "rural tourism + e-commerce" development. The simulation results are shown in Fig. 3 and Fig. 4.

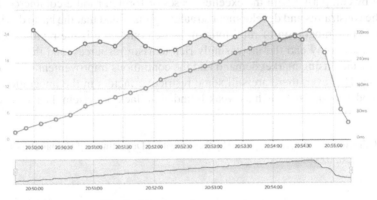

Fig. 3. Number of concurrent users

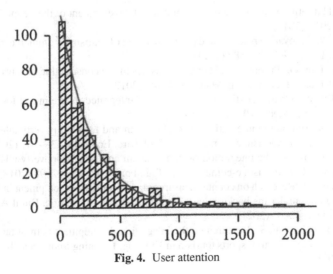

Fig. 4. User attention

The establishment of rural tourism e-commerce platform provides a large number of diversified information services for rural tourism industrialization, and provides timely and accurate practical information of rural tourism industrialization resources, market, production, policies and regulations, relevant technical personnel, disaster prevention and mitigation for rural tourism producers, operators and managers. At the same time, it provides online trading platform for enterprises and villagers, Reduce the threshold for enterprises and villagers to engage in e-commerce, cultivate and support rural e-commerce enterprises [24].

8 Conclusion

Based on the perspective of the integration of rural tourism and e-commerce, this paper combs and summarizes the development status of this mode at home and abroad, and summarizes the problems existing in the integration of rural tourism and e-commerce in China by comparing with the excellent cases of foreign rural e-commerce, and discusses the constraints and development strategies. It is found that the basic driving force to promote the integration of rural tourism and e-commerce is the rapid development of China's economy and the increasingly rich tourism resources, and the core driving force is the increasing market demand and the continuous improvement of information technology. However, there are still some restrictive factors in the integration of rural tourism and e-commerce, such as weak foundation, lack of system, lack of talents and funds.

Acknowledgements. Project supported by the Education Department of Hainan Province, project number: Hnky2020ZD-28.

References

1. Statistical bulletin of national economic and social development of the People's Republic of China in 2019 (2019)
2. Yu, P., Fu, Y.: innovation research and platform design of tourism e-commerce mode in China. Comput. Modern. **07**, 179–182 (2011)
3. Zhang, Y., Lan, G.: Problems and countermeasures in the construction of tourism informatization in China. Modern. Shop. Malls **18**, 47–48 (2012)
4. Lei, C., Hu, W.: Characteristics and Strategies of integrated marketing of tourism industry using new media. Media **10**, 80–82 (2016)
5. Sun, J.: Research on the integration of rural tourism and targeted poverty alleviation under the background of "tourism e-commerce+." Natl. Circ. Econ. **33**, 138–139 (2019)
6. Lin, N.: Research on the integration of rural tourism and targeted poverty alleviation under the background of "tourism e-commerce+." Tour. Forum **11**(06), 92–98 (2018)
7. Xu, Y., Song, Z.: Research on countermeasures of rural tourism development in Saiyang town of Jiujiang City under the background of targeted poverty alleviation. Rural Agric. Farmers (B Edition) **03**, 42–43 (2021)
8. Zhao, Y., Lan, T., Cai, S.: Research on the integration development path of targeted poverty alleviation policy and rural sports tourism in Xinjiang. Liaoning Sports Sci. Technol. **43**(02), 25–28 (2021)

9. Liu, Y., Gao, S.: Problems and countermeasures of rural tourism development from the perspective of targeted poverty alleviation. Rural Sci. Technol. **12**(02), 44–45 (2021)

10. Yuan, W.: Integrated development path of "rural tourism + targeted poverty alleviation." Contemp. Tour. **18**(35), 40–41 (2020)

11. Hui, X.: Current situation and countermeasures of rural tourism targeted poverty alleviation. Agric. Econ. **05**, 91–93 (2020)

12. Ma, J.: Research on the development of rural tourism in Shanxi Province from the perspective of targeted poverty alleviation. Mod. Market. (Next Issue) **12**, 117–118 (2019)

13. Wang, Y.: Problems and countermeasures of rural tourism development in Henan Province from the perspective of targeted poverty alleviation. Rural Sci. Technol. **34**, 43–44 (2019)

14. Tian, C., Sun, L.: Exploration on the integrated development of rural tourism and targeted poverty alleviation. Shanxi Agric. Econ. **20**, 77+79 (2019)

15. Zhai, Y.: Analysis of rural tourism development countermeasures from the perspective of targeted poverty alleviation – taking Guanyang business village as an example. J. Shanxi Econ. Manage. Cadre Coll. **27**(03), 61-63+67 (2019)

16. Liu, F., Cao, Z.: Research on the integration of folk sports and rural tourism in Shaoguan from the perspective of targeted poverty alleviation. Contemp. Sports Sci. Technol. **9**(20), 201–204 (2019)

17. Zhao, Y.: Research on the integrated development of sports culture and rural tourism in Chizhou under the background of targeted poverty alleviation. Shanxi Agric. Econ. **12**, 117–118 (2019)

18. Yan, J.: Research on the integrated development mode of rural tourism and tourism targeted poverty alleviation in Hengnan County. Tour. Overview (Second Half) **12**, 175–176 (2019)

19. Wang, D., Li, G., Jing, X.: Integrated development of rural tourism and targeted poverty alleviation. Resour. Hum. Settle. **03**, 10–14 (2019)

20. Fu, X.: Research on the integration development of Guangxi ethnic sports ecological culture construction and rural tourism under the background of targeted poverty alleviation. Comparat. Study Cult. Innov. **3**(01), 44–45 (2019)

21. Liu, Q.: Research on the integration of rural tourism development and ethnic sports cultural creativity in Guangxi under the background of targeted poverty alleviation. J. Baise Univ. **31**(06), 99–102 (2018)

22. Zhu, A.: Research on targeted poverty alleviation of rural tourism in Taihang mountain area of Hebei Province under the coordinated development of Beijing Tianjin Hebei. Tour. Overview (Second Half) **14**, 145 (2018)

23. Yang, L., Guo, Z.: Financial support for rural tourism development in Guang'an under the background of targeted poverty alleviation. Tour. Overview (Second Half) **14**, 164-165+167 (2018)

24. Gao, L.: Integrated development path of rural tourism and targeted poverty alleviation in contiguous destitute areas: a case study of Jingyuan County, Gansu Province. Tour. Overview (Second Half) **14**, 168 (2018)

System Design and Development of Logistics Management Talents Training Program in Colleges and Universities Under the Background of Big Data

Na Yin[✉]

School of International Education, Wuhan 430056, China
ccshcc@ccshcc.cn

Abstract. Combined with the basic positioning of logistics personnel training in modern colleges and universities, this paper discusses the system design and development ideas of logistics management personnel training scheme in Colleges and Universities under the background of big data, and formulates a set of feasible personnel training scheme, so as to realize the direct correspondence with the actual professional posts of small and medium-sized logistics enterprises and realize the personnel training objectives.

Keywords: Big data background · Logistics major · Personnel training · Program

1 Introduction

The 12th Five Year Plan of the people's Republic of China clearly puts forward that the logistics industry is one of the top ten revitalizing industries, which indicates that the development of modern logistics industry has been placed in a very important position. It points out that we should vigorously develop modern logistics industry, accelerate the establishment of socialized, specialized and information-based modern logistics service system, vigorously develop third-party logistics, and give priority to the integration and utilization of existing logistics resources, Strengthen the construction and connection of logistics infrastructure, improve logistics efficiency and reduce logistics cost.

Guangdong Province also attaches great importance to the development of modern logistics industry, which is close to the national development plan [1]. In the 12th Five Year Plan of Guangdong Province, it points out that it is necessary to focus on the development of modern logistics and other modern service industries, and clearly requires to take the national planning goal as the overall goal, speed up the construction of logistics infrastructure and network, and create a convenient and efficient modern logistics network with the Pearl River Delta as the main body.

M. A. Jan and F. Khan (Eds.): BigIoT-EDU 2021, LNICST 392, pp. 320–329, 2021.
https://doi.org/10.1007/978-3-030-87903-7_40

2 Current Situation of Logistics Talents' Ability and Quality Demand

The Pearl River Delta is the earliest region to develop logistics industry in China. According to relevant data, the gap of logistics employees in Guangdong will reach 3 million. Different types of logistics enterprises, the nature of logistics jobs, different industries and different regions have different demands for logistics talents [2, 3]. According to the basic orientation of logistics personnel training in higher vocational colleges, it is necessary to master international trade, transportation, logistics geography, as well as solid English ability and logistics operation related knowledge. In order to obtain accurate data, we have completed the Research Report on the demand for logistics talents in Guangdong Province, which analyzes the demand for logistics talents of small and medium-sized export-oriented manufacturing enterprises and small and medium-sized micro logistics enterprises in Guangdong region, especially in the Pearl River Delta region, and lays a foundation for the development of training standards for high skilled talents. The specific jobs are shown in Fig. 1.

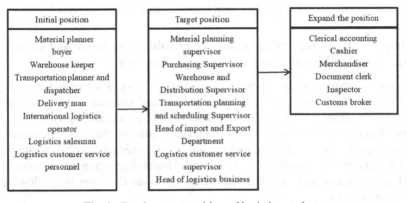

Fig. 1. Employment position of logistics students

Generally speaking, the ability and quality requirements of small and medium-sized logistics enterprises in Guangdong Province for higher vocational logistics talents can be summarized as good communication ability, practical operation ability and team cooperation ability. Professional theoretical knowledge requirements mainly include logistics management knowledge, financial knowledge, professional English, computer information system knowledge, etc. The most effective way is to strengthen the construction of logistics discipline and specialty in Colleges and universities, and improve the knowledge and skills of students majoring in logistics in higher vocational colleges [4–6]. In the Pearl River Delta region, it puts forward specific requirements for the talent training of Logistics Specialty in higher vocational colleges. How to formulate a talent training plan suitable for the development of logistics enterprises is the top priority for the development of Logistics Specialty in higher vocational colleges.

3 On the Construction of the System Design and Development of Talent Training Program

3.1 Establishing the Training Plan Oriented by Professional Ability

Logistics specialty or department should establish professional ability requirements and knowledge modules and build a new curriculum system based on the professional ability requirements of logistics management post group, the job requirements of small and medium-sized logistics enterprises, and the logistics workflow [7]. The curriculum system must highlight the cultivation of post ability and professional quality, and pay attention to the cultivation of students' professional ethics and entrepreneurial ability. In addition, the curriculum system should dilute the boundary between professional basic theory courses and professional training courses, extend the scope of cooperation with enterprises, and jointly develop a number of practical training courses and professional quality courses with the combination of work and study in the curriculum design [8]. At present, we have made great efforts to list the courses of "warehousing and distribution practice", "production operation and management", "logistics and transportation practice" as excellent courses development projects, with the purpose of building the integration of courses and certificates and training students' professional ability.

3.2 Improve the Construction Quality of Training Base Inside and Outside the School to Meet the Needs of Practical Teaching

The logistics specialty or department of the college should strengthen the practical teaching link and highlight the skill training. Students' practical training includes not only on campus training, but also post practice in off campus training base [9–11]. The premise of combining learning with working and improving students' vocational skills is to have a good practice and training base inside and outside the school. Strengthen the data management and control, study the data tracking and detection differences between related radar targets, and calculate the difference value data. Set the following data calculation formula:

$$R_i = \frac{\sum\limits_{k=1}^{n_i} R_{ik} V_{ik}}{\sum\limits_{k=1}^{n_i} V_{ik}} \tag{1}$$

According to the target position of radar tracking target, the data foothold of position function is analyzed. At the same time, the data difference value between different regions is reasonably planned. According to the information range of difference value, the simplification of algorithm is strengthened, the operation process is simplified, and the operation time is shortened, so as to obtain more efficient tracking and detection data results, Using the selected index parameters, the paper designs the transformation model formula to accurately control the test object of the radar target:

$$K = \sqrt{S - T^3} + \frac{p - 0}{N} \tag{2}$$

4 Design and Development of the Construction Path of "Professional + Entrepreneurial" Mode

According to the construction idea of the system design and development of the new talent training scheme, this paper puts forward the construction path of system design and development based on the special industry * enterprise mode and the "professional + entrepreneurial mode" [12–14]. Through this mode, we can cultivate talents suitable for small and medium-sized logistics enterprises in Guangdong Province, solve the gap of skilled talents in enterprises, and find a breakthrough and direction for the society to solve the severe employment situation, Enhance students' entrepreneurial passion and ability to lay a solid foundation for future entrepreneurship.

4.1 Jointly Study and Discuss to Determine Talent Training Objectives and Specifications

Through expert argumentation, group discussion and other ways, the training objectives and specifications of logistics management professionals are accurately positioned, the logistics management professional construction Steering Committee is established, and the revision opinions and suggestions of talent training program are discussed at regular meetings (see Fig. 2).

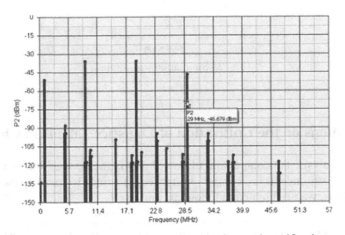

Fig. 2. Simulation of talent training objectives and specifications

4.2 Joint Development of Modular Curriculum

System and specific courses according to the requirements of the new talent training program, the curriculum system of logistics management major adopts the curriculum system of combining work with study module to strengthen skills and ability training [15, 16]. The curriculum system consists of basic professional quality module, professional positioning internship module and entrepreneurial ability training module. The preliminary designed and developed curriculum system consists of professional practice courses (platform practice courses, professional technology practice courses, broadening training), professional theory courses (platform courses, professional technology courses, broadening courses), public basic courses and directional elective courses [17]. Mainly in the platform practice courses, it offers logistics cognition training, international trade training, basic management training, market research, accounting practice, etc., In the professional and technical practice courses, there are warehousing and distribution training, logistics transportation training, customs declaration comprehensive training, procurement and supply management training, chain enterprise logistics training, business negotiation training, etc. in the outward bound training courses, there are supply chain management training, e-commerce training, entrepreneurship scheme design, graduation thesis design, graduation practice, etc. (see Fig. 3).

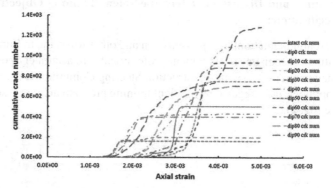

Fig. 3. Simulation with strain

5   Suggestions on the Cultivation of Logistics Talents in China

In the final analysis, the competition of enterprises is the competition of talents. The training of logistics talents is a complex system engineering. To speed up the training of logistics talents, we need to do the following work well.

5.1   We Should Create a "Soft Environment" for the Development of Logistics and Form a Correct Understanding of "Logistics"

In a sense, in the development of modern logistics, the gap between us and developed countries is not only in equipment technology and capital, but also in concept and knowledge. For a long time in China, there are still many people who do not know enough about the importance of circulation under the condition of socialist market economy, especially the importance of modern circulation mode and industry, which will greatly restrict the development of China's advanced productivity and competitiveness [18]. We should pay attention to the propaganda and guidance of logistics, so that people can understand modern logistics and accept the concept of modern logistics. In order to form the idea of big logistics system in the whole society, we must let more people understand and accept it. It is not enough to build some hard facilities. We should organize experts and scholars to introduce the basic knowledge of logistics, publicize the idea of logistics to the society, create a kind of atmosphere of logistics, make more public understand logistics, attract excellent talents to participate in this field, and create a "soft environment" for the real development of logistics in China.

5.2 Training Logistics Talents Through Various Ways

In view of the current situation of logistics education and personnel training in China, logistics education in China should focus on the logistics discipline education. The development of various forms of education and training should be carried out simultaneously. Through academic education, vocational training, training organized by industry associations and short-term training organized by enterprises, logistics talents should be cultivated through academic education, To improve the education and training system of logistics talents, new disciplines are in the process of continuous research and improvement. After the rapid increase of logistics demand, the lag of logistics discipline construction has become a problem that cannot be ignored. At present, China has not yet formed a complete logistics education system. Although a number of colleges and universities have set up undergraduate education of logistics specialty, they are still in the stage of self planning and design of curriculum and practice, lack of standardization in curriculum setting, textbook selection and training direction, and the number of talents cultivated is uneven [19]. We should refer to the foreign logistics personnel training system, according to the requirements of market demand and discipline theory, fully consider the current situation of logistics education resources, and do a good job in logistics discipline construction.

5.3 Set Up the Correct Goal of Logistics Personnel Training

It is the basic premise of logistics education to set up a correct goal of logistics personnel training in multi-level logistics. The goal and mode of logistics personnel training should be based on the needs of the field and the characteristics of the discipline, and the current situation of logistics education resources should be fully considered, Comprehensive design of modern logistics personnel training different levels of talents, take different training methods, logistics postgraduate education mainly train logistics researchers, teachers of higher education and enterprise logistics managers, they should have a solid logistics research method, be able to solve the major theoretical and practical problems in the modern logistics industry, undergraduate education should focus on quality, knowledge and coordinated development [20]. For enterprises, we should cultivate logistics management talents with solid theoretical foundation, wide knowledge, strong adaptability, logistics operation and management, logistics system planning and design ability, as well as logistics professional and intermediate logistics management talents with basic theory of related disciplines, so that they can directly manage and operate the whole process of logistics operation. Junior college and higher vocational education are the basic logistics education, which should focus on operation, skills and practice, and mainly cultivate practical operators with professional skills and engaged in logistics process management.

6   Training Logistics Talents Through Vocational Training

6.1   Improve Quality

In order to improve the quality of logistics personnel, we should pay attention to the vocational education of in-service personnel and strengthen the training of existing logistics personnel. According to the survey, about 92% of the logistics employees in the United States have bachelor's degree, 41% have master's degree, and 22% have obtained the formal qualification certificate. To improve the professional quality of these employees, we can only complete it through continuing education. At the same time, with the development of social economy and the progress of science and technology, even the logistics practitioners with higher education will face the problem of knowledge updating and self enrichment. Therefore, in addition to academic education, continuing education is a necessary supplement and extension of logistics personnel training [21]. As shown in Fig. 4. For the job, the starting point should be high managers, and the forms should be diversified. Teaching, research, discussion and communication can be adopted, and different levels should be set to meet the needs of different personnel. Short, targeted, practical to be strong, to play a practical function and role in a short time, to be able to solve problems. Long term training should be considered in the long run, not practical.

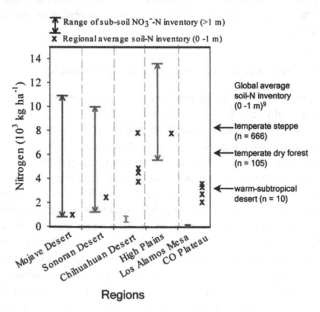

Fig. 4. Main quality improvement models

6.2 Unify the Qualification Certification of Posts and Improve the Training System of Logistics Talents

In order to improve the standardization of continuing education and integrate with international practice as soon as possible (professionals need to pass the evaluation and assessment of authoritative institutions to obtain qualification certification), it is particularly important to unify the post qualification certification. In addition, it has also introduced the qualification certification system of the Royal Logistics Association of China and Beijing. However, because there is no unified and authoritative post qualification certification standard in China, the existing continuing education and qualification certification have not been generally recognized by the logistics industry. Therefore, China should take the lead of the government, organize universities and industry associations to jointly form an expert group, and work out a logistics training standard and post qualification certification system that is not only modern logistics concept, but also in line with China's specific reality, so as to promote the development of China's logistics continuing education and improve the logistics talent training system.

6.3 Carry Out Special Research on Logistics Education to Promote the Cultivation of Logistics Talents

Logistics experts should show logistics research and education topics, focus on the growth law of logistics talents, the contribution of logistics talents to logistics development and the comparison of logistics education systems in various countries, support the research and innovation activities of universities and scientific research institutions in logistics, improve the integration of logistics theory and technology in China, and encourage the development and application of Cooperative Logistics Technology in universities and research institutions [22]. Experts in the field of logistics also write training materials to provide sufficient logistics support for colleges and universities in China.

7 Concluding Remarks

To build a talent training program of logistics management specialty, which is supported by "industry + enterprise" mode and systematically designed and developed by "specialty + entrepreneurship" mode, to realize "three highlights, three optimizations and three strengthening", that is, to highlight the improvement of ability and quality, the cultivation of professional ethics and the training of professional skills; to optimize the curriculum system, teaching methods and methods; to strengthen the training of logistics management talents [23, 24]. We should strengthen the teaching of foreign languages and computers, the teaching of Humanities and Social Sciences, and the teaching of entrepreneurship theory, so that students can "learn by doing" and "learn by doing", so as to realize the standard of "high quality, wide knowledge and strong ability" for senior skilled talents and obtain the recognition and acceptance of social enterprises.

Acknowledgements. Research on the cultivation of applied talents for undergraduate logistics management (sino-foreign cooperation education) based on national teaching quality standards.

References

1. Wang, H., Li, X.: Curriculum system construction of logistics management specialty in higher vocational colleges based on working process systematization. Educ. Career **36**, 35–36 (2011)
2. Liu, Y.: Research on the development of work process systematization course for logistics management major in higher vocational colleges. Harbin Inst. Technol. **6**, 23–25 (2012)
3. Wang, X., Jiao, L., Wang, X.: Curriculum development and implementation of higher vocational logistics management specialty with systematic working process. Vocat. Educ. Explor. **7**, 166–167 (2010)
4. Cao, J., Chen, X.: On the development of logistics management curriculum system based on working process. J. Liaoning High. Vocat. Coll. **2**, 45–48 (2010)
5. Zhang, T., Hu, J., Yang, Y., Zhang, H., He, Y.: Exploration of "PCA double spiral ladder" talent training mode under the background of modern apprenticeship – taking logistics management specialty as an example. China market **09**, 91–93 (2021)
6. Li, H., Lin, C., Lan, Z.: Construction of "customer fresh load" mode of rural fresh e-commerce under shared logistics: a case study of Guanzhuang she Township in Shanghang County, Fujian Province. J. Yunnan Agric. Univ. (Social Science) **15**(02), 67–72 (2021)
7. Chen, Y., Yu, B., Liu, B.: Design of evaluation system for talent training program of logistics management specialty in higher vocational colleges. China Bus. Theor. **06**, 180–181 (2021)
8. Wang, G.: Research on the cultivation of cross border logistics talents in Henan Province under the background of digital trade. J. High. Educ. **10**, 151-154+159 (2021)
9. Tian, N.: Research on the reform of logistics personnel training mode in local colleges and Universities – taking Zhengzhou as an example. Today's Fortune **06**, 192–193 (2021)
10. He, Y.: Research on the training of logistics talents in secondary vocational schools based on market orientation. Manage. Technol. Small Medium Sized Enterpr. (Zhongxunjiao) **03**, 154–155 (2021)
11. Zhu, J., Zhao, L.: Research on the relationship between Xingtai logistics industry and national economic development under the background of new normal. J. Xingtai Univ. **36**(01), 48-52+86 (2021)
12. Xu, X., Tang, J.: Exploration on curriculum system construction of higher vocational logistics management major integrating emergency logistics content. Logist. Eng. Manage. **43**(03), 185–188 (2021)
13. Wu, Y., Zeng, H.: Development status and Prospect of logistics industry in Guizhou Province under the background of big data+. Logist. Eng. Manage. **43**(03), 57–59 (2021)
14. Gao, H.: One of the "one belt, one road" background of independent institute international logistics curriculum teaching reform. Logist. Eng. Manage. **43**(03), 194–196 (2021)
15. Liu, S., Liu, Y., Gui, T.: Practical research on talent training mode of production and education integration in higher vocational colleges – taking logistics management major of Shanghai communications vocational and technical college as an example. Logist. Eng. Manage. **43**(03), 179-180+196 (2021)
16. Qi, S.: Problems and countermeasures of experiential teaching from the perspective of applied talents training – taking logistics management as an example. Mod. Bus. Ind. **42**(12), 161–163 (2021)
17. Liu, X., Xie, S.: One belt, one road, international logistics park's strategic evaluation and development conception – based on the double circulation perspective in the post epidemic era, 2021 monthly (03): 59–71
18. Sun, X.: Suggestions on e-commerce logistics personnel training in vocational colleges. China Train. **03**, 46–47 (2021)

19. Fu, Y., Lin, H.: Exploration and practice of the "three education" reform of the "1 + X" certificate of logistics management major based on "integration of competition and courses" – taking Hainan vocational and technical college of economics and trade as an example. Logist. Sci. Technol. **44**(03), 158–160 (2021)
20. Ma, L.: One of the "one belt, one road" strategy for training logistics application-oriented talents. Manage. Technol. SMEs **03**, 108–109 (2021)
21. Liming. Relieving the "talent bottleneck" is the meaning of the logistics industry planning in the 14th five year plan. Modern Logistics News, 2021–03–03 (A04)
22. Zhong, Y.: Exploration on the realization path of cultivating high quality logistics talents under the background of intelligent logistics. China Logist. Procurement **05**, 60–61 (2021)
23. Wang, B.: Research on modern apprenticeship training of secondary vocational logistics management specialty. China Logist. Procurement **05**, 65 (2021)
24. Wang, L.: Discussion on the teaching reform of e-commerce logistics management in higher vocational colleges. J. Jiamusi Vocat. Coll. **37**(03), 147–148 (2021)

Teaching Analysis of Digital Marketing Mode Under the Influence of Big Data

Jinhong Zhao[✉]

The College of Arts and Sciences, YunNan Normal University, Kunming 650222, China

Abstract. With the rapid development of the global Internet, online shopping and online shopping has become a fashion. At the same time, network Huina is a new model suitable for the era of big data in the new era. Next, this paper first analyzes the basic concepts of big data and digital marketing, and then analyzes the methods and Strategies of digital marketing mode in the era of big data for reference.

Keywords: Big data · Network marketing · Ant colony algorithm

1 Introduction

With the advent of the era of big data, all enterprises begin to implement the snow sale mode on the network, and regard the network snow sales mode as the mode of enterprise sales development. The establishment of the network snow sales mode under the data is the precondition for some enterprises to carry out the network sales business [1]. The network is updated every moment, which also makes the network snow sales mode in enterprises constantly change and update. Because some decision makers don't pay enough attention to network marketing, they can't choose their own marketing strategy. For example, China's Alibaba bought part of the shares of sina Weibo, which to a large extent shows a very important battle Luo plan of Alibaba's sales. Therefore, a comprehensive understanding of the network marketing model is now an important topic for many enterprises.

2 Big Data and Network Marketing Mode

2.1 Big Data

Big data literally means a large amount of data. Its real meaning is to use modern advanced computer technology to deal with a large amount of data which is difficult to be processed by human or conventional processing technology. It is very difficult to make effective statistics of these resources for the traditional technology processing, and thus become useless resources. Due to the accelerating process of modernization, a large number of data are produced every day. With the passage of time, the data will be more

M. A. Jan and F. Khan (Eds.): BigIoT-EDU 2021, LNICST 392, pp. 330–338, 2021.
https://doi.org/10.1007/978-3-030-87903-7_41

and more large. According to statistics, there are 40zb data generated by the Internet every day.

In the modern network marketing, many enterprises in the process of marketing do not lack of data resources, the most important problem is that too much data, many times difficult to deal with. Enterprises in the economic business, need to carry out statistics on all aspects, but also on the customer, market data centralized statistical analysis, these data statistics together formed a large number of data, how the enterprise carries on the comprehensive and effective management and utilization of such large data, for many enterprises, there is no doubt that it is a very big problem and challenge. In the era of Internet, network marketing needs a lot of data, and makes use of big data to make a choice for internal marketing. Therefore, computer big data processing technology is very important.

2.2 Analysis of Big Data Density Peak Clustering Algorithm

In the density peak clustering algorithm, each point has two attributes, one is the density attribute, the other is the exclusion attribute. The higher the density, the more clustered the data at this point, and the possible density center of the data. The larger the exclusive group value is, the more likely it is to be a new cluster. For the points distributed in the center of the database, clusters and clusters are mostly distinguished by low density [2]. In short, the larger the exclusion value, the greater the probability of forming new data clustering. In the data set to be clustered by big data processing, the calculation formula of local density ρ_i is as follows:

$$\rho_i = \sum x(d_j \cdot d_c) \tag{1}$$

$$x(x) = \begin{cases} 1 \\ 0 \end{cases} (x_1 + x_2) \tag{2}$$

3 Research on Network Marketing Mode Under Big Data

With the development of science and technology, computer technology has been applied in the network marketing mode, which can summarize a large number of useful information and make it a big data network marketing model. In the big data network marketing mode in the Internet era, we need to constantly explore new network marketing mode if we want the marketing to run normally.

3.1 Product Association Mining Marketing

Commodity association mining marketing has great promotion value. There are many successful examples of this kind of marketing in network marketing, such as the classic examples at home and abroad, beer and diapers. The sellers sell the goods to the supermarkets, and the supermarket operators put the beer and diapers together. This practice can make the sales of the two commodities exceed the expectation to a great extent. Many people don't think that there is a great correlation between the two commodities, but a careful study shows that there is a great correlation. Because most of the American women are housewives, they have no time to go out to buy things. They have to wait for their lovers to pick them up after work. Their children use a lot of diapers every day, Many people buy beer when they buy diapers. In this case, diapers and beer form a certain relationship. Therefore, in this marketing model, we need to use big data as the mining basis to discover the association between individual data. No matter how much single data, no matter how high the value is, there is no correlation between them. Therefore, it is necessary to analyze the original data and establish the relationship between the data.

3.2 Social Network Marketing of Big Data

The data generated by the network marketing mode of social data is relatively large, such as network friends, QQ friends and twitter, etc. these information are used for social network marketing. For example, Mengniu yogurt is combined with the network resources of renren.com, plus the star effect, and "sour milk music dream" is taken as the theme in various places and colleges, At the same time, the young people in Colleges and universities are striving for their dreams and vigorously promoting Mengniu, so that people can drink Mengniu milk and enjoy high-quality life [3]. For example, the red rice machine advertising in the QQ space to forward and vigorously promote, so that the sales of red rice greatly increased, exceeding the pre-sales expected sales; and last year's popular network "where the object is strongly sought after by netizens and so on. These seemingly simple things are actually the media and propaganda of social network marketing by the initiators using big data.

3.3 Big Data User Behavior Analysis Marketing

Big data user behavior analysis marketing mode. This marketing mode is mainly to record and objectively analyze the user's online data, analyze the valuable customers, and then carry out reasonable marketing plan, for example, cloud letter, a social tool developed in marketing. This kind of social tool has a strong ability of self analysis. It can analyze consumers' attitude towards products according to consumers' evaluation of products and social history, then establish a model of consumers' purchase desire, and conduct systematic analysis on it. Then, according to the results of analysis, targeted customers can be selected to provide reliable customers for future online marketing.

3.4 Personalized Recommendation Marketing of Big Data

Big data personalized recommendation marketing is also a very important mode in network marketing. In many social activities, some large-scale social platforms, such as forums, communities and microblogs, allow users to build their own social circle and establish their own circle of friends. In their own circle of friends, they can freely publish the information they want to publish. This information has great value for advertising enterprises, through which the psychological needs of consumers can be analyzed. This personalized marketing model makes use of the speed of network communication and huge social groups. This form is a very important part of online sales. Therefore, some analysts are constantly appearing in our field of vision. This kind of analysis tool is different from ordinary analysis tools. It mainly carries out targeted marketing through personalized algorithm.

3.5   Big Data Analysis Marketing of Modern Communication

Modern communication data analysis marketing application is also very extensive, for example, Taobao quantum constant channel statistics. Quantum constant channel statistics mainly includes two aspects in different functions: quantum constant channel website statistics and quantum constant channel store statistics. Quantum Hengdao website statistics is mainly for users and third-party data and some specific statistics of content, such as blog visits and third-party statistics users. To carry out comprehensive data monitoring and system analysis. Using the statistical analysis of data in the Internet to analyze the rules of users' use in the network, and use the analysis results to adjust the corresponding network marketing plan. The shop statistics of quantum Hengdao is generally used in Taobao stores for real-time data statistics.

4   Micro Blog Marketing Case Analysis Based on Big Data

Now the marketing of micro blog is very hot, which is due to the rise of micro blog. There are tens of thousands of users in microblog, and there are a large number of common good people gathered together. Therefore, advertising dissemination is very easy, and the strength of dissemination is very wide. In this circle, advertising data publicity is very important. Each character in microblog represents a meaning. For example, a small circle represents a user, a small red circle represents a big V user in the microblog, and a green small circle represents a core data hole user, while a small blue country is a combination of green small country and red small circle. If in a certain star circle, there is a "big customer's remarks will be transmitted in the star circle through various ways. For example, if the star wants to promote a kind of "cosmetics", then the star will carry out a big advertisement on the" big "and get a very good effect [4]. However, if you want to further promote the "cosmetics" to other circles, then it needs "core structure hole" users to connect with each other in two different circles, so as to promote the advertisement to each circle and achieve the purpose of network marketing [5]. The following Fig. 1 shows the micro blog marketing simulation.

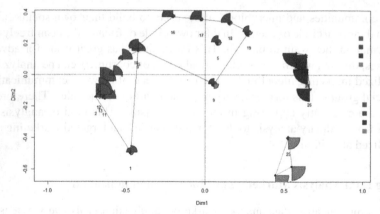

Fig. 1. Simulation of Micro blog marketing

5   BF Network Marketing Strategy and Planning in the Era of Big Data

5.1   Marketing Strategy of BF Network in the Era of Big Data

Keeping up with the development trend of Internet and mobile Internet, BF network aims to improve its own operation ability, explore innovative marketing mode, and provide precision marketing services for brands and customers. BF network has transformed from the traditional website construction into a diversified online news release platform, including website, mobile official website, mobile client, official microblog and wechat of media. It also fully integrates offline activities and services, links resources, and carries out o2o closed-loop marketing mode of online and offline combination, that is, fully invoking online resources of Internet and mobile internet terminal, Effective combination of landing activities and offline services enables users to obtain information online, participate in interaction, and experience the marketing of brand merchants in the o2o closed-loop mode of personal participation offline [6].

BF network is not only a network media, but also a fourth media news website affiliated to the Propaganda Department of the municipal Party committee. As a news website, it has its own incomparable characteristics - strong government background and social credibility and influence, which is the incomparable advantage of commercial websites. As a local news portal, news has become one of the important ways to obtain social credibility and influence [7]. It is not only the mission of news websites to stick to the position of public opinion, but also one of the important chips to occupy more market share and better market resources with commercial websites in the market environment. Combined with the news website's own government resources and social influence, we can better transform the credibility and influence into marketing driving force and marketing conversion rate; we can better play our own advantages and create a high-quality brand marketing service system. With the rapid development of the Internet and mobile Internet, and the arrival of the era of big data, BF network has been transformed into a new media group to build an all media matrix. With the influence and credibility of BF network as the core, BF network has been established in websites, microblogs, wechat and clients. The website has BF, its channels and frontier client sites. BF and its channels have their own microblog corporate accounts and wechat public accounts, forming a three-dimensional social new media matrix. The mobile terminal has BF mobile phone official website, BF app client and frontier client, In the era of all media, BF network has constructed a necessary for the development of media convergence - the four carriages (namely website, microblog, wechat, client) pattern. All of these contribute to the diversification and three-dimensional development of BF's marketing system. From the previous online publicity on the Internet, combined with the planning and implementation of offline activities, the online and offline activities can interact effectively, and the resources of different channels can be combined and integrated. In the process of obtaining news and information, netizens can extend to news and information services and marketing services, and participate in offline activities, With a sense of experience and interaction, and a sense of participation and sharing effect in the process, the marketing mode of BF network is more abundant, and the audience can obtain marketing service and marketing experience in different experiences [8, 9].

5.2 The Marketing Mode of "Media + Activities"

"Media + The marketing mode of "activities"", It is one of the most important manifestations of O2O marketing mode, With a long marketing time, online and offline marketing effect, Precisely achieve the characteristics of the target population: It can integrate multi platform and cross platform resources, And we can get more effective information from users through offline activities, Let users increase the sense of experience, belonging and interaction, To achieve precision marketing. BF network develops a new mode of network interaction, Using network channel to build personalized interactive platform, By organizing colorful activities, While sharing the network feast with tens of millions of netizens, it also creates unlimited business opportunities for brand businesses. The marketing mode of "media + activities" is shown in Fig. 2.

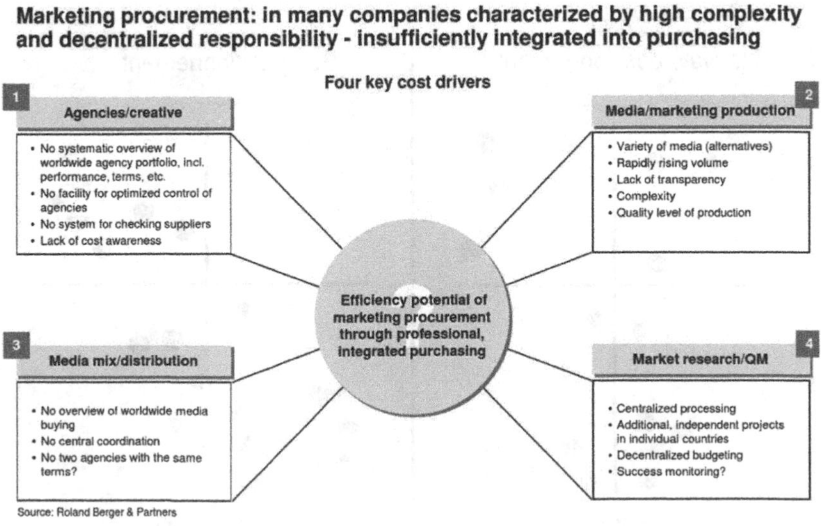

Marketing procurement: in many companies characterized by high complexity and decentralized responsibility - insufficiently integrated into purchasing

Four key cost drivers

1 **Agencies/creative**
- No systematic overview of worldwide agency portfolio, incl. performance, terms, etc.
- No facility for optimized control of agencies
- No system for checking suppliers
- Lack of cost awareness

2 **Media/marketing production**
- Variety of media (alternatives)
- Rapidly rising volume
- Lack of transparency
- Complexity
- Quality level of production

Efficiency potential of marketing procurement through professional, integrated purchasing

3 **Media mix/distribution**
- No overview of worldwide media buying
- No central coordination
- No two agencies with the same terms?

4 **Market research/QM**
- Centralized processing
- Additional, independent projects in individual countries
- Decentralized budgeting
- Success monitoring?

Source: Roland Berger & Partners

Fig. 2. The marketing mode of "media + activity"

5.3 The Marketing Mode of "Media + Service"

The marketing mode of "media + service" realizes one-stop marketing solution [10, 11]. For brand businesses, BF can help them establish official website and official media matrix (including official microblog, wechat and mobile client of mobile phone version) according to their needs, and realize personalized requirements of brand businesses with its own technical strength. For example, establish a brand official website and official media matrix for a brand, so that Internet users can find the information aggregation of the brand on the Internet and mobile Internet multi terminals, obtain all the brand information at the first time, and provide continuous information update, news maintenance and in-depth cooperation with news websites, so as to realize the transformation from news website users to brand users, Realize the transformation from news website to brand official website. The marketing mode of "media + service" can also be targeted at local government departments. With the implementation of government affairs activities with the help of the Internet, government departments urgently need safe and reliable partners to build government affairs Internet portals, establish Internet windows and enhance the government brand image. Government information release should also take into account the interaction with the people. BF network uses its own high-quality government background and resources to establish good cooperative relations with Tianjin Municipal People's Congress, CPPCC, district and county Party committees and bureaus. It not only provides website construction and maintenance for government affairs in the form of services, but also promotes the development of Internet-based government affairs in new media, and creates and maintains official microblog, wechat public accounts, APP services and other services for the cooperative party and government departments, We should establish channels for the interaction between the government and the people. Through the form of "media + service", not only the government resources, credibility and influence of the media are deepened and enlarged in the cooperation with the government, but also the income generating benefits of the media operation are realized. The marketing mode of "media + service" is shown in Fig. 3.

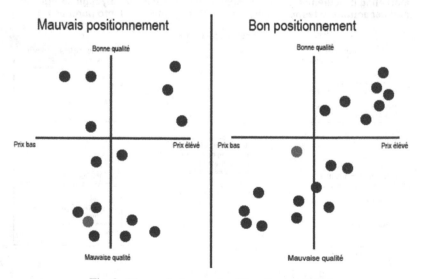

Fig. 3. The marketing mode of "media + activity"

6 Marketing Mode of Brand Promotion

The marketing mode of brand promotion usually uses "influence" marketing, also known as label and evaluation marketing. For the marketing mode of brand promotion, it is necessary to fully base on the advantages of localized resources, deeply tap localized resources and benchmarking enterprises, deeply cooperate with localized industry flagship, high-quality brands and excellent enterprises, enhance the influence of industries, enterprises and brands through the media's own influence, and build a win-win marketing mechanism [12, 13]. Through the authority of news websites, we can improve the exposure and attention of brands, create added value of brand value, let media influence drive the influence of enterprises or brands, and bring value-added effect to enterprises or brands, so as to promote the marketing of their products and services. Through user feedback, it also helps brand businesses adjust their brand quality in time, pay attention to the direction of user needs, develop more effective and more accurate marketing methods, and make the most limited influence of localization the most valuable influence. For example, in cooperation with a certain brand, in addition to online advertising, special content marketing, or building an all media publicity platform for its official website through multi terminal brand businesses on the Internet and mobile Internet, we can also use bf.com to have a number of high-quality self-made columns, such as news reception hall, zero distance between the government and the people, the first scene, entrepreneurial life, to carry out column naming cooperation, So that the brand can not only be exposed on the Internet advertising, but also be fed back through the host's oral broadcast of thanks, the film's tail falling, and other feedback, such as recording background, computer back stickers, wheat logo and other materials will be displayed at the same time; in addition, it can also be selected in the BF 315 consumer trustworthy brand, through the support and feedback of Internet users, Use data to reflect the brand value and brand influence [14]. The shortlisted brands can not only obtain the honor of "most trusted brand" with influence, but also pay attention to the areas to be improved according to user feedback. The award of "the most trusted brand" can be used as brand goodwill to build up the reputation of the brand in the interaction of online and offline marketing promotion, so as to achieve better reputation marketing and attract the attention and transformation of more potential consumer groups at the same time of brand promotion [15].

Media has a strong attribute of information aggregation, and the marketing mode of "media + public opinion" can produce strong guidance and direction for marketing activities through market feedback and user attention. In the Internet era, information security is very important, whether it is government agencies or brand businesses, the demand for public opinion information is more and more urgent. To understand the concerned topics, hot events and audience feedback, all need to be applied through public opinion analysis [16]. The marketing mode of "media public opinion" is to deeply excavate public opinion information and attention guidance through the media's keen insight and news perspective, so that brand businesses can make targeted marketing strategies and take the audience's needs as the starting point of marketing strategies.

7 Conclusion

Extensive network marketing can greatly reduce the cost caused by advertising, so that enterprises can get better benefits and greater profits. Big data has penetrated into people's lives. To make full use of big data, enterprises should carry out reasonable marketing according to users' Internet habits and content. This not only makes the user's network data leak to a certain extent, but also reduces the user's security. Some network information contains the user's real information. Therefore, we should update the loopholes in technology to make up for the loopholes and protect the basic privacy of users. To a large extent, network maintainers are required to carry out comprehensive maintenance. This protection is not only a technical challenge, but also needs to be protected by relevant legal measures in reality. The behaviors of non-compliance on the network should be thoroughly checked and severely attacked.

Acknowledgements. Yunnan Provincial Education Department of the sixth batch of university science and technology innovation team project construction "Yunnan Province university wisdom tourism science and technology innovation team".

References

1. Li, X.: Analysis of new marketing model in the era of big data. National Circulation Economy (11), 9–10 (2020)
2. Qiu, Y.: A new marketing situation in the era of big data Tiktok's marketing model analysis. Modern Business Trade **39**(24), 46–47 (2018)
3. Lin, X.: Analysis of new marketing model in the era of big data. China's Strategic Emerging Industries (12), 145 (2018)
4. Zhang, G.: Analysis of network marketing mode based on big data era. Modern Commerce (32), 59–60 (2014)
5. Ma, J.: Enterprise marketing strategy research. Contemporary Economy (15), 13–14 (2014)
6. Sheng, H.: Research on the current situation and trend of China's media marketing. Master's thesis, Dalian University of Technology, Dalian (2007)
7. Huang, W., Huang, W.: Web marketing analysis based on Web3.0. China Business (2009)
8. Lin, Y.: On advertising precision marketing and communication strategy in the era of big data. Jin Media (10), 76–77 (2014)
9. Rogers, E.: Diffusion of Innovation. China Compiler Press, Beijing (2002)
10. Long, S.: Internet thinking in the transformation of BBC. Media (4), 5–8 (2012)
11. Xu, J.: Analysis of precision marketing. Information Network (8), 16–17 (2006)
12. Howard, C.: Measuring the conversion rate: the science and art of marketing optimization, pp. 79–85. Electronic Industry Press, Beijing (2014)
13. Lippmann, W.: Public Opinion, p. 15. Macmillin, New York (1965)
14. Wilberschram, Porter, W., Li, Q., et al.: Trans. Introduction to Communication, p. 29. Xinhua Publishing House, Beijing (1984)
15. Tang, X.: Data mining technology based on data warehouse, pp. 76–79. China Machine Press, Beijing (2014)
16. Ma, Q.: Internet web technology innovation leads to the change and thinking of marketing mode. Science Technology Industry (02), 25 (2011)

The Application of Big Data Analysis in Model of Financial Accounting Talents in Colleges and Universities

XiuJuan Zhang(✉)

Yunnan Technology and Business University, Yunnan 651700, China

Abstract. In the era of knowledge economy, the biggest challenge to the cultivation of talents in Colleges and universities is to make knowledge-based and creative talents. The cultivation of modeling talents has become the main body of value creation in Colleges and universities, and then has become the key point that colleges and universities must grasp.

Keyword: Center of gravity · Key points

1 Introduction

Talents have always been valued by all countries in the world, and education is the main means to cultivate talents. Higher education direct. Then we will transport all kinds of senior professionals to the society, promote economic, technological and social development, and enhance international competitiveness. Therefore, in the face of the global new technological revolution and the comprehensive challenges of the new century society, all countries in the world are predicting and studying the future, paying increasing attention to and accelerating the reform of higher education, and putting forward new and higher requirements for the quality of talents. "There can not be a unified, fixed and rigid standard model for the quality of talents in the 21st century. Instead, we should proceed from the actual situation of various countries and vary from time to time and from country to country. We should base on China's national conditions and proceed from China's actual situation.

It puts forward the strategic requirements of training and bringing up high-quality creative talents to Chinese education in the new century. With the deepening of the reform of colleges and universities, the living environment of colleges and universities has undergone profound changes, and the competition brought by the market economy has become increasingly fierce [1]. How can colleges and universities win in the market and obtain sustainable development.

2 Journals Reviewed

2.1 Training to Meet the Needs of Society

According to the relevant data of China Journal Network and Beijing Library, the standard of education in the University of science and technology is very important. On

M. A. Jan and F. Khan (Eds.): BigIoT-EDU 2021, LNICST 392, pp. 339–345, 2021.
https://doi.org/10.1007/978-3-030-87903-7_42

the issue of human standard, many experts in higher education have different answers, some even have different opinions. On the contrary. But we all have a common answer, that is, the goal of education. Almost everyone in Zhunzhong middle school mentioned that it is necessary to cultivate people who can adapt to the needs of society and have "high competence". As for "high". Some people say that they should have a high degree, while others say that they should have rich practical experience and a strong sense of responsibility. Most people emphasize the working background of famous companies, but there is still no unified and clear answer however. One thing is common. We all think that creative talents in the new century are the same direction. The result of simple data collection is very surprising. Imagine that there are so many higher education experts, each of them. There is such a big difference in people's ideas and views. If everyone makes educational policies according to their own ideas. Then we can imagine the quality of talents trained by our colleges and universities. As a result, there is only "high". Without a clear definition and standard, the requirement of "competency" still can't help colleges and universities to carry out high-quality training.

2.2 Construction of Victory Characteristic Model

The construction of competency model can solve these puzzles for us. From the research of competency model abroad, in the middle and late 20th century, McLellan of Harvard University. Professor and a series of follow-up research results make people see a new dawn of modern human resource management theory, It provides a brand-new perspective and a more favorable tool for the practice of human resource management in enterprises, namely. This paper makes a comprehensive and systematic study of personnel, and analyzes the competency of comprehensive evaluation from explicit characteristics to implicit characteristics. Law. This method can not only meet the requirements of modern human resource management, but also build a certain post competency. The qualitative model has a clear statement about the competency and the combination structure of a person who should take on a certain job. It can also become an important scale and basis for personnel quality evaluation from explicit to implicit characteristics, so as to provide a reference for personnel quality evaluation [2]. The reasonable allocation of human resources provides a scientific premise. The research on Competency Model in international academic circles started earlier, and began to study and develop since 1980s. And practice the competency model based on business strategy. Armstrong (1991).

3 Analysis on the Current Situation of Talent Cultivation in Colleges and Universities

Colleges and universities are highly concentrated industries of knowledge, technology and capital, and education economy is an important part of state-owned economy. With the development of higher education and the high-speed development of socialist market economy, technology is becoming more and more important. The speed of technology renewal will be faster and faster, and the corresponding is to have high intelligence and high efficiency to adapt to the social development. This is also the only way for colleges and universities to enter the market and meet the challenges. But it goes with it the current situation of talent cultivation, especially the cultivation of talent competence, is worrying.

3.1 On the Particularity of the Cultivation of Talents' Competence in Colleges and Universities

There are obvious differences between the competency training of university talents and that of enterprise talents. In addition to the knowledge and skills necessary for talent cultivation, especially in the cultivation of personality, creativity and innovation. It has its own special requirements in the cultivation of spirit, practical ability and entrepreneurship.

3.1.1 The Individualized Requirements of the Competence of University Talents

Comrade Jiang Zemin put forward at the third national education working conference: "we must strengthen the national innovation. Ability refers to the height of the rise and fall of the Chinese nation Today's world is developing by leaps and bounds. The era of knowledge economy has come. Looking back on history, every progress in human life is associated with creation and innovation. Together, the history of mankind is a history of constant innovation. Innovation is the main line of social development. It is also the soul of today's era of knowledge economy, in which the favorable personality development of college students is to cultivate students' comprehensive ability. The premise of quality is always accompanied by innovation. It can be seen that "strengthening the good personality of college students".

Exhibition is an important part of cultivating innovative spirit and strengthening quality education. "①1In the era of knowledge economy, innovation has become the decisive factor of social development. And innovation ability comes from personality development. In the era of knowledge economy, people's innovative spirit and creative ability are required more strongly [3]. Zhang of human personality. Yang said that the generalization of human creativity will be the requirement of the innovation of the times. Psychological research and the research on gifted children in different countries.

3.1.2 The Practice of Training Children

The practice of training young children has proved that the spirit of innovation is the most important, dynamic and effective quality of human beings. It is also a part of human heredity. It is a common and universal thing for human beings. Maslow once pointed out that creativity is "possessed by any child and lost by most people when they grow up." This shows that the improper education in the day after tomorrow will obliterate the unique and unique personality in nature. He made it popular and lost his creativity. The development of people's creativity is closely related to the development of personality. It includes people's world outlook, philosophy of life, way of life, ethical standards, mode of thinking and so on of course, Personality development here refers to people with sound and good personality. Quality education is to recognize the individual.

On the basis of differences, encourage the development of personality. Personality cultivation is the need to develop one's own potential and create a better future. Therefore, the purpose of personality cultivation is to help every teenager find their own potential life in the future. Living on the road to fully show the growth point of talent. The person with strong innovation ability must get good personality development. Without the development of personality, it is impossible to form and realize their creative ability.

3.2 Country's Educational Circles have Ignored the Cultivation

Personality, and even avoided talking about personality, which is fundamentally against the law of talent growth and education. Comprehensive development and personality development are inseparable. Quality education is just the right way. On the basis of recognizing individual differences, we should encourage the development of personality and cultivate students' good personality. That's what we need. Classroom teaching changes the previous indoctrination education method, induces students to think independently at any time, and encourages students to speak boldly. Timely put forward their own opinions and views. At the same time, in order to make students form creative thinking in the process of education. The school and the teacher must give the students a free environment and avoid repressing fantasy and forcing compliance can. In other words, only when people's subjective consciousness, independent personality and personality are fully developed, can they be more consciously and fully developed. Improve the comprehensive quality of the individual, increase and play their creative ability.

4   The Cultivation of Creativity and Innovative Spirit of Talents in Colleges and Universities

4.1   The Talents Cultivated in Higher Education Should be Creative Senior Professionals. The High School of the People's Republic of China

According to Article 5 of the equal education law, "the task of higher education is to cultivate high-quality talents with innovative spirit and practical ability. We should develop science, technology and culture and promote socialist modernization The 21st century is a century in which the knowledge economy with modern science and technology as the core gradually takes the leading position." The essence of knowledge economy is creation. New. With the rapid development of science and technology, the cycle of scientific and technological achievements transforming into productive forces and the life cycle of products and commodities. The cycle is getting shorter and shorter. Innovation has become the most important activity of human beings and the basic driving force of survival and development. create. New advantages can make up for the disadvantages in resources and capital. As shown in Fig. 1. If we accelerate innovation, we can take the initiative in market competition. Right. Therefore, cultivating creative talents should be the basic task of education, especially in higher education. "In the era of knowledge economy, science and technology advance by leaps and bounds, the speed of knowledge renewal and the process of high-tech industrialization. Increasingly, knowledge economy is a new economic form based on the latest technology and the essence of human knowledge. It's just beginning. In this era, the country's innovation ability includes knowledge innovation ability and technological innovation ability.

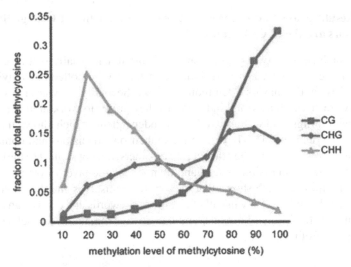

Fig. 1. Main trend of innovative talent training

4.2 The Key to the Future of the Family and the Rise and Fall of the Nation

As one of the main bodies of the national innovation system, colleges and universities bear the responsibility of knowing. At the same time, it shoulders the responsibility of cultivating creative talents. Building knowledge oriented economy. The national innovation system in the economic era requires colleges and universities to cultivate a large team of high-quality talents. The development of College Students' creativity resources is the basis of the sustainable development of rejuvenating the country through science and education. Therefore, it is necessary to cultivate and develop universities. Students' creativity has great practical significance and far-reaching influence.

$$f(y|\theta) = \sum_{k=1}^{K} \pi_k \phi(y|\theta_k) \tag{1}$$

$$Q\left(\theta, \theta^{(i)}\right) = E_z[\log f(y, z|\theta)|y] \tag{2}$$

5 Analysis on the Characteristics of College Students' Creativity

Creativity and the ability to express in a more general way. The ability to solve problems, acquire knowledge and create fantasy. Creativity can't be simply classified into a certain category. A kind of ability, creativity is composed of many kinds of abilities, is a kind of comprehensive ability. Huang Xiting and others believe that the cultivation of College Students' creativity involves six factors: first, a strong sense of well-being. The second is the uniqueness and novelty of Association; the third is the independence of individuals; the fourth is the independence of knowledge. Five is not afraid to make mistakes; six is correct values. They use the sixteen personalities of cartel.

5.1   The Results Show that the Personality Characteristics of College Students' Inventors are the Main Factors

Bias: one is high sensitivity, which is sensitive, emotional, usually soft hearted, easily moved and more feminine, Love art, indulge in fantasy, in collective activities, their unrealistic views and behavior often reduce the set. Secondly, they are not happy with others. They are silent, lonely and indifferent. They prefer to work alone, and they are not happy with things. Third, they are highly independent, which shows that they are independent and decisive [4]. They often make their own decisions and complete their own work plans on their own. They do not rely on others, are not subject to the constraints of public opinion, do not dislike others, and do not need the favor of others; The fourth is high self-discipline, which shows that they know themselves and the other, and they are strict in self-discipline. They usually have the same words and deeds, can reasonably control their own feelings and actions, and can always maintain their self-esteem and win the respect of others."

5.2   College Students are in the Period of Study, They Can Not Be Asked to Make Earth Shaking Creation

Although the creative activities of college students have also produced new and meaningful products, their creative activities are not so good. Most of things in human history are not the first creation, but in terms of personal history, they have the significance of creation. Therefore, the creativity of college students, on the one hand, has the general characteristics of human creativity; on the other hand, due to the particularity of the environment, the creativity of college students has its particularity.

The establishment of the competency model of talent cultivation in Colleges and universities provides a broader perspective and new direction for talent cultivation in Colleges and universities. Technology plays a fundamental and guiding role in personnel training activities, thus promoting personnel training.

6   Concluding Remarks

The development of theory and practice is the new starting point of modern higher education personnel training. Guiding college students to carry out career planning and helping them to realize career development is the core of modern human resources. It is a basic concept of development. It is also a basic requirement of humanistic management in Colleges and universities. The essence of humanistic management is that university organizations should view their own goals and objectives from a systematic point of view. We should respect and balance the interests of all stakeholders, and lead and motivate them in a humanized and personalized wayl. Encourage college students to promote the realization of their reasonable wishes and dreams as the starting point of management, in respect, truth. In the environment of honesty, trust and support, the common development of colleges and universities and college students can be realized, so that college students can have a better understanding of their future.

References

1. Zhang, H.: Talent recruitment and selection based on competency model. Hum. Resour. (04), 84–85 (2021)
2. Wang, K.: Research on the cultivation of reserve cadres in t property insurance company based on competency model. Nanjing University of Posts and Telecommunications (2020)
3. Zhang, X.: Research on structured interview based on post competency model under project risk method. J. Hubei Open Vocational College **33**(23), 130–131 (2020)
4. Wang, X.: Research on connotation and extension of competency. Quality Market (19), 79–81 (2020)

The Development of Intelligent Technology in Preschool Education Strategy

Xiaoli Wang(✉)

School of Teacher Education, Weifang Engineering Vocational College, Qingzhou 262500, Shandong, China

Abstract. Early childhood is not only the enlightenment stage of children's intellectual development, but also the key period of children's life development. Good early childhood education can lay a good foundation for children's life development. It is of great significance to put forward the strategy of sustainable development in preschool education.

Keywords: Intelligent technology · Preschool education · Personalized recommendation · Collaborative filtering

1 The Current Situation of Preschool Education

At present, the development of preschool education in China is as follows:

1. The condition of running a school is not satisfactory. Preschool education in China is not a compulsory education stage, the state does not have a fixed financial allocation for preschool education, which to a certain extent leads to the education leaders ignore the importance of preschool education, and rarely arrange special kindergarten and employment funds for preschool education. Based on this, the school running conditions of preschool education institutions in some areas are generally poor, which are mainly reflected in the following aspects: insufficient infrastructure and backward teaching equipment; lack of some health facilities and large outdoor game facilities. The school running of preschool education institutions in remote areas often only includes a few rooms, a few tables and a few stools. The environment for children is very monotonous, which is not conducive to the development of early childhood education [1].

2. The teachers of preschool education are insufficient, and the quality of most preschool education is not high. As we all know, good teachers are an important material guarantee to ensure the quality of education, but there are some areas, especially the underdeveloped county areas, which are limited by the backward economic conditions. Their monthly income of preschool education is often lower than that of urban preschool education, and the school environment is difficult. In some areas, preschool education teachers are generally lacking. Because of this, teachers who

M. A. Jan and F. Khan (Eds.): BigIoT-EDU 2021, LNICST 392, pp. 346–355, 2021.
https://doi.org/10.1007/978-3-030-87903-7_43

are willing to teach in preschool education institutions in remote areas generally do not meet the high school standard, and some even graduate from primary school, which seriously does not meet the national standard. This is also a major reason for the low quality of preschool education in most colleges and universities.

2 Problems in the Development of Preschool Education

(1) The number of pre-school education teaching institutions is small, the teaching equipment is backward, China's rapid economic development, people's living standards are constantly improving, so people also put forward high requirements for the teaching equipment of term education, but because the number of pre-school education institutions is very limited, the number of government-owned kindergartens is very few, leading to the high cost of teaching institutions. These problems make institutions far from being able to meet the needs of parents for their children's preschool education. In addition, the only educational institutions have a limited number of teaching equipment. Some institutions even have the problems of non-standard basic equipment and seriously backward equipment, which has a great hidden danger for children's safety.

(2) The preschool education system is not perfect. Because the development of preschool education industry in China is not long, preschool education has not formed a perfect system. Although many institutions claim to be "preschool education", the teaching content tends to be primary school. They think that preschool education is to let children feel the learning atmosphere of primary school in advance and learn the teaching content of primary school in advance. Therefore, all aspects are implemented according to the standards of primary schools, including allowing children to do a lot of homework, carrying out the so-called mid-term and final examinations, and only paying attention to the learning of Chinese, mathematics and other contents. In addition, this kind of exam oriented education leads to children's early learning weariness, which is very harmful to children's later learning career [2].

(3) Teachers are not strong enough. Due to parents' attention to preschool education, the demand for preschool education teachers is increasing, and many preschool education graduates and teachers do not have a precise understanding of the preschool education system. Teachers have misunderstandings about preschool education, which leads to problems in students' education. Children in preschool not only have a desire to explore the world, but also have a desire to know, At the same time, their outlook on life, values and world outlook are also in a blank period, while the former teachers only pay attention to the teaching of Chinese and mathematics knowledge to students, and do not actively guide them to form correct ideas and learning habits.

Euclidean distance. It is the most widely used distance calculation method in the process of cluster analysis:

$$d(x_i, x_j) = \|x_i - x_j\|_2 = \sqrt{\sum_{k=1}^{q} (x_{ik} - x_{jk})^2} \tag{1}$$

Manhattan distance. It takes the absolute value of the difference between the features in each data and accumulates them:

$$d(x_i, y_j) = \sum_{k=1}^{q} |x_{ik} - x_{jk}| \tag{2}$$

Chebyshev. Which means to take the maximum absolute value of the difference between features in each data. The Chebyshev distance formula is shown in Fig. 3

$$d(x_i, x_j) = \max_k (|x_{ik} - x_{jk}|) \tag{3}$$

3 Countermeasures to the Problems in the Development of Preschool Education

3.1 Promote the Development of Preschool Education Institutions

Preschool education plays an irreplaceable role in the education system. The study in this period has a great impact on children's life. Children are the future of the motherland, receiving a good education is the premise of children's good growth, and preschool education is the basis of education [3]. Therefore, the state should attach importance to the construction of pre-school education institutions, so that they can be included in the national education planning. In addition, we should provide financial support for existing educational institutions, encourage and support them to improve teaching equipment, improve their own education level and teaching ability, and meet parents' needs for preschool education from multiple channels and perspectives. At the same time, educational institutions should improve their own rules and regulations, such as the establishment of a sound management mechanism for preschool teachers, innovation and development of more adaptive preschool education model. In addition, parents should also pay attention to the children's preschool education, let the children study in a better educational institution, provide a good environment for the children, cultivate good learning habits.

3.2 Guide the Whole Society to Establish a Correct Awareness of Preschool Education

Parents are the main decision-makers of children's education. There are still some parents who lack proper understanding of pre-school education, which leads to children's lack of proper pre-school education. Therefore, the state and pre-school teaching institutions can publicize the importance of pre-school education, such as holding seminars, distributing publicity materials and making various advertisements, so as to guide parents to establish correct concepts and awareness of pre-school education. In addition, preschool teachers in preschool teaching institutions should also organize more parents to visit or parent-child activities, so that they can deeply understand the importance of preschool teaching and truly understand the content of preschool teaching. For children, parents are the first teachers. Therefore, parents should not only place their hopes on preschool teaching institutions, but also imperceptibly establish a good learning environment for children, cultivate children's excellent learning habits, and set an example for children to form correct three outlooks. Parents should cooperate with pre-school teaching institutions and guide their children in the right way.

3.3 Improving the Teaching System of Preschool Education

With the development of quality education, educational institutions and teachers should change their teaching concepts and teaching purposes, and truly realize the importance of preschool education and the learning content that preschool education should have. In terms of teaching content, we should let children accept the quality teaching they should receive at their age, meet their curiosity about the world, and carry out more courses such as art, sports, music, etc., so that students can understand the world from the perspective of colorful colors and wonderful music. In addition, from the perspective of teaching mode, teachers can carry out more extracurricular activities. On the premise of ensuring students' safety, children can freely explore and feel the world, which has a great positive impact on children's life development. At the same time, teachers can actively guide students to develop correct learning habits in the term education stage, which also plays an important role in children's later learning career (see Fig. 1).

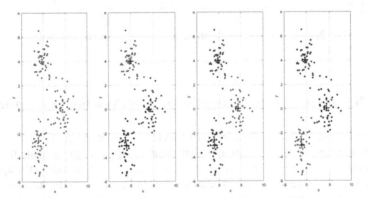

Fig. 1. Improving the teaching system simulation analysis

3.4 Improving Teachers' Professional Quality

In addition to providing financial support for preschool education, the state should also pay attention to the cultivation of preschool education teachers and the improvement of their professional quality. For example, we can set the assessment system to ensure the quality of teachers, to ensure that children get the quality of teaching. As the "imparter" of teaching content, teachers play an important role. Only by accumulating excellent professional knowledge, can they freely use what they have learned to educate and guide the development of students. All pre-school education institutions should also improve the threshold of applying for teachers, through continuous professional training, to improve teachers' teaching ability, so as to improve work efficiency [4]. Teachers should also be strictly controlled to ensure the teaching quality and teaching level of teachers, so as to ensure the development of students. Teachers should innovate and develop their own teaching forms, and use information-based teaching mode or situational teaching mode to stimulate children's interest in learning. Children are in the age of curiosity, lively and active. Teachers should make good use of children's characteristics, improve their interest in learning, and make them develop good learning habits (see Fig. 2) [4].

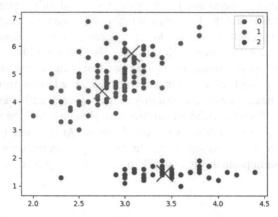

Fig. 2. Improving teachers' professional quality simulation analysis

4   Application of Artificial Intelligence in Preschool Education

The application of artificial intelligence in education has been widely studied [5]. As a basic part of the education system, it is particularly important to integrate the application of artificial intelligence into preschool education. On the one hand, the application of artificial intelligence in preschool education will improve the infrastructure environment of preschool education, help teachers improve their teaching level, help parents accompany their children more scientifically, and ultimately improve the comprehensive quality of the educated; on the other hand, the improvement of the quality of preschool education will enhance the development level of preschool education in China, To further promote the self-renewal and development of China's education, the application of artificial intelligence in preschool education mainly includes intelligent counseling, intelligent kindergarten management and intelligent wearable devices. At present, the main application products of artificial intelligence in preschool education are intelligent robots. The intelligent robots represented by alpha egg series provide high-quality educational resources for children, assist teachers to enrich teaching methods, and finally provide different solutions for different children to realize personalized teaching. The intelligent robot will discover children's potential in a certain aspect in advance from the aspects of children's learning ability, cognitive thinking and personal interest, so as to help teachers and parents better understand children and achieve the effect of teaching students in accordance with their aptitude [6]. The application of artificial intelligence in preschool education is shown in Fig. 3.

Artificial intelligence also has important applications in kindergarten management, including campus security management and early warning, The intelligent wearable devices can monitor children's health and safety in real time, Daily Mobile Smart parts

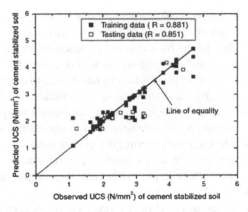

Fig. 3. Application of artificial intelligence in preschool education

monitoring data analysis to get the health status of the body, output the most reasonable diet, send to parents on time. In kindergartens, intelligent wearable devices can monitor children's physical state and timely reflect abnormal conditions to teachers. Classroom environment control devices can adjust the brightness, air, temperature and humidity of the classroom according to children's state to create the most comfortable nap environment. At present, the application of artificial intelligence products in preschool education stage is single, mainly intelligent robots, with simple functions. Scholars have little research on the application of artificial intelligence in preschool education [7]. The application of artificial intelligence in preschool education is still in the early stage of exploration. The universality of various intelligent products is low, but the potential is huge. Therefore, the application of artificial intelligence in preschool education should learn from the achievements of artificial intelligence in other education stages, expand the scope of application, enrich intelligent products, and promote the development of preschool education [8].

5   Reform of Preschool Education in China

In the era of artificial intelligence, the development concept of preschool education needs some changes. We can start from three aspects, that is, the transformation from learning style to carrying style, the transformation from curriculum teaching method to project teaching, and the transformation from short supplement teaching to long development teaching. It is a change from learning style to carrying style. It is urgent for China's preschool education to get out of the dilemma of learning orientation. At present, China's preschool education is still learning oriented. The growth and development of children's body and mind is more urgent and important than the simple "behavior change or potential for change" (that is, learning). The main purpose of preschool education is to let children grow up. We must cherish children's world and children's life, and fully understand and respect children's ideas. Preschool education activities should be carefully arranged and precisely planned, and should be child-centered and child centered, consciously making use of children's original ecological life, or even partially retaining children's natural original ecological life. Preschool education should gradually weaken the learning orientation and give full consideration to life orientation or growth orientation [9]. The second is the change from curriculum teaching method to project teaching method. The traditional curriculum teaching method is guided by the teaching content, which aims to instill knowledge into children by teachers, and children passively accept knowledge. For some children, it is difficult to concentrate and the learning effect is poor. Project teaching is a kind of active learning and cognitive process starting with children's interest, which can promote children to carry out interesting, personalized and meaningful exploration in their own world [10]. It originates from children's interest, can balance skill learning and children's interest, pay attention to children's learning goals, and let children participate in purposeful and meaningful learning activities. In this way, it is an important direction for the development of preschool education in the future to promote the enthusiasm of children's exploration, deepen sensory understanding and achieve the best learning effect. Preschool education teaching can gradually infiltrate the project-based teaching method, integrate knowledge into the process of project exploration, and let children actively excavate and discover new knowledge, so as to achieve the teaching purpose. The third is the change from short form teaching to long form teaching. Everyone in the education of knowledge is not all necessary, so from early childhood, how to teach them the necessary ability is the most important, according to the children's talent, potential and various, interest to design personalized curriculum, try to make up for the shortcomings of education and promote the combination of education. At present, education is mostly short education, especially in the process of children's education, focusing on all aspects of training, as well as mining and making up for the shortcomings, the process is hard and easy to hit the enthusiasm and self-confidence of children's learning. The education of developing one's strong points enables teachers, parents and even children to constantly tap their own potential, make themselves more confident and actively explore the unknown. In preschool education, it can improve children's nature, potential and creativity [11].

6   Application Trend of Artificial Intelligence in Preschool Education

To explore the application of artificial intelligence in school education, firstly, select appropriate teaching activities according to the teaching methods, contents and objectives in different periods of preschool education [12]. Then, with the help of the functions and characteristics of artificial intelligence technology, get the combination point of artificial intelligence products and teaching activities. Finally, construct the application scenarios of artificial intelligence in different periods of preschool education. The application of artificial intelligence in preschool education will start from the current content of preschool education teaching activities, namely campus education, family education, preschool special education, combined with the development needs of rural preschool education, preschool education resource allocation, preschool education specialty, preschool education for special children. The application of artificial intelligence in preschool education will start from the current content of preschool education teaching activities, namely campus education, family education, preschool special education, combined with the development needs of rural preschool education, preschool education resource allocation, preschool education specialty, preschool education for special children.

6.1   Application of Artificial Intelligence in Preschool General Education

To explore the application of artificial intelligence in preschool education, first of all, select appropriate teaching activities according to the teaching methods, contents and objectives in different periods of preschool education; then, with the help of the functions and characteristics of artificial intelligence technology, find out the combination of artificial intelligence products and teaching activities; finally, construct the application scenarios of artificial intelligence in different periods of preschool education [13].

Gifted children, more emphasis on early intellectual development and enlightenment, these children in life ability and other aspects do not need too much help, the content is relatively simple. For problem children, it aims at early intervention and prevention, more is psychological counseling. According to the psychological index parameters measured by artificial intelligence, using the appearance characteristics and functional characteristics of artificial intelligence products, we patiently and gradually carry out education and guidance. For children with disabilities (deaf mute, hearing-impaired, visually impaired, mentally retarded, autistic and physically disabled), it mainly includes early rehabilitation training, education and compensation for all-round development and correction of defects, involving the early stage of 03 years old and preschool education of 4–6 years old.

6.2   The Application Direction of Artificial Intelligence in Preschool Education

The application of artificial intelligence in preschool education will start from the current development needs of preschool education, focusing on rural preschool education, the inclusive nature of kindergartens, the allocation of educational resources, preschool education specialty and preschool education for special children [14].

7   Problems in the Application of Artificial Intelligence in Preschool Education

At present, there are three problems in the application research of artificial intelligence in Education: there are few application scenarios of artificial intelligence in the process of education and teaching, lack of systematic description and evaluation system of application scenarios of artificial intelligence in campus, lack of emotional training and emotional interaction in the process of interaction between artificial intelligence products and educates. Taking this as a reference, the application of artificial intelligence in preschool education can start from three aspects.

One is to increase the research on the application scenarios of artificial intelligence in the process of preschool education. A small number of artificial intelligence education products and application scenarios are difficult to meet the "teaching" and "learning" needs of education intelligence, and are not enough to promote the development of education intelligence. Only by proposing a series of subdivided application scenarios and implementation measures, can we meet the needs of the development of preschool education and promote the development of intelligent chemistry preschool education.

The second is the systematic description and evaluation system of the application scenarios of artificial intelligence in preschool education. The systematic description of AI application scenarios in campus should include: teaching tasks, teaching objectives, teaching modes, strategies and tools, teaching process design, teaching evaluation and so on. Perfect systematic application scenario description is an important reference to further implement the application of artificial intelligence in the field of education, which guides teachers, students, education administrators and designers of educational intelligence products to jointly realize the application of artificial intelligence in teaching. In addition, the evaluation of the application of artificial intelligence in preschool education is an important way to promote its self-renewal and development. Building and improving a scientific and comprehensive evaluation system is an important research content to promote the application of artificial intelligence in the field of education at this stage.

Children's demand for emotional care is higher, emotional training is related to children's mental health and character training is particularly important. Therefore, teachers and parents should participate in the research and development of artificial intelligence products and scene construction, and take the initiative to carry out emotional interaction and emotional cultivation with children. In addition, the change of preschool education mode should be the basis for the construction of application scenarios of artificial intelligence in preschool education. The change of education mode should be fully considered in the design of teaching content in application scenarios. It makes preschool education give full play to the advantages of life oriented or growth oriented learning style, project-based teaching method and long-term teaching.

8   Conclusion

Nowadays, preschool education has been widely concerned by the society. The development of preschool education determines the education level of children. Therefore, the development of preschool education has a long way to go. Starting from the actual situation, it is very important for the educated children to improve the teaching quality of preschool education and develop the educational hardware equipment of preschool education, which also has a positive impact on the development of preschool education.

References

1. Wang, Y.: Qualitative Research on the Development of New Rural Public Parks Under the Background of Three-Year Action Plan of Shaanxi Preschool Education. Shaanxi Normal University (2014)
2. Zhang, Y.: Community education is the necessity of the development of preschool education. In: Collection of Academic Papers Commemorating the Centennial of China's Early Childhood Education, vol. 397. Jiangsu Education Press, Nanjing (2004)

3. Jin, G., Zhao, M.: Advantages and disadvantages of diversified forms of early childhood education: a case study of Tengchong County in Yunnan Province. Preschool Educ. Res. (09), 1922 (2011)
4. Zhang, R.: Current situation and Countermeasures of agricultural preschool education. J. Shanxi Radio TV Univ. **16**(06), 24–26 (2011)
5. Yu, Y., Zhang, B.: Achievements and prospects of preschool education in China in 40 years of reform and opening up. Chin. J. Educ. (12), 18–26 (2018)
6. Wang, Z.: Application of artificial intelligence technology in education. Comput. Sci. (2), 33–34+16 (1984)
7. Zou, M.: Artificial intelligence and its application in music education. Northern Music **38**(15), 254–255 (2018)
8. Sun, Y.: Application of Artificial Intelligence in Language Training Guidance and Correction. Beijing University of Posts and Telecommunications, Beijing (2018)
9. Cao, Y., Liu, Z.: The value, Dilemma and Countermeasures of artificial intelligence in sports. Sports Cult. Guide (1), 31–35 (2018)
10. Huang, Y.: Application of artificial intelligence technology in high school mathematics teaching. Math. Learn. Res. (15), 64+66 (2018)
11. Sun, Z.: On the influence of artificial intelligence on Distance Education under the new situation. J. Jilin Radio TV Univ. (09), 64–65 (2018)
12. Zhang, D.: Design and Development of Children's Earthquake Safety Education Game Based on Artificial Intelligence. Yunnan Normal University, Kunming (2017)
13. Li, P.: The application of artificial intelligence in the management of primary and middle schools and preschool education. Mod. Commun. **01802**, 171–172 (2018)
14. Qiu, C.: Application of artificial intelligence technology in primary school English Teaching. Huaxia Teach. (31), 56–57 (2018)

Application of Open Source Data Mining Software Weka in Marketing Teaching

Zhengteng Hao[✉]

School of Economics and Management, Qinghai Nationalities University, Xining 810007, Qinghai, China

Abstract. Marketing is a professional basic course for economic management students. In the process of teaching, we should not only teach students theoretical knowledge, but also teach students to integrate into the real economic management activities. In order to stimulate students' interest in learning and improve students' practical ability, this paper discusses the teaching method of introducing open source data mining software Weka into classroom teaching, and gives practical teaching examples to improve the teaching effect. The training quality of the marketing course has been a useful attempt, and the classroom effect is good.

Keywords: Weka · Marketing teaching · Apriori correlation method · Shopping basket analysis

1 Introduction

Marketing is a professional basic course for students majoring in economic management. With the rapid development of information technology, both commodity information and customer information are massive. How to mine useful marketing information in big data and then apply it to our marketing is a problem that marketing students need to solve. Data mining technology in computer science is to solve the problem of automatically analyzing and discovering useful information in large databases. Apriori association algorithm is mainly used to discover meaningful connections hidden in large data. This paper attempts to explain the application of Apriori association algorithm in supermarket data analysis by taking open source software Weka as an example in marketing, In order to guide students to use data mining methods to solve practical problems [1].

2 Data Mining Weka

2.1 Open Source Software WEAK

The full name of Weka is Waikato environment for knowledge analysis. Its source code can be downloaded from http://www.cs.waikato.ac.nz/ml/WEKA Get it. At the 11th acmsigkdd International Conference, the Weka group of Waikato University won the

M. A. Jan and F. Khan (Eds.): BigIoT-EDU 2021, LNICST 392, pp. 356–365, 2021.
https://doi.org/10.1007/978-3-030-87903-7_44

highest service award in the field of data mining and knowledge exploration. Weka system has been widely recognized and has become one of the more complete data mining tools. As an open source data mining platform, Weka integrates a large number of machine learning algorithms that can undertake the task of data mining, including data preprocessing, classification, regression, clustering, association rules and visualization on the new interactive interface [2–5]. In modern business society, the data of enterprises are generally massive. If students can use the advanced software Weka to analyze the marketing data management and dig out the hidden relationships from the massive data, it will certainly be of great benefit to the mining and utilization of marketing data and the discovery of business opportunities.

2.2 Definition and Research Significance of Software Defect Prediction

With the increasing dependence on computer software, how to effectively improve the quality of software has become the focus and difficulty in the field of software engineering. The traditional software quality assurance methods (such as static code review or dynamic software testing) are inefficient, which need a lot of manpower and material, and need additional modification operations such as code instrumentation. Software defect prediction technology has gradually become one of the research hotspots in the field of software engineering - 4, 5, 6, 8, 9, 10.1, 12 software defect prediction, generally refers to the analysis of software source code or development process, design the degree element that has correlation with software defects, and then mine and analyze the software history warehouse to create defect prediction data set [6]. Based on the defect prediction data set, a specific modeling method (machine learning) is used to build the defect prediction model, which can predict and analyze the potential defect modules in the subsequent versions of the software. 13 software defect prediction can greatly reduce the manpower and material resources required for testing, and there is no need to modify the code. After the development of software department and module is completed, the defect prediction can be carried out in time, and the potential defect modules can be identified in advance, so that the project director can optimize the allocation of test resources and improve the test efficiency and software product quality [8–10]. Therefore, the research on software defect prediction technology, the establishment of software defect prediction model, and the early prediction and identification of software defects are not only of great research significance, but also of great application value.

2.3 Data Mining Software Weka

With the rapid development of science and technology and social economy, computer technology has been widely used in various industries. Computer software (system) is more and more closely related to people's work and life. The impact of software quality and system reliability on the efficiency and safety of production and management activities is also growing. However, with the continuous growth of people's demand for software, the scale of software is becoming larger. With the increase of complexity, software development and maintenance become more and more difficult. Software with hidden defects may lead to software failure or system collapse in the process of operation. Serious software defects will bring huge economic losses to enterprises, and may even lead to casualties. Unclear user requirements, non-standard software development process, lack of experience and ability of software developers and other reasons will lead to software defects. In order to ensure the quality of software, software testers will check software defects through certain software testing methods, and then inform developers to deal with software defects. However, affected by the actual factors such as the development schedule and cost control of software projects, software test engineers can not completely test all software modules.

3 Software Defect Prediction Method

According to the different methods of software defect prediction, it can be divided into static prediction and dynamic prediction. Static prediction mainly quantifies software code into static features, makes statistical analysis on these features and historical defect information, mines the distribution law of historical defects and constructs a prediction model, and then forecasts new program modules based on the prediction model. Dynamic prediction is to analyze the occurrence time of software defects, mining the relationship between software defects and their occurrence time [11]. At present, with the wide application of machine learning algorithm, static software defect prediction has achieved good prediction results, which has attracted more attention of researchers.

3.1 Software Defect Prediction Process

The process of software defect prediction can be divided into four stages. The software history warehouse is mined and analyzed, from which the program modules are extracted. The granularity of program module can be set as file, package, class, function, etc. After the modules are extracted, these program modules are marked as defective or non defective modules respectively. Design metrics. Based on the software static code or software development process, the corresponding metrics are designed to measure the software of program modules, and the defect prediction data set is constructed. Build defect prediction model. After the necessary data preprocessing of the defect prediction data set, the software defect prediction model is constructed with the help of specific modeling methods (such as machine learning method). Defect prediction. The software defect prediction model is used to predict the new program module, and the program module is predicted to be defect free and defect freep [12]. The prediction target can also be the defect number or defect density contained in the program module.

3.2 Metric Meta Design

The design of metrics is the core problem in the research of software defect prediction. Typical metrics include the number of lines of code, McCabe loop complexity and Halstead scientific metrics. Taking the complexity of the McCabe loop as a metric, the complexity of the control flow of the program is mainly considered. The assumption is that the higher the complexity of the control flow of the program module, the higher the possibility of containing defects. The assumption is that if there are more operators and operands in the program module, the more difficult it is to read the code, and the more likely it is to contain defects. Some researchers design metrics based on software development process. 2021 based on software development process mainly considers software project management, developer experience, code modification characteristics and so on. Researchers have conducted a comparative study on which way to design metrics is more effective in building defect prediction models. Graves 22, moser21 and others believe that the degree element designed based on the development process is more effective [13]. Menzies23 and others believe that the metric element of static code can build a high-quality defect prediction model. It can be seen that there is no obvious difference between the two. Reasonable design of metrics can achieve good prediction results.

3.3 Construction and Application of Defect Prediction Model

Before building the defect prediction model, it may be necessary to preprocess the defect prediction data set according to the quality of the data set. This is because the data set may have problems such as noise, dimension disaster and class imbalance. Preprocessing can improve the quality of the data set. 2 after preprocessing the data set, the defect prediction model is constructed with the help of certain algorithms. As the core of artificial intelligence and teaching data science, machine learning has been widely used in software defect prediction, which is selected as the algorithm to build software defect prediction model [14]. Machine learning methods commonly used in software defect prediction can be divided into classification methods and regression methods. Classification methods mainly include classification regression tree, naive Bayes, k nearest neighbor, support vector machine, ensemble learning and cluster analysis; regression methods mainly include linear regression, polynomial regression, stepwise regression and elastic regression.

4 Analysis Method of Curriculum Relevance

Pearson product distance correlation is mainly used to calculate the correlation between continuous variables. The Pearson correlation coefficient is meaningful only when the population of variables is normal distribution or can be approximately regarded as normal distribution, and the number of samples is not less than 30.

(1) The formula for calculating Pearson product distance correlation coefficient is as follows 1:

$$r_{AB} = \frac{\sum_{i=1}^{n} (a_i - \overline{a})(b_i - \overline{b})}{\sqrt{\sum_{i=1}^{n} (a_i - \overline{a})^2 \sum_{i=1}^{n} (b_i - \overline{b})^2}} \tag{1}$$

(2) The test statistic formula is shown in Fig. 2:

$$t = \frac{r_{AB}\sqrt{n - 2}}{\sqrt{1 - r_{AB}^2}} \tag{2}$$

In terms of technical implementation, the system is developed based on spring MVC web framework and mybatis architecture [15]. The platform can be divided into two core parts, one is data transmission function, the other is data mining function, each part is the complete framework of MVC theory.

The overall architecture of the system is shown in Fig. 1:

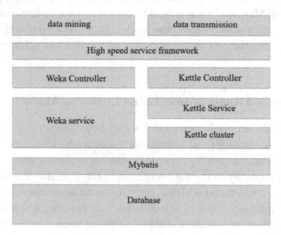

Fig. 1. Platform technical architecture

The technical architecture of the system is mainly divided into two functional applications, both of which are realized by three-tier MVC structure. The two functional applications are data transmission application and data mining application. The two applications implement the method call through the framework, which makes the application relatively independent and convenient for development.

5 Application of Open Source Software Weka in Marketing Teaching

In the process of teaching, we choose the shopping basket analysis experiment. The shopping basket analysis is to apply the correlation technology to the transaction process, especially to analyze the supermarket cashier data, and find out the commodities that appear in groups. For marketers, this is the main source of sales information for data mining. For example, after automatic analysis of the cashier data, it is found that customers who buy beer also buy potato chips. This discovery is of great significance to supermarket managers [16]. This information can be used for a variety of purposes, For example, planning the location of shelves, selling only one of the products that will be purchased at the same time at a discount, and providing product coupons that match the products sold separately. Businesses can also identify special customers from customers' purchase behaviors, not only analyzing their historical purchase patterns, but also providing a new way to identify the special customers, Moreover, it can accurately provide special purchase information that may be of great interest to potential users.

In the teaching experiment, we use a supermarket shopping basket analysis data set of Weka, the file name is supermarket.ar. This data set is collected from a real supermarket in New Zealand. There are 217 attributes and 4627 instances in the data set. It is very suitable for shopping basket analysis experiments. First, we use the preprocessing panel of Weka's Explorer interface, Load the supermarket.arf. In the current relation sub panel, we can see the basic information of the dataset. Because the dataset has many attributes and a large amount of data, students will click the Edit button on the top of the preprocessing panel to open the viewer window of the dataset and view the data file. Let students understand the properties and structure of data through proper explanation.

Among the ten association rules obtained from the calculation results, many commodities appear many times, and the total amount is very high. We can conclude that: first, customers who buy Biscuits, frozen food and other fast food will buy fruits and vegetables to supplement their body's vitamins; second, customers who buy Biscuits, frozen food, fruits and vegetables will buy bread and cakes; third, customers who buy the above food will buy a large amount at a time, and the total amount will be very high; fourth, transactions with a high total amount, They usually buy bread and cakes and so on. If the information is provided to the supermarket, we can rearrange the shelves, rearrange the supermarket, provide fast payment channels, arrange delivery and other additional services according to the knowledge, so as to enhance the market competitiveness.

For data mining, data is the core. The quality of data can directly affect the results of data mining. However, the actual data is very heavy. The so-called forest is big, there are all kinds of birds, and any problems may appear when there are more data, such as incomplete data, default attribute values, noisy data in the data set, wrong data, different coding formats or naming rules in the data set [17]. This will bring extra difficulties for data mining, even poor data will lead to the result of the error, so the data preprocessing is a step that can not be omitted.

If a data mining tool wants to have good performance, preprocessing is essential. In the process of mining, the preprocessing process will take up most of the time. There are four steps in preprocessing: data cleaning, integration and transformation, rule constraint and concept layering. Data cleaning includes: incomplete data processing, processing noise points and inconsistent data (see Fig. 2).

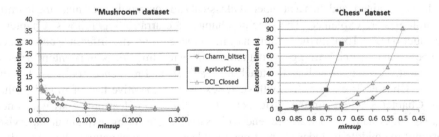

Fig. 2. Results of simulation with open source data mining (a)

K-means is a classic clustering algorithm. The basic principle of kmeans is: after determining the number of clusters K, k points are randomly selected as the center of the cluster. According to the Euclidean distance between each instance point and the center point in the data, each instance point is assigned to the cluster where the nearest center point is located. Then we get the cluster through the previous step, get its centroid point, redefine the centroid point as the center point, and then repeat the whole process [18–20]. In the continuous iteration process, the center of the cluster will continue to change, until in several successive iterations, the center of each cluster is exactly the same as the center of the previous round, which means that the cluster has been allocated, and the kmeans algorithm has been successfully implemented (see Fig. 3).

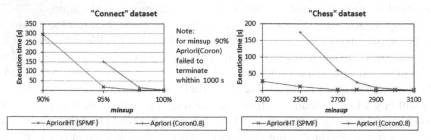

Fig. 3. Results of simulation with open source data mining (b)

6 Teaching Objectives and Contents

The purpose of this course is to enable students to understand the basic direction of machine learning; to master the basic algorithm of machine learning; to master the method of using Weka platform to realize machine learning algorithm; to understand the relevant research ideas of machine learning, from which to learn some methods of pioneers to solve problems; to further understand the usage and performance of learning algorithm through experiments, As shown in Fig. 4. In order to achieve these goals, on the basis of fully referring to the existing classic machine learning textbooks, the course team designed the following teaching content and syllabus: This course will take the classification task in data mining as an example, First, the evaluation of classification model is explained, and then a number of classical and commonly used machine learning technologies are explained. The specific chapters are arranged as follows: Chapter 1: introduction [21]. Explain the definition of machine learning, the difference and connection between machine learning and data mining, the teaching ideas and content arrangement of this course, as well as the teaching materials and reference books used in this course. Chapter 2: explain the method, index and comparative test of model evaluation. Chapter 3–9: explain the basic technology of machine learning: start with linear regression, explain linear learning: end with K-means clustering, explain unsupervised learning; the middle includes support machine learning, neural network learning, decision tree learning, Bayesian learning and nearest neighbor learning. Chapter 10–13: introduce the advanced technology of machine learning, including ensemble learning, cost sensitive learning, evolutionary learning and reinforcement learning.

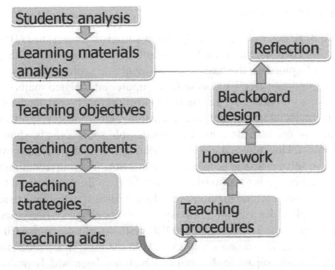

Fig. 4. Shows the teaching objectives

7 Teaching Evaluation and Assessment

7.1 Machine Learning

For machine learning, an elective course of computer major, which attaches great importance to the cultivation of hands-on ability, classroom theoretical knowledge teaching is of course important, but the more important thing is extracurricular practice, so that students' practical ability can be improved, practical application problems can be solved, and truly useful machine learning algorithms can be written [22]. In order to achieve this goal, the course team designed eight practical training and assessment problems after class, which are mainly used to consolidate and review the theory of machine learning algorithm taught in class, and master the implementation of machine learning algorithm by using the international machine learning open source experimental platform Weka. Through the experiment, we can further understand the usage and performance of machine learning algorithm, Improve the programming ability of machine learning algorithm: (1) implementation and experimental test of linear regression algorithm (10 points); (2) implementation and experimental test of logistic regression algorithm (10 points); (3) implementation and experimental test of SMO classification algorithm (15 points); (4) implementation and experimental test of BP classification algorithm (15 points); (5) implementation and experimental test of D3 classification algorithm (15 points); (4) implementation and experimental test of BP classification algorithm (15 points); (5) implementation and experimental test of D3 classification algorithm (15 points); (6) NB classification algorithm implementation and experimental test (7) KNN classification algorithm implementation and experimental test (10 points).

7.2 All Aspects of the Course

Implementation of K-means clustering algorithm and experimental test (10 points) students submit AVA code file based on Weka platform and screenshot of test results on any data set with Explorer as the basis of assessment and scoring. Machine learning is a new and very important elective course for computer and related majors. It has been widely used in many fields, such as image recognition, intelligent medical treatment, market analysis, financial investment, fraud screening, environmental protection, scientific research and so on, and has achieved considerable social effects, showing a good application prospect. How to teach this course well in University and improve students' practical ability to solve practical problems with machine learning technology is a problem that every teacher needs to seriously consider [23]. In this paper, from the teaching objectives and content, teaching methods and characteristics, to teaching evaluation and assessment, a complete description of all aspects of this course. After years of efforts, this course has been opened twice on the MOOC platform of love course China University, with 12913 and 12671 students respectively. In the third time, 2702 students have been pre selected to participate in the course, which has been widely praised by other college students and social learners, and has been selected into the first batch of excellent online open courses for undergraduates in Hubei Province [24]. The future work mainly includes to further supplement and improve the teaching content of the course, put forward new teaching methods, and improve the content and scoring standard of teaching evaluation and assessment according to the feedback of students.

8 Conclusions

In order to stimulate students' interest in learning marketing course, we try to apply an open source tool Weka in the teaching process, and with the help of its visual environment and typical algorithm, we demonstrate a practical problem-solving process for students in class. Through these teaching steps, students can gradually understand the open source software Weka and master the use of typical algorithm. We use Weka to process and analyze business data, improve data processing ability, and mine valuable information for marketing. At the same time, Weka software is open source software. For students with programming foundation, they can analyze the principle of the algorithm, and also optimize the algorithm through their own programming to further improve their ability to solve problems.

References

1. Chu, H.: Design and Implementation of University Data Exchange Platform Based on ETL. University of Electronic Science and Technology, Chengdu (2014)
2. Zhang, C.: Design and Implementation of Enterprise Data Exchange Platform Based on BTL. Harbin Institute of Technology, Heilongjiang (2016)
3. Xu, J., Pei, Y.: Review of data ETL. Comput. Sci. 38(4), 15–20 (2011)
4. Zhang, R.: Review of ETL data extraction, software guide. 9(10), 164–165 (2010)
5. Zhang, S.: The application of experiential teaching in the marketing teaching of secondary vocational education. Mod. Vocat. Educ. (30), 95 (2017)

6. Jiang, Y.: Application of marketing simulation training software in marketing course teaching. Contemp. Tourism (Golf Travel) (10), 193+211 (2017)
7. Weifang: The application of case teaching in marketing teaching. China Natl. Expo (10), 71–72 (2017)
8. Zhang, Y., Li, X.: A study on the teaching reform of flipped classroom mode in marketing teaching in Colleges and universities. J. High. Educ. (19), 121–123 (2017)
9. Wang, W.: Application of task driven teaching method in marketing teaching. China Market (28), 232–233+237 (2017)
10. Jin, Z., Zheng, C.: PBL teaching method in marketing teaching in Colleges and universities. Era Educ. (19), 160–161 (2017)
11. Liyaqin: The application of participatory case teaching in Marketing Teaching: based on constructivism learning theory. Contemp. Teach. Res. Cluster (10), 20 (2017)
12. Xialiping: Application of inquiry teaching method in the course of tourism marketing. Knowl. Econ. (20), 112–113 (2017)
13. Luo, X.: The application of experience teaching mode in automobile marketing teaching of Higher Vocational Colleges. Tax Payment (27), 120 (2017)
14. Liyifang: The application of mixed teaching mode in marketing course teaching. New Campus (Reading) (09), 24–25 (2017)
15. Wang, D.: The new application of case teaching method in marketing course teaching. New Campus (Early Ten Days) (09), 50 (2017)
16. Zhao, J.: Explore the application of interactive teaching mode in marketing teaching of higher vocational education. China Market (25), 227–228 (2017)
17. Xu, C.: Application analysis of project teaching method in marketing planning course teaching. Mod. Bus. Trade Ind. (25), 164–165 (2017)
18. Wang, J.: The application of project teaching method in the teaching of network marketing in secondary vocational school. Mod. Vocat. Educ. (24), 51 (2017)
19. Zousha: The application of project-based teaching method in tourism marketing teaching. Mod. Mark. (Next Issue) (07), 76 (2017)
20. This paper discusses the innovative application of SPOC in the teaching of marketing courses in Higher Vocational Colleges. China Mark. (22), 112+117 (2017)
21. Chen, H.: On the application of situational teaching method in marketing major teaching. Sci. Educ. Lit. (Next Issue) (07), 41–42 (2017)
22. Yangzhaoyun, Liyiting, Chenrainbow, Zhangdechun: Analysis of marketing teaching reform of Applied Undergraduate. Mod. Econ. Inf. (14), 421 (2017)
23. Hu, M.: The application of micro course in marketing course teaching in Higher Vocational Colleges. Educ. Mod. 4(29), 242–243 (2017)
24. Wang, Y.: Application of case teaching method in marketing teaching. Mod. Shopping Malls (13), 251–252 (2017)

Data Mining for Quality Analysis of College English Teaching

Zhaoli Wu[✉]

Shandong Institute of Women's Studies, Jinan 250000, Shandong, China

Abstract. This paper will introduce the factors that affect the quality of college English teaching in detail. Through professional research and investigation, the researchers take the quality of English teaching in a certain university as an example. The practical application of data mining technology in college English teaching quality is demonstrated by confirming the test object, test method, sample form, test result and result analysis, in order to optimize the teaching effect with the help of network technology.

Keywords: Data mining technology · College English · Teaching quality

1 Introduction

With the rapid change of English teaching form, the factors affecting the quality of college English teaching are gradually increasing. In order to accurately evaluate the actual effect of English teaching reform, researchers have introduced data mining technology in due course. Through the reasonable application of this technology to find out the actual impact of various elements on English teaching. The composition of traditional teaching quality indicators is often based on the experience of the makers. There are many subjective and qualitative components in the indicators, and the weight distribution of indicators is mostly based on subjective experience, even the same. Many different disciplines and different types of courses adopt the same teaching quality indicator system. In fact, College English teaching has its own characteristics (such as emphasizing the interaction between teachers and students, diversified teaching modes and conditions, etc.), many factors that affect the quality of teaching are often hidden in the teaching process, environment and conditions and are not easy to be found. Only by showing these factors can we design teaching quality indicators that are not only in line with the characteristics of College English teaching, but also scientific, reasonable and avoid subjective [1–4].

2 Experimental Design

2.1 Object and Purpose of the Experiment

From April to June in 2007, 1132 non English Major Freshmen in a university were selected as the subjects of the survey. Taking the making notable progress (MNP) of

© ICST Institute for Computer Sciences, Social Informatics and Telecommunications Engineering 2021
Published by Springer Nature Switzerland AG 2021. All Rights Reserved
M. A. Jan and F. Khan (Eds.): BigIoT-EDU 2021, LNICST 392, pp. 366–374, 2021.
https://doi.org/10.1007/978-3-030-87903-7_45

students as the objective evaluation index, this paper analyzes the factors influencing MNP by CHAID (chi squared automatic interaction detector) algorithm. (Note: in this experiment, the improvement of students' scores by more than 5% under the same test difficulty is regarded as the standard of students' obvious progress [5, 6].

2.2 Experimental Methods

This paper uses CHAID algorithm to analyze the factors that affect MNP. CHAID is an analysis method based on target optimization, which has the functions of target selection, variable selection and clustering. It is suitable for classification and rank data analysis. According to the given response variables and the selected explanatory variables, the optimal segmentation of samples is carried out. According to the significance of chi square test, the automatic judgment grouping of multiple contingency table is carried out. Generally, it has good effect on the automatic classification of discrete data sets [7].

2.3 Sample Characteristics

In order to systematically observe the teaching factors that affect students' learning quality, we designed 13 attributes that may affect the teaching quality, such as "learning purpose", "teaching means", "teaching mode" and "teacher type" [8]. Some of these attributes reflect the subjective factors in students' learning process, such as "learning purpose", "overall evaluation of teachers" and "entrance English achievement", while others reflect the objective factors such as teaching conditions and environment, such as "teaching mode", "teaching means" and "teacher's degree", The complete sample statistics are shown in Fig. 1.

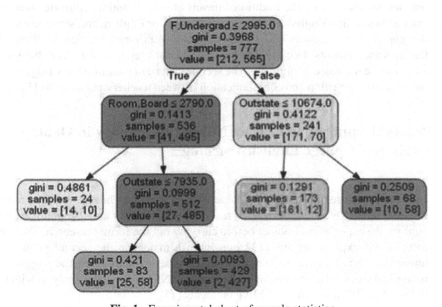

Fig. 1. Experimental chart of sample statistics

3 Elements Affecting the Quality of College English Teaching

(1) Purpose of study

There are many factors that affect the quality of college English teaching. The most important factor is the students' learning purpose. Specifically, for college students, they have studied English for a long time and have a certain understanding of the subject. The important factor affecting its development and progress is learning purpose, that is, interest. When students are interested in English learning, their subjective will to learn this subject will be stronger, which not only effectively enhances classroom participation, but also improves English learning thinking. By perfecting listening, speaking, reading and writing in English courses, the passion of language learning will be aroused, and the quality of English teaching will be guaranteed [9].

(2) Teaching model

The teaching mode of college English will also affect the teaching quality of this subject. Because foreign language teaching has strong practicality, both students and teachers should strengthen English practice in time compared with other courses. Use listening and speaking in English teaching to master the connotation of the subject and improve the practical application of English subjects [10]. In college English classroom, teachers should take the content of teaching materials and the actual level of students as the basis, and increase the communication and practicality of English classroom through the change of English teaching form. On the basis of ensuring students' English theoretical knowledge, teachers should improve their English literacy with practice and promote the overall development of teaching.

(3) Students' emotional factors

Cultivate students' emotional factors. When teaching English to college students, teachers should give up the traditional means of indoctrination, cultivate students' interest in learning English and students' feelings for English, and strive to narrow the gap between teachers and students, establish an equal relationship between teachers and students. In class, teachers should not keep a straight face, but smile and increase the sense of intimacy. The second sister can use more body language to increase the interaction and communication between teachers and students [11–14].

4 Practical Application of Data Mining Technology in Quality Analysis of College English Teaching

(1) Test subjects

In order to explore the quality of college English teaching, the relevant researchers use data mining technology to analyze the phenomenon in an all-round way, and confirm the experimental object before carrying out the formal research. The subjects of this experiment are 1132 non-English majors in the second grade of a university. The objective evaluation criteria are designed by using the progress of individual scores of each student. The index is MNP, professional analysis with the help of CHAID algorithm and MNP elements [15].

(2) Test methods

Through the CHAID algorithm, the researchers analyze the MNP elements in detail. In this algorithm, the clustering function, variable selection and target selection can be used to analyze and judge the order grade data and classification. CHAID algorithm can optimize and segment the samples according to the selected explanatory variables and reaction variables, and automatically form groups related to significance according to chi-square test. Discrete class data sets are used to show the actual effect of automatic classification. The data mining algorithm is shown in Fig. 2.

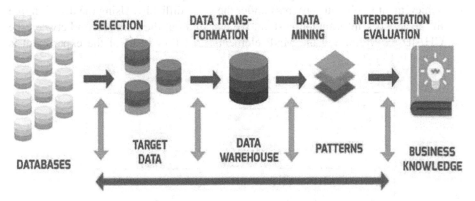

Fig. 2. Data mining algorithm

(3) Sample morphology

In order to better test the quality of college English teaching and find out all kinds of factors that affect students' English learning, relevant personnel should design a set of appropriate questionnaires, the main contents of which can be set as teacher type, teaching mode, teaching means and learning purpose [16].

For example, the questionnaire for the teaching model can cover a variety of elements. The specific data are shown in Table 1. The corresponding effects brought to students when teachers use different English teaching models can be seen from the data in Table 1.

Table 1. Survey data related to teaching patterns

Teaching model	Happy English	College English	New Vision English
Number of samples (s)	422	387	110
% of total (%)	37.3	34.2	9.8
Learning status	Optimal	Optimal	Difference

After completing the questionnaire survey, the relevant researchers can put the data information they obtain into the network information system, through the efficient fusion of data mining technology and decision tree, timely collate, analyze and study the impact of various data on the quality of college English teaching, and then increase the scientific nature of the experiment [17].

(4) Test results

With the exploration of data mining technology, relevant researchers have found that this element has brought changes to the quality of English teaching. According to the relevant data, there are 52 nodes with more than 95% confidence level in the decision tree. The teaching or learning quality in this study is mainly the ratio of individual achievement to obvious progress under the same difficulty. Using the decision tree in the network information system, we can find out the relationship between the CHAID algorithm and the MNP elements. The flow chart of the experiment is shown in Fig. 3.

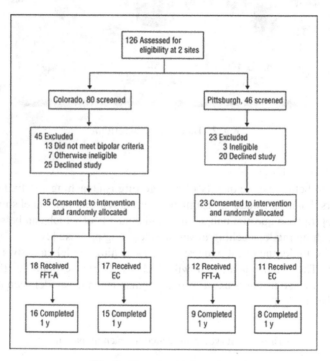

Fig. 3. Experiment flow chart

In the current sample of sophomore students, the total number of students is 1132, compared with the last examination, there are 208 students who have made more than 6% progress, accounting for 18% of the total sample. According to the data in the decision tree, the most important factor affecting the quality of students' English learning is "learning purpose", which is very appropriate. Learning purpose will affect the subjective initiative of college students. As a second language, if students' participation is not strong, it will not only affect their learning state, but also reduce the overall quality of English learning. In addition, teaching mode and teaching method also affect students' English learning quality. Compared with the subjective elements of learning purpose, this objective element also always affects students' English learning style and result.

(5) Analysis of results

After understanding the main factors affecting students' English learning, the relevant researchers can analyze the influence degree of learning purpose, teaching mode and teaching method in detail [18]. Specifically, the purpose or motivation of learning mainly represents the interest of learning, that is, the teacher's teaching in English classroom or the attraction of the subject itself. With the rapid development of network information technology, teachers need to bring this technology into English classroom, that is, to enhance the attractiveness of English classroom by means of multimedia technology. For example, in normal English learning, the content of teaching materials covers four items of listening, speaking, reading and writing, while in practical teaching, teachers pay more attention to reading and writing, and ignore the two items of listening and speaking. The main purpose of English is communication tools. If students are difficult to communicate in English, it will greatly weaken their interest in English learning, It also reduces its learning effect. If multimedia technology is used in formal English classroom, it will not only stimulate students' hearing and vision, but also increase the sense of classroom participation and improve the practical effect of English learning. Figure 4 is the result analysis chart:

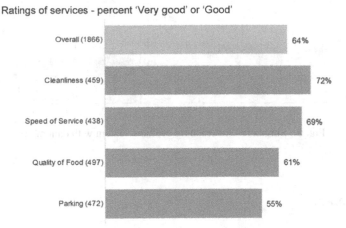

Fig. 4. Result analysis chart

At the same time, the teaching mode and teaching method will also affect the English learning quality of sophomore students. Because the students in this stage already have a certain English foundation, teachers should actively change teaching methods in English classroom teaching. Strengthen the English level of each individual student with various teaching forms. Under the influence of learning attitude and learning ability, students' English scores will be greatly different. Teachers should always adhere to the concept of teaching according to their aptitude, design more appropriate learning methods for students with different abilities, and strengthen the effectiveness and pertinence of teaching. Through the improvement of learning purpose, teaching method and teaching mode, teachers and students can make progress together to improve the overall quality of English teaching [19].

Moreover, with the help of data mining technology and decision tree, relevant researchers can also find out the correlation between CHAID algorithm and MNP elements. Generally speaking, each node can fully reflect the results of MNP elements. That is, we can see the teaching quality of college English teaching in different environments and elements, such as teacher education, teaching means, teaching mode and so on. Through the data index researchers in each node, we can find out the actual influence of each element on the effect of English teaching. For example, 44 samples and 13 MNP elements are included in 52 nodes. The purpose of learning includes going abroad or looking for a job (see Fig. 5).

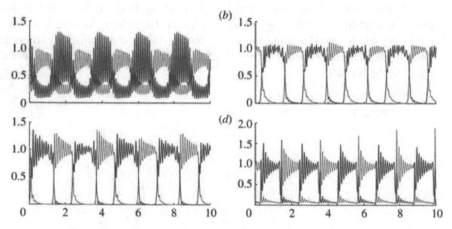

Fig. 5. Simulation for teaching evaluate system with data mining

5 Summary

To sum up, through the practical research on the quality of college English teaching, relevant researchers can find out many elements that affect the quality of English teaching, and use data mining technology to accurately find out the importance of all kinds of elements. Make college English teachers understand the current teaching situation, its teaching innovation also has stronger pertinence.

The analysis method of College English teaching quality based on data mining provides a new technical route for the construction of teaching quality index system.

This method uses data mining technology to find hidden, difficult to find but valuable information of teaching quality influencing factors from a large number of data covering College English teaching process, conditions and environment. The classification order of CHAID decision tree reflects the importance of the factors that affect the quality of teaching. Through data mining, it is found that the most important factors affecting the learning progress rate of the experimental sample group are "learning motivation", "teaching method" and "teaching mode". This method can make up for the subjective defects of the traditional teaching quality indicators, and provide a more objective basis for the construction of teaching quality system.

This method is helpful to provide targeted improvement strategies for teaching quality. Each node of Pmnp decision tree is a specific classification group, which reflects the quantitative effect of various factors on Pmnp. It allows analysts to directly observe the quality of English Teaching under different factors and environments (such as teaching mode, teaching means, teachers' education background), so as to compare the advantages and disadvantages of various teaching measures or their applicability to different student groups, It provides an intuitive and solid basis for proposing targeted improvement measures.

References

1. Gao Shu, G.: A study on the cultivation of talents in the eco-model of college English curriculum under the of Big Data Old brand marketing (01), 141–142 (2021)
2. Shan, B.: Analysis on the main factors and countermeasures affecting the quality of college English teaching. Shanxi Youth (06), 201 (2020)
3. Tang, S., Zhang, Y., Wu, Y., Zhang, X.: Digital transformation and engineering data governance of engineering enterprises in BIM environment. Oper. Manag. (04), 73–77 (2021)
4. Hong, Y.: Book borrowing analysis based on association mining. Libr. Res. Work (04), 75–79 (2021)
5. Song, X.: Cultivation of students' autonomous learning ability in junior middle school English Teaching. Learn. Weekly (13), 99–100 (2021)
6. Ye, X.: How to cultivate students' autonomous learning ability in Higher Vocational English Teaching. Knowl. Base (07), 138–139 (2021)
7. Zhang, X.: Strategies for cultivating English intercultural communication ability. Cult. Ind. (09), 38–40 (2021)
8. Zhang, Y.: On the cultivation of intercultural communicative competence in high school English Teaching. Cult. Ind. (09), 41–42 (2021)
9. Jia, X.: Cultivation of high school English autonomous learning ability. Coll. Entrance Examination (12), 49–50 (2021)
10. Mao, H.: Practical research on output oriented method in Chemical English Teaching. Thermosetting Resins **36**(02), 75 (2021)
11. Zhong, W.: Significance of Chemical Engineering English professional reform in the new era. Thermosetting Resins **36**(02), 80 (2021)
12. Liu, X.: Exploration of English grammar teaching strategies in Higher Vocational Colleges. Invention Innov. (Vocat. Educ.) (03), 12–13 (2021)
13. Yang, L.: Practice of oral English and writing teaching in Senior High School under unit theme. Invention Innov. (Vocat. Educ.) (03), 45–46 (2021)
14. Ke, W.: Qing novel coronavirus pneumonia during the epidemic prevention and control period of practice and reflection of online teaching. Sci. Educ. Union (Next 2021) (03), 174–175 (2021)

15. Jiang, Y.: A preliminary study of emotional teaching in primary school English culture time. Prim. Sch. Students (Next Issue) (04), 1 (2021)
16. Yang, Y.: Optimizing teaching strategies for oral English Teaching. Prim. Sch. Students (Next Issue) (04), 30 (2021)
17. Han, Q.: Analysis of English Discourse Teaching in primary schools based on schema theory. Prim. Sch. Students (Next Issue) (04), 41 (2021)
18. Geng, B.: Application of interactive performance method in primary school English Teaching. Prim. Sch. Students (Zhongxunjiao) (04), 31 (2021)
19. Gu, L.: On the cultivation of English language sense and oral English Teaching. Prim. Sch. Students (Next Issue) (04), 54 (2021)

Exploration and Practice of University Students Health Education Promotion Model Under Big Data Information

Wanjun Chen(✉)

South China Institute of Software Engineering,
Guangzhou University, Guangzhou 510990, China
cwj@sise.com.cn

Abstract. This paper introduces the promotion mode of College Students' health education based on big data information, summarizes the methods and means of the promotion mode of health education in teaching, and puts forward that we should make full use of modern educational technology and health education resources, guide students to carry out health education, prevent and treat diseases under the new medical mode, strengthen school health education, and explore new ways of College Students' health education And practice.

Keywords: Big data information · College students' health education · Exploration · Practice

1 Introduction

The core content of health education in higher schools is to spread health knowledge, help college students show its limitations and establish modern health awareness. At present, as an independent subject with the change of medical model, some subjects related to psychological factors and social factors, health education in Colleges and universities has been attached great importance by the governments of various countries. For hundreds of years, "medical model" has been based on biomedicine for hundreds of years, such as vision decline, chronic pharyngitis, tinnitus, sudden ear. The incidence rate of deafness and malignant tumor has been on the rise. The formation of these diseases is based on biological factors to prevent and treat diseases [1]. The occurrence and development of diseases are difficult to solve under the biomedical mode. With the development of medical education and medical science, the biomedical model has become obvious. In 1984, health education for college students began to be implemented in China's colleges and universities. At the same time, the emergence of the "biological psychological social" medical model made people not satisfied with treatment, but required to focus on prevention and health care, Make your body and mind in a better state of health.

© ICST Institute for Computer Sciences, Social Informatics and Telecommunications Engineering 2021
Published by Springer Nature Switzerland AG 2021. All Rights Reserved
M. A. Jan and F. Khan (Eds.): BigIoT-EDU 2021, LNICST 392, pp. 375–384, 2021.
https://doi.org/10.1007/978-3-030-87903-7_46

2   Proposed Work

2.1   Personalized Recommendation Information Transmission

According to the correlation features, the user's personal consumption data is extracted, and the fuzzy decision function is obtained. The nonlinear mapping $\phi : n \in R^n \to Q$ is used to represent the user's personalized grooming space. The data information is combined with the decision function, and the intelligent algorithm is used to map to the sample set of quotient recommendation. The hypothesis is that represents the model input vector as the target test value, and the number of N tables is used to calculate the personalized recommendation objective function:

$$\min imize \frac{1}{2} \|w\|^2 + \overset{n}{\underset{i=1}{B}} (j_i + j_i^*) \tag{1}$$

$$subject, m_i - (w\phi(a_i) + b) \le \varepsilon - j_i \tag{2}$$

$$(w\phi(a_i) + b) - m_i \le \varepsilon - j_i \tag{3}$$

In the era of traditional economy, data is a low-energy thing, from data, information, knowledge to wisdom is more and more valuable. But in the era of big data, data has become extremely powerful. Big data not only promotes big knowledge and wisdom, but also promotes big thinking and big pattern. As a new strategic resource, big data has attracted great attention from the industry, academia and political circles. Developed countries have launched plans to develop big data. Big data is rapidly changing the society we live in and the way we think. Whether you realize or feel it or not, it has become an indisputable fact that big data has entered people's field of vision. It is influencing the existence and operation of various social systems in its own way. Similarly, big data also enters and influences college students' mental health education system in an invisible way.

2.2   Significance and Characteristics of Health Education for College Students

2.2.1   The Significance of Health Education

The Central Committee of the Communist Youth League "on Further Strengthening and improving the health education of college students" shows that the Party Central Committee and the government are concerned about the physical and mental health of college students. According to the health education data of Xi'an Jiaotong University Hospital and Shaanxi Province's brother colleges, since 2016, the number of college students with mental illness, psychosomatic disease and vision decline has increased to varying degrees. In the face of such a serious situation, according to the characteristics of college students in the late adolescence, they have greater plasticity, It is a long-term work and task for colleges and universities to carry out health education in order to help them establish modern health awareness, increase health knowledge, and improve comprehensive disease prevention ability of physical and mental health, which is related to the future of our country. Practice has proved that college students' health education plays a positive role in overcoming diseases [2–6]. It is the need for college students to maintain a healthy state during the period of school, and it is also the need to ensure college students' physical and mental health during the period of school and become the national qualified talents. Nowadays, the extensive attention of Chinese people to health also indicates the coming of a new era of health education in Colleges and universities in China. Figure 1 shows the health education level of college students from 2014 to 2020 based on big data.

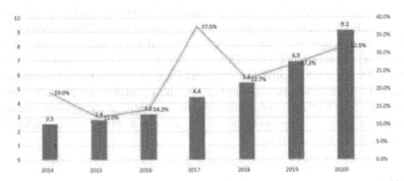

Fig. 1. The health education level of college students from 2014 to 2020 based on big data

2.2.2 Characteristics of Health Education

College Students' health education is based on health education. It can enhance college students' self-protection awareness and develop good health behaviors, so that college students are not only physically healthy, but also psychologically and morally healthy, and have good social adaptability. Early intervention should be carried out on the psychological diseases and psychosomatic diseases of college students during their study.

College Students' health education is aimed at the students at the later stage of puberty. The subject content includes not only basic medical treatment and clinical application technology, but also psychology, psychosomatic medicine and public health diagnosis and treatment skills. Therefore, in the teaching of health education for college students, more attention should be paid to guiding the students to change to the medical model of "biology psychology society" to understand and prevent diseases [7]. This will help students master the knowledge of health education, understand the ultimate goal of health education, spread the knowledge of health education to social groups, realize the right of everyone to enjoy health care, and serve the health of the whole people.

2.3 Ways to Promote Health Education

According to the characteristics of students in different areas, different environments and different living habits, we should arrange targeted teaching contents, such as teaching freshmen about environment and adaptation, environment and health, sleep and diet hygiene and psychological development characteristics of college students, so as to make them adapt to the study and life in the new environment as soon as possible. The teaching method adopts the teaching method of focusing on key problems and case teaching method [8–10]. The common problems encountered by freshmen are discussed collectively.

2.4 Teaching

Using new teaching methods to stimulate students' interest in learning and improve teaching effect. Questioning teaching is a new teaching mode. The attending doctors take 50 health education problems encountered in real life as teaching materials, apply specific cases and adopt problem-oriented method to constantly stimulate students to think and research, so that college students can learn to analyze problems and improve their ability to solve problems. The new teaching mode has obvious advantages over the traditional teaching methods in cultivating students' scientific thinking mode, stimulating their interest in learning, improving their ability of self-study and language expression, communication and cooperation.

2.5 Lecture

Different and flexible teaching methods were used to carry out health education lectures. According to the characteristics of sophomores who have basically adapted to the study and life of the University, we give special lectures on "hygiene and protection of eyesight", "food hygiene and nutrition", as well as female physiological health knowledge to female students. Among them, "health and protection of vision" is the focus of health education. Poor eyesight is one of the most common diseases detected by students in our country. Most of them are myopia. Myopia limits students' attention and reduces their ability to identify distant and fine targets, which seriously affects students' learning. The purpose of the special lecture on "food hygiene and nutrition" is to enable students to master the types of nutrients, develop good eating habits and prevent food poisoning. In women's physiology and health education, according to the special structure and physiological characteristics of women's body, women are taught menstrual health and health knowledge, such as breast pain and health care, menstrual health care, women's sports health, etc. [11]. In the teaching of modern social disease prevention and control knowledge, such as AIDS prevention and control, premarital sexual behavior and morality, heuristic teaching method is used to let students choose 100 health education topics before class, and find materials, learn and think after class.

3   Opportunities and Challenges of College Students' Mental Health Education in the Era of Big Data

In the era of traditional economy, data is a low-energy thing, from data, information, knowledge to wisdom is more and more valuable. But in the era of big data, data has become extremely powerful. Big data not only promotes big knowledge and wisdom, but also promotes big thinking and big pattern. As a new strategic resource, big data has attracted great attention from the industry, academia and political circles. Developed countries have launched plans to develop big data [12, 13]. Big data is rapidly changing the society we live in and the way we think. Whether you realize or feel it or not, it has become an indisputable fact that big data has entered people's field of vision. It is influencing the existence and operation of various social systems in its own way. Similarly, big data also enters and influences college students' mental health education system in an invisible way.

3.1 Challenges

The era of big data will bring many breakthroughs in college students' mental health education. However, just like everything else, college students' mental health education also faces many challenges in the process of using big data, such as data awareness, data collection, data storage, data talent and data privacy, which will restrict the role of big data. First, China has not yet upgraded big data to the national strategic level, and the industry generally lacks big data awareness, which is the key factor hindering the implementation of big data technology in various industries, and will also hinder people from patiently and seriously studying the origin and mechanism of big data. Second, in many colleges and universities, college students' mental health education has become a matter for a small number of mental health educators or student counselors, and other departments have become "bystanders". As data resources are scattered in different departments, there is a common phenomenon of "data island" and "fragmentation".

3.2 Four in One, Improve the Guarantee Mechanism of College Students' Mental Health Education

In order to make good use of big data in college students' mental health education and give full play to its due value, we must establish and improve the corresponding guarantee mechanism in order to meet the challenges of the era of big data [14]. First, a big data leading group should be set up. Colleges and universities should gather all departments involved in college students' mental health education and relevant personnel engaged in the development and utilization of big data to establish a strong leading group for big data application. Team members should not only have firm belief and be willing to devote themselves to the reform action, but also understand and support the application of big data, which can strengthen the integration of all kinds of information resources, provide top-level design and lay a solid foundation for the application of big data in college students' mental health education.

4 Adaptation and Innovation of College Students' Mental Health Education in the Era of Big Data

Big data is starting an important transformation of the times. It has become a strategic resource as important as natural resources and human resources. It is another disruptive technological change after cloud computing and Internet of things. Its storm is sweeping all industries with massive data [15]. Facing the opportunities and challenges brought by big data, college students' mental health education should actively integrate into the wave of big data, use data resources to improve education effect, promote service innovation, and improve the scientific level of College Students' mental health education.

4.1 Concept First, Set Up the Data Consciousness of College Students' Mental Health Education

With the advent of the era of big data, for the first time, human beings have the opportunity and conditions to widely and deeply obtain and use complete data, explore the laws of the real world, and obtain knowledge that was impossible to obtain in the past. As the most populous country in the world, China produces a huge amount of data. However, according to Wu Hequan, an academician of the Chinese Academy of engineering, more than half of the data in China have not been properly protected. Therefore, it is imperative to first advocate and strengthen data awareness in Colleges and universities. Both college administrators, teachers and mental health educators should update their ideas, understand the relevant knowledge of big data, and fully realize that big data is a valuable resource for mental health education. Secondly, mental health educators in Colleges and universities should keep pace with the times, Finally, mental health educators in Colleges and universities should establish the consciousness of data precipitation and data application in daily practice and scientific research, Pay attention to the collection and storage of mental health education related information, lay a solid data foundation for mental health education, and truly play the value of big data in college mental health education. After the opening of the era of big data, human society is undergoing a profound change. Like the invention of the Internet, the big data wave is not only a revolution in the field of information technology, but also a sharp tool to launch transparent government, accelerate enterprise innovation and lead social change in the world. It will bring new ways of thinking and management change to all walks of life. The advent of the era of big data has also imperceptibly affected the ways and behavior habits of college teachers and students in exploring the unknown world. How to actively respond to the opportunities and challenges brought by the era of big data and promote the development of College Students' mental health education with the times is an important issue for college counselors and other mental health educators to think deeply.

4.2 Conform to the Trend and Innovate the Research Paradigm of College Students' Mental Health Education

At present, influenced by many factors, college students have different degrees of interpersonal barriers, study and work pressure, emotional love confusion, Internet Dependence and other psychological problems. In the past, the research on these psychological problems mainly adopted the way of empirical research and questionnaire sampling survey, and achieved a lot of valuable research results [16]. However, due to the limitations of sample selection and data analysis, it is often difficult to draw systematic and profound conclusions from the research results. Mental health educators increasingly feel powerless in practical work, which has a great relationship with the lack of scientific development of mental health education. In the era of big data, everything is digitalized, and many disciplines have been deeply integrated with information technology. It is a general trend to use data to study college students' mental health education. In practice, it is urgent to promote the transformation of research paradigm of mental health education. Combining point with area to find out the way of College Students' mental health education.

At present, college students' mental health education lacks effectiveness and effectiveness. One of the important reasons is that mental health education is divorced from the changing real social life. Social existence has changed, but the ways and means of mental health education and the design of education system are still in the past, lacking the consideration and follow-up of the times. In the era of big data, everything can be quantified. Mental health educators can analyze the psychological roots of college students from their various activities and family and social relations, timely and accurately grasp their psychological status, seek the path to solve mental health education with big data, and realize the refinement of mental health education service.

5 Effect of Health Education

In order to understand the effect of health education through various ways, the hospital of Xi'an Jiaotong University issued questionnaires to 1000 college students who had participated in health education in 2012, The results show that 92% of the students who have participated in the health education knowledge study have obvious knowledge and improvement on the prevention and control of infectious diseases and AIDS, 82% of the students improved their bad living habits and behavior significantly [17]. They generally believe that behavior is closely related to health. A good behavior can promote people's mental and physical health, otherwise it will harm health. At the same time, 300 college students who did not participate in the study were given questionnaires and compared with those who had participated in health education. The results showed that the students who had participated in health education had obvious ability to prevent and control infectious diseases, modern social diseases, tuberculosis and AIDS, while those who had not participated in health education had obvious ability to prevent and control infectious diseases, tuberculosis and AIDS, The awareness of health education and the ability of disease prevention and treatment are still limited to the level of knowledge in middle school, and there are significant differences between the two groups of students.

For the health effect based on big data, we only selected the data from 1990 to 2018, which is increasing year by year from the effect point of view, such as the red line in Fig. 2, and the other two lines represent people's cognition of big data. As we can see from Fig. 2, we still have a normal attitude towards this change. So it doesn't change much.

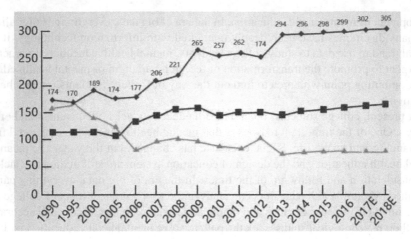

Fig. 2. The health effect based on big data

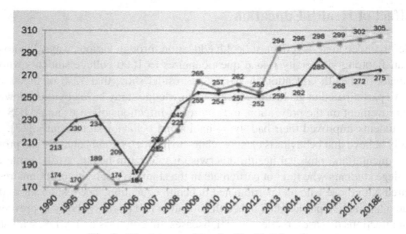

Fig. 3. Big data cognition for Health Education

We can also see from Fig. 3 that people's health education on the use of big data is on the rise. In the past, people only mentioned big data, but now people use big data. So the effect is different.

6 Conclusion

Health education for college students has been carried out for nearly 30 years in China's colleges and universities. Through the comprehensive application of various teaching methods and means, we have improved the teaching effect and enhanced the students' ability of disease prevention and treatment. Although we have gained some experience in the exploration, as a new subject, there are many imperfections in the teaching, In the future, we will actively learn from the successful experience of health education for college students at home and abroad, and constantly improve and explore the effective ways and modes to adapt to the health education and teaching of college students in China. In the era of big data, all changes are accelerating [18–23]. If we don't understand this change, we will become blind actors and can't keep up with the pace of social progress. Only by further strengthening the awareness of big data and actively tapping the inherent potential of big data application, can colleges and universities effectively promote the transformation, innovation and development of College Students' mental health education, make it quickly keep up with the changes of the times, and embark on the healthy road of self-improvement, self-improvement and self transcendence.

References

1. Zhengliang's, Q.: Health Education Course in Colleges and Universities. Xi'an Jiaotong University Press, Xi'an, vol. 89 (2010)
2. Zhengliang, Q.: Health Education for College Students. Shaanxi Science and Technology Press, Xi'an, pp. 11–13 (2010)
3. Mei, W., Ping, L., Qiongyu, W., et al.: Exploration and practice of skin laser cosmetic medicine education. Chin. Med. Educ. Technol. **29**(3), 316–318 (2015)
4. Zhengliang, Q.: Health Education for College Students. Shaanxi Science and Technology Press, Xi'an, vol. 4143 (2010)
5. Yingxin, L., He, H.: Health management model based on national health and health management big data platform. Int. J. Biomed. Eng. **40**(005), 307–314 (2017)
6. Hao, W.: Building medical big data platform to promote healthy China Construction. Modern Health (2018)
7. Zesen, Y., Xiao, W.: Health control methods, devices, media and electronic devices of big data platform, cn108733532a
8. Ren, D., Zhigang, D., Juan, H., et al.: Design and implementation of provincial health care big data platform. Chin. J. Health Inform. Manag. **014**(001), 31–34 (2017)
9. Jiantao, C.: Design of big data platform for user health service and key technologies of analysis and processing. Nanjing University of Posts and Telecommunications (2016)
10. Jie, L., Jinwen, Y., Xuanchen, Y., et al.: Design and application of provincial health care big data platform. Journal of Medical Informatics (2018)
11. Gang, L.: Research on the cultivation of contemporary college students' healthy political psychology. Shanxi Normal University (2019)
12. Xuejiao, F.: Moral education countermeasures of College Counselors in the perspective of caring ethics. Legal Syst. Soc. **17**, 174–176 (2019)
13. Two works of our students won the prize in the national: 2019 scientific popularization competition of College Students' health education. J. Fujian Med. Univ. (Soc. Sci. Edn.) **20**(02), 66 (2019)
14. Lan, Z., Zemata, D.: Research on the methods and paths of health education and management of college students. J. Jinzhou Med. Univ. (Soc. Sci. Edn.) **17**(03), 73–75+79 (2019)
15. Huan, G.: Investigation on current situation of national health education for college students. Advances in Social Sciences, vol. 8, no. 5 (2019)
16. Liming, C.: The relationship between College Students' health literacy and health status. Zhejiang University (2019)
17. Rui, Z., Yanyi, H., Jin, L.: Health education problems and strategies of higher vocational college students. South Vocat. Educ. J. **9**(03), 55–59 (2019)
18. XuXi, L.: Research on the cooperative education of College Students' health education and ideological and political education. Chongqing Medical University (2019)

19. Miao, L.: The research on the education of College Students' healthy online lifestyle development. Huazhong Normal University (2019)
20. Fen, Y.: Research on health support system of university students in Kunming. Yunnan University (2019)
21. Sai, L.: Shaping the healthy personality of college students in the context of Ideological and political education. Time Rep. **04**, 62–63 (2019)
22. Xiangling, L.: Survey on the demand of health education service in Colleges and universities. Health Educ. Health promotion **14**(02), 156–158 (2019)
23. Guojing, Z.: The significance of aesthetic education to college students' mental health: an exploration practice. Emotional Reading **11**, 11 (2019)

Research on Management Mechanism of Higher Education Based on Big Data Analysis

Zhili Ni[✉]

Wuhan Railway Vocational College of Technology, No. 1 Canglong Avenue, Jiangxia District, Wuhan 430205, China

Abstract. With the development of the era of knowledge economy, higher education in various countries is facing more and more fierce competition and challenges. In order to improve the quality of higher education and the pursuit of high performance of higher education, all countries intend to introduce total quality management into higher education, so as to improve the quality of Education and management efficiency. European and American countries applied TQM to higher education earlier, and have formed a more mature and perfect management system. The overall quality management of higher education should be oriented towards students' satisfaction, adopting the management method of all staff participation and forming a culture atmosphere of continuous improvement.

Keywords: Higher education · Total quality management · Management mechanismm

1 Introduction

With the development of knowledge-based economy, the rapid updating of scientific and technological information, and the increase of international cultural exchanges, higher education in the 21st century is facing new and more intense global challenges. At the same time, due to the importance of higher education in the social economy, all countries in the world take it as a consensus to improve the management of higher education. In recent years, in order to effectively improve the quality of higher education and pursue high performance of higher education output, all countries intend to introduce the concept of total quality management into higher education management, which is regarded as a new opportunity for the development of higher education. However, we need to point out that the dilemma of the application of total quality management in higher education lies in the difficulty of clearly defining the autonomy of academic staff and the needs of students. First of all, employee empowerment is an important dimension in the quality management system. However, the academic staff's participation in university decision-making and management is still low. Secondly, students' needs are more and more diversified and personalized, which also brings challenges to the design of curriculum and training methods [1]. Therefore, the university should combine the resources, conditions and characteristics of the University, formulate quality standards

M. A. Jan and F. Khan (Eds.): BigIoT-EDU 2021, LNICST 392, pp. 385–390, 2021.
https://doi.org/10.1007/978-3-030-87903-7_47

and performance indicators, comprehensively enhance the quality and competitiveness of university education, and achieve excellence.

2 Confidence K-means Clustering Algorithm for Structure Recognition of Confidence Rule Base

Belief rule reasoning (rmer) is an expert system based on evidence reasoning, decision theory and production rule reasoning under uncertain information. Belief rules can effectively model uncertainty and nonlinearity, and ensure reasonable accuracy and the interpretability of language.

2.1 Belief Rule Reasoning

2.1.1 Main Introduction

In traditional methods, the structure and parameters of the confidence rule base are determined by experts or decision-makers in advance according to experience knowledge or other original models. However, it is very difficult and unreliable to use only expert knowledge to determine the structure and parameters of the confidence rule base in the case of large-scale confidence rule base and rapidly changing patterns. The small difference between rule weight and attribute weight of belief rule will bring great changes to the performance of belief rule reasoning. For this reason, Yang et al. Proposed several single objective and multi-objective nonlinear optimization models (OM tbrbs) for training confidence rule base and applied them to graphite composition detection. The remarkable feature of these models is that the input and output information can only be given partially, such as incomplete or fuzzy, mathematical or judgmental, or mixed.

2.1.2 Application Algorithm

The production rule reasoning model can be represented by the following four tuples:

$$R \leq X, A, D, F > \tag{1}$$

By extracting the evaluation level of the antecedent variables of each rule from the historical data, the reasonable structure of the confidence rule base is obtained. The optimal clustering of the algorithm is proportional to the distance between any two adjacent evaluation grades, and ensures the shortest distance between the historical data and the nearest evaluation grade.

2.1.3 Belief Rule Reasoning Algorithm Based on Evidential Reasoning

Since the independence of belief rules meets the requirement of recursive evidential reasoning (RER), Yang et al. Proposed a belief rule reasoning algorithm based on recursive evidential reasoning. In order to generate the excitation weight of each confidence rule, the relationship between the current input and the evaluation level reference value of the antecedent variable of each rule must be determined before reasoning [2]. The main idea is to check the evaluation level of each rule's antecedent variable to determine the matching degree of the current input corresponding to each evaluation level.

$$S(X_i) = \{(A_{i,j}, \alpha_{i,j}); j = 1\} \tag{2}$$

The function mapping between input and output of belief rule reasoning and the distributed form of reasoning output can be expressed as follows:

$$S(X_i) = \{(D_n, \beta_n); j = 1\} \tag{3}$$

2.2 Advantages of Belief Rule Reasoning

When the belief rule reasoning is applied to the problem of only input variable historical data, it is necessary to put forward the corresponding algorithm to mine information from the input variable historical data to identify the structure of the confidence rule base because of the volatility and nonstationarity of the antecedent variables. In this paper, based on K-means 1 and fuzzy c-means 1, a confidence K-means clustering algorithm is proposed to dynamically identify the structure of the confidence rule base for system control. Each data in K-means belongs to only one cluster, and each data in fuzzy c-means belongs to each cluster in the form of membership function. In the framework of evidential reasoning, each historical data in the confidence K-means belongs to two adjacent clusters (evaluation grades) in the form of confidence. By extracting the evaluation level of the antecedent variables of each rule from the historical data, the reasonable structure of the confidence rule base is obtained. The optimal clustering of the algorithm is proportional to the distance between any two adjacent evaluation grades, and ensures the shortest distance between the historical data and the nearest evaluation grade. These two characteristics show that the algorithm can improve the identification degree of the optimal structure and the accuracy of identification and reasoning.

3 Application Pattern and Structure Recognition of Belief Rule Reasoning

3.1 Confidence Rule Base Process

The application of confidence rule reasoning can be divided into system approximator and system controller. The application mode of confidence rule reasoning as system approximator and system controller can be shown in Fig. 1. As a system approximator, the training and adjustment of the confidence rule base is driven by the error of observation output and reasoning output [3]. The reasoning output of the confidence rule is the prediction of the actual system output. As a system controller, the training and regulation of the confidence rule base is driven by the system performance, and the reasoning output of the confidence rule is the control or decision variable in the actual system operation process. In the application mode of system approximator and system controller. For the application of the confidence rule base shown in Fig. 1 as a system approximator, the historical data is given in the form of input and output data pairs. Based on the definition of the random utility of the confidence rule, Zhou et al. Proposed the sequential algorithm of adding rules and deleting rules to adjust the structure of the confidence rule base. letter.

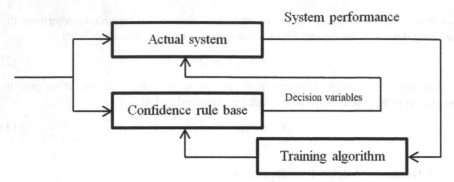

Fig. 1. Application mode of confidence rule reasoning as system controller

3.2 Structure Recognition of Confidence Rule Base

The structure recognition in this paper is to mine information from the historical data of input variables to determine the evaluation level and utility of each rule antecedent variable under the condition of determining the logical relationship between the antecedent variables of the rule base and the cascade hierarchical framework of the confidence rule base. In order to build the rule base, it is necessary to determine the evaluation level of each preceding variable and the number of evaluation levels [4]. The structure of the confidence rule base and the number of its rules depend on the position and number of the evaluation level in the input variable domain. To find the optimal structure of the confidence rule base is to find the most appropriate number and location of the evaluation level set, which can ensure that the confidence rule base can form the best fit to the sampling data and system pattern. The antecedent of each confidence rule is composed of all antecedent variables and each evaluation grade.

4   Quality Assurance System of Higher Education

4.1   Improving the Quality of University Education

The pursuit of excellence is not only the goal and ideal of the University, but also the spirit and soul of the University. The pursuit of excellence in the quality of university education has become the mainstream concept of the development of higher education in various countries. Higher education quality management is developed according to the quality management and quality assurance standards of the international organization. In the past two decades, the concept of quality has replaced the concept of "efficiency" and become the primary goal of higher education. As Green said, "the 1980s is the era of efficiency, while the 1990s is the era of attaching importance to quality. Especially in the 1980s, the government's investment in universities is decreasing, the public management movement is rising, the society and the public demand for university performance is higher and higher, and the global international competition is intensifying. Universities need to effectively evaluate their own school running performance, so as to be able to cope with internal and external challenges and improve the overall quality of university running.

4.2 Management Mechanism Adopted Abroad

In 1997, the quality assurance agency for Higher Education (QAA) was set up to audit the teaching quality and academic quality of universities, so as to reduce the duplication of institutions and waste of resources. At the same time, it ensures the tradition of university autonomy and effectively combines external audit with self-management. Since 90% of French higher education funds are borne by the government, it is an important responsibility of the government to audit and evaluate the quality of higher education. French higher education quality assurance is mainly responsible by its national evaluation committee. In the declaration of "quality and diversity of higher education", Australia proposed to independently audit the policies and procedures of university quality assurance, recommended to the government the allocation of annual school running funds, and established the Australian university quality management as a national institution to supervise and audit the quality of higher education in Australia. So far, American universities have basically adopted total quality management to ensure and improve their education quality.

5 Total Quality Management Mechanism of Higher Education

5.1 On the Quality of Higher Education

5.1.1 Cardinal Principle

For the country, the quality of higher education is the guarantee to enhance the competitiveness of the country. For the society, receiving higher education is the driving force to realize social mobility and the catalyst to promote economic development. For enterprises, in the era of global competition, high-quality talent training and supply is the basis for enterprise innovation and R & D, Moreover, high-quality talent output can also improve the popularity of the University in the world, which can improve the quality of students and form a good and orderly cycle. Therefore, from the above point of view, the quality management of higher education is the key factor of school competitiveness and performance management.

5.1.2 Student Satisfaction Oriented

Education is one of the important products in global service trade. According to the classification of WTO, education has been listed as "service sector" together with 14 departments, such as business services, communication, sales and finance. It is an important service product in Global trade. Students and parents are regarded as customers. "The development of higher education based on customer orientation attaches great importance to the response of external customers (parents and students) and timely corrects the education service. The improvement of students' satisfaction will also lead to the improvement of students' loyalty. The United States is one of the countries with the highest degree of marketization of higher education, and its student satisfaction is a very important evaluation index of higher education [5]. Paying attention to the needs of students and using student satisfaction assessment to continuously improve and improve the quality of education are the key factors for the success of American higher education.

5.2 Adopt the Management Method of Full Participation

The implementation of total quality management must be supported and participated by all employees. Some studies show that about 70% to 75% of the failure of organizational change is due to the failure to meet the expectations of key stakeholders. The choice of goals in an organization should be recognized and committed by all members, and enable members to work hard to achieve the goals. As UNESCO put forward in the global initiative for quality assurance capability, "quality assurance is effective only when all stakeholders understand and accept the challenges and benefits they face. To form an atmosphere of attaching importance to quality requires strong and firm management stakeholders from higher education leaders to participate in quality management, which is the value demand of quality management. The stakeholders here include internal and external stakeholders, including staff, students, alumni and community, etc. In the knowledge society, the purpose of organization is to make knowledge more productive.

6 Conclusion

The optimal confidence k-clustering ensures that the sampling points are close to the evaluation level with the minimum distance, that is, the input variables are close to the antecedents of the confidence rules as much as possible, which improves the accuracy of recognition and reasoning of the confidence rule base. The algorithm is also suitable for the application of confidence rule base as system approximator. In this paper, the number of evaluation grades in the confidence rule base is still given by the decision maker. The next step is to propose the corresponding data mining algorithm, which not only dynamically determines and adjusts the position of evaluation grade utility, but also identifies its number.

References

1. Dong, C.: Eight Principles of Quality Management. Guangdong Economic Publishing House, Guangdong (2008)
2. Yang, G.: Specific strategies for improving the quality of Higher Education. In: Proceedings of the Second Cross Strait Forum on Higher Education: Higher Education Quality, School Running Mode and Development Strategy. South China University of Technology, Guangzhou (2006)
3. Lin, T.: Total quality management supported by IBM. World Educ. Inform. **4**, 32 (1995)
4. Yang, C.: Challenge, transcendence and excellence of high energy education in Taiwan. Educ. Mater. **44**, 1–28
5. Jinli, S.: New development of higher education quality assurance system in UK: Practice of QAA institutional review. Educ. Res. Dev. **1**, 173–206 (2006)

Research on Teaching Reform of College Chinese Based on Data Analysis

Donglei Zhang[✉] and Bingxian Li

Weifang Engineering Vocational College, Weifang 262500, Shandong Province, China

Abstract. The rapid development of vocational education speeds up the pace of education and teaching transformation in higher vocational colleges. However, due to the influence of internal and external environment in the actual operation process, there are some problems in College Chinese curriculum in higher vocational colleges, which need to be solved and improved urgently. Therefore, this paper analyzes the current situation of Chinese teaching in higher vocational colleges, and analyzes the reasons for the problems. Combined with the needs of educational development, this paper puts forward the Countermeasures of College Chinese teaching reform, in order to provide theoretical reference for educators.

Keywords: Higher vocational colleges · College Chinese · Teaching innovation

1 Introduction

With the rapid development of computer network, communication technology and Internet, distance education based on Internet has become a new teaching method, and more and more people pay attention to it. Because online teaching has many characteristics such as timeliness, sharing, interaction and individualization, it has incomparable advantages compared with traditional teaching mode. It creates a new teaching mode, breaks the limitation of traditional teaching mode in time and space, and adopts advanced teaching means and methods, which can greatly improve teaching efficiency and teaching effect, Teaching activities have been promoted to a new level. The booming online education, with its new high-tech teaching means, provides a solution with less investment, quick effect, high quality and high efficiency to solve the contradiction between the serious shortage of resources and the growing demand for teaching advice. Fortunately, it will become an important direction of the development of teaching advice in the future.

College Chinese is a public comparative course of higher vocational education, and it is also the transfer of cultivating students' Humanistic ability. Under the background of deepening the collection of funds from the government and the army, the problems existing in the articles of combatting Tai and contending for life by the Supreme People's court are increasingly realized. In order to solve the existing problems, the teaching method of "Li Yue Bei Si Meng" has been developed rapidly, which meets the requirements of

© ICST Institute for Computer Sciences, Social Informatics and Telecommunications Engineering 2021
Published by Springer Nature Switzerland AG 2021. All Rights Reserved
M. A. Jan and F. Khan (Eds.): BigIoT-EDU 2021, LNICST 392, pp. 391–397, 2021.
https://doi.org/10.1007/978-3-030-87903-7_48

modern students. It is the focus of College Chinese educators in higher vocational colleges. Based on this, under the condition of understanding the actual teaching situation, this paper puts forward new ideas for the reform of College Chinese curriculum, and provides a reference for College Chinese to get out of the teaching dilemma.

2   Analysis on the Current Situation of College Chinese Teaching in Higher Vocational Colleges

2.1   Fuzzy Orientation of Curriculum

Influenced by the traditional education concept, most higher vocational colleges put more energy into the professional courses in the process of setting up the education system. They think that when the students are mainly studying the basic courses of certificate, they should reduce the time of Chinese teaching as much as possible, or even do not set up the College Chinese course, which makes the College Chinese course mere formality and difficult to achieve the overall education, Influenced by the external education environment, the lack of communication with the outside world in College Chinese teaching research in higher vocational colleges, the blocking of teaching information in various vocational colleges, make teachers unable to timely understand the innovative teaching methods and ideas, teaching mood is increasingly low. The emergence of this situation will increase the randomness of teachers' teaching, or completely copy the teaching ideas of undergraduate colleges, ignoring the thinking of students' learning ability and needs. Or follow the traditional single teaching mode, the classroom learning atmosphere is boring, aggravating students' long-term learning fatigue. As a result, the effect of College English teaching is not good, and the transformation and innovation of teaching mode can not be completed under the background of the comprehensive implementation of education reform.

2.2   Students Do Not Pay Enough Attention to the Curriculum

Compared with the ordinary undergraduate students, the difference of vocational college students is reflected in their own literacy and learning ability, and the scientific and cultural foundation is relatively weak. From the present situation, although vocational college students have learning motivation and good learning attitude, their own communication ability and expression ability are also obvious. Influenced by the traditional teaching concept of our country, students are in a passive learning environment, their creativity and learning enthusiasm are weakened, and their attention to basic education is insufficient. The prejudice against basic education in higher vocational colleges has not completely disappeared, and the students' utilitarian tendency is serious. Many students think that basic subjects such as Chinese are dispensable. They can't concentrate in the classroom, which not only wastes the time and energy of learning, but also reduces the students' humanistic quality. This kind of negative emotion will reduce the enthusiasm of students to actively understand and read literary works, make them unable to accept the influence of literary education, unable to experience the charm of literary works, and realize the improvement of their own humanistic realm.

3   The Causes of Problems in College Chinese Teaching

3.1   The Teaching Content of College Chinese is Too Old

As the key to attract students, teaching content is the basis of realizing the teaching goal of College Chinese and has direct connection with students. At the same time, the choice of teaching content also determines the final effect of teaching, especially for the college language of humanities, it is particularly important to choose the teaching response which is in line with the students' learning ability. Only students' curiosity about the content of the curriculum can they cooperate with teachers to complete the exploration and research of the curriculum content, and then successfully copy the relevant knowledge and skills. However, from the current situation, the College Chinese teaching materials can not meet the requirements of education development, carry out the innovation of teaching content, but also exist that the content of the teaching materials is too old and the content of the white text textbooks is composed of literary works, many works flow far away, with a deeper connotation. It is not suitable for the modern social situation that they reflect the social problems which have been applied to the society with strong era atmosphere. At the same time, it is not suitable for the aesthetic standards of the young students. Students are easy to understand the situation in the process of reading, which leads to the loss of the original educational value of Chinese teaching content.

3.2   Insufficient Investment in Curriculum Construction

First, the construction of Chinese teachers in higher vocational colleges is insufficient. With the rapid development of vocational education, many vocational colleges pay more attention to the expansion of mechanics and ignore the construction of teachers. The contradiction between the number of students and the level of teachers is becoming increasingly prominent. As a result, most of the teachers of College Chinese course are employed or part-time, and most of the education funds they receive are special funds for professional courses, It will affect the quality and efficiency of curriculum construction. Third, the lack of teaching infrastructure. As a part of vocational education, higher vocational colleges focus on training high skilled talents. Therefore, the investment in professional courses is more, and the cost of human and material resources is larger. For the humanities of Chinese characters, many higher vocational colleges think that the content of teaching materials has a certain richness, and there is no need to invest too much resources in discipline construction, which leads to the lack of Chinese mathematics infrastructure, and the overall effect of carrying out Chinese mathematics activities in the central environment will be affected.

4   Description of Intelligent Group Paper Model

4.1   Interpretation of Test Paper

Test paper generation is a typical multi-objective optimization problem. Test paper evaluation criteria include total score, cursory distribution, question type distribution, difficulty of test paper, test time and discrimination of test paper, Set the evaluation function y - "(total score, cursive distribution, question type distribution, difficulty of the test paper, and the best degree of closeness between the generated test paper and the target, so as to judge the relative quality of the test paper. Suppose there are m questions in the test paper, and each question has n parameters, including score (A1), chapter (A2), question type (A3), difficulty (A4), time (as), and discrimination (A6). The total score of the test paper is s. Figure 1 shows the spectrum of disturbance voltage at the end of intelligent components. The distribution of chapters is C (= 1.2 The total number of chapters (P), the distribution of questions (R2) The total test time is t, the total discrimination of the test paper is Q, and the test paper model is the objective matrix X of mxn, which should be close to the following constraints.

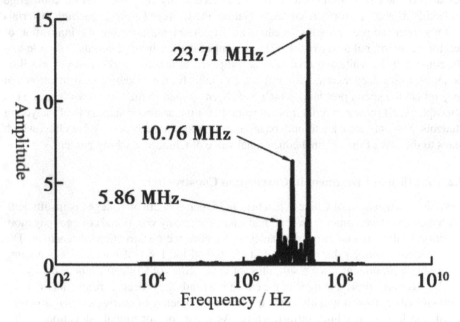

Fig. 1. Spectrum of disturbance voltage at the end of intelligent components

4.2 Algorithm Explanation

It takes a long time to generate a test paper, and it has a large complexity. At the same time, it doesn't always return effective test questions, which is inefficient. Generally, it is suitable for the examination system with less constraints. Backtracking heuristics is to return to the last recorded state type when the search fails, change the new state according to certain rules, and continuously backtrack to test paper generation or return to the starting point. This algorithm has a high success rate, Generally, it is suitable for the examination system with a small number of questions. When the number of questions is large, the complexity of empty questions is large, the program structure is extremely complex, and the time required for generating papers is long, The efficiency of genetic algorithm is low, and genetic algorithm is a probability search algorithm. It is a randomized search and optimization algorithm that simulates the biological natural selection and genetic mutation mechanism to solve complex problems. The algorithm has the characteristics of adaptive optimization, intelligent search technology and good convergence, so it can effectively solve the constrained optimization problem in test paper composition, and improve the speed and success rate of test paper composition.

The sum of the scores of all the questions is the total score:

$$S = \sum_{i=0}^{i=m} a_{ij} \tag{1}$$

According to the proportion of a question score in the total score, the sum of the difficulty is the same as the total difficulty of the test paper:

$$D = \sum_{i=0}^{i=m} (a_{i4} a_{il}/S) \tag{2}$$

According to the proportion of the score of a question in the total score, the sum of the discrimination is the same as the total discrimination of the test paper:

$$Q = \sum_{i=0}^{i=m} (a_{i6} a_{il}/S) \tag{3}$$

5 System Function Design

5.1 Making Clear the Overall Teaching Goal and Establishing the Concept of Professional Education Service

College Chinese is an important part of vocational college education system, which is a course with comprehensive education function. College Chinese takes Chinese as the main carrier, and integrates ancient and modern Chinese literature works as the main teaching content. It has distinctive instrumental and humanistic characteristics. Relevant education results show that through the study of College Chinese curriculum, students can constantly enhance their literary literacy, and improve their expression ability and communication ability. Especially for the graduates who are about to enter the job, they have a solid knowledge of mother tongue and can complete the job transition more quickly to adapt to the requirements of the job. Therefore, higher vocational colleges should pay attention to the construction of College Chinese course star, and set the teaching goal to cultivate students' Chinese ability as the core, and to promote students' all-round development as the purpose.

5.2 Pay Attention to the Construction of Teaching Materials and Improve the Sense of the Times of Teaching Content

Under the background of the comprehensive implementation of the reform of teaching sound, the teaching reform of College Chinese in higher vocational colleges should not stop at the construction of a complete teaching system, but also pay attention to the cultivation of students' intellectual and cultural way of thinking, combined with the needs of contemporary young students, write scientific and reasonable teaching materials, and endow the course with new vitality, In addition to the traditional contents, inspirational books at home and abroad can also be introduced into the teaching materials, such as the works of Mo Yan, Yu Hua, Liu Xuyun and other modern popular writers, so as to better adapt to the way of thinking of contemporary students and reflect the humanistic characteristics of Chinese. Through Li Zhihe's teaching experience, students can establish correct values and learning attitude, so as to help teachers complete the task of Chinese curriculum education.

In a word, with the rapid development and transformation of education and the change of social requirements for talents, high skilled and high-quality talents have become the favorite of the market and posts. Under this background, higher vocational colleges should realize the urgency of education reform and actively adopt innovative teaching mode based on the actual teaching content of College Chinese. And bold attempts to constantly enhance the timeliness of Chinese teaching, so as to better complete the goal of College Chinese teaching reform.

6   Summary and Prospect

The algorithm is applied to the CA system for automatic test paper generation. The algorithm can meet the requirements of automatic test paper generation, and the test paper generation is close to the predetermined constraint conditions. It shows that the algorithm can basically meet the requirements of teachers. At present, it has been applied to the network-based customized CA software in our college, and it provides a good auxiliary role for teachers in the teaching process, It is well received by teachers. With the rapid development of information technology and network technology, computer aided instruction (CA) has become a hot issue in modern teaching and reform. It integrates a variety of teaching auxiliary functions, makes the CA system have good interactive control, plays a very excellent auxiliary role in teaching, and comprehensively improves the teaching effect, The most outstanding is the online examination function of CA system. It can use computer to realize intelligent test paper generation, automatic test paper modification and statistics, which not only reduces the workload of teachers, but also realizes the separation of test and teaching, and makes the examination more standardized. Therefore, intelligent test paper generation is one of the most important tasks of online examination system, and its essence is according to the specific selection constraints, It is a typical multi-objective optimization problem. The efficiency and quality of the problem depends on the design of the algorithm. Therefore, choosing the appropriate algorithm is conducive to improving the success rate of test paper generation and the quality of test paper generation.

References

1. Fengli, Z.: Path selection of College Chinese practice teaching mode reform. Chin. Charact. Cult. **20**, 15–16 (2020)

2. Yueping, W.: Research on the reform of College Chinese teaching based on improving the Chinese ability of undergraduate application-oriented talents. Chin. J. Multimedia Netw. Teach. **10**, 117–119 (2020)
3. Yan, Z.: Exploration of Chinese classroom teaching reform in Higher Vocational Colleges from the perspective of student center. J. Xiangyang Vocat. Techn. Coll. **19**(05), 81–84 (2020)
4. Fei, W.: Practical research on creative writing in College Chinese teaching reform in art vocational colleges. Popular Lit. Art **17**, 214–215 (2020)

Research on the Application of Data Statistics Technology in the Training Course of Service and Management for the Elderly

Mengmeng Sun[✉]

Shandong Institute of Commerce and Technology, Jinan, China
sunmengmeng1985@163.com

Abstract. With the significant improvement of living standards and medical conditions, the elderly service robot has become an important research direction of service robot and has broad application prospects. This paper introduces the development status of the elderly service robot at home and abroad, and introduces the key technologies of the elderly service robot: autonomous mobile technology, perception technology, intelligent control technology and communication technology. The main problems and solutions are summarized.

Keywords: Old clothing major · Core curriculum · Teaching reform

1 Introduction

The survey shows that many developed countries are facing the problem of aging population. With the significant improvement of people's living standards and the continuous improvement of medical conditions, a variety of service robots for the elderly have appeared in our lives. In the face of empty nest elderly families and single elderly residents, how to use the home service robot to carry out rehabilitation training, psychological care, life support and daily monitoring for the sick or disabled elderly has become the focus of the current research on the elderly service robot.

2 Research Status at Home and Abroad

2.1 Current Situation Abroad

Based on the research of traditional service robots, service robots for the elderly are gradually developed, which take the elderly as the service object. Shigeki sugano, a professor at Waseda University in Japan, has developed twindy-0, a robot for the elderly. The service robot can bend down to pick up objects and lift 35 kg with both hands. The skin of the robot is soft and it can handle objects of various shapes. ROBOSOFT, a French robot company, has launched a robot called kompal, which is mainly used to help the elderly and the disabled. Robots can speak, understand conversations, and navigate autonomously [1]. South Korea launches Robot Companion for the elderly. South Korea has developed the robot to provide a playmate for the elderly and prevent dementia.

M. A. Jan and F. Khan (Eds.): BigIoT-EDU 2021, LNICST 392, pp. 398–404, 2021.
https://doi.org/10.1007/978-3-030-87903-7_49

2.2 Domestic Situation

In China, the research of elderly service robot started late. But in recent years, there has been a certain development. For example, the service robot researched by Harbin Institute of technology of China can recognize drugs, tea cups and other objects, and can sing. The Chinese Academy of Sciences has launched the subject of "robot for the elderly service" in the knowledge innovation project "robot for thousands of households". And in 2012, a life size "service robot for the elderly" was displayed. Generally speaking, the development of domestic elderly service robots started late, and there is still a certain gap with developed countries, and most of them are low-end products.

3 Key Technology

3.1 Autonomous Mobile Technology

Autonomous mobile technology is one of the key technologies of service robot. Robot navigation technology is an important aspect of autonomous mobile technology. The mobile robot can sense the environment and its own state through sensors, and realize the target oriented autonomous motion (navigation) in the environment with obstacles. The current indoor robot navigation technology mainly includes: RFID navigation, magnetic navigation, ultrasonic and radar navigation, voice navigation and visual navigation. Tong Feng et al. [2]. Designed a mobile robot ultrasonic navigation system for indoor structure environment. Berns proposed a navigation technology based on RFID. Hu Zhijun and others designed a visual guidance service robot for the elderly. Li Xinde proposed a robot vision navigation method based on general object recognition and GPU acceleration technology algorithm, and studied the vision navigation algorithm of service robot.

3.2 Perception Technology

The elderly service robot is a special service robot for the elderly. Its sensing system is a sensor network composed of various sensors. It includes pressure sensor, which is used to sense touch on the elderly accompany robot; smoke and harmful gas sensor, which is used to sense the indoor environment; photoelectric sensor, which is used to sense the intensity of indoor light, speed sensor, which is used to measure the moving speed and distance; proximity sensor, which is used for short-distance accurate mobile positioning; voice sensor, which is used to realize man-machine dialogue, and complete voice instructions: visual sensor, Perception of indoor space environment, better complete the object recognition, positioning and grasping.

3.3 Intelligent Decision Control Technology

Intelligent decision-making and control is an automatic decision-making and control technology that can drive intelligent machine to make intelligent decision and achieve control objectives without human intervention. The intelligent decision-making and control of the elderly service robot includes automatic and intelligent decision-making and control in the process of autonomous movement, accurate positioning, recognition and grasping objects, human-computer interaction, network control and so on. Intelligent decision control includes: fuzzy control, neural network control, artificial intelligence control, humanoid control, chaos control and so on. Intelligent decision-making and control is mainly used to solve the complex control problem that the control object can not be accurately modeled, which has the characteristics of nonlinearity. In the navigation trajectory control, Zhang Lixun and others studied the sliding mode trajectory tracking control of service robot based on global vision.

4   The Pain Points of Traditional Expansion Methods

4.1   Value Evaluation System

Due to the increasing demand of network quality and network development, the pressure of network operation is increasing. However, due to the limited investment in network development and the adoption of user perceived capacity guarantee scheme in network construction, high-value users and high-value scenarios can not get better network quality, which leads to the low efficiency of network development, It can't maximize the investment benefit. The expansion cycle is long, and the expansion lags behind, which affects users' perception. The method of daily expansion is early construction and late evaluation [3]. Although multi-dimensional analysis can be carried out by optimization means, the expansion work is late.

4.2   User Perception

That is to say, after finding out the capacity problems or poor user perception problems, the soft expansion or supplementary expansion of the site will take a long time from the expansion application to the equipment distribution and the completion of the expansion, which will affect the user experience. Moreover, due to personnel flow and other reasons, the area after expansion is not a hot spot area, and the expansion can not achieve the ideal effect, and the user perception is poor, so the unified expansion standard is adopted, The traditional expansion method of high value scenario classification and expansion standard is one size fits all, without considering the different business model of different scenarios and different network quality, resulting in the different critical point of each cell traffic suppression. Adopting the same standard will lead to the situation of no expansion, wrong expansion and no expansion.

5   Identification of Cell Traffic Depression based on Clustering Algorithm

There are many factors that lead to traffic depression, such as different social scenarios, different number of people, different wireless environment, different time periods, different business types, different user behaviors, etc. How to comprehensively consider these factors and predict the suppressed traffic has become the business pain point of wireless network operators. This paper studies machine learning algorithms such as clustering algorithm, linear regression and association rules, and uses four steps of classification, fitting, suppression identification and suppression prediction to complete the suppression traffic prediction for each cell, So that operators in the new station, expansion station can complete the site planning and expansion according to the actual user needs.

5.1 Cell Scene Clustering based on Kmeans Algorithm

5.1.1 Brief Introduction of Kmeans Clustering Algorithm

The so-called clustering algorithm refers to the method of automatically dividing a pile of unlabeled data into several categories, which belongs to the unsupervised learning method. This method should ensure that the data of the same category have similar characteristics [4]. According to the distance or similarity (affinity) between samples, the more similar and less different samples are clustered into one class (cluster), and finally multiple clusters are formed, so that the samples within the same cluster have high similarity and the differences between different clusters are high. The algorithm steps are as follows: a) randomly select k sample points as the initial starting point; b) calculate the distance from all points to the initial point, And the closest point set is regarded as the same class. C) recalculate the center of gravity of the same class (using Euclidean distance) d) recalculate the distance from all points to the new classification center, and take the center of gravity as the new starting point. E) go back to step b) to re divide the classification and the center of gravity.

5.1.2 Effect Evaluation of Community Clustering

There are two elbow methods to choose the best number k of kmeans clustering: elbow method and contour coefficient method. Elbow method measures the rationality of clustering number k according to the sum of square error, and its definition is as follows:

$$SSE = \sum_{p \in C_i} |p - m_i|^2 \tag{1}$$

The contour coefficient of a sample point x is defined as follows:

$$S = \frac{b - a}{\max(a, b)} \tag{2}$$

Is the average distance between sample point X and other samples in the nearest cluster. The nearest cluster is defined as follows:

$$C_i = \arg\min_{c_k} \frac{1}{n} \sum_{p \in c_k} |p - x_i|^2 \tag{3}$$

Using the above principle, the visualization of the sum of square error contour coefficient under different K values is realized, as shown in Fig. 1. From the SSE curve (blue) of various k values in the clustering of traffic and user number in Fig. 1, it can be found that the curvature is the highest when $k = 3$, and the contour coefficient (red dot) is the second largest when $k = 3$. According to elbow analysis method, the best value k of this clustering number is 3.

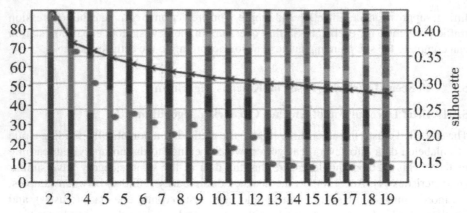

Fig. 1. Selection of optimal K according to elbow method and contour coefficient method(Color figure online)

5.2 Identification and Calculation of Flow Depression

Overview of fpgrowth algorithm the motivation of association rules is shopping basket analysis. By finding the relationship between different products that customers put into the "shopping basket", this paper analyzes customers' shopping habits. The discovery of this association can help retailers to understand which products are frequently purchased by customers at the same time, so as to help them make better marketing strategies. Fpgroftl algorithm consists of an FP Tree and an item head table. Each item points to its position in the tree through a node chain. The process of establishing FP Tree is as follows: a) the start node is empty. b) First, insert the first row of keywords. C) insert the second row of keywords. If there is a duplicate prefix path, the nodes on the path + 1D) after inserting all the data, the FP Tree and linked list are also built.

6   The Current Situation and Reform Significance of the Core Curriculum of the Elderly Service Major

6.1   The Current Situation of Core Curriculum Setting for Senior Service Major

The development of enterprises in the field of old clothes is in its infancy, and the industry lacks a unified standardized management system, which can be used for reference, There is also a lack of uniform talent training standards and core curriculum system in Colleges and universities to divide the courses involved in the major of old clothes into six categories: physical health service, mental health service, leisure and entertainment service, elderly service management, social work and other six categories. Among these courses, more than 80% of the professionals believe that "16 courses such as life care for the elderly" are the core courses of the elderly service and management major. Among them, 9 courses belong to the category of physical health services, 3 courses belong to the category of elderly service management, 2 courses belong to the category of mental health services, and 2 courses belong to the category of leisure and entertainment services. These courses mainly cover nutrition diet, life care, disease care, rehabilitation services, exercise guidance and many other services for the elderly. This is consistent with the first three contents of the meals, nursing, rehabilitation and psychological services clearly stipulated in the basic standard for the social welfare of the elderly.

6.2 The Significance of the Core Curriculum Reform of the Elderly Service Major

The professional education of elderly service is one of the most important contents of vocational education, and it is an important way to cultivate skilled talents of elderly service management and service technology. The purpose of the core curriculum reform of the elderly clothing specialty is to "deepen the cooperation between industry and education, the combination of school and enterprise, the combination of education and training, and improve the education mechanism of combining morality and technology, and the combination of work and study". Guided by the systematization of the working process, the carrier and effective path of the professional technical skills training are explored through the matching of the professional post group needs and the actual service work, so as to build and extend the students' professional core competence. With the purpose of strengthening the technical skills training, the teaching content is reorganized, the learning situation and task unit are orderly, and the teaching module design and project system simulation exercise are carried out, Focus on the organic integration of the core curriculum content and the actual service management technology to achieve seamless docking.

7   Conclusion and Prospect

With the development of society and science and technology, the technology of service robot is constantly improving, and people's requirements for the quality of life are constantly improving, which promotes the rise and development of elderly service robot for the elderly group, and will eventually form an industry. In order to better develop the elderly service robot, and finally apply it to the family, researchers on the one hand will apply the latest technology to the project, at the same time continue to strengthen the contact with potential customer groups, and finally develop intelligent, modular, networked, low-cost and lightweight elderly service robot, and integrate it into our society as soon as possible. To solve the aging population, improve the quality of life of the elderly and other social problems, better serve mankind.

References

1. Feng, D., Gong, D., Sheng, L.: Problems in modern apprenticeship system of elderly service and management specialty in Private Higher Vocational Colleges -- Taking Zhejiang Dongfang vocational and Technical College as an example. Educ. Modernization 7(02), 122–124 (2020)
2. Yan, C.: Exploration and practice of talent training mode of "school enterprise joint education" for elderly service and management major -- Taking the construction of elderly service and management major of Shaanxi national defense industry vocational and Technical College as an example. J. Shaanxi Nat. Defense Ind. VocaT. Tech. Coll. 29(01), 3–5 + 35 (2019)

3. Qi, C., Ying, S.: Exploration and practice of university government enterprise cooperation in building the specialty of elderly service and management – Taking Tianjin Sino German University of applied technology as an example. J. Tianjin Sino Ger. Univ. Appl. Technol. **03**, 106–109 (2018)
4. Guangjun, X.: On the innovation of training mode of elderly service and management talents in Liaoning Province – Taking the elderly service and management specialty of Liaoyang vocational and Technical College as an example. J. Liaoning Teach. Coll. (Soc. Sci. Edn.) **05**, 131–132 (2017)

Research on the Application of Information Data Classification in Employment Guidance for Higher Vocational Students

Fang Fang[✉]

Zhejiang Tongji Vocational College of Science and Technology, Hangzhou 311231, China

Abstract. In order to further explore and guide the employment of higher vocational students, the author makes extensive statistics on the decision-making factors of graduates' employment in recent years. Based on the classification of information data, this paper analyzes the impact of information data classification on the application of graduate employment guidance. The main significance of this study is to improve students' knowledge structure, improve their comprehensive quality and social competitiveness.

Keywords: Information data classification · Higher vocational students · Application research

1 Introduction

The school running idea of higher vocational colleges is "employment oriented", and the employment rate has always been an important indicator to measure the level of higher vocational colleges [1]. At the same time, we have noticed that with the development of society, the high-level and low-level indicators to measure the employment quality of graduates are becoming more and more important. How to improve the employment quality of students under the premise of maintaining a high employment rate is a problem of great concern to decision makers in higher vocational colleges. It is of great significance for higher vocational colleges to improve employment guidance, increase employment rate and employment quality by using the employment data of students over the years to mine useful knowledge and provide it to decision makers. Data mining technology can transform some existing data into useful knowledge and mine valuable information.

2 C4.5 Algorithm

C4.5 algorithm is an improvement of ID3 algorithm. Different from the sub-3 algorithm, C45 algorithm selects the node attribute of each node in the decision tree based on the data gain rate. By default, the attribute with the highest data gain rate under the current

M. A. Jan and F. Khan (Eds.): BigIoT-EDU 2021, LNICST 392, pp. 405–414, 2021.
https://doi.org/10.1007/978-3-030-87903-7_50

branch node is selected as the test attribute of the current node [2–4]. The C4.5 algorithm has this characteristic, which greatly reduces the amount of data needed to classify the samples in the data mining results, and can accurately reflect the minimum randomness or "impure" of the partition. This theoretical method minimizes the number of expected tests required for an object classification, thus designing a simplest decision tree. For the convenience of the research, the related terms in the algorithm are defined below [5].

Definition 1: let S be a set of s data samples, and the Category attribute can take M different values corresponding to m different categories. Suppose S is the number of samples in category C, and the amount of information required for classifying a given data object is called the entropy before S partition.

$$I(s_1, s_2, \cdots, s_m = -\sum_{i=1}^{m} P_i \log_2 P_i) \tag{1}$$

Attribute A is used to divide the information needed for the current sample set, and the entropy after S partition is used, i.e.

$$E(A) = -\sum_{j=1}^{v} \frac{s_{1j} + \cdots + s_{mj}}{s} I(s_{1j}, \cdots, s_{1j}) \tag{2}$$

3 The Current Situation and Problems of Employment Guidance System in Higher Vocational Colleges

3.1 Current Situation Analysis

At present, the first mock exam system and employment guidance service system in higher vocational colleges include eight modules: employment policy and regulations, employment preparation and job hunting skills, employment process innovation and 1 industry education; application training for real life, simulation interview; changing roles to adapt to workplace safety education. Some secondary colleges add some innovation points according to their own characteristics, and set up resume production competition to improve students' application ability. But we can see that all the guidance is around the job before and in the job, lack of career guidance after the job.

3.2 Existing Problems

According to the 2014 "Jiangsu Province college graduates employer survey report", employers are faced with outstanding problems in the recruitment process, such as graduates' lack of understanding of the applied units and positions, poor communication skills, and stereotyped resumes. This makes us think about whether the uniform employment guidance course can adapt to the diversified development of students' employment. The employment guidance system has many repetitive contents, lacks innovation, and does not really play the role of guiding students' employment.

(1) The connotation and importance of the employment guidance system are not well understood in Colleges and universities. According to the characteristics of college students and the needs of social employers, the employment guidance for college students should correctly teach them to find a job that can give full play to their potential. In reality, the understanding of the employment guidance in Colleges and universities mostly lies in that the employment guidance is enough to provide some employment information choices for students, so that students can find jobs, and the employment consultation system in Colleges and universities is equal to the employment guidance of the export part of school education. In the preparation of the entrance end of college work, such as enrollment plan, major setting, curriculum arrangement and so on, it is ignored that the improvement of College Students' employability is a growth and long-term process, rather than a short-term education can get growth.

(2) The curriculum system of employment guidance is lack of timeliness and pertinence [6, 7]. At present, the curriculum system of employment guidance in higher vocational colleges is outdated in design, lack of innovation, lack of pertinence in teaching content, relatively single teaching method, difficult to quantify and scientific in teaching syllabus and teaching plan, difficult to realize scientific guidance, lack of perfect educational content and comprehensive teaching method, which is lack of flexibility, consistency and professionalism, There is a huge difference between the desire of college students and enterprises, which has little effect on the graduates to participate in the fierce employment competition. In view of the lack of substantive counseling content in the current employment form of entrepreneurship and further education, the psychological counseling course is only superficial, and there is no specific guidance in the aspects of cultivation, professional ethics education and employment psychology.

(3) The function of the employment guidance department has some deviation. Now all colleges have set up employment consulting agencies, but its function is more emphasis on administration, and the function of serving employment and education employment is missing. In addition, some schools blindly pursue the employment rate, ignore the construction of the employment guidance system, lack of research and Research on the employment curriculum system, and need to change the employment guidance department from passive communication to active exploration of the construction of employment guidance.

4   Changing the Employment Guidance System and Promoting Students' High Quality Employment

4.1   The Concept of Employment Guidance

At present, the employment guidance of Chinese style is mostly seasonal guidance, that is, the employment guidance course is only opened half a year or one year before graduation. We should realize that the employment guidance system is a dynamic and changing process. Move forward the employment guidance and run through the employment guidance before and after job hunting, forming a dynamic guidance system. Schools should change from "helping students get employment" to "teaching students to get employment", reset the concept of employment service, build a dynamic system to enhance students' successful employment ability, so that students can not only get employment, but also have high quality.

4.2 The Curriculum System of Employment Guidance Changes to the Trend of Diversification

(1) Diversified teaching forms, schools will change from offline courses to online courses, using MOOCS and other forms to enrich the mode of employment guidance courses and enhance students' interest. The employment guidance system for college students should be more in line with the reality, follow the rules and needs of the modern market, and build a more professional and humanized employment guidance curriculum system with new technical means.

(2) Diversification of psychological education at present, the phenomenon of delayed employment among graduates of higher vocational colleges is mostly due to the psychological reasons of students. In the traditional employment guidance course, psychological guidance is only used as a module to explain the theory, without realizing that students' employment psychological problems are a gradual process [8–10]. Schools should integrate employment psychological counseling into ordinary daily psychological counseling, and transform classroom teaching into targeted face-to-face counseling mode. In addition, more psychological counseling should be given to students with employment difficulties and disabilities.

(3) In recent years, the state has vigorously developed innovation and entrepreneurship education, and schools comply with the national requirements to increase entrepreneurship education in the employment curriculum system. However, as far as the current situation is concerned, no matter teachers, textbooks or teaching methods are lack of a scientific planning, more is the propaganda of entrepreneurship policy, theory is more than practice.

4.3 Transformation of Teachers and Functional Departments to Specialization

The "knowledge structure, professional ideas, cultural cultivation and working methods" of the employment guidance staff are very important to the construction of the employment guidance system. Higher vocational colleges should pay more attention to it, improve the construction of career guidance teachers, and improve the human resource training system of employment guidance. Professional teachers should be drawn into the team of career guidance teachers to meet the requirements of students' professional knowledge. Career guidance teachers need to improve their professional ability, actively participate in the training of career guidance and entrepreneurship guidance, and enrich their knowledge, To improve the ability to serve students, the functional departments should change their identities and coordinate the employment work of the whole college [11]. The employment work not only needs the policy service or the statistics of the employment rate of graduates, but also appears in the whole learning process of college students, so as to establish a long-term system centered on cultivating the employment ability of college students. The employment guidance department needs to give full play to the collaborative ability, strengthen the cooperation with other functional departments, and transform the employment work to the whole process, full staff, specialization and informatization. At the same time, we need to make full use of network media, such as microblog, wechat and mobile phone client to push employment information in time to serve students' employment [12]. To sum up, higher vocational colleges must actively adapt to the actual needs of society and market economy, supplement and improve the employment guidance system for college students, so as to better adapt to the diversified development trend of employment, In terms of professional communication ability, team cooperation and formal professional etiquette, it helps college students to establish the awareness of professional people and realize the zero distance connection between school and society, learning and employment.

5 Changing the Employment Guidance System and Promoting Students' High Quality Employment

5.1 Changing the Concept of Employment Guidance

At present, higher vocational colleges have more employment guidance skills, less puzzles, more information, less ideas, more jobs and less dedication. This leads to the poor effect of employment guidance in higher vocational colleges. The employment guidance system is not fully competent for the function of employment guidance. It is necessary to change the concept of employment guidance and provide students with professional, market-oriented, personalized, information-based and diversified guidance mode.

5.2 Higher Vocational Schools Should Actively Take Various Measures to Promote the Smooth Employment of Graduates

And other vocational schools should actively implement relevant employment reform policies for graduates, increase investment in human, material and financial resources, and take various measures to bridge the gap between graduates and enterprises and institutions. (1) Set up or adjust major and related courses according to market research. (2) It is necessary to set up management and service institutions for graduates' employment, implement the system of top leader responsibility for graduates' employment, and invest certain human, material and financial resources to implement it. (3) We should fully implement the spirit of the central ministries and commissions on setting up vocational guidance courses, and provide students with comprehensive, scientific and systematic vocational guidance. (4) Actively build a production and learning platform to achieve in-depth cooperation between schools and enterprises. Schools can jointly run schools with enterprises to achieve order based training, provide talent reserves for enterprises, and adjust teaching plans and syllabus according to the needs of enterprises, so as to solve the contradiction between supply and demand between enterprises and schools. (5) All departments of the University cooperate closely to establish a pre employment mechanism that combines internship with recommended employment. If the graduates are arranged to work directly in the production post of the enterprise and practice according to the requirements of the enterprise employees, it can not only make the graduates apply what they have learned, but also make them develop a high sense of professional responsibility, the spirit of hard work and the consciousness of abiding by the professional ethics.

5.3 The Employers Should Change the Concept of Employment, and the Parents of the Graduates Should Change Their Own Concept

(1) Some employers will put forward some unreasonable requirements when recruiting graduates from higher vocational schools. On the one hand, it needs the relevant government departments to investigate and punish this phenomenon and order them to correct it. On the other hand, it also needs the employing units to change their concept of employment and create all kinds of opportunities for these graduates in a fair, open and just way in recruitment to help them grow up. (2) Some parents of higher vocational school graduates want their children to work stably and have a rich income. They also have the idea of looking after their face and unwilling to let their children stay away from themselves and work in other places, which also affects the smooth employment of graduates. Parents should give good advice on the employment of their children, encourage their children to start their own businesses, and support their children to go out to hard areas and jobs.

6   Data Acquisition

Using C4.5 algorithm for data mining and analysis needs to establish specific and verifiable research objects, so we should systematically and carefully collect and sort out the predictable potential factors that may affect students' employment before establishing the algorithm analysis template [13–15]. The accuracy of data acquisition samples directly affects the reference value of algorithm analysis results.

According to the needs of research and analysis, this paper mainly collects students' basic information from the "basic information service" interface in the student status management system of Nanjing vocational and Technical College of information, and derives the "student performance information" of 2016 graduates from the "Learning Center - score query service" interface. Colleagues from the employment guidance center under the Student Work Committee of Nanjing Institute of information technology provided us with the "employment status information" of relevant graduates in 2016. As shown in Fig. 1. The author uses random sampling method to intercept 600 relevant records from nearly 5000 data records as the object of this study and analysis. 400 pieces of data are arranged to form the training data set from the 600 pieces of information of graduates intercepted, and the remaining 200 pieces of data are allocated to the test data set.

From the "basic information service" interface of student status management system of Nanjing vocational and Technical College of information, the "basic information of students" is collected, which mainly includes the following contents: Department, major, class, name, student number, gender, ability, specialty, political health, reward and punishment, training experience, social practice activities, etc. In addition, the interface also shows factors such as ethnic origin [16–20]. ID number, etc., which are not related to graduates' employment choice or are subject to anti discrimination and anti local protection policies, which have little influence on Graduates' employment.

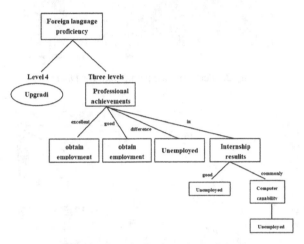

Fig. 1. Decision tree analysis chart.

7 Simulation for Generating Classification Rules

The process from root node to each leaf node in the decision tree with excellent practical ability can be seen from the above classification principles that graduates with rich social practice experience and high English application ability account for the vast majority of better enterprise employment samples, while graduates with relatively weak social practice experience who have won awards above provincial and municipal level, Basically, they can be employed in poor state-owned enterprises, general foreign enterprises and better private enterprises; graduates with general social practice ability and without high-level rewards can only mix with poor enterprises(see Fig. 2 and Fig. 3).

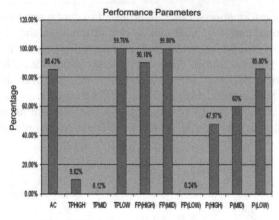

Fig. 2. Performance parameters for Data

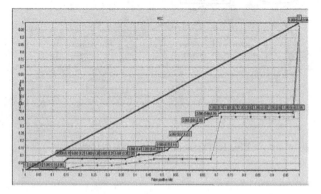

Fig. 3. ROC Curve for Class

8   Conclusion

According to the discrete characteristics of the potential data attributes that may affect the employment of university graduates, this paper uses the C4.5 algorithm of decision tree to mine and analyze the target data, constructs the analysis model of the influencing factors of College Students' employment, and establishes the classification rules, Data mining analysis has high reference value and practical significance. The classification rules of influencing factors of graduates' employment analyzed by C4.5 algorithm model has a guiding role in improving the comprehensive quality of college students, and can also provide ideas for the transformation of employment guidance work in Colleges and universities. Data mining technology has been successfully applied in other business fields, but it is still a new attempt in the field of education. In this paper, the decision tree method of data mining technology is introduced into the employment data analysis of higher vocational colleges. The analysis results will help to improve the pertinence of talent training programs and improve the employment quality of school students. Of course, in the actual analysis operation, select a small data training sample to get the analysis result persuasive is limited, at this time we will face thousands of data, data attributes will reach 10 or more. Therefore, we can not simply use manual calculation, we need to seek the help of computer software [21–24]. Sasanswer tree is two kinds of decision tree analysis software which are widely used.

References

1. Mai, X., Jia, P., Weng, J., et al.: Application of decision tree based on multi-scale rough set model in university employment data analysis. J. South Chin. Normal Univ. (Nat. Sci. Edn.) **46**(4), 31–36 (2014)
2. Chang, Z., Wang, L.: Application of a new decision tree model in employment analysis. Comput. Eng. Sci. **33**(5), 141–145 (2011)
3. Zhang, J., Wang, Q.: Application of C4.5 algorithm in graduate employment information database. Inform. Technol. **11**, 31–33 (2009)
4. Lv, X.: Data Mining Method and Application. Renmin University Press, Beijing (2009)
5. Pan, J., Tai, H.: Innovation research on employment guidance mode of Higher Vocational Students under the background of "Internet plus". Modern. Educ. **4**(38), 16–17 (2017)
6. Dai, H.: Exploration on the effectiveness and long-term effectiveness of employment guidance for higher vocational students. New Campus (Reading) **09**, 192 (2017)
7. Gan, Y.: Employment problems and Countermeasures of higher vocational students from the perspective of stable matching theory. Educ. Career **17**, 61–64 (2017)
8. Ni, Q.: Research on Vocational College Students' employment guidance based on peer counseling mode. Educ. Modern. **4**(32), 294–295 (2017)
9. Jiang, G., Rao, D., Feng, J.: Research on personalized employment guidance for students from poor families in Higher Vocational Colleges – a case study of Guangdong Construction Vocational and technical college. Employ. Chin. Coll. Stud. **15**, 39–44 (2017)
10. Hua, Y.: Thinking on using physical education in Higher Vocational Colleges to assist employment guidance. Contemp. Sports Sci. Technol. **7**(22), 107–116 (2017)
11. Nie, P., Hu, J.: Analysis on the current situation of employment guidance and management system in Higher Vocational Medical Colleges -- Based on the employment management system of Cangzhou Medical College. Shanxi Youth **14**, 229–230 (2017)
12. Ren, F.: On how to promote the development of vocational guidance in Colleges and universities. Shandong Text. Econ. **07**, 59–61 (2017)
13. Liang, X., He, L., Chen, J.: Research on employment guidance for chemical engineering students in Higher Vocational Colleges. Shandong Text. Econ. **07**, 53–55 (2017)
14. Li, X.: On the optimization path of employment education in Higher Vocational Colleges. Zhiyin Lizhi **13**, 22 (2017)
15. Han, B., Qu, S., Wang, D.: Research on employment guidance for art vocational college students. Sci. Chin. **20**, 193 (2017)
16. Fang, X.: Research on the satisfaction of employment guidance course in Higher Vocational Colleges. Health Vocat. Educ. **35**(13), 16–17 (2017)
17. Tang, F.: Psychological analysis and employment guidance strategies of vocational college students [J / OL]. China Training: 1–2 [2021–04–10]. https://doi.org/10.14149/j.cnki.ct.201 70622.007
18. Jiang, Y.: Problems and Countermeasures of Ideological and political education in Vocational College Students' Employment Guidance. Lab. Soc. Secur. World **18**, 9–10 (2017)
19. Zhang, L.: Research on Counselors' employment guidance for students based on 2 + 1 teaching mode. Modern Econ. Inform. **11**, 434 (2017)

20. Lin, W.: Reflections on the work mode of College Students' Employment Guidance under the background of wechat – Taking Guangzhou Urban Construction Vocational College as an example. Tomorrow Fashion **10**, 225–226 (2017)

21. Wang, W., Chen, X.: Problems and Countermeasures of employment guidance for students majoring in finance and economics in Higher Vocational Colleges. Chin. Foreign Entrepreneurs **14**, 200–203 (2017)

22. Song, X.: Problems and Countermeasures of Vocational College Students' employment guidance in the new era. J. Higher Educ. **09**, 182–183 (2017)

23. Yerkenbuick, M.B.: Problems and countermeasures in employment of higher vocational students-Taking Yili vocational and technical college as an example. Chin. Training **09**, 59–60 (2017)

24. Liao, W.: Analysis on the current situation and Countermeasures of Vocational Students' Employment Guidance. Teaching Manag. Educ. Res. **2**(08), 19–20 (2017)

Study on the Application of Fuzzy Algorithm in English-Chinese Translation System in Vocational College Education

Fen Han[✉]

Xi'an Vocational and Technical College, Xi'an 710077, Shannxi, China

Abstract. In E-C translation system, fuzzy algorithm can extract the best context and feature words and improve the accuracy of translation. This paper studies the English Chinese translation system based on fuzzy algorithm, and compares the translation effect of English Chinese translation system based on fuzzy algorithm. The simulation results show that, compared with the traditional E-C translation system, the E-C translation system using fuzzy algorithm has higher accuracy and recall rate, and higher translation accuracy.

Keywords: Fuzzy algorithm · English Chinese translation

1 Introduction

In recent years, with the deepening of globalization and the increasingly extensive international cooperation, English is becoming more and more important. As a tool of communication between English and Chinese, English Chinese translation system has become a research hotspot. There are two traditional English Chinese translation systems: rule-based and case-based. However, these systems can not extract the optimal context and feature semantics in the process of translation, resulting in the low accuracy of English Chinese translation. In addition, with more and more information at present, how to obtain the optimal translation from massive data becomes the key of English Chinese system [1]. Based on this, this paper proposes to introduce fuzzy algorithm into English Chinese translation system, and construct language mapping model by selecting optimal feature semantics, so as to achieve optimal translation of English Chinese translation system and improve translation accuracy. However, the commonly used fuzzy algorithm is easy to lead to the loss and lack of language, which affects the accuracy of translation. The improved fuzzy algorithm becomes the key to the research of English Chinese translation system.

2 Statistical Machine Translation

2.1 Introduction to Statistical Machine Translation

Machine translation is a very difficult job. According to the common understanding, to translate a sentence, we should first understand the source text, understand the meaning

M. A. Jan and F. Khan (Eds.): BigIoT-EDU 2021, LNICST 392, pp. 415–426, 2021.
https://doi.org/10.1007/978-3-030-87903-7_51

and sentence structure of each word in the source text, then select the translation of each word, and reorganize the sentence according to the grammatical structure of the translation. The traditional rule-based machine translation (RBMT) method is usually written by human language experts in various forms of rules, and then the computer completes the operation according to the rules written by human. This approach is effective to some extent, but when the number of rules reaches a large scale, the system becomes difficult to control. Moreover, in the face of all kinds of complex language phenomena, it is difficult to cover all the situations with manually written rules. Therefore, the rule-based machine translation system is often difficult to continue to improve after reaching a certain level [2–4]. Therefore, in early 1990, IBM company proposed a new machine translation method, called statistical machine translation method. The basic idea of this method is to collect a large number of bilingual sentences and the size of sentence pairs, and then use some algorithm to automatically learn translation knowledge from these corpora, and finally use these knowledge to translate, usually reaching 1 million.

2.2 Advantages of Statistical Machine Translation

2.2.1 Statistical Machine Translation Can Make Full Use of the Existing Translation

The basic working method of statistical machine translation is to establish statistical model, formulate algorithm rules, and use a large number of high-quality bilingual parallel idioms to train the system. In today's Internet era, there are a large number of bilingual texts accessed in home machine-readable format. After aligning and processing the bilingual texts, the statistical machine translation engine 18 can read and learn them, thus improving the construction efficiency and translation quality. For example, coge company imported 20 billion words of official documents from the United Nations into its translation engine for training, thus improving the quality of the translation engine in a relatively short period of time, making the translation engine of big Google able to catch up with others and ranking among the best in the evaluation RO of NST. have to Google's translation engine can be ranked in. The evaluation in NIST. They are at the top of the list.

2.2.2 Statistical Machine Translation has Higher Translation Quality

The traditional working mechanism of rule-based machine translation is realized through the in-depth study and continuous revision of language rules, and the machine transformation at the lexical, syntactic and other levels. The research of this transformation is complex and arduous, but the transformation result is stiff and unnatural, which can not be compared with manual translation. Statistical machine translation (SMT) is a new way to train students with high-quality manual translation, which makes the corpus in the translation system more authentic and more in line with the norms of manual translation. The earliest statistical translation system is based on the word translation model, which only considers the linear relationship between words, without considering the sentence structure and context information [5, 6]. When the word order of two languages is quite different, the effect is not good. With the continuous efforts of researchers, a phrase based translation model has emerged. The basic idea is to extract a large number of aligned phrase fragments from large-scale corpus, and use these phrase fragments to match and combine the sentences to be translated. Due to the restriction of phrases, the choice of words is more accurate, and it is helpful to the translation of some common words and idioms, and the word order of translation is relatively determined, which makes the translation results more in line with the characteristics of the target language.

2.2.3 Higher Construction Efficiency and Translation Quality

The traditional rule-based machine translation (RBMT) is more complex in the study of language rules, which requires long-term work, huge development costs and repeated debugging. The representative of this aspect is the American SYSTRAN translation engine. Compared with RBMT, the construction and training of statistical machine translation system can be completed in weeks or even days, which greatly saves time and cost and improves development efficiency. Experts believe that if the same amount of money is invested in the machine translation system, the translation quality of the statistical machine translation system will be higher. (ER, 200518-21) Candide, a system developed by IBM in a short period of time by using statistical machine translation method, has a performance comparable to that developed by the famous machine translation company SYSTRAN after decades of development [7]. At present, the Institute of computing of Chinese Academy of Sciences is cooperating with Harbin Institute of technology and Xiamen University to develop SLK road statistical machine translation system. It is believed that it will make remarkable achievements in the near future.

2.3 Static Binary Compiler

FX! 32 contains the following interactive components: agent (transparent agent), loader (loader), runtime and emulator (runtime and emulator), binary translator (translator), server (server) and manager (Manager). Agent belongs to Windows NT process, which is used to detect whether there are x86 programs that need to be loaded and executed. If a program is detected to be executed, the loader is called to load the x86 code [8–10]. If the source program is executed for the first time, the emulator interprets the x86 code and gets its profiling information. Profiling contains the information of the static translator: the address of the call instruction to be interpreted, the source/destination address pair to jump indirectly, and the instruction address to perform the non aligned memory access. After the above program is terminated, FX! 32 server calls static binary translator and applies the collected profile information for static translation. At the same time, the generated profile information is stored in the database for subsequent execution of similar programs. Using this method, the source x86/WinNT program is translated to alpha machine incrementally [11]. The main drawback of FX! 32 system is that the first time execution is very slow, and it relies on pure static profiling, and can not dynamically optimize the code according to the change of program execution.

3 Fuzzy Algorithm

3.1 Fuzzy c-Means Clustering Algorithm

Fuzzy FCM algorithm is one of the most perfect algorithms in theory, and it is also very common in practical application. In essence, it belongs to the category based on partition algorithm, but it is a kind of flexible fuzzy partition, which introduces fuzzy theory on the basis of clustering, and then uses membership function to determine the classification of a data object.

An algorithm based on the set of FCM is a set of data $X = \{x_1, x_2, \ldots x_n\} \subset R^s$ is divided into C groups, and its cluster center is represented by C_i.

$$J(U, c_i, \ldots c_c) = \sum_{i=1}^{c} J_i = \sum_{j=1}^{n} u_{ij} d_{ij} \tag{1}$$

Where u_{ij} is between 0 and 1, C_i is the cluster center of the i class, $d_{ij} = \|c_i - x_j\|$ is the Euclidean distance between the ith cluster center C_i and the j-th data object x, and $m \in [1, +\infty)$ is a weighted index to control the fuzzy degree of clustering. The higher the m value, the greater the degree of ambiguity.

In order to improve the translation accuracy of English Chinese translation system, an English Chinese translation model system based on human interaction and feature extraction is proposed. Based on the overall qualitative understanding and logical reasoning ability of human-computer interaction, the translation similarity model is introduced into the English Chinese translation system model, and the translation results are obtained by calculating the translation similarity between two semantic loops in the same language space. Compared with SOA, SCA and SLA, the proposed algorithm has higher accuracy, accuracy, precision η and recall rate RecA, which provides a new method and approach for English translation.

3.2 The Concrete Steps of FCM Algorithm

The rule-based recognition method is mainly obtained from corpus automatically or compiled by experts, which has the advantages of easy to understand, but it is not universal, time-consuming and easy to produce ambiguity; the method based on statistics transforms the problem of noun phrase recognition into the problem of tagging similar words, which is simple and flexible, and does not rely on specific language model. It is a popular translation algorithm in the past; However, this method is based on a large number of sample data, which is prone to over fitting problems. With the rise of artificial intelligence and machine learning methods, neural network-based noun phrase recognition method has been applied to English noun phrase recognition. However, due to the complexity of English grammar, the accuracy of English noun phrase recognition needs to be improved. In order to improve the translation accuracy of English Chinese translation system, an English Chinese translation model system based on human interaction and feature extraction is proposed. Based on the qualitative understanding of human interaction and the ability of logical reasoning 3, the translation similarity model is introduced into the English Chinese translation system model, and the translation results are obtained by calculating the translation similarity between two semantic vectors in the same semantic space. Compared with SOA, SCA and SLA, the English Chinese translation based on human-computer interaction and feature extraction has higher accuracy, accuracy, precision η and recall rate, which provides a new way for English translation. In semi supervised clustering, tag data acquisition needs a certain cost. Therefore, in the actual application of data sets, the amount of label data is relatively small, at this time SFCM algorithm can not reflect its advantages [12–14]. Therefore, this paper redefines the iterative formula of SFCM algorithm, and puts the parameters representing the weight of labeled data points in the iterative expression of cluster center, so as to adjust the influence of supervision information. Secondly, the iris data set in machine learning library is tested and verified by MATLAB programming. Finally, the accuracy of the improved ssfcm algorithm is analyzed and compared with the previous basic algorithm. The experimental results show that the improved algorithm has good performance.

4 Application of Computer Aided Translation Technology in Simultaneous Interpreting

This part introduces the application and main problems of computer assisted translation technology (hereinafter referred to as "technology") in the simultaneous interpreting of two stages, namely, pre preparation and on-site translation.

4.1 Technology Assisted Multimodal Pre Translation Preparation

Before the start of on-site translation activities, multi-modal pre translation with technical assistance can be carried out. The technologies used include online translation, machine translation, voice generation, etc. the technologies used include daofan translation, SDL studio17, Microsoft edge, etc. among the speakers of international conferences, some of them have strong accents, such as non-native speakers of English and native speakers of English speaking countries such as India and Singapore, which are often difficult to recognize. In the process of SDL studio17 automatic translation, you can choose to use Google translation, baidu translation, Sogou translation and other automatic translation engines for automatic translation, and choose online or local translation memory, Save translation results for future invocation. In addition, SDL Studio17, after the addition of multilingual control, the automatic generation of alphabetic terminology is easy to recite and find [15]. The simultaneous interpreting of medical conferences involving the author involves a lot of medical knowledge and terminology. It is difficult to learn and memorize in a short time. For this reason, the author consults relevant academic papers according to the speaker's information and the theme of his speech. Under the guidance of the concept of "multimodal interpretation preparation before translation", I watched the related topic speeches, classroom teaching videos and operation videos on domestic and foreign video websites, and analyzed the knowledge points, vocabulary, expression and sentence patterns. The Department also used the reading function of Microsoft edge browser and Baidu translation to input I English words into the browser, watch the words and listen to the reading, In terms of the subject matter of the related topics of stimulating storage, the author found that most of the words, including terminology, are clearly pronounced in the Xiangya English Chinese Medical Dictionary and the Zhongshan medical dictionary, and most of the three pronunciations are very standardized.

4.2 Technology Assisted Live Simultaneous Interpreting Practice

Simultaneous interpreting is simultaneous interpreting. The translation is transmitted from the same device headset to the interpreter, and the translator is translators through the interpreter workbench microphone. Then, speech recognition technology, the translator's target language recognition into text, and projected on the front of the central screen. Speech recognition technology company arranges personnel on site to manually correct or delete Chinese words and sentences with serious recognition errors. All people at the meeting can choose to use headphones or watch subtitles to get the main idea of the speech. The translated words can be saved and used for meeting minutes, news reports, translation quality evaluation and control. It is worth noting that during the second session (2018) of the simultaneous interpreting, the speech recognition technology provided the company with the initiative to propose that the source language be transferred to the only display screen of the interpreter's workroom for reference. However, the author found that due to the excessive use of medical lingual and jargon, the effect of writing is not good [15]. Moreover, the simultaneous interpreting room and the conference hall are separated by walls, and can not directly observe the scene. If the only display screen is used for the display, it can not be used to broadcast the slide show on the spot. After weighing the pros and cons, the author and his partner agreed that the display screen should be used to play live slides. In the process of live translation, the author and his partner also used Google translation, Youdao translation and other online translation engines to translate unfamiliar words, phrases, sentences and paragraphs in the temporary speaker's Fantasy for reference. The author found that the translation and automatic pronunciation of most medical terms are very accurate. In the field translation, the author found that the intervention of technology has also brought some new problems, which can be summarized and analyzed from the perspective of speech recognition system and human factors.

5   Synonymous Translation

Synonymy is the most important relation in word. If the two expressions replace each other in the language text without changing their meaning, the two expressions are synonymous, and sy η set means synonymy [16–18]. Antonymy is not the basic organizational relationship of WordNet. It is a morphological relationship rather than a semantic relationship between concepts, such as the antonym pair heavy/light". As for the antonymy of adjectives, the semantic organization of descriptive adjectives is completely different from that of nouns. It adopts n-dimensional hyperspace structure instead of tree structure. Antonymy is the basic semantic relationship of descriptive adjectives, so synset of adjectives is connected through antonymy.

5.1 Technical Literacy and Psychological Quality of Interpreters in the Age of Technology Assisted Simultaneous Interpreting

In the context of technical assistance translation, the ability to use technology should be an important component of the quality of simultaneous interpreting. Technology, especially language technology, is a double-edged sword for translators. Various technologies help translators improve the quality of translation and the efficiency of pre translation preparation. For example, speech recognition can transcribe the target language produced by the interpreter, which is convenient for the interpreter to summarize, analyze and improve after translation. IFLYTEK was invited to attend the conference of simultaneous interpreting and interpreting at the end of 2018. At the meeting, iFLYTEK demonstrated its interpreter assistance software. This software can not only recognize the speech, but also highlight the numbers and proper terms in a special way, and provide the reference translation of terms. If this software constantly reduces the workload of simultaneous interpreting, it can even reduce the burden of memory and note taking. On the other hand, the mode of speech recognition and simultaneous interpreting has brought great psychological pressure to the interpreter. In the process of traditional simultaneous interpreting, the target language will only be kept in the brain in the form of voice [19, 20]. If the interpreter is careless, he will not be particularly careful. In the new mode of simultaneous interpreting and subtitle translation, the long time of omission in translation is displayed on the screen. This undoubtedly brings more psychological pressure to the interpreter. The interpreter needs better compression ability. In addition, after the introduction of speech recognition, the pronunciation is not standard enough and the effect of target language recognition is poor, which may be eliminated by the market. With the continuous progress of machine translation technology, a number of translators with poor translation quality have been pushed to the edge of unemployment.

5.2 How to Protect Intellectual Property Rights and "Innocence" of Interpreters in the Era of Technology Assisted Simultaneous Interpreting

Over the years, the protection of interpreters' knowledge has aroused great concern. Many kinds of intellectual property works of interpreters may be infringed and abused. For example, after the target language translated by the interpreter is recognized by the software, it will continue to be used by the organizer to add subtitles, write reports, and release technical materials. The excellent translation engine belongs to the secondary translation of the interpreter's knowledge, so that the vocabulary produced by the interpreter in the preparation of translation can be shared with the technology company and the host office for the high-quality and future translation quality. When I accept the task of translation, I agree with all parties about the intellectual property rights of the translation products, and the items and products that need to be paid for should be specified in the agreement. Except that the root cause of low translation is not always the poor level of the interpreter. Sometimes factors other than the interpreter, such as the on-site technology and physical environment of the speaker's speech, will affect the translation effect [21]. However, if the audience does not grasp the spoon (or pay for bilingual listening), they can not or have no time to consider and accurately analyze the reasons for the poor translation quality. Under the traditional simultaneous interpreting mode, the translator will not only enhance the scapegoat transfer technology but also increase the omission of the interpreter. How to determine the responsibility of the interpreter becomes more complex. Translators should be strict with themselves and be good at using technical means to collect evidence and correct their names.

5.3 Technology Assisted Simultaneous Interpreting May Not Improve the Communication Effect of Speakers

Speech recognition and assisted simultaneous interpreting can help speakers generate speeches, so that they can be processed and edited in the later stage, and can be transmitted in text, or for the speakers or organizers to review and reflect after the meeting. However, it should be noted that some English speakers are not only very different from the typical standard native speakers in phonetic level, but also have great differences in the use of words, phrases, grammar, sentence patterns and text structures. The machine recognition of the corpus produced by them is difficult, and the effect of on-site translation and recognition is also poor. In addition, the use of simultaneous interpreting translation screen shows that this "information delivery" mode may lead speakers to be unable to attract audience attention. Voice transcribing technology providers_ A staff member described a typical scene to the author: a "heavyweight speaker is standing on the left end of the huge conference hall to make a report, and the live voice transcription screen is set on the right end of the conference hall. During the meeting, people pay attention to the right end of the screen, while the speaker is "in the cold", and the nonverbal communication with the audience is blocked.

6 An Analysis of English Chinese Translation System

6.1 System Architecture

The system mainly includes interface layer, implementation layer and data layer. The interface layer is mainly for the user to classify the translated text through the user interface; the implementation layer mainly completes the realization of each text classification process, designs the memory data storage, and provides the interface for calling the interface layer; the data layer is mainly used to complete the data exchange between the interface layer and the implementation layer. The details are shown in Fig. 1.

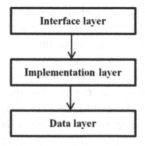

Fig. 1. System architecture of English Chinese translation system

In order to achieve user operation, the user interface is classified by MFC dialog box, which is divided into four modules: path selection, classification, preprocessing and classification change. TF-IDF and LDA are used for feature extraction. For the classifier design, according to the phenomenon that each category has unique feature words, the classification lexicon is designed. According to the frequency of each category of feature words in the document, the document category is first judged, and then other classification design is carried out by using Sm classification method [22]. According to formula (2), when Eq. (2) is satisfied, the document is directly set as the category, otherwise SVM classification is carried out.

$$\frac{Count(i)}{Total(N)} > 80\% \tag{2}$$

Where $Count(i)$ represents the number of feature words in the current Classified Thesaurus; $Total(N)$ represents the total number of feature words in the document set.

7 Analysis of Simulation Experiment

In order to ensure the accuracy and effectiveness of the analysis, the experimental parameters are set uniformly, including phrase translation amount of 500 characters, short text translation amount of 600 characters, translation rate of 15 kbps, semantic recognition rate of 25 kbps [23]. The only difference between simulation experiments of English Chinese translation system is that feature extraction algorithms are of different types. Tf-df feature extraction algorithm, LDA feature extraction algorithm and no feature extraction algorithm are used to process the preparation rate, recall rate and accuracy rate.

The system simulation results show that: compared with the English Chinese translation system without feature extraction algorithm, the τ f-df feature extraction algorithm has higher accuracy, recall and accuracy, and has higher translation accuracy; LDA feature extraction algorithm has higher accuracy, recall and accuracy, and better translation accuracy than τ f-df feature extraction algorithm. Theoretically speaking, the τ f-df feature extraction algorithm is realized by calculating document similarity when classifying documents. However, this method does not consider the semantics of feature words or sentences, resulting in two sentences with the same meaning or causal relationship being divided into different categories, and LDA feature extraction algorithm can identify [24]. Therefore, compared with the τ f-df feature extraction algorithm, LDA feature extraction algorithm has higher translation accuracy, which are shonw in Fig. 2 and 3.

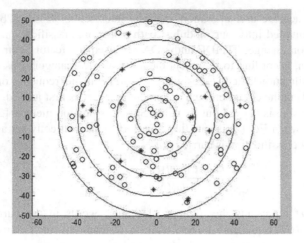

Fig. 2. Fuzzy algorithm with translation system

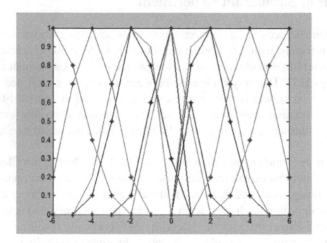

Fig. 3. Fuzzy evaluation simulation

8 Conclusion

This paper studies the English Chinese translation system based on feature extraction algorithm, and studies two types of feature extraction algorithms through simulation experiments. In the English Chinese translation system, feature extraction algorithm can be selected through feature semantic optimization, and construct semantic ontology mapping model to achieve the optimal solution and achieve the best English Chinese translation. Compared with τ f-idf feature extraction algorithm, LDA feature extraction algorithm has higher accuracy, recall and accuracy, and has higher translation accuracy.

Acknowledgements. 2020 Scientific and Research Program Funded by Shaanxi Provincial Education Department (20JK0366).

References

1. Ma, J., Huang, D., Liu, H., et al.: Research and application of English part of speech tagging for machine translation. China Commun. **3**, 66–75 (2012)
2. Wang, Y.: Part of speech tagging in English Chinese machine translation system. Comput. Eng. Appl. **46**(20), 99–102 (2010)
3. Zhang, Z., Yang, H., Zhao, Y., et al.: Topic text network construction method based on Gibbs sampling results. Comput. Eng. **43**(6), 150–157 (2017)
4. Guo, L.: Design of computer intelligent scoring system for English translation based on natural language processing. Mod. Electron. Technol. **42**(4), 158–160 (2019)
5. Fan, H.: Lack of subjective translation ability in translation teaching and countermeasures. Road Success **08**, 40–41 (2021)
6. Li, H.: On the importance of ethnic Chinese translation in cultural development. (05), 16–17 (2021). Huang He. Huang Huang Zi. Huang Zhong Ren
7. Song, D., Lin, Y.: Translation strategies of poetry and verse in the original version of a dream of Red Mansions. Chin. Cult. Stud. **01**, 144–157 (2021)
8. Li, X.: Legalization of local normative legal documents translation from the perspective of "rule of law". J. Nanyang Normal Univ. 1–6 (2021). http://kns.cnki.net/kcms/detail/41.1327.Z.20210311.1710.008.html
9. Li, Y.: The spread of Confucianism overseas under the influence of Confucian Classics Translation: a case study of Finglett's Confucianism research and its inheritance. Chin. Cult. Stud. **01**, 170–180 (2021)
10. Xi, W.: Translation strategies of conceptual metaphors in English and Chinese culture loaded words from a cross cultural perspective. Mod. Vocat. Educ. **11**, 98–99 (2021)
11. Li, Z., Wang, S.: Research on the cultivation of applied translation talents in nonferrous metallurgy – a review of metallurgical English. Nonferrous Met. (Smelting) **03**, 195 (2021)
12. Changshan: Translation strategies of modern agricultural science and technology English: a review of modern agricultural science and technology English listening and reading course. Plant Quar. **35**(02), 93 (2021)
13. Mao, H., Wu, L.: The red translator of Ningbo going out. Ningbo daily, March 2021. (009)
14. Wang, X., Zhang, J.: A study of "creative treason" in Chinese translation of Martin Fiyero. Sci. Educ. Wenhui (First Issue) **03**, 188–190 (2021)
15. Zhang, S.: Research on the ways to improve the core literacy of translators based on social needs. Foreign Lang. Teach. **42**(02), 55–59 (2021)
16. Geng, J.: Problems and countermeasures in English translation of imported and exported food packaging in China. Foreign Trade Pract. **03**, 59–62 (2021)
17. Liu, X., Wang, M., Zhang, Y.: The present situation and Prospect of translation studies on the biography of the arched hero – based on the analysis of CNKI literature resources. Cult. Ind. **07**, 60–61 (2021)
18. Zhang, Y.: The influence of cultural factors on subtitle translation from the perspective of Skopos theory: a case study of crazy animal City. Cult. Ind. **07**, 64–65 (2021)
19. Sang, D.: Research on garment English teaching and translation. Cotton Text. Technol. **49**(03), 94–95 (2021)
20. Palidan, A.: A study of conversion techniques and types in Chinese Uyghur translation. Media Forum **4**(05), 94–95 (2021)
21. Wen, Q., Guan, C.: A tentative exploration of intercultural communication in Translation Classroom Teaching: a review of pragmatics in Translation Classroom Teaching. Chin. J. Educ. **03**, 112 (2021)
22. Zhao, X.: A study of biblical translation by Chinese in the late Qing Dynasty and the Republic of China. J. Ningbo Univ. (Hum. Ed.) **34**(02), 8–16 (2021)

23. Sun, B.: Translation principles of Xixia officials and titles. J. Southwest Univ. Nat. (Hum. Soc. Sci.) **42**(03), 60–71 (2021)
24. Yan, D.: College English translation teaching based on intercultural thinking: a review of College English teaching and intercultural competence training. Chin. J. Educ. **03**, 131 (2021)

Teaching System and Application of Preschool Education with Big Data

Jiayun Xie[✉]

Building 3, Qiangwei Garden, Junfa City, Panlong District, Kunming 651700, Yunnan, China

Abstract. Online education service not only covers millions of Internet users in small and medium-sized cities, but also has the advantage of low cost compared with offline preschool education. Therefore, online education is the development direction of China's future education. Network education platform, has been more and more widely used in education and teaching, network teaching platform is mainly divided into several categories, but most of them are developed by computer professionals, whether in the pre production of the website, or in the post maintenance of the website, are time-consuming and laborious. For a school, the cost of using courses is relatively high, and the computer level of personnel is relatively high. As an open source software, Moodle is more convenient to use, and teachers with general computer level can also use it. Therefore, in recent years, its application in the field of education is more and more favored by educators.

Keywords: Preschool education · Moodle · Teaching model · Teaching design

1 Introduction

Today, with the rapid development of science and technology information, the network has brought great changes to our social life. The application of network in the field of education has led to great changes. Network learning has become a new way of learning in school education and a second classroom for students to learn and communicate. The rapid and direct network interaction makes people's communication jump out of the limitation of time and space. People are trying all kinds of things, such as online job hunting, online trading and online education [1]. As the forefront of the new generation of education, kindergarten should also keep pace with the times, so that children can take the first step on the starting line of keeping up with the trend as soon as possible. Piaget's cognitive construction theory makes us realize that the most fundamental thing for children to understand the network and use the network for learning is how to promote their interaction with the "network education environment", because only in this way can children really feel and appreciate the charm of the network, stimulate more desire for exploration, and develop their own more potential. Therefore, in order to achieve a better effect of preschool education, it is necessary to establish a network teaching platform which is similar to classroom teaching. It can not only meet the needs of teachers' teaching, but also facilitate the implementation of home cooperation, home co education and home interaction, so that busy parents can better understand their children's school situation after work and know their children's shortcomings [2–4].

M. A. Jan and F. Khan (Eds.): BigIoT-EDU 2021, LNICST 392, pp. 427–435, 2021.
https://doi.org/10.1007/978-3-030-87903-7_52

2 Related Research Status

In China, in 2003, Pang Lijuan and others talked about the importance of preschool education to people's development (including people's personality and cognitive development) and to education, family and society. In 2009, Liu Yan analyzed several problems of preschool education in China, and proposed that the rapid development of preschool education in the future can not do without the support of the government. In 2013, Zhang Na put forward new ideas on the curriculum model of preschool education. In view of many disordered phenomena in kindergarten curriculum, blind imitation has made the curriculum rigid and inflexible [5]. This paper puts forward that the new curriculum mode should be applied with open thinking: first, the development mode of combining foreign curriculum with localized curriculum; second, the mode of combining curriculum theory with practical curriculum practice; third, the mode of combining inheriting tradition with insisting on development. The curriculum should be diversified, be able to build its own curriculum model according to the actual situation, and understand the curriculum model with an open attitude [6].

In foreign countries, Agnes of Belo University uses sound reading teaching in early childhood education. Early childhood is a time network full of imagination and fantasy. Therefore, it is very natural at this stage. Through literacy, folktales, nursery rhymes and other materials, education should cultivate children's imagination. The purpose of this paper is to stimulate the psychology that children's success depends on the preparation they receive when they are young. They insist that children's full preparation for life requires their ability to read and write; Anthony of the University of Florida in the United States also thinks that preschool children should urgently develop early literacy and reading skills; KAMs Khan of the University of Turkey is making a comparative analysis of the early education and the current situation of preschool education in Australia and Turkey. In Turkey's early childhood education, women's active participation in daily life is a key development. The importance of early childhood education is emphasized. It can be seen that preschool education is not compulsory in these two countries [7].

In this experiment, accuracy, precision, recall and macro F1 are used to measure the experimental effect. The definition is as follows:

$$Accuracy = \frac{TP + TN}{P + N} \tag{1}$$

$$Precision = \frac{TP}{TP + FP} \tag{2}$$

$$Recall = \frac{TP}{TP + FN} \tag{3}$$

For preschool teachers, most of the computer professional knowledge is not strong, for the establishment of preschool education platform has no operation control authority, so the content of the website can only be passively accepted. For the local pre-school education website, especially the website adapted to the internal use of kindergartens, we should strengthen the pertinence of the content of the website, at the same time, we also hope to have a platform suitable for teachers to update their operation at any time and in time.

3 Problems in Practical Teaching of Preschool Education Major

3.1 Practice Teaching has not been Effectively Implemented

In the practical teaching process of preschool education major in some colleges, although the overall teaching objectives and contents are set, and the theoretical teaching and practical teaching hours of preschool education major are specifically arranged, there are still some problems in the actual teaching, such as the implementation of some courses is not in place, However, it does not involve the practice teaching of children's psychological development and early childhood education activities. The teaching mode is still regular and lacks practicality. Due to the lack of conservation curriculum arrangement in preschool education major of junior colleges, it is difficult for students to learn conservation practice, and it is difficult for teaching content and teaching methods to innovate effectively with the development of the times. Teachers still occupy a dominant position in the teaching process. Teachers require students to learn how to turn students into passive learning, and it is difficult to plan all kinds of children's practical activities independently [8–11].

3.2 Lack of Practical Teaching Conditions

Due to the lack of sufficient hardware and software facilities, it is difficult to effectively meet the diversified learning needs of students. Although some colleges have basic handicraft room, dance training room, calligraphy room and piano room, there is no lack of powerful training room or skill training classroom. For example, infant care skills, swimming guidance skills, environmental facilities skills and so on, but these professional skills training rooms are not perfect. In addition, some junior colleges lack professional experience in preschool education, which only has a theoretical basis. Most of the teachers enter the work after graduation, which makes it difficult to effectively master the practical activities of kindergarten. During the University, they also learn theoretical knowledge, lack of practical teaching experience of children, and are limited to talking on paper in the teaching process, which makes it difficult to improve the practical teaching ability. At present, the practice teaching time of preschool education major is too short in most colleges in our country [12]. The school has arranged the usual skills training and students' classroom time, and the rest of the time has not been reasonably arranged, so it is difficult to improve the students' comprehensive ability [13].

3.3 The Management of Practice Teaching is Not Enough

The practice teaching process of preschool education should be a gradual process, but some teachers just regard the practice teaching as a teaching task, and do not pay attention to whether students participate in the practice teaching activities, and do not carry out targeted teaching according to the students' learning situation, lack of practice teaching management. At the same time, the school did not pay attention to the practice teaching management, lack of specific staff and special management organization, it is difficult to strengthen the practice implementation process [14].

4   The Construction of Teaching Ecosystem

(1) Building a harmonious and equal relationship between teachers and students
Teaching is mainly taught by teachers, students passively accept knowledge into teacher guidance, students take the initiative to explore learning, fully respect the dominant position of students. Teachers are no longer superior, but explore and cooperate with students [15]. Teachers and students should increase more effective interaction, so as to stimulate students' learning initiative. Teachers' personal emotion and teaching style can also exert a subtle influence on students, and students' positive feedback can promote the emotional communication between teachers and students. Teaching should not be a cold and boring teaching of knowledge and skills, but an emotional and happy learning process.

(2) Making open and diversified teaching objectives
In the era of "education informatization 2.0", the construction of home education platform, education software app, website and we media operation have sprung up one after another, which requires kindergarten teachers to master the use of information technology to serve teaching, optimize classroom experience, and improve the ability of curriculum construction. The teaching goal of information course is to cultivate students' information literacy. In addition to knowledge and skill goals, teachers should pay more attention to process goals and emotional goals [16]. For example, in "data processing in excel", let students collect data from life, process data, and feel the whole process of data processing, so that they can know how to apply Excel to life. Through the case of "city temperament from reading index" in the classroom, we can feel that the society attaches great importance to reading and imperceptibly tell students that learning is endless. At the same time, we should cultivate students' big data thinking and feel the efficiency and convenience of information technology. As an information technology teacher, we should be able to combine the teaching content, create a meaningful teaching situation, consciously infiltrate the information technology emotional goals, cultivate sentiment, stimulate emotion, and cause students' emotional resonance.

(3) Renew the teaching content in keeping with the times
The information courses of preschool education major mainly include computer application and multimedia production. With the emergence of information products such as artificial intelligence, Internet of things, cloud computing and big data, the curriculum should keep up with the development of the times and closely integrate with the needs of kindergartens. Some kindergartens have used flipped classroom, online courses, vr virtual reality for teaching, and set up 3D printing courses, robot education, children's interesting programming, etc. In order to train talents to adapt to the development of preschool education industry, the curriculum should also follow the latest research in the field of information, try to add digital, scientific and technological curriculum elements, and even research and cooperate with excellent third-party education software companies on preschool education content, increase information knowledge, and broaden the scope of employment of students [17, 18].

(4) Explore the teaching method of online and offline integration

As a mature online learning platform, our super star Fanya platform provides rich curriculum resources and services, establishes a channel for interaction between teachers and students, and effectively connects school education and family education [19]. The combination of online courses and online courses is conducive to promoting students' fragmented learning and lifelong learning. Teachers should be good at using big data, cloud computing and other technologies to develop personalized training programs for students, so that cloud education, digital learning and other high-quality network resource sharing become more convenient and efficient. As the information classroom is oriented to social needs and students' professional needs, it requires strong practicality and skills. Therefore, in online and offline teaching, it should be advocated to carry out project-based teaching combined with kindergarten, with discussion and exploration as the main way of learning, so that students can actively participate, so as to improve the ability of information teaching and sustainable development [20].

(5) Create a high quality and efficient information technology teaching environment
The teaching of information technology course must be guaranteed by information technology facilities. The school should increase the investment in software and hardware equipment, pay attention to the maintenance and update of multimedia classroom, network classroom, campus network and other facilities, and improve the utilization rate of students' facilities. In addition, the school should also pay attention to the investment in educational software resources, to establish a stable network resource platform, to support the communication and interaction between teachers and students with the network, and through various evaluation activities, encourage teachers to accumulate multimedia materials, courseware and teaching cases and other teaching materials, at the same time, increase cooperation with kindergartens and other colleges and universities, and enrich a large number of teaching materials Learning resources are added to the platform to provide convenient learning channels for teachers and students at any time and realize resource sharing [21].

(6) Construction of multi dimensional teaching comprehensive evaluation system
Teaching evaluation has the functions of diagnosis, encouragement and guidance. Teachers should be good at making use of multi-dimensional and three-dimensional evaluation, not only for students' knowledge ability, but also for their practical ability, information consciousness, innovation ability and other aspects of process evaluation. Teachers can use big data and cloud computing for diagnostic evaluation, and make targeted training programs according to students' interest and potential to teach students in accordance with their aptitude. In teaching, teachers use big data to collect students' process materials, push learning materials that meet students' current needs according to students' feedback, and adjust teaching in time. Teachers can also encourage students to use mobile app to make interesting works, share the results through social platforms, and finally include the likes, comments and votes in the assessment. Multi dimensional evaluation can bring more valuable evaluation basis in many aspects, so as to motivate students and promote teaching.

5   System Construction

5.1   Preparation and Arrangement of Supporting Materials

(1) Hand related pictures and knowledge selection. According to the requirements of preschool education manual course teaching practice, based on the basis of learners, according to the teaching objectives, the front-end analysis is carried out to determine the appropriate manual course related knowledge, pictures and operation steps, so as to pave the way for the later case design and implementation [22].

(2) Material collection and processing. Combined with children's manual knowledge, pictures and operation steps, combined with the actual teaching needs, the materials needed for scientific organization and classification.

(3) The selection of children's manual knowledge, pictures and operation steps. Based on the specific teaching objectives and learning activities of preschool education manual course, according to the teaching standards, integrate the relevant knowledge and materials, and reasonably plan and organize the learning unit [23].

5.2   System Design Ideas

According to the structure of Moodle course platform system, the teaching system of preschool education manual course program selects two basic page types: test page and branch table page. Among them, the test page is used to present questions to the learners. After the learners learn, test and submit the answers to the test questions, the program teaching system provides real-time diagnostic learning feedback to the learners, and dynamically guides the learners to enter the next knowledge point or return to the original knowledge point to learn again. In order to reflect the overall performance of learners, the test page is set up to record sub items and summarize them into the overall performance of learners. The branch table page is mainly used to set up learning branches. Each branch is relatively independent, based on the learning objectives, and does not set up separate items. Correct and wrong answers will only affect the learning progress, but will not affect the learners' performance. When all the learning branches are completed, you can enter the next stage of knowledge learning. In Moodle platform, teachers can fully combine the advantages of the test page and the branch table page to complete the docking, feedback and jumping between the corresponding learning links. If learners complete the answer within the specified time, they can view their own learning results and test point explanations. Moodle platform can show each learner's answer process and results in detail in the form of a report on the teacher's side, and the display content mainly includes learning scores, answer time and so on, As well as the highest score, the lowest score, the average answer time, the most time and the least time, and other classified summary information, and even can be refined. The system design idea is shown in Fig. 1.

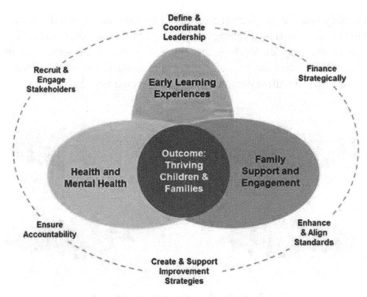

Fig. 1. System design ideas

5.3 System Construction Process

This paper describes the design process of the paper-cut technology section, which mainly includes three basic processes: adding content directory and content display page, adding test branch table page and setting the jump target of each test question/branch.

(1) Add content directory and content display page. Click the "new content page" hyperlink to fill in the page title, edit the page in the content, and add "basic knowledge" and "basic skills" and other test items in the content page. After learning, learners can click the relevant content to complete the learning test of knowledge and skills.

(2) Add clusters and test questions. According to the test content of "basic knowledge" and "basic skills" added in the content page, build the "basic knowledge" and "basic skills" cluster, add all the test questions of the knowledge point in the cluster, and finally add the cluster end page.

(3) Jump to goal setting. When all the test feedback questions and branches are added, open the "test questions/branches list" page. According to the requirements of the corresponding course teaching plan, click the "Edit" icon in the "behavior" to find the corresponding test feedback answer or description item. According to the branch direction, in the jump list after each item, select the target question or branch to jump to, and then complete the program teaching activity addition.

Login with the learner's account, you can see the learning examples of program teaching activities, and easily jump to the page of program teaching activities. The remaining time of program teaching system learning is displayed under the directory of program teaching system, which is used to help learners grasp the learning progress of program teaching system; The learning score of program teaching activities is displayed at the bottom of the program teaching system, and the interactive module is displayed at the bottom of the program teaching system. According to the settings of these jump buttons and question pages, the page Jump and interaction of program teaching activities are completed, so as to realize personalized learning (see Fig. 2).

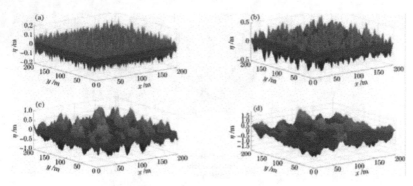

Fig. 2. Simulation with education system

6 Conclusion

Today's society is a society with popularization of network education and rapid updating of information. With abundant network resources, Moodle provides a new tool for preschool education to expand education, broaden vision and communicate. As parents, they can see their children's performance in kindergarten, and communicate with teachers to give timely feedback; as children, they can learn more and more interesting knowledge from the platform; as teachers, they can learn more about each child's situation according to online learning communication, so as to teach students according to their own characteristics. Therefore, it is of great significance to establish an interactive platform for preschool education based on Moodle.

References

1. Zhu, D.: Research on the teaching application of Moodle tutorial activity module. Shaanxi Normal University (2016)
2. Deng, M., Liang, C.: Application of programming instructional design based on Moodle platform. Modern Primary and Secondary Education, no. 12, pp. 44–48 (2015)
3. Li, J.: Innovation and application of handwork in preschool education. Art Educ. Res. **06**, 107 (2018)
4. Zhang, B.: Research on the construction of handmade curriculum in kindergarten teachers. Shandong Normal University (2017)
5. Mo, Q.: Construction and implementation of practical teaching system of preschool education major at junior college level. Mod. Vocat. Educ. **17**, 117 (2018)

6. Jiang, X.: Construction of practical teaching system at junior college level for preschool education major. J. Chongqing Univ. (22), 43 (2020). Qingfeng
7. Zhao, Y.: A preliminary study on the practical teaching system at the junior college level of preschool education. Life **17**(20), 48 (2017)
8. Tao, C., Lin, C.: Research on the design of "whole course" practice teaching curriculum system of preschool education major at junior college level. Sci. Technol. Innov. Guide **13**(31), 123–125 (2016)
9. Wang, L.: On the construction of practical teaching system at junior college level of preschool education major. Curriculum Educ. Res. **28**, 35–36 (2015)
10. Hao, H.: Construction strategy of practice teaching system of "Kindergarten Activity Design" courses for preschool education major in Higher Vocational Colleges – taking Harbin preschool teachers college as an example. J. Lanzhou Inst. Educ. **31**(05), 149–151+153 (2015)
11. Wang, X.: Research on the construction of the whole process practice teaching system of preschool education specialty at junior college level. Educ. Explor. **05**, 145–147 (2013)
12. Gao, R.: Research on the construction of practical teaching system in Normal Universities – taking preschool education major of Jiaozuo Teachers College as an example. University (Acad. Ed.) (06), 61–64+56 (2012)
13. Liang, Z.: Construction of the whole process practice teaching mode of preschool education specialty at junior college level. Preschool Educ. Res. (05), 30–33 (2011)
14. Fan, X.: Constructing the practical teaching system of preschool education major with professional ability as the core – taking Qiongtai Teachers College as an example. J. Hainan Radio Telev. Univ. **12**(01), 77–82 (2011)
15. Guo, Y.: Research on practice teaching system of kindergarten activity design courses for preschool education majors. Hunan Normal University (2010)
16. Xu, Y.: A preliminary study on the practical teaching system at the junior college level of preschool education major. Preschool Educ. Res. (z1), 101–102 (2006)
17. Zhang, Z.: On the opportunities and challenges of marketing under the background of big data. J. hechi Univ. (10), 173–174 (2021). Shang Xun
18. Song, B., Yunzhen, L.V.: On the influence of data mining on comparative education research. Heilongjiang High. Educ. Res. **38**(04), 87–90 (2020)
19. Deng, Q., Chen, S.: Discipline construction of higher education in China under the background of education big data. J. Ningbo Radio TV Univ. **17**(04), 122–128 (2019)
20. Huang, L.: Ideas and thoughts on building big data science of ideological and political education. China Train. **14**, 220–221 (2016)
21. Hu, J. Research on college ideological and political education in the era of big data. University of Electronic Science and technology (2016)
22. Yang, L.: Analysis on the transformation of comparative education research methodology from the perspective of post structuralism. Northeast Normal University (2015)
23. Li, X.: Collaborative research on ideological and political education. Lanzhou University (2016)

Research on the Innovative Application of Big Data Financing Model in University Infrastructure Projects

Fangyan Yu[✉], Xiong Na, Liang Chen, and Yanping Xu

Nanchang Institute of Technology, Nanchang 330044, China

Abstract. In this paper, the author starts with the theoretical analysis, based on the theory of educational economics and project management, comprehensively summarizes the experience of BOT financing mode at home and abroad, and on the basis of fully understanding and grasping the guidelines and policies issued by the State Council, the Ministry of education and governments at all levels BOT financing mode is a successful mode in general enterprises. It actively explores the necessity and feasibility of its application in university infrastructure, discusses its implementation steps, scope and other specific operation problems, and puts forward some feasible suggestions for the problems and difficulties that may be encountered. This paper attempts to explore a new way for the popularization of higher education and the socialization of logistics construction in Colleges and universities, and also hopes to improve the theory of financing diversification in Colleges and universities and expand the scope of project management financing theory.

Keywords: University · BOT financing mode · Infrastructure construction

1 Introduction

Since the rise of BOT investment and financing in the 1980s, it has gradually become the focus of international investment boom. BOT means The English abbreviation of build operate transfer is a project financing mode. Its typical mode is: "the government signs a concession agreement with a project company of non-governmental departments for an infrastructure project, and grants the project company the responsibility for investment, financing, construction, operation and maintenance of the project. Within the term of the concession agreement, the project company will be granted the right to undertake the project The project company shall charge appropriate fees to the users of the facilities, so as to repay the debts, recover the costs used for the construction, operation and maintenance of the project and obtain profits. After the expiration of the concession period, the project company must transfer the ownership of the project to the local government departments free of charge according to the agreement." [1].

Published by Springer Nature Switzerland AG 2021. All Rights Reserved
M. A. Jan and F. Khan (Eds.): BigIoT-EDU 2021, LNICST 392, pp. 436–446, 2021.
https://doi.org/10.1007/978-3-030-87903-7_53

①After China's accession to the WTO, it has more and more close ties with other countries in the world, and the cooperation and exchanges with other countries are becoming more and more frequent. Besides the cooperation in economy and other aspects, the cooperation and exchanges in education are also increasing. China's higher education industry presents the international situation, from elite education to mass education, and the trend of marketization is increasing. Because the vast majority of colleges and universities in our country are public institutions, which are of public welfare nature, single financing mode and short of channels, almost all major colleges and universities in our country are facing a serious shortage of funds in infrastructure construction. This also forces colleges and universities to raise funds from the society through various channels and ways to alleviate the pressure of fund shortage. But after all, colleges and universities are institutions whose main task is to cultivate talents. They can't only pursue economic benefits and take great risks. So what kind of financing methods should colleges and universities take to solve the problem of shortage of funds for infrastructure construction in Colleges and universities, and ensure that it can be carried out in a relatively safe and stable environment has become a series of problems to be solved. In order to solve these problems quickly and efficiently, we can introduce BOT financing mode into colleges and universities. This is not only conducive to the construction and development of university infrastructure, but also conducive to the development of social economy. So can colleges and universities introduce this kind of project financing method which is usually used in enterprises? What are the similarities and differences between the BOT mode in enterprises and that in Colleges and universities? What kind of operation steps should be followed in the specific operation process and what problems should be paid attention to in the application of BOT financing mode in Colleges and universities are the focus of this paper.

2   BOT Mode

2.1   The Definition, Characteristics and Forms of BOT Financing Mode

(1)   The definition of BOT financing mode

BOT is the abbreviation of Build Operate Transfer, which means build operate transfer. Also known as "public works concession". According to the international practice, the connotation of the typical BOT investment mode is that the government and the private sector project company sign a contract, and the project company is responsible for financing, designing and constructing the infrastructure service project. After the completion of the project, the project company will repay the project debt and obtain the return on investment through the operation of the project within the concession period specified in the concession agreement signed with the government. At the end of the concession period, the property right of the project will be transferred to the government of the country where the project is located free of charge. Specifically speaking, the so-called B, that is, build (construction), refers to the project company to raise funds for the construction of the project. The project company is generally composed of construction company, finance company, equipment and raw material supply company, international syndicate, etc. As for the qualification of the project company in BOT investment mode, if the credit rating of the project sponsor is relatively high, the investor shall form the project company alone; if the credit rating of the project sponsor is relatively low, the project sponsor and the investor shall form the project company together. The so-called "O", namely "operate", refers to that the project company, through the operation and management of the project, recovers the investment, repays the loan and gains profits within the specified franchise period [2]. The so-called "t", namely transfer, refers to that the project company needs to transfer the property right of the project to the government of the country where the project is located free of charge after the expiration of the franchise period. To sum up, BOT is a financing mode to attract capital investment in government or collective project construction. It was proposed by Turkish Prime Minister Ozar in 1984, and then formed in North America in the late 19th century. In the 1980s, with the rapid development of social economy in developing countries, the demand for infrastructure construction of the state and government has been increasing. Obviously, their development is restricted by backward technology and shortage of funds. Therefore, attracting private capital and private capital has become a possible and inevitable choice, and BOT financing mode emerges as the times require. After more than 20 years of development, BOT financing mode has become an effective way to solve the problem of national public infrastructure construction funds. At home and abroad, there are many built and under construction projects have introduced BOT mode. For example: Sydney Harbour Tunnel in Australia, channel tunnel in England and France, Erythrina bridge in Quanzhou, Fujian Province, etc. In the construction of 2008 Olympic facilities in China, more than 30 Olympic venues and the main "bird's nest" gymnasium project adopt the typical BOT financing mode (Figs. 1 and 2).

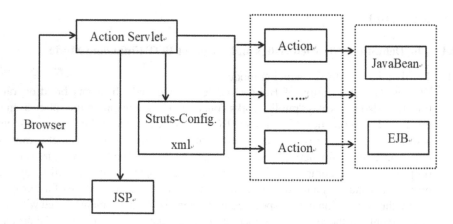

Fig. 1. Struts frame structure

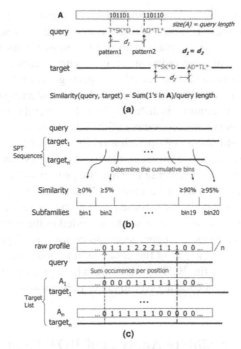

Fig. 2. Query algorithm

2.2 Characteristics of BOT Financing Mode

BOT financing projects usually have a large amount of one-time investment. This is because BOT projects are generally used in large-scale public works projects, such as power plants, urban water supply systems, highways, airports, etc., and the investment of these projects is often very large. Therefore, the construction and financing of large-scale projects usually combine the traditional equity investment and project financing, in which equity investment usually accounts for less than 30% of the total investment, and most of the remaining funds come from project financing. BOT is one of the more typical and effective project financing methods.

There are many participants in BOT financing projects. Generally, a BOT project must include the project sponsor, government departments, project construction units, users, equipment and raw material suppliers, among which the project company is the most important participant and executor. But all the participants may affect the progress of the project. The operation cycle of BOT financing project is long. Because of the large scale of BOT financing projects, the time span from project bidding negotiation, feasibility study to the end of the business cycle and the final realization of project transfer often takes more than ten years, decades or even longer.

BOT financing project has great risk. The investment scale, the number of participants and the business cycle of BOT project jointly determine that there are various risks in the whole implementation process, such as the risk of government policy change, the risk of financial market interest rate or exchange rate and institutional change, engineering construction risk, irresistible risk, raw material or equipment supply and price risk, tax risk, etc. However, there is a basic principle in the risk allocation of BOT project, that is to allocate the risk to the party who has the most ability to control or bear a certain risk. BOT financing project has strict and complex contract provisions. Due to the above characteristics of BOT projects, BOT projects generally have a set of long-term, complex and strict contract arrangements for avoiding, transferring risks and facilitating management. This kind of contract is different from the general economic contract. First of all, there must be a concession agreement of the government or the public sector. This concession agreement stipulates the operation period of the project company, which is very important in BOT financing projects, that is, the concession period. The project company is only allowed to operate and manage the project within this period.

BOT financing projects must have predictable and stable sources of income. The investors of BOT project usually get income by charging the facility usage fee after the completion of the project, so the price must be included in the contract.

The basic content of financial supervision and control. BOT project must be transferred free of charge. After the end of the concession period, unless both parties intend to continue cooperation on the BOT project, the whole property right and the project itself shall be transferred to the local government or public management department free of charge.

3   Necessity and Feasibility Analysis of BOT Financing Mode in Colleges and Universities

(1) The present situation of infrastructure construction in Colleges and Universities
In recent years, the rapid development of China's economy has also brought the development of higher education into a rapid development era. In 1998, the number of colleges and universities in China was 1020, and the number of colleges and universities in this college was only 3.41 million. By 2006, there were 1867 universities, more than twice that in 1998. However, the number of students in this college reached 17.388 million, which was 5 times that of 1998. However, due to the relative shortage of the overall education funds, the conditions of running a university to different degrees are tight, which seriously affects the expansion and growth of the scale of the school. This is mainly reflected in the following aspects:

1. The shortage of funds hinders the rapid development of Higher Education
Since the reform and opening up, the state has paid more and more attention to education. In the meantime, the national product has been increasing year by year due to the rapid development of national economy, and the state has increased its investment in education cost. From 1978 to 1985, the average annual added value has been increased every year The annual growth rate of RMB 1.694 billion is 6.31 billion yuan, which is nearly 9 times that of 1980 in 1994. ⑤Although the scale of higher education is expanding rapidly, the public finance investment of governments at all levels has not increased in the same proportion in time, so the average education funds of students are still in a downward trend. For most universities, especially non key universities, the relative shortage of state funding, coupled with the rapid

increase of the number of universities, the surge in the number of students and the rise in prices, the university funds are still very tight. The statistics announcement of the implementation of national education funds of Ministry of education, the State Bureau of statistics and the Ministry of Finance in 2005 pointed out that in 2005, the expenditure on the public service expenses in the national average budget of ordinary colleges and universities was 5375.94 yuan, down 3.18% from 5552.50 yuan last year. The expenditure of public funds in the national average budget of higher education is 2237.57 yuan, down 2.65% from 2298.41 yuan last year [3].

2. The dormitory and classroom are short and the teaching equipment is old
Due to the rapid growth of the number of students in school, the school building conditions have declined. According to the statistics of the development and research center of Shanghai Academy of education and Sciences, the three-year expansion of Chinese universities and colleges and universities, the statistics show that: "in 1999, the national ordinary university schools occupied an area of 626000 mu, the building area of the school buildings was 174.5 million square meters, and the average teaching administrative room was 12.3 m^2. The average student dormitory area is 5.4 m^2. In 2000, the national ordinary university schools occupied an area of 76000 mu, the building area of the school buildings was 204.4 million square meters, and the average teaching administrative room for students was 11.2 m^2. The average student dormitory area is 5.4 m^2. In 2001, the national ordinary university schools occupied an area of 906000 mu, the building area of the school buildings was 2557 million square meters, and the average teaching administrative room for students was 11.1 m^2. The average student dormitory area is 5.8 m^2. ⑥At the same time, many colleges and universities lack of classroom quantity or the equipment in the classroom is old, which can not meet the needs of modern teaching. On the one hand, the total value of teaching equipment in many universities can not meet the relevant national standards, on the one hand, many old and outdated experimental equipment can not be updated and scrapped in time. All of these have seriously affected the improvement of the quality of higher education and the development of higher education.

3. Low efficiency of educational funds
For a long time, the higher education funds in China are facing a very awkward situation. On the one hand, the investment in higher education funds is relatively insufficient, on the one hand, the efficiency of the use of educational funds is very low. For example, the utilization rate of teaching facilities, sports equipment and experimental equipment in the school is too low. According to the official statistics released by the Ministry of education in 2007, by the end of 2006, the number of staff and staff in general colleges and universities in China was 1872601 and 1179168 full-time teachers in general universities. Then, the number of administrative logistics in ordinary universities was 693433, accounting for 37% of the total number of teachers and staff. The above data can be seen clearly that due to the cross and replacement of institutions, the proportion of managers is relatively large, which causes the proportion of the expenditure of management personnel in the funds of colleges and universities is too large, and the redundancy of personnel also increases the friction in daily management and reduces the efficiency.

The scale of higher education has accelerated its expansion at the turn of the century. In 1999, the State Council decided to expand the scale of college enrollment. For a period of time, the scale expansion has become the theme of the development of higher education in China, and the size of college students has grown rapidly. By 2006, the number of Chinese higher education students in school reached 25 million, and the gross enrollment rate of higher education was 23%. The number of students enrolled in the college in China has increased from 1080000 in 1998 to 54.46 million in 2006. The size of students in the college has increased from 34.41 million in 1998 to 17.39 million in 2006. The number of graduate students in school has also increased from 200000 in 1998 to 111 million in 2005. ⑦From the quantity, China has become the largest country in the world higher education, and has successfully achieved the leap from elite education to higher education popularization. But in the aspect of logistics facilities in Colleges and universities, the following problems are also faced:

(1) Logistics base conditions are relatively poor

With the rapid development of higher education in China, the elite education in the past is also transforming into mass education. The main problem faced by this process is insufficient funds, and the logistics basic conditions seriously restrict the development of colleges and universities. In recent years, the state and universities continue to promote the process of socialization of university logistics, although the number of new student apartments and canteens in Colleges and universities has increased more than in the past, however, the number of students relatively expanded, there are still more students in the accommodation conditions have not been significantly improved [4]. This phenomenon is caused by the expansion of enrollment in Colleges and universities, the increase of students; on the other hand, the lack of funds in Colleges and universities in China is a long-standing historical problem, too much debt, and the total amount of logistics facilities in many universities is far from the quantity required by the state. However, the limited funds can only maintain the basic operation, and cannot increase the construction and development of logistics facilities. The construction of infrastructure lags behind the requirements of higher education development in China.

(2) Unclear property rights

The property right of the legal person of the logistics group is caused by the unclear property right. In law, it cannot operate independently, bear the profit and loss and bear the relevant civil liability. This causes the school as the parent of the logistics group to bear a great joint and several liability; because the logistics group is not using the full cost accounting method, the latter is not used for the accounting of the full cost Qin group is not a company in strict sense, so it can not participate in market competition according to the requirements of market economy, and follow the internal development law of the enterprise to participate in market competition, and give full play to the characteristics of the enterprise, and it is not in line with the development law of accounting system and market economy. As the property of the school, logistics facilities are leased unconditionally to the logistics group for operation, but at the same time, the logistics group does not reflect such assets in their accounts, which makes the equipment being used have no loss and depreciation on the book, and the funds in use have no interest, and the collective and national ones are suffering from the losses.

(3) The substantive separation of ownership and management right
"The separation of school and enterprise, the separation of ownership and management right" is the basic principle of socialization of logistics in Colleges and universities in China. This requires that colleges and universities should simplify the functions of institutions, transform and independent institutions, and avoid administrative functions and business services.

The interweaving and confusion of the functions of affairs. However, in the actual operation process, due to human reasons, ownership and management rights are often integrated, weakening or even ignoring the economic contract relationship between school and enterprise, which makes the financial unclear, difficult to manage, and return to the difficulties at the beginning of reform.

(2) Analysis of the current situation of University financing
Facing the reality of insufficient funds, the state, society and universities are trying to find ways to raise funds through various channels to try to solve the problems that colleges and universities are facing. There are several practical ways to do so:

1. Play the basic role of government financial allocation.
For a long time, in view of the nature of university service, it has been to obtain financial allocation from the state as an institution, how much rice to cook in a large pot. However, with the beginning of the reform of colleges and universities, colleges and universities try to implement charge education to increase their income to promote the development of colleges and universities, and end the previous era of "big package". However, with the rapid development of our economy, science and technology are changing day by day, people's living standards are increasing, and the requirements for education are also increasing. This puts forward objective requirements for the human and facilities of colleges and universities. However, due to the lack of funds, the development of colleges and universities is difficult. However, according to the relevant statistics of UNESCO, the proportion of higher education funds in China in the national education funds has ranked the top in the world. In 2001, the financial education funds of China broke through the 300 billion yuan for the first time, reaching 305.7 billion yuan. On this basis, the annual increase of education investment was maintained. By 2005, the financial investment in education had exceeded RMB 500 billion, reaching 516.11 billion yuan, doubling in five years, with an annual growth rate of 15%. Therefore, the main channel of the source of funds is still the government financial allocation, so it must play its basic role in the financing of colleges and universities.

2. The income of tuition and miscellaneous expenses is another source of college funds.
In the past long time, compulsory education in China has no obligation and is still charging. But higher education in the non compulsory education stage is not only free of all costs, but also arranged by the state. This not only distorts the stage of compulsory education, but also is not conducive to the development of education

in China, but also to the long-term and rapid development of our economy. From the mid-1980s, China began to try to implement part of the students pay fees to attend school. Besides the public fee students in the plan, the students who are self-employed and the students appointed by the employing units need to pay for the tuition. Then, the reform of "parallel track" began in the early 1990s, and by 1997, all the students in Colleges and universities had to pay for school. The charge is based on the principle of reasonable sharing of education cost, comprehensively measuring the interests and bearing capacity of the country, society, University and family, and adopting the upper limit of the university charge according to the education subjective departments in the region where the university is located. The specific standards are the policies determined by the university itself, which makes the university charge not only reasonable but also feasible. The proportion of tuition and miscellaneous expenses in the financing of colleges and universities has increased significantly, among which central universities rose from 10.4% in 1998 to 19.9% in 2006, while local universities rose from 17.9% in 1998 to 40.1% in 2004. By 2006, the total income of ordinary colleges and universities in China was 85.75 billion yuan, accounting for 29.2% of the total income. It can be said that the majority of students' families have made great contributions to make the tuition and miscellaneous income become another important source of financing and the development of colleges and universities.

4   Research on the Application of BOT Financing Mode in Colleges and Universities

The application of BOT financing mode in the infrastructure construction of colleges and universities should include project determination, bidding and tendering, signing of franchise agreement, construction, operation and transfer.

(1) Determination of project

The infrastructure construction of colleges and universities should be based on the development plan of the University, and select a number of projects from the projects that may or need project financing for feasibility demonstration. Based on the approval of the feasibility study of the project, this paper focuses on the feasibility of adopting BOT financing mode to meet the needs of the project, and makes corresponding decisions to determine the project to implement BOT financing.

(2) Project bidding

After passing the project feasibility study, the project bidding notice shall be released to the whole society, and the bidding documents shall be issued to the independent bidders or bidding consortia who have passed the qualification examination. The bidding documents are generally composed of instructions to bidders, franchise agreement, technical requirements, reference materials, etc. Colleges and universities should open their bids at the deadline. At the time of bid opening, information such as the submission time of the bid, whether the seal of the bid is in good condition, the name of the bidder (list of members of the bidding consortium), the unit price of the construction facilities charge bid, whether the bid guarantee meets the requirements of the bidding documents, and the formal written authorization of the legal representative or the consortium will be disclosed and recorded. Before bid evaluation, colleges and universities should set up a bid evaluation committee according to relevant laws and regulations. The bid evaluation committee evaluates the bidding documents provided by the bidders and recommends the successful candidates to the bidders. Finally, the Management Committee of university logistics department decides the successful bidder.

(3) Sign franchise agreement

After the successful bidder is determined, colleges and universities shall issue a letter of acceptance to the successful bidder. At that time, both parties shall conduct further consultation and negotiation to confirm the contents of the tender and sign the franchise agreement according to the provisions of the bidding documents. The agreement mainly includes the time and method of the investor's capital investment, the construction period, the content and period of the investor's management and operation, and the allocation and undertaking of risks. The successful bidder shall complete the financing delivery and the registration of the project company according to the commitment period in the tender document, and both parties shall formally sign the franchise agreement after the design documents are approved.

(4) Project construction

After the franchise agreement is formally signed, the project company will carry out the design, construction and completion of the construction tasks in an organized way on the premise of complying with the relevant laws and regulations.

(5) Project operation

This phase will last until the expiration of the concession agreement. In this phase, the project company will operate the project directly or by signing a contract with the operator in accordance with the standards of the project agreement, the loan agreements and the conditions agreed with the investors. During the whole operation period of the project, the project company shall maintain the project facilities in accordance with the requirements of the agreement. In order to ensure that the operation and maintenance can be carried out in accordance with the requirements of the agreement, universities have the right to supervise and inspect the project company or operators and put forward rectification suggestions.

5 Conclusion

Education industry is a basic industry related to the future development of the country, and it is a sunrise industry with broad market and development prospects. But the traditional financial allocation based education funds can not meet the rapid development of the university construction funds. With China's accession to the WTO and the further development of China's education, a large number of domestic and foreign capital has poured into the higher education market in China. We must strengthen our efforts to carry out reform and innovation if we want to develop education, build a well-off society in an all-round way, and stand invincible in competition. How to deal with the pressure of expanding enrollment to the basic school conditions and logistics service facilities of ordinary universities; how to solve the problems of the difficulty of covering investment and financing for the infrastructure of colleges and universities; how to solve the problems of low efficiency and waste in the operation of university infrastructure; in this context, BOT financing mode is presented in front of us with its unique advantages.

Through the necessity and feasibility analysis of BOT financing mode of university infrastructure, we can see that we can not only build the BOT financing mode of university infrastructure with its own characteristics, but also deeply feel the vitality and great impetus it brings to the leapfrog development and logistics reform of our country. The BOT of university infrastructure can be seen The mode of investment and financing not only solves the problem of difficulty in the investment and financing of university infrastructure, but also activates the whole logistics market of colleges and universities, which makes it go into the process of enterprise, marketization and commercialization. It provides a comprehensive guarantee for the development of higher education. But according to the current situation, the BOT mode of infrastructure in Colleges and universities is still in the initial stage, and there are still a series of problems. This paper, through the theoretical research on BOT project investment and financing mode in the construction of University Infrastructure in China, and at the same time, it also applies BOT to colleges and universities Some problems of financing mode are analyzed and corresponding countermeasures are put forward, which try to provide reference for university decision-making, and hope to provide valuable reference for the reform and innovation of the investment and financing system in Colleges and universities.

Acknowledgements. Science and Technology Research Project of Jiangxi Education Department in 2019, Project name: innovative Application Research of BOT financing Model in University Infrastructure Project, Project number GJJ191018.

References

1. Zhang, D.: Problems and suggestions on management of sporadic infrastructure projects in colleges and universities. Development **12**, 47 (2006)
2. Hu, C.: Cause analysis and Countermeasures of influencing project schedule. Sci. Technol. Plaza **10**, 170–173 (2011)
3. Yu, J.: Review of budget and final accounts of university infrastructure projects. Archit. Technol. Dev. **36**(12), 63–64 (2009)
4. Wu, S.: Thinking and improvement of audit practice of university infrastructure projects. J. Anhui Inst. Archit. Technol. (Nat. Sci. Ed.) **05**, 65–68 (2007)

Application of Data Mining Technology in Economic Statistics

Yao Chen[✉]

Yunnan Technology and Business University, Kunming 651701, Yunnan, China

Abstract. At present, the domestic social and economic development is constantly improving, and during this period, there are many economic data reflect the uncertainty. Relevant departments make a classification of these data by means of statistical calculation, and then make a summary of the economic development in a period of time, so as to plan the next economic activities. However, in the unified calculation of these data, we must use a reasonable trial method, and the current unified calculation method has certain uncertainty, unable to get accurate results. If the data mining technology is used reasonably, the above problems will be solved. This paper discusses the application of data mining technology in economic statistics.

Keywords: Equipment support · Audit information · Data mining · Quick recommendation

1 Introduction

In the context of the rapid development of domestic economy, the statistical work of economic data of personnel in relevant units is becoming more and more complex, and there are more and more data categories. However, there is obvious uncertainty when using the traditional statistical method to sort out and calculate these data. This brings great difficulty for the following economic activities such as data analysis. With the help of data mining technology, we can improve the traditional way, not only to calculate the accurate data, but also to expand the depth and breadth of data statistics. In this way, we can get the practical data, and provide the guarantee for the orderly development of economic activities [1].

The so-called data mining technology is formed through the effective integration of multiple related professional knowledge. We can sort out and analyze the complex data, and through a series of sorting and analysis process, we can calculate the data with practical effect. It can be said that data mining technology belongs to the unified computing technology that can accurately process complex data.

M. A. Jan and F. Khan (Eds.): BigIoT-EDU 2021, LNICST 392, pp. 447–453, 2021.
https://doi.org/10.1007/978-3-030-87903-7_54

2 The Advantages of Data Mining Technology in Economic Statistics

2.1 With Strong Comprehensive Ability

In today's era when the domestic economy is greatly improved, it is necessary to accurately analyze and calculate the economic data. However, the standards for data projects are different in different working groups of relevant departments. In the real environment of various economic activities, usually involves a lot of management departments. During the period of making some choices and decisions, different management departments must take accurate data as the basis. However, different management departments will reflect different management methods, so the perspective of data requirements will be different. If the data mining technology is used reasonably, the data can be processed from different angles, so as to meet the different needs of different management departments for data, and can account for different departments. Summary work to provide the basis. For example, in the financial department of an enterprise, when accounting, the accounting personnel need to summarize the data. In this process, we must use data mining technology to format the data stored in the computer. This is the basis of accounting and summary. And this advantage is reflected by data mining technology. It can be said that data mining technology can play a comprehensive role in economic activities [2].

2.2 The Effect of Actual Statistics is Strong

Reasonable use of data mining technology can make a deep analysis and collation of complex data, in order to calculate the data with practical effect. For example, the application of data mining technology to a large number of complex data collation can make a reasonable and accurate calculation and collation of a large number of data. In order to make the complicated data more clear. Improve the efficiency of management. And this technology can also make a deeper and wider collation of the current data, and fully guarantee the reliability of the statistical data [3].

2.3 The Applicability of Data Mining Technology is Strong

To make accurate accounting and summary of economic data, the main purpose is to provide basis and guidance for economic activities in various real environments. However, different economic activities have different statistical angles. If the data mining technology can be used effectively, it can further improve the traditional calculation method. And can meet the statistical work of many units. It can be said that data mining technology belongs to the unified calculation method with the function of sorting out. The reasonable application of this technology to the economic data calculation can guarantee the integration of high-value information, so as to meet the needs of different angles of different economic activities. It can be said that this technology has good applicability.

3   Data Mining Technology

3.1   Basic Concepts of Data Mining

With the change of communication mode and the development of storage technology in recent years, a large number of data are generated and recorded in every bad section and every moment of life and production. From the state to enterprises, they gradually realize the value of data, which will become a symbol of national strength and wealth in the future. Any of our activities will produce data, and these intangible data, like historical relics, can discover the value of cultural relics in specific fields by using effective archaeological discoveries. The process of value exploration and discovery of data as "historical relics" is data mining. 2. Just as people can predict some familiar affairs according to their long-term work experience, In the information age, data mining is an important link to find, extract or capture knowledge from the daily accumulated massive data [4].

According to the results of the requirement analysis of the equipment support information system, this chapter will carry out the specific design and implementation of the system. This project uses BS architecture, Java background language, bootstrap foreground language and the corresponding data mining technology to design the system, so as to meet the functional requirements of remote access, friendly interaction and quick recommendation. The main contents of equipment support information system design include: system overall framework design, database design, core function module design and user support information recommendation design. In the process of core function module design and user support information recommendation design, the business process is expanded and described, and the function interface is implemented.

Equipment support information table is used to store the questions and answers submitted by users, which is the most important data in the system. In order to ensure the security of the system and introduce the security mechanism in the future, the picture content is stored in the table in Base64 format instead of separately. Equipment support information status includes saved, submitted, unanswered, answered, failed and passed. At the same time, with activiti5 calling process progress ID, the equipment support information flow is realized. User support data is the core data of data mining in the system, which should have the required user information, equipment information and other data content, such as the number of user organization, equipment number, support object number and equipment parameters. Equipment parameter is the type of equipment parameter corresponding to special character segmentation. Local audit table is used to store local audit information, including user number, user name, user organization number, log type and operation time. When more data is accumulated, it is helpful to achieve faster query recommendation through system upgrade and time series data mining.

3.2   Common Algorithms of Data Mining

According to the data mining tasks in different application scenarios, the data mining algorithms are generally different, and sometimes the mining algorithms need to be modified according to the research data objects.

Logistic regression analysis can explain the relationship between the discrete dependent variable and the independent variable. Generally, the binary variable, i.e. the dependent variable, has two values. For example, Y(1) represents the positive result and Y(0) represents the reverse result. At this time, the independent variable x is a group of explanatory variables of the dependent variable, and the logarithm of the approximate response ratio of X to the dependent variable $Y = 1$:

$$\text{Logit}(P) = \ln\left(\frac{P}{1-P}\right) = \alpha + \beta \tag{1}$$

In the formula, a is the intercept and β is the slope. In the multi classification relationship, the value of dependent variable is not limited to two kinds, that is, there are multiple values of Y.

Typical discriminant function is based on Bayes theory. It analyzes and calculates the correlation between the observed values of each category, and then establishes the corresponding classification rules, so as to realize the re classification of the original sample. By comparing the original class with the predicted class, the classification and discriminant accuracy of the initial sample are finally determined. It can be expressed as:

$$X^i = \left(x_1{}^i, x_2{}^i, \ldots x_m{}^i\right), i = 1, 2, \ldots k \tag{2}$$

Based on the deficiency of multiple logistic regression prediction, the classification method based on discriminant function is used for further classification prediction.

4 Application of Data Mining Technology in Economic Statistics

4.1 The Method of Data Integration Lays a Foundation for the Accurate Statistics of Economic Data

With the continuous development of social economy, the relevant data is bound to increase correspondingly, and the categories are also more abundant, which makes the data statistics work more difficult. If the data mining technology is reasonably used, the data in the computer database can be checked and sorted, which ensures the efficiency of data identification and provides guarantee for the orderly development of data statistics work.

The equipment support information system is not only the data provider, but also the data demander. The acquisition, storage and utilization of all kinds of data in the support process is very important to the system. Combined with the characteristics of all kinds of important data, this paper focuses on the following factors to design the database of the system: 1) considering the scalability and adaptability of the system, it is necessary to fully reserve the field and table relationship. 2) Because the data in the system needs to connect with other systems or equipment, it is necessary to ensure that the user support data meets the standard audit data format proposed by this topic, which is the data requirement for data mining of user support data, and also the basis for establishing the standard equipment support information database. 3) Equipment support information is sensitive. In order to facilitate the subsequent adoption of data encryption and other security measures, increase the redundancy appropriately, and store the text and pictures in the support information in the form of text in the database, so as to ensure the consistency of data and avoid the mismatch of graphics and text in equipment information call.

4.2   The Pretreatment Method Should Be Applied in It

The reason why data mining technology can do efficient unified calculation for complex data is mainly due to its intelligent function, so when there is deviation in the initial unified data, it will inevitably lead to the uncertainty of the unified data. But in the specific work, the initial statistics data are often uncertain and imperfect, so with the help of preprocessing, these uncertain and imperfect data can be completely deleted to ensure the accuracy and integrity of the data by means of data mining technology. The pretreatment method should be used, as shown in Fig. 1.

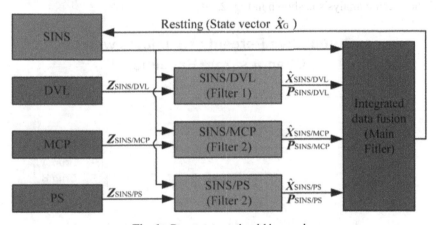

Fig. 1. Pretreatment should be used

4.3   Regression Analysis Method Can More Clearly Reflect the Data Statistics

Regression analysis method can show the relationship between variables more clearly. This method can more directly analyze the current commodity market share, sales and other data statistics. Taking the simple linear regression analysis method as an example, the qualitative variable is set as X, the dependent variable is set as y, and a is the intercept and B is the correlation coefficient. At this point, we can get the linear equation as $y = a + bX$. At this time, the results of different dependent variables are substituted into the formula, which can intuitively analyze the variables.

The system uses Java language to complete the main development of the application. User and role management, equipment information management, equipment support information management, business process management, statistical analysis and data management constitute the basic functions. Role management provides the functions of creating, editing, deleting and permission control of roles in the system. User management provides users with various roles to add, edit and delete. Equipment information management provides the addition and editing of equipment attribute parameter types and corresponding items. Equipment support information management provides support information creation, answer, approval and recommendation.

Timeliness of data. The interface for obtaining and importing the user support information of different equipment systems is reserved in the system, which is convenient to obtain the audit log in time and ensure that the data in the system is kept up-to-date. The choice of data mining algorithm. The core function of the system is to find valuable information in the user support data, then use the system filtering algorithm to analyze the user similarity, and use the association rules to analyze the support area between equipment. According to the characteristics of user information, equipment information, audit information and other support data, this paper improves the traditional Apriori algorithm to meet the requirements of application scenarios.

Regression analysis method can more clearly reflect the statistical nature of the data, and the specific analysis is shown in Fig. 2.

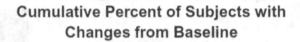

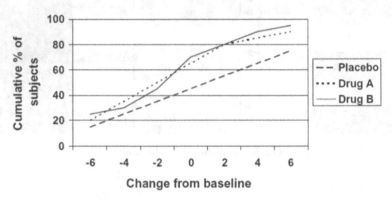

Fig. 2. Regression analysis method reflects the statistical nature of data

5   Conclusion

Data statistics can play an irreplaceable role in the social and economic development. The traditional statistics can not meet the needs of social and economic development. With the help of data mining technology, the traditional statistical method can be improved, which provides a guarantee for the realization of accurate data. It can be said that the use of data mining technology can make a clear arrangement and analysis of the current economic data, so as to obtain accurate and reliable data, and provide the basis for future economic activities.

References

1. Wang, S.: Wang Shuan's Theory and Practice of Foreign Military Information Construction. PLA Press, Beijing (2008)
2. Jiawu, H., Qiuxing, G.: Support measures and characteristics of US military equipment package in Iraq war. Foreign Mil. Sci. **8**, 40–42 (2003)
3. Yao, H.: Overview of the development of US military logistics information system. Mod. Mil. (05), 87–91 (2016)
4. Zhao, J., Deng, J., Liu, Y., Jungle, T.: Digital construction of information support for ordnance equipment. Logist. Technol. **36**(10), 142–145 (2017)

Multi-objective Evolutionary Algorithm Based on Data Analysis and Its Application in Portfolio Optimization

RuiWu Jiang[1(✉)] and Jing Tang[2]

[1] School of Mathematics and Statistics, Zhaotong University, Zhaotong 657000, China
ztjrw2003@tom.com
[2] Yongfeng Middle School, Zhaoyang District, Zhaotong 657000, China
ztxystxyjrw@tom.com

Abstract. The rapid development of computer technology has expanded the scope and field of use of this technology. At present, through the application of computer in optimization problems, it can effectively meet various specific objectives. Based on this, this paper discusses the multi-objective evolutionary algorithm and significance based on DEA, and analyzes the application of the algorithm in combination for reference.

Keywords: DEA model · Multi-objective evolutionary algorithm · Portfolio

1 Introduction

Because portfolio optimization problems are often NP difficult to solve, it is difficult to get the optimal solution in polynomial time, but DEA model is used as the model to evaluate at the optimal level. The portfolio optimization problem can be effectively solved by integrating it into multi-objective evolutionary algorithm. Therefore, it is necessary to analyze the algorithm and its application.

2 Discuss the DEA Based Multi-objective Evolutionary Algorithm and Its Advanced Significance

The so-called DEA, mainly refers to the production leading edge obtained by data envelopment analysis model based on known data, and the decision unit (DMU) with multi-input and multi-output is evaluated to obtain the optimal solution. At present, the DEA model mainly includes FG model, CCR model, ST model and BCC model. At present, there are many effective portfolio models which are limited by the relevant conditions to transform the problem into NP difficult solution. In general, the initial solution of intelligent optimization algorithm is random. Tabu search and other algorithms are more dependent on the initial solution, which leads to the initial search state is not

M. A. Jan and F. Khan (Eds.): BigIoT-EDU 2021, LNICST 392, pp. 454–458, 2021.
https://doi.org/10.1007/978-3-030-87903-7_55

ideal. However, the DEA model evaluation is based on the optimal level and uses the difference operator to solve the interdependence of subproblems, which can effectively avoid falling into the local optimal solution and enhance the diversity of the algorithm. Therefore, by using DEA to improve the multi-objective evolutionary algorithm, it is not necessary to determine the index weight coefficient and any form of relational expression, but also to reduce the time complexity of the algorithm and solve the problem of portfolio optimization. The specific optimization methods are as follows:

1. **The decomposition MOP, artificially decomposes it into multiple single-objective optimization problems.**

MOEA/D algorithm does not treat MOP as a whole, but as a single-objective optimization problem. When decomposing it, it mainly uses Chebyshev aggregation method. Then the optimal solution is obtained by modifying the weight vector [1]. In this case, each generation of population is picking the optimal solution and then forming the set of optimal solutions, which is different from the MOEA/D algorithm can only optimize the adjacent subproblems, which can find the optimal solution in all populations.

2. **The initial population is generated using the DEA model, which has an initial population efficiency value of**

1. DEA evaluation idea is to analyze whether the DMU input and output data are relatively effective or invalid by judging the input and output data. In the concrete operation, the input and output of the DMU should be fixed, and the DMU should be projected by mathematical planning. After that, the relative validity is evaluated according to the deviation between the DEA and the two. DEA general model is that there are n DMUs, and the input and output elements are m and s respectively x_{ij} That means j i input to the DMU, y_{rj} The r output representing the j DMU, and the input > 0, output 0. The input weight in i is v $.0_{ij}u$ is the r output weight$_{rj}$ $X_j = \left(x_{1j}, x_{2j}, ..., x_{mj} \right)'$ $Y_j = \left(y_{1j}, y_{2j}, ..., y_{sj} \right)'$ $v = (v_1, v_2, ..., v_m)'$ $u = (u_1, u_2, ..., u_s)'$ $h_j = u'Y_j/v'X_j$ and the weight ≥ 0. In this case, the efficiency evaluation indicators i the evaluation decision-making unit are as follows: to satisfy the h by v and u, the appropriate choice coefficient$_j \leq$ 1. Since the CCR model is a fractional linear programming problem, it can be transformed into an equivalent linear programming problem when it is solved, and the CCR model is used when the initial population is generated.

Third, use difference operator. The whole operation of differential evolution algorithm is relatively simple and stable, and it has high optimization efficiency. After applying differential operator to multi-objective evolutionary algorithm, it can produce new individuals in the algorithm. At the same time, it can also combine the optimization algorithm based on scalar method. For the DEA model to optimize the multi-objective evolutionary algorithm, the efficiency and performance of differential operators should be improved. From the current research situation, polynomial mutation operators can be used to disturb the solution, thus increasing the local search ability of the algorithm.

Fourth, the solution that does not fully conform to the release constraint is repaired. At the same time, it is very likely that the solution that does not fully conform to the release constraint appears, so it needs to be repaired. That is, directly modify the gene bits beyond the range in the test function. At the same time, in the portfolio problem, in order to make the newly generated individual satisfied and, the last coding position of the individual should be. i when $-1 \leq i_j \leq 1 \sum_{j=1}^{N} i_j = 1 \, j_N = 1 - \sum_{j=1}^{N-1} j_j$ NWhen > 1, this means that the sum of individual values is too small. For this reason, the minimum coding position needs to be regenerated with a range of [0, 1], and then verified again until the assumed conditions are satisfied; when the i is met$_N$ < -1, the maximum coding position needs to be regenerated in the range of $[-1]$.

Fifth, algorithm flow and framework. 1 Population initialization. The Euclidean distance between two weight vectors is calculated, and the nearest weight T each weight vector is found, that is, the X is $i = 1, ..., N$ $B(i) = \{i_1, ..., i_T\}$ i $X^{i1}, ..., X^{iT}$ T nearest weight vector is. Then, the initial population is generated in feasible space by DEA, and the reference point is initialized. Finally, the initialization EP is empty. 2 Population update. A new solution is obtained by using the difference operator to make and produce a new solution, and the polynomial is used to disturb it. Finally, it is produced, repaired and improved y, so that it is within the range of the solution. Then update the neighborhood solution and delete the dominant vector in the EP. If there is no dominant vector, add it to the EP. If the added result does not meet the stop condition, continue to update the population. If the stop condition is satisfied, stop and output the EP $B(i)klx_kx_lyy'zF(y')F(y')$.

To sum up, the improved algorithm mainly improves the multi-objective evolutionary algorithm (MOEA/D) by DEA the initial solution, and then provides a new way for the initial solution of the algorithm. That is, the optimization of the multi-objective evolutionary algorithm with the DEA model can effectively improve the quality of the initial solution, guide the later iteration of the algorithm, and strengthen the local search ability of the whole algorithm by using the difference operator as the crossover operator [2].

3 Analysis of the Application of DEA Based Multi-objective Evolutionary Algorithm in Portfolio Optimization

To judge the optimization effect of DEA based multi-objective evolutionary algorithm in portfolio, the MOEA/D is compared with the optimized algorithm. The application is as follows.

3.1 Application of Classical M-V Models

Select the yield data for 10 stocks from 2012 to 2016, Carry out simulation experiment using Matlab. Set the initial parameter values of the MOEA/D algorithm and the DEA-MOEA/D algorithm: the target population size is 50, Variance probability and crossover probability are 0.6 and 0.5, respectively, Through random experiments on these 10 stocks, The generation distance and diversity index of each algorithm are obtained. Specific data are shown in Table 1.

Table 1. MOEA/D comparison of portfolio optimization results under DEA-MOEA/D algorithm

Mean-variance problem	Generation distance		Diversity indicators	
	Mean	Variance	Mean	Variance
MORA/D	0.0143	0.0591	0.0011	0.0003
DEA-MOEA/D	0.0115	0.0478	0.0009	0.0002

The above table shows that the optimized DEA-MOEA/D results are better than the MOEA/D algorithm, whether the generation distance or diversity index.

3.2 M-V Models with Cardinality Constraints

Still choosing yield data for 10 stocks from 2012 to 2016, Carry out simulation experiment using Matlab. Set the initial parameter values of the MOEA/D algorithm and the DEA-MOEA/D algorithm: the target population size is 50, Variance probability and crossover probability are 0.6 and 0.5, respectively, The number of iterations is 500, The scaling ratio is 0.2. Under this model, Using MOEA/D algorithm and DEA-MOEA/D algorithm to select 3, 4, 5 cardinality, Random experiments were carried out respectively. Take one of these random experiments, The results are shown in Tables 2 and 1.

Table 2. Comparison of portfolio optimization results with cardinality constraints under MOEA/D and DEA-MOEA/D algorithms

		Generation distance		Diversity indicators	
		Mean	Variance	Mean	Variance
K = 3	MOEA/D	0.0038	0.00001	0.9482	0.0258
	DEA-MOEA/D	0.0031	0.00000	0.9170	0.0314
K = 4	MOEA/D	0.0029	0.00001	0.9322	0.0517
	DEA-MOEA/D	0.0024	0.00001	0.8865	0.0840
K = 5	MOEA/D	0.0029	0.00008	0.8997	0.0511
	DEA-MOEA/D	0.0022	0.00005	0.8931	0.0416

Table 2 and Fig. 1 show that although the two algorithms are not ideal in solving the problem of cardinality constrained portfolio, the DEA-MOEA/D algorithm is closer to the front surface than the MOEA/D algorithm, but there are some solutions that do not converge. This is mainly because the function model of portfolio with cardinality constraint is discontinuous, so the result is reasonable.

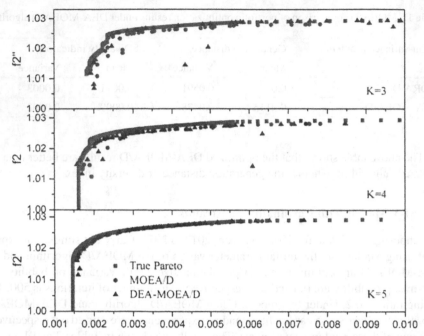

Fig. 1. Comparison of portfolio optimization results with cardinality constraints under MOEA/D and DEA-MOEA/D algorithms

Through the application of the multi-objective evolutionary algorithm based on the DEA model to the portfolio, it can be clearly seen that compared with other algorithms, the final solution of the DEA-MOEA/D algorithm is better, and the effectiveness and diversity of the front surface are enhanced. Although the application effect in portfolio with cardinality constraint is general, as shown in the figure, the performance of the algorithm will gradually increase with the increase of cardinality. All in all, by using the DEA model to optimize the multi-objective evolutionary algorithm and applying it to portfolio optimization, the optimal results can be obtained with the help of the enhanced convergence speed and the solution of diversity.

Conclusion: to sum up, DEA based multi-objective evolutionary algorithm has strong practical significance for portfolio optimization. Therefore, the DEA model should be used to optimize the multi-objective evolutionary algorithm to improve its convergence speed and increase the diversity of solutions, so as to obtain the optimal solution in the portfolio problem.

References

1. Xie, C., Long, G., Cheng, W., et al.: Advances in large-scale multi-objective evolutionary optimization algorithms. Guangxi Sci. **27**(06), 600–608 (2016)
2. Wang, Z., Li, H.: A multiobjective evolutionary algorithm using polynomial variation strategy and decomposition method. Microelectron. Comput. **38**(01), 95–100 (2021)

Research and Development of Green Pottery Under the Cooperation Platform of University and Enterprise Under Big Data Analysis

Jian Zheng[✉]

Panzhihua College, Sichuan 617000, China

Abstract. In order to further understand the local national culture, this paper takes the school enterprise cooperation as the platform, and uses genetic algorithm to study the R & D of Huili Qingtao. Through the simulation study, it is found that the pottery there is named "huilvtao" because of its special chemical composition and mineral composition, so it is used to make various pieces of green pottery. It is also because of its unique geographical location that Hanbao has created a unique ethnic culture in the border area of Hanbao, which is different from Bayu and Chu.

Keywords: Genetic algorithm · School enterprise cooperation · Huili green pottery

1 Introduction

Huili green pottery is a famous product in Sichuan. Because of the use of "malachite green stone" ingredients and named. The product is gem green, fresh and elegant. There is a clear sound of metal. Glaze does not contain lead, non-toxic, tasteless, high temperature, acid and alkali resistance, unique in Sichuan pottery. In the early 1980s, with the help of Sichuan Academy of fine arts, many kinds of glaze were developed, such as raindrop glaze, iron red glaze, pink blue glaze and so on. The number of products has increased from more than 10 to more than 200, including dragon and phoenix wine sets with rich national style, small and exquisite deformed animals, vases, flower cuttings, wall hanging, table lamps, tea sets, pen washing, etc.

The main raw material of Huili green pottery is white clay, which is made by special technology. The glaze color is made from natural malachite, then added with rice bran mortar and green slurry. The green glaze prepared by experienced glazes is green, crystal clear and extremely bright after firing. In a variety of temperatures, can appear dark green, green and also known as "green vegetables" and other different colors. Because of its green color and unique characteristics, it has been called "green pottery" since ancient times. [2] The Trademark Office of the State Administration for Industry and Commerce announced that "Huili LvTao" was approved for the registration of China's geographical indication certification trademark. This indicates that "Huili green pottery" has a sign,

M. A. Jan and F. Khan (Eds.): BigIoT-EDU 2021, LNICST 392, pp. 459–465, 2021.
https://doi.org/10.1007/978-3-030-87903-7_56

and its unique "ID card" with specific humanistic skills and natural resource value has far-reaching significance for promoting the development of Huili green pottery.

With the continuous enrollment expansion of colleges and universities in our country, graduates are facing many difficulties in employment. In order to solve this problem, in addition to policy guidance and creating more employment opportunities, it is of great significance to actively guide and support college students with innovation and entrepreneurship to carry out individual or team entrepreneurship, it is beneficial to create employment opportunities, improve the employment rate and improve the comprehensive quality of talents. The school enterprise cooperation mode can enrich the social experience of college students and improve their practical skills, so as to further stimulate the innovation and entrepreneurship passion of college graduates, Therefore, we must carry out systematic analysis, and then design and develop a college students' innovation and entrepreneurship experience platform.

2   Research on Innovation and Entrepreneurship Mechanism of College Students Based on Multidimensional Dynamic Innovation Model

The innovation and entrepreneurship mechanism of college students involves the government level, the social level, the university level and the enterprise level, which is a system engineering with considerable complexity. The multi-dimensional dynamic innovation model (mdmi) with the dynamic adjustment characteristics of multi factor integration can be used for systematic analysis of the innovation and entrepreneurship mechanism of college students [1]. The multi-dimensional dynamic innovation model can be divided into three sub levels (see Fig. 1): the first level is the entity level, which includes government entity, University entity, enterprise entity and social entity, and is in

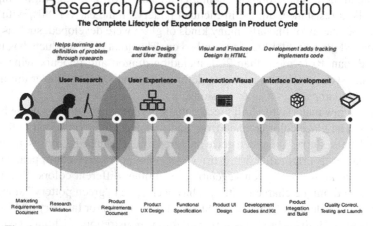

Fig. 1. Research framework of innovation and entrepreneurship mechanism

the dominant position in the whole model; the second level is the influencing factor level, which includes the elements that have great influence on the innovation and entrepreneurship mechanism of college students, and is uncertain, The third layer is the driving layer, including the external factors that drive the innovation and entrepreneurship mechanism of college students, and the driving layer also has uncertainty.

3 Research on the Importance of Influencing Factors of College Students' Innovation and Entrepreneurship Based on Multivariate Support Vector Machine

The multidimensional dynamic innovation model established above analyzes the innovation and entrepreneurship mechanism of College Students under the influence of multiple factors from a qualitative point of view, An importance model of College Students' innovation and entrepreneurship influencing factors based on multivariate support vector machine is established. The original SVM model is extended to binary classifier to solve the multivariate classification problem, and the multivariate classification problem is decomposed into multiple binary self classification problems. In order to solve the binary sub classification problem between class J and class k and maximize the boundary between the data, the soft marginal objective function is as follows:

$$\min_{w_j, w_k \in R^d} \frac{1}{2} \| w_j - w_k \|_2^2 + C \sum_{y_i \in \{j,k\}} \xi_i^{jk} \tag{1}$$

$$\frac{1}{2} \sum_{j=1}^{c} \| w_j \|_2^2 + \frac{1}{2} \sum_{j=1}^{c} b_j^2 \tag{2}$$

$$\tilde{y}_i = \arg \max_j w_j^T x_i + b_i \tag{3}$$

4 Platform Design and Implementation

4.1 Platform Requirement Analysis

The principle of requirement analysis needs to meet the following aspects: S1: convenient for users to set the original data and keep it unchanged for a period of time to ensure the stability of the system; S2: ensure that the role is reasonable and unique; S3: the system should have certain pressure resistance and robustness to deal with attacks or data pressure; S2: ensure that the role is reasonable and unique; S4: regular maintenance and update. As shown in Fig. 2. The main function modules of Pinghe include project declaration module, expert audit module, fund management module, project progress tracking module, etc. the project declaration module is the basis of the whole system, responsible for project declaration, budget analysis, etc.; the expert audit module is the core module of Pinghe, responsible for project audit, rationality detection, etc.; The fund management module is mainly used for information interaction between project sponsors and investment users. The project progress tracking module is mainly used to track and promote the implementation status of the project, and is mainly responsible for the final implementation report of the project.

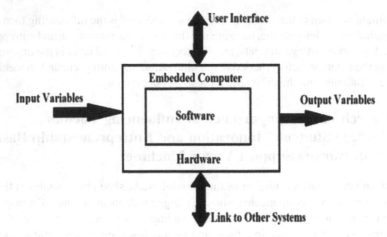

Fig. 2. Platform requirement analysis

4.2 Overall Design of the Platform

The function module of student innovation and entrepreneurship experience. The database adopts the database LDK mode, and the key point is to design the data layer 3.3 platform to realize the model reporting, expert review, fund management module in the project. Due to the limited space, only part of the system implementation is displayed. The main frequency of the system is 30 gh. The running memory of the system is 16 GB, and the storage space is 8. It is the system platform for checking the external books. The CPU of the hardware device is Intel Core 5 fineness section audit result B, and the network bandwidth is 20 m [2]. The system data storage software is MSSQL server2015 the servers are connected by LAN, one of which is the master node and the other two are slave nodes. On each Hadoop node (i.e. the server), the first step is to create the user and set the corresponding password; the next step is to modify the/etc./hosts file; the next step is to install and configure SSH, In order to realize the master node's control over the slave node, the last step is to install the Java running environment and Hadoop. After the installation of Hadoop, you need to configure Hadoop. First, you need to configure the environment variables and the Java path. Next, you need to configure the configuration file core- site.xml , hdfs- ite.xml and so on.

4.3 Implementation of Project Declaration Module

"The project declaration module is one of the important modules based on the school enterprise cooperation innovation and entrepreneurship platform, which is mainly composed of declaration, modification and other functions. The declaration page includes three parts: the left submenu, the top menu bar and the left submenu of the main page display the corresponding functions, and the page of each module can be opened by clicking the expand button; The top menu bar displays the home page, user management, login / exit buttons; the home page provides information input channels for project applicants. 3.3.3 system login interface users need to log in to the college students' innovation and entrepreneurship experience platform to carry out the corresponding operation. Users need to enter the user name and password on the login page, The background verifies the validity of the user name and password [3]. After successful verification, it will go to the main page. "Profit seeking" or the pursuit of profit is the natural attribute of enterprises, which is also the internal driving force for enterprises to participate in school enterprise cooperation and form a long-term cooperation mechanism. The author thinks that the opportunity cost of enterprises participating in school enterprise cooperation should be compensated clearly in the revision of vocational education law. The "opportunity cost" here includes two parts. The first is the total amount of enterprise investment that can be estimated by money, such as capital, equipment, technical personnel and posts. The second is the income that can be obtained if these inputs are not used for school enterprise cooperation, but for normal product production. The total amount of compensation should be 120% of the opportunity cost of the enterprise (assuming that the average social profit of the enterprise or industry after calculation).

5 Analysis on Actor Network of School Enterprise Cooperation in Higher Vocational Colleges

After determining the main actors and their relationships in the actor network of school enterprise cooperation, we analyze the translation process according to the problem analysis framework proposed by carlon, which includes four key links: problem solving, benefit giving, recruitment and mobilization, Describe how different heterogeneous actors are "translated" and maintained in the actor network of school enterprise cooperation under the conditions and roles set by the core actors.

5.1 Problematization

The premise of who is the core actor is to make clear who is the core actor. As mentioned above, the tasks of the core actors are: (1) to determine the common goals; (2) to define the categories and interest demands of the actors that may be included in the network; (3) to resolve the contradictions and conflicts of heterogeneous actors; (4) to propose OPP solutions that can be recognized and accepted by different actors. Core actors should have higher authority, coordination and management ability than other actors. From the current domestic school enterprise cooperation, higher vocational colleges often take the initiative to find enterprises to carry out school enterprise cooperation by science and Technology Department, science and technology industry department or school enterprise cooperation office. As a result, many colleges and researchers often refer to the phenomenon of "uneven hot and cold", "one hot" and so on. The root of this phenomenon lies in the fact that higher vocational colleges do not have the authority, coordination and management ability of core actors, which directly leads to the failure to propose OPP solutions that can be recognized and accepted by different actors, thus forming an interdependent network alliance. In this case, the Ministry of education and local education departments and bureaus (some scholars also proposed that the Ministry of education, the Ministry of human resources and social security, the Ministry of Finance and relevant ministries and commissions jointly set up the top management organization "school enterprise cooperation committee") should become the school enterprise Cooperation Bank.

5.2 Benefit Endowing

The role of laws and regulations in giving benefits is actually to establish an interest coordination mechanism among different actors, and it is also a means for core actors to ensure that other actors play their roles without "betrayal" behavior, so as to maintain the healthy operation of the whole network. The common practice of European and American developed countries to promote the development of local vocational education is to formulate laws and regulations. Germany promulgated the Vocational Education Law (Basic Law) in 1969 and the Vocational Education Promotion Law (supporting law) in 1981. In 2005, a new vocational education law was promulgated, including "General Provisions", "Vocational Education", "organization of vocational education", "vocational education research, planning and statistics", "Federal Institute of vocational education", "punishment rules", "transitional provisions and bridging provisions", with a total of 103 articles. This is the right way to cultivate a large number of skilled people.

The vocational education of "school enterprise cooperation, work study combination" is the legal standard. In 1996, China promulgated the "Vocational Education Law" and in 2005, the State Council promulgated the "decision on vigorously developing vocational education". At the same time, local governments at all levels also formulated and issued many relevant documents, especially in the "Ministry of education, Ministry of Finance on the implementation of the national Model Higher Vocational Colleges Construction Plan" in 2005, After the promulgation of the "opinions on speeding up the reform and development of Higher Vocational Education", it has focused on supporting 100 higher vocational colleges to carry out key construction, including exploring the school enterprise cooperation system. At the same time, it has also introduced specific policies, including tax incentives, financial support, employment, technology research and development, donations, incentives, etc., to encourage school enterprise cooperation.

6 Attach Importance to the Role of Vocational Education Group

From the practice of Vocational Education in China in recent years, vocational education group is conducive to the realization of scale, integration and intensification of vocational education. Vocational education groups have recruited many vocational colleges, secondary vocational schools, enterprises, industry associations, intermediary agencies and relevant government departments, at least in form, initially realizing the complementary resource advantages of heterogeneous actors. What should be done next is to translate the roles, status and interests of different actors in the framework of the new vocational education law, and put forward action plans and development strategies acceptable to all parties. The actor network of school enterprise cooperation needs a vocational education group including all kinds of stakeholders to support, and the selection and recruitment of allies is an important link in the construction of actor network of school enterprise cooperation [4]. Only a sufficient number of allies or member units endowed with appropriate interests can give full play to the attraction and cohesion of actor network. Usually, the network alliance of school enterprise cooperation actors in Vocational Education Group has the characteristics of circle structure, which is divided into core actors, main actors and outer actors. This requires that in the process of constructing or optimizing the existing vocational education group, the core actors should have good identification ability to other actors, and recruit the appropriate actors through interest endowing, carry out accurate role positioning, and promote the stable operation of the whole school enterprise cooperation actor network.

7 Epilogue

Based on the school enterprise cooperation mode, this paper uses multidimensional dynamic innovation model (MDM) to analyze the problems and development trends in the process of College Students' innovation and entrepreneurship activities from multiple perspectives, and constructs a platform for college students' innovation and entrepreneurship experience, which has good performance. The importance of influencing factors of College Students' innovation and entrepreneurship is analyzed by using multiple support vector machine algorithm, the platform is stable, practical and functional. It can better meet the requirements of university innovation and entrepreneurship experience platform.

References

1. On the optimization of College Students' innovation and entrepreneurship education system under Zhang Hua's school enterprise cooperation mode. J. Xuchang Univ. 34(4), 144–146 (2015)
2. Yi, L., Ming, C.: Thinking and exploration of college students' innovation and entrepreneurship practice based on school enterprise cooperation mode. Sci. Educ. Guide 15, 178–179 (2016)
3. Hengliang, W.: research on the construction of school enterprise cooperation mode for college students' innovation and entrepreneurship in Tibet University. J. Tibet Univ. 31(4), 134–139 (2016)
4. Yuanzheng, W., Jie, N., Yuting, D.: Innovation and entrepreneurship promotion strategy of college students based on multidimensional dynamic innovation model. Lab. Res. Explor. 35(2), 205–210 (2016)

Applications of Information Technology

Research on Electronic Finance Comprehensive Experimental System Platform Based on Information Technology

Haiping Yu$^{(\boxtimes)}$

Anhui Institute of International Business, Hefei 231100, Anhui, China

Abstract. In this paper, for the information technology e-finance comprehensive experimental system platform research, for the practical courses of information technology finance, students have no way, and almost impossible to carry out classroom learning and practice in the real social business system. At the same time, due to the relevance of the financial system and society, students can not complete in a concentrated period of time, in fact, it needs a lot of time it takes a long time to experience the operation practice, and these actual business transactions themselves are completed through the network. Therefore, for online banking and other financial courses, it is more suitable for the practice teaching through the virtual experimental system.

Keywords: Spring framework · Experimental system platform · Information technology · e-finance

1 Introduction

With the rapid development of computer technology, the Internet has become the information network with the largest coverage, the largest scale and the richest information resources in the world, which has greatly promoted the development of world scientific research and demonstrated the great role of information network in promoting the development of teaching and scientific research. In the modern information society, scientific research projects are becoming more and more complex and the scale is expanding, and many projects need large-scale interdisciplinary cooperation to be effectively solved; modern scientific research activities cost a lot, and the experimental equipment is large-scale and expensive, so it is urgent to share the experimental equipment to reduce the research cost; The globalization of information network and the new achievements of communication and computer technology have greatly enhanced the ability of interaction, cooperation and resource sharing of scientific researchers, provided more effective means for scientific researchers in different fields to carry out cooperative research, and better faced the challenges brought by the surge of information. In this context, the concept of "virtual experiment" came into being, showing the development trend of teaching and research environment and methods in information society [1].

M. A. Jan and F. Khan (Eds.): BigIoT-EDU 2021, LNICST 392, pp. 469–474, 2021.
https://doi.org/10.1007/978-3-030-87903-7_57

Electronic finance and traditional finance have different forms. It is a financial activity existing in the electronic space. It has the characteristics of virtual form and network operation mode. It is the product of the combination of information technology and modern finance. "From the perspective of financial service audience, e-finance provides financial related services. In order to improve the quality of personnel training, many colleges and universities in China have built financial or e-finance laboratories, which can be divided into two types according to the laboratory functions, namely, professional or on-the-job training oriented laboratories and financial analysis oriented laboratories. The financial teaching simulation software, which is mainly based on training, carries out transactional and operational training for students. In this way, the students can only become ordinary operators, and can not undertake the background research work with high skills and high quality requirements; P financial analysis laboratory is located in financial data analysis and decision support, It is mainly used in the courses and research of investment analysis, risk prediction and income prediction, and basically ignores the cultivation of financial practice skills, which will also bring some problems. Students can not understand the business process of financial practice and the related content of financial products. It can be seen that the current financial laboratory in Colleges and universities has a single function, which can not fully meet the needs of personnel training.

2 Design of Electronic Finance Comprehensive Experiment System Platform

2.1 Struts Framework Structure

Experimental teaching is based on the whole talent training system, with the cultivation of practical ability and innovation ability as the core, deepening the reform of experimental teaching content system, introducing teaching and scientific research achievements at home and abroad, constructing an experimental teaching system suitable for the cultivation of students' exploration spirit, scientific thinking, practical ability and innovation ability, and providing high-quality services for the learning and scientific research activities of teachers and students [2].

The experiment is convenient and interactive. We should overcome the shortcomings of the traditional practice teaching, such as large investment, limited space and insufficient staffing. Students can carry out experimental activities through the computer terminal at any time. Effectively meet the needs of students to improve their professional practice ability.

The experimental content is comprehensive and extensible. Due to the wide range of teaching contents in the field of economy and finance, including banking, insurance, financial management, accounting, trade finance, logistics, finance, financial information, etc., the system can not only integrate the commonly used economic and financial models, According to the teaching needs, we can further develop new economic and financial models, integrate them into the system, expand the experimental teaching content, and make it a virtual experimental teaching system with wider coverage and higher technical content.

There should be a database for experiment and it is easy to access. The purpose of the economic and financial experiment is to process and integrate the data on the basis of a large number of business data, economic and financial data and other aspects of data, transform them into thematic information resources, and provide them to the majority of teachers and students for experimental analysis, as the basis for analysis, prediction and decision-making of economic and financial activities. Therefore, there must be a rich database of data resources for the experimenters Call.

2.2 Spring Technology

Spring framework is a lightweight framework based on J2EE, which includes AOP, IOC/Di, MVC and other applications, and can be flexibly chosen according to the actual situation of the project. The framework is a layered application development framework, unlike struts, hibernate and other open source frameworks dedicated to a certain layer. Spring is committed to the whole application level architecture program, so that all levels of the program can be coordinated, so as to maximize efficiency. In addition, spring can integrate other excellent open source frameworks to form a coherent and unified architecture.

Spring provides a lightweight solution for application development. It includes the core mechanism of dependency injection, the integration of AOP's declarative transaction management and various persistence layer technologies, and the excellent webmvc framework. Spring provides an excellent solution for the presentation layer [3], business logic layer and data persistence layer of J2EE application. In addition, POJO management is also supported, and the objects of each layer of J2EE application are "welded" together. The framework consists of 7 modules, as shown in Fig. 1.

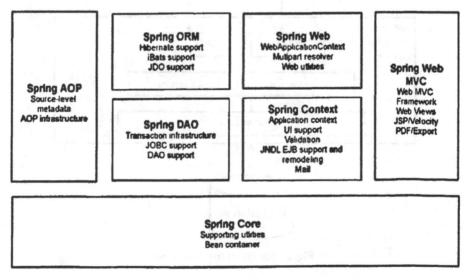

Fig. 1. Struts frame structure

3   Data Structure of Electronic Finance Experiment System

The system adopts a typical J2EE three-tier structure (Fig. 2), namely presentation layer, persistence layer and business logic layer. The three-tier system puts business rules, data access and validity verification in the persistence layer. The client does not directly interact with the database, but establishes a connection with the persistence layer through components, and then the middle layer interacts with the database.

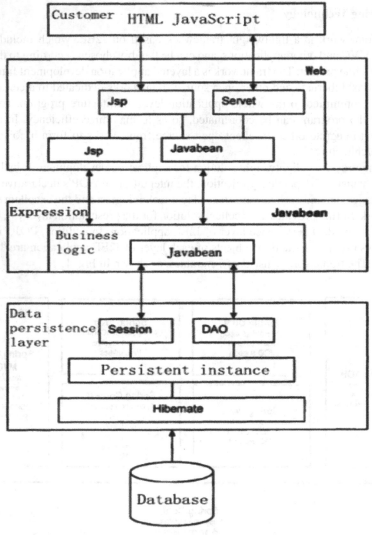

Fig. 2. Structural frame drawing

The reflection function of Java allows programs to dynamically call methods in instances by passing information such as class object instances, method names and parameter sequences. The following is the system to achieve dynamic call program structure [4].

```
public Object invokeMethod(Object owner, String methodName, Object[] args)throws Exception{
    Class ownerClass = owner.getClass();
    Class[]argsClass = new Class[args.length];
    for(int i = 0, j-args.length: isj; i++)
    {argsClass[i] = args[i].getClass();
    }
    Method method = ownerClass.getMethod(methodName,argsClass);
    return method.invoke(owner,args);
}
```

4   Network Structure Design

The design of the experimental platform is divided into two parts, which are network structure and soft resource configuration. The school laboratory can not be allowed to use the financial private network, so through the use of LAN technology, routing technology, three-layer switching technology, wireless local area network technology to build a simulated financial network, and according to the needs of teaching and scientific research, the network interconnection is designed as a dual connection mode, supporting users to access the simulated financial network and the Internet at the same time.

(1) To prevent information leakage and ensure the security of data transmission between the bank and its customers through the public network; (2) to prevent illegal access and restrict the access of illegal users to the internal network of the bank; (3) to protect the integrity of information and to protect the integrity of data by using technical means such as digital certificate; (4) To prevent fake users and provide authentication and authorization methods; (5) to ensure the availability of the system, the client can reduce the possibility of attacking the intranet system by authorizing the use of open services; (6) customer experience of financial services and products to meet the access needs of various clients to financial services, reflecting the characteristics of Internet-based and mobile financial services.

At present, the development of financial services and products and the application of management information lag behind the speed of information infrastructure construction and business development. Financial information technology software investment is obviously insufficient compared with hardware construction investment. F in the future, the demand for intelligent management talents of financial system development and system operation will increase greatly. The soft resource allocation of the platform should highlight the dual functions of talent training and social service, and support experimental teaching and financial research. Therefore, the soft resource allocation of the platform should be considered from two aspects: one is software resources, simulation financial simulation system, data management and analysis software, system development software and system operation intelligent management software; the other is data resources, financial database and data warehouse. The choice of the two kinds of resources should be advanced and redeveloping.

5 Conclusion

At present, the experimental teaching is basically carried out in the laboratory, which has the disadvantages of large investment, large loss, low efficiency, long cycle, difficult maintenance, slow equipment update and so on. The e-finance experimental teaching integrated platform designed and constructed by this research institute is for the development of e-finance system. With the development and popularization of network technology, virtual experiment emerges as the times require. Compared with traditional experimental teaching, virtual experiment has many advantages, such as easy to use, low cost, easy to realize repeated experiments under the same conditions, good sharing under the network environment, wide range of benefits and so on.

Acknowledgements. Provincial Quality Engineering——Financial Management Specialty Group of Anhui institute of International Business, 2020zyq20.

References

1. Guoping, H., Xiaoying, X.: Problems and countermeasures of financial interest rate construction in China's banking industry. J. Banker **12**, 114–117 (2011)
2. Qianli, M., Haotian, X.: Application and innovation of data warehouse in the information construction of commercial banks. Gansu Finance **11**, 27–29 (2012)
3. Jin, C., Jinhong, C.: Warm Discussion on Electronic Finance, 2nd edn. Zhejiang University Press, Jizhou (2010)
4. Jin, C., Jinhong, C.: Introduction to Electronic Finance, 2nd edn. Zhejiang University Press, Jizhou (2010)

Research on Information Media Technology Design of Information Collection and Release System

Rui Wang(✉)

Huali College, Guangdong University of Technology, Guangzhou 511325, China

Abstract. With the development of newspaper competition, the scale of newspaper business continues to expand, and the newspaper media industry is gradually moving towards the road of collectivization and informatization. The timeliness of news is the primary factor of news report. With the development of computer, network and digital media technology, news information collection and release system based on digital media will become an important part of modern newspaper collectivization. Therefore, the main research object of this paper is the new production flow model of newspaper news collection and press release under the new all media production management workflow system (platform) based on digital media, in order to solve how to better ensure the timeliness of newspaper news and enhance the competitiveness of the newspaper industry under the fierce competition.

Keywords: Digital media · All media · News timeliness · Newspaper news gathering · Newspaper news release

1 Introduction

Digitalization makes the whole traditional newspaper group face a new grim situation. In April 2006, Tsinghua University Social Science Literature Publishing House jointly published the 2006 media blue book, which pointed out that with 2005 as the "inflection point", Chinese traditional newspapers have stopped rising for many years, and advertising turnover has fallen sharply, with an average decline of more than 15%. At the same time, Blue Book believes that after more than ten years of exponential growth, all kinds of emerging media represented by the Internet have approached the "critical point". In the next two or three years, new digital media such as Internet media will also present explosive development. In this new form of media reform, the strong position of traditional newspaper groups has been fundamentally shaken [1]. Blue Book Analysis believes that the newspaper industry's economic downturn is due to the national macro-control on the surface, and is mainly impacted by the network and other emerging media on the deep. The deeper reason is that the transformation period of newspaper media form has come.

M. A. Jan and F. Khan (Eds.): BigIoT-EDU 2021, LNICST 392, pp. 475–479, 2021.
https://doi.org/10.1007/978-3-030-87903-7_58

2 All Media Concept and Related Technical Standards

2.1 Definition of All Media

According to Baidu Encyclopedia, "all media" refers to the media organizations and operators using text, graphics, images, animation, web pages, sound and video and other media means of expression (Multimedia), through radio, television, audio-visual, film, publishing, newspapers, magazines, websites and other media forms (business integration), Through the integration of radio and television network, telecommunication network and Internet Network (three network integration), we can provide users with TV, computer, mobile phone and other terminals (three screens in one), and receive any media content by anyone, at any time, at any place and in any way.

2.2 Information Collection Algorithm

In the process of clustering, it is easy to generate many local minimum values, which affect the generation of the best number of clusters, and it is easy to misjudge the experimental results. As a good index to judge the clustering effect, the contour coefficient not only reflects the cohesion between data, but also realizes the difference between data. Therefore, this paper will use the contour coefficient as the evaluation index of clustering effect.

$$Sil(X_i) = \frac{B(X_i) - a(x_i)}{\max\{a(x_i), b(x_i)\}} \tag{1}$$

$$a(x_i) = \frac{1}{|C_k|} \sum_{x_j \in C_j} Dist(x_i, x_j) \tag{2}$$

2.3 Characteristics of Omnimedia

"All media" does not only exclude a single form of traditional media, but also integrates various forms of media, so that we still attach great importance to the traditional value of a single form. Moreover, the single form of unified media is regarded as an important part of "all media". 3. "All media" does not reflect the simple connection between the media in the "cross media" era, but the whole network media, whose online media is fully mixed. "All media" has the most complete coverage, the most comprehensive technical means, the most complete media carrier, the most complete audience communication. In the field of media market, "all media" is large and comprehensive, but it shows super segmentation for the overall performance of the audience [2]. For example, through "all media", we can have many forms of expression for the same information, but at the same time, we can also balance and adjust the forms of media expression according to the individual needs and focus of individual audience groups. For example, when displaying the information of a real estate, we can use graphics and text to display the information of the objective description of the house type map and the building book; through the use of audio and video, we can display the dynamic information more intuitively.

3 Requirement Design and Analysis of Newspaper News Information Collection and Release System

3.1 Understanding of System Project Requirements

Modern newspaper industry has become the integration of paper media, online media and mobile media, and the integration of traditional media and new media. Facing the new situation, China's newspaper industry should fully integrate the new and old media, share resources, and realize "one generation, multiple release" of news information [3]. At present, the main problems faced by the newspaper industry are the aging of newspaper readers, the decline of social influence, the emergence of non-traditional competitors (Internet), the change of the positioning of important competitors, the emergence of new customers, the loss of market share in key market segments, the loss of customers, the pressure from the profit margin and so on.

3.2 System Construction Design Principle and Construction Content

The system design should fully inherit the software and hardware of the current information construction of Dazhong daily. At the same time, the system architecture design and the construction of each subsystem should meet the needs of building digital media system, data scale and load. Advanced technology is an important goal pursued by system construction. From content management platform to network information integration and search service, we must achieve a higher level of current media system construction, and meet the direction of IT technology development and progress, based on flexible user definable template publishing technology. According to the specific needs, application environment and future development plan of Dazhong newspaper group, the basic way to build the system is to use software products that meet the needs, are advanced in technology, mature and stable, and are flexible to expand. On the basis of product implementation and deployment, in the corresponding development environment, with the help of the development interface provided by software products, personalized customization and secondary development are carried out according to the specific needs of the project, so as to achieve the technical and functional objectives of the project. The system constructed with mature software modules has the stability and reliability of operation, which can effectively shorten the project implementation cycle, avoid technical risks, and ensure the realization of the technical and functional objectives of the project. As can be seen from the content, the system can realize six complete functions, and the functional modules of the system are designed according to these functions, as shown in Fig. 1 below.

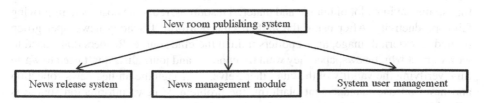

Fig. 1. Function module diagram of news release system

4 UML Basic Model of System

4.1 System Use Case Diagram

Before you create a use case diagram, you first need to identify the actors. In the news release system, it needs the participation of users and information publishers. Users can browse news topics such as information topics, news Hotspots or picture topics; all information content published on the website can be retrieved by keyword, title, full text, author, source, release time, release period and other ways; in addition, users can download pictures and other related materials from the website. As the leader of website updating, information publisher uses this system to edit, publish, modify and delete information. The system needs a special manager for daily maintenance and management, so it needs the participation of the system administrator.

4.2 System Management Time Sequence

The system administrator manages the system sequence diagram. The system administrator can add or delete the publisher through the interaction with the management window. The specific operation is completed by the interaction between the management window and the database. The results of the administrator's operation will be displayed on the page. It shows the time sequence diagram managed by the system administrator in the news release system. Sequence diagram involves five objects: Super administrator, operation interface, control class, account class and account data [4]. The super administrator of the system puts forward the request of adding an account in the operation interface and inputs the new account information at the same time. The control class is responsible for checking and obtaining the application information. After confirming the information, the account is established and the account information is set by the account class. Finally, the account information is added to the account data and the adding result is displayed in the operation interface.

5 Concluding Remarks

The system introduces the ability of all media material collection and editing, which provides the foundation for the group's all media reform. It makes it possible for data of various media to be collected, processed and released. So that the group can run a complete multimedia news process without the help of other organizations. Because all publishing media use the same collection platform, the collection of content can be shared with each other, and the staff will not be sent to the same news scene for many times, thus saving a lot of human and material resources, reducing costs, and improving labor productivity. After using the unified clue platform, Dazhong newspaper group unified its external image, and reporters unified the clue source. Readers don't need to worry about which newspaper they want to report to, and journalists don't need to waste time verifying the validity of the clues themselves. Through the unified news collection and news release management platform, the problem of news information source is effectively solved, and the timeliness is enhanced.

Acknowledgements. Research on the interaction between the "two sessions" of the party media news client of the Guangdong Provincial Education Department —— Taking the people's Daily mobile phone client as an example, the project number 2017 WTSCX133; Guangdong university key platform and scientific research project (innovation talent class) Guangdong government Weibo WeChat information release and public communication research, project number :2017 WQNCX177.

References

1. Wu, C., Zhang, L.: The way of newspaper process reform. China Reporter **4**, 72–75 (2007)
2. Shi, L.: business model construction of digital transformation of newspaper industry. J. Southwest Univ. National. (Human. Soc. Sci. Edn.) (6) (2010)
3. Jiang, J.: Creating an all media working platform to realize the transformation of media production mode. Ind. Eng. Manage. (2009)
4. Wu, F., Yao, X.: Digital newspaper industry: the development direction of traditional newspaper industry. Media Observ. (2) (2007)
5. Long, Y.: Design and implementation of Internet public opinion information collection system. University of Electronic Science and technology
6. Zhong, M.: Design and implementation of microblog information collection and analysis system. University of national defense science and technology
7. Yan, F.: News information collection and release system based on digital media. Shandong University
8. Song, X.: Design and implementation of information data acquisition and analysis center of civil aviation news of China. University of Chinese Academy of Sciences (School of engineering management and information technology) (2016)
9. Zhang, Z.: Design and implementation of a hotel digital information publishing system. Sci. Technol. Inf. **22**, 262–263 (2013)
10. Huang, Z.: Design and implementation of social intelligence information collection system of Zhongshan Public Security Bureau. University of Electronic Science and Technology (2016)

Research on the Application of Digital Technology in Civil Engineering Survey Management System

Xiaowen Hu$^{(\boxtimes)}$, Ronggui Liu, and Zhongjie Jia

Nantong Polytechic College, NanTong 226002, JiangSu, China

Abstract. With the continuous development of science and technology, computer digital measurement technology develops rapidly. In engineering measurement, digital measurement technology plays a very important role, It is an important guarantee for the smooth development of engineering survey. This paper analyzes the requirements of civil engineering survey management information system, aiming to lay the foundation for the computer digital development of civil engineering survey management information system, and build a multi-agent, multi module, multi-stage computer digital engineering survey management system. So that the managers of engineering construction project fully realize the important influence of engineering survey management on project quality, progress and cost, and fully realize the importance of computer digital civil engineering survey management information system. Only by paying attention to the development of engineering survey management information system, can we promote its continuous improvement and give full play to the maximum benefits.

Keywords: Computer digitization · Civil engineering · Engineering survey

1 Introduction

Measurement plays a very important role in civil engineering. The inspection before construction and the measurement in the construction process also need to be carried out when installing equipment. After the completion of the project and the later inspection and maintenance work need to be measured. The role of measurement is to predict and measure. Through the collection of data to test, we can ensure that we can master the operation process of the whole project after checking. In the whole life cycle of civil engineering, there will be a large amount of and diversified engineering information. Managers need to use modern computer information technology and effectively use and manage the engineering information to realize the effective management of the whole life of the project. The engineering survey management is an important daily project management work. The effect of engineering measurement management has an important influence on the construction quality, cost control, construction period organization and management efficiency of the project [1]. In the early stage of the project, the project construction program and after the completion and acceptance of

M. A. Jan and F. Khan (Eds.): BigIoT-EDU 2021, LNICST 392, pp. 480–486, 2021.
https://doi.org/10.1007/978-3-030-87903-7_59

the project, the engineering measurement is essential. It can be said that the engineering measurement is necessary in the construction measurement The project has the ability of all aspects and full cycle influence. The work content contained in engineering survey is closely related to other contents in project management, and is connected with each other and supported. It is a technical control of project duration, quality and cost, and an important guarantee for the smooth realization of all objectives of project construction.

At the same time, the continuous development of computer digital information technology has constantly promoted the change of people's life and working style, and its influence in the construction of engineering projects is also more and more extensive, which promotes the continuous improvement of the digital information level of engineering project management computer, and plays an important supporting role in project management, even in some enterprises and some projects The development of information technology in the process has played a transformative role in the management, which is more systematic, comprehensive and integrated, and more attention to management efficiency, which is of great significance to the measurement management of civil engineering [2].

2   Civil Engineering Survey Technology

Engineering survey is a very important link in the whole construction project. It is the first step for an engineering project to proceed smoothly. The improvement of Surveying and mapping technology ensures the quality of engineering survey. So in today's social development, we should continue to study the modern measurement technology in engineering measurement, and better apply it [3]. In the continuous development of science and technology, measurement technology is also improving, and more widely used, such technology is applied to engineering measurement, so that its quality is enhanced at the same time also has a guarantee, at the same time greatly improves the efficiency of engineering measurement, so that it can be carried out very smoothly, to provide data support for the future engineering construction, it can be seen that the new technology of Surveying and mapping is of great significance to engineering measurement Quantity management plays an important role in better development.

3   Computer Digitization and Civil Engineering

3.1   The Influence of Computers

One of the most outstanding and greatest scientific and technological inventions of human beings in the 20th century is computer. The birth of computer has opened a new page for the history of human science and technology, especially in its indispensable civil engineering, which has a great impact on the development of human society, marking the unprecedented information age for mankind. Digital information management is the most widely used field of computer at present. Because of the massive storage of computer, a large amount of data can be input into the computer for storage, processing, calculation, classification and sorting.

3.2 Digital Engineering Survey Technology

With the rapid development of science and technology, measurement technology and methods are constantly changing. From the beginning of manual recording, manual calculation, to manual recording calculator calculation, until today's electronic recording, electronic calculation, greatly improve the efficiency and accuracy of measurement data processing. The comparison between traditional measurement method and computer measurement technology is shown in Table 1.

Table 1. Comparison between traditional measurement method and computer measurement technology

Data processing method	Using tools	advantage	shortcoming	Conditions of use	Development prospects
Manual calculation	Manual calculation	Low cost	The calculation efficiency is low and the error probability is high	A small amount of calculation	Auxiliary calculation
The computer combines many kinds of software	All kinds of adjustment software	The calculation efficiency is high and the error probability is small	High cost	Large amount of data	It has wide prospect and wide application range

Modern measuring instruments are mainly developing in the direction of automation and digitization. The most representative instrument in engineering measurement is called total station, which is the product of the combination of electronic theodolite and range finder. Total station has many functions, including electronic angle measurement, mobile recording, storage and so on. It has very high efficiency in engineering. With the progress of science and technology, the total station is more and more developed, and gradually has a lot of new functions, such as automatic focusing can aim at the target, and can measure the operation inside the software, and so on [4].

At this stage, modern measuring instruments have gradually realized the informatization and digitization, which makes the detected information more intuitive and can better analyze the information. Because there is a large memory in the measuring instruments, the main purpose is to record the engineering measured data more timely and accurately, and to prevent the huge loss caused by the loss. Now there is no memory in the engineering measurement instruments, mainly using the computer to record the detected data, and can also be processed and transmitted in time. The most important thing of modern measuring instruments is to move forward in the direction of informatization. Because the realization of informatization brings great convenience for engineering measurement, it can realize the integration of information processing and transmission, which provides great guarantee for the smooth progress of engineering construction, and also improves the construction efficiency of engineering construction, and has certain guarantee for its quality.

The experimental results are analyzed.

In this paper, 100 000 English words are randomly selected from the thesaurus and divided into two groups. Within the specified time, the system in this paper is used to carry out the comprehensive query of the vocabulary respectively. The query results are shown in Table 1.

4   Application of Computer Digital Civil Engineering Survey

4.1   GPS Positioning and RTK Measurement Technology

The GPS positioning system takes the satellite system as the core, and the ground monitoring system and the terminal signal receiving system realize the positioning analysis of the object. The current positioning technology for static positioning and dynamic positioning is divided into the following two types: one is differential technology, which can improve the positioning accuracy by determining a reference point, setting up a base station, setting up a receiver, calculating the signal difference of the point and correcting the positioning results [5]. The second is 0 measurement technology, Base station, mobile station and radio communication equipment constitute T system. The base station is set up with the observation building position as the reference point. Observation points are set around the base station. Combined with the satellite signal received by the signal receiver and the time and position data, the time history curve is established, and the three-dimensional coordinates, velocity and other parameters are obtained to realize the dynamic monitoring of the observation point displacement.

4.2   Ground 3D Laser Scanning Technology

The three-dimensional laser scanning technology is suitable for the field measurement, and the three-dimensional laser scanning technology is relatively fast. The application principle of this technology is to use laser to carry out distance measurement, take 3D laser scanner as the main measurement tool, complete data acquisition and processing in working state, match the corresponding color gray with the reflected laser intensity, and obtain the three-dimensional coordinates of the measuring point in X, y and Z directions.

4.3   UAV Tilt Photogrammetry Technology

UAV barrel photogrammetry technology mainly uses UAV equipped with infrared camera and digital camera to collect ground information through low altitude photography to ensure the mapping accuracy. The realization of this technology includes the following five processes: one is camera calibration, which recovers the position relationship between the photo and the horizontal head by means of pre calibration. A calibration plate and a calibration field are set in the room to obtain the distortion coefficient according to the azimuth elements of the main distance change, so as to realize the camera calibration and ensure the acquisition of the real ground image; the other is image positioning, which is automatically completed by fploonn software, The third is to make the line plan, import the specific point coordinate value in the artificial fish swarm period method and 10 software, determine the flight height, speed, line, shooting angle, shooting frequency and other parameters, so as to plan the optimal line and generate the boundary value of the flight area with high accuracy. The fourth is the image acquisition, which can be used for image processing The fourth rotation UAV system is introduced to improve the negative judgment ability of the platform, support a variety of high-precision navigation equipment, optimize the ground holding rate and other data, and improve the quality of image acquisition. The fifth is the surface 3D reconstruction and fine product generation. After the confirmation of exterior orientation elements, the scene is reconstructed based on automatic national image matching technology, and the point cloud data is output and the point cloud density is adjusted, With the image matching algorithm, feature points are automatically selected to ensure the generation of accurate point cloud.

5 Computer Digital Engineering Management

The quality of civil engineering management and management measures, to a large extent, determines the level of project management efficiency, as well as the survival and sustainable development of construction enterprises. And the normal organization and development of personnel, equipment, materials and technology in the construction process must be realized through good project management.

Civil engineering projects generally have the characteristics of long periodicity, high mobility and many open-air operations. In addition, the construction process often needs multi process cross construction, and the comprehensive application of multi project construction, which determines the complexity and particularity of civil engineering project construction. Therefore, scientific project management measures must be formulated and implemented to ensure the smooth progress of engineering construction, so as to achieve the expected quality standards, functional requirements and cost control requirements of the project.

With the rapid development of China's civil engineering industry, all kinds of new construction technology, new technology, new materials and new equipment are constantly popularized and applied in civil engineering projects. In recent years, engineering projects generally present the trend of large-scale structure, complex function and novel decoration, which put forward higher and newer requirements for engineering management.

Therefore, digital management based on computer is of great significance to civil engineering. Computer management system is a human oriented system that uses computer hardware, software, network communication equipment and other office equipment to collect, transmit, process, store, update, expand and maintain information. The application of management information system is to realize the management, adjustment and control of various activities of the organization. A lot of valuable organization information, management information, economic information, technical information and regulatory information will help to choose a variety of possible schemes during the decision-making period of the project, which is conducive to the project objective control during the implementation period of the project, and also conducive to the operation after the completion of the project. The objectives and meanings achieved are shown in the Table 2 below.

Table 2. The purpose and significance of computer digital engineering management

Achieve the goal	Significance
Digitalization and centralization of information storage	It is conducive to the retrieval and inquiry of project information
Programming of information processing and transformation	It is helpful to improve the accuracy of data processing
Digitalization and electronization of information transmission	It can improve the fidelity and confidentiality of data transmission
Convenient access to information Improve information transparency Information flow flattening	It is conducive to information exchange and collaborative work among project participants

6   Conclusion

This paper focuses on the development and application of measurement management in civil engineering and the implementation and optimization of project management measures from three aspects of computer digitization, engineering measurement and engineering management. In order to improve the level of project management and realize the informatization of project management, it is necessary to use computer and network technology to realize project management. Through the collection, storage and processing of relevant data, the project management information system is established to serve as the basis for the project management planning, decision-making, control and inspection, so as to ensure the smooth implementation of the project.

Acknowledgement. Guiding projects of Nantong municipal science and technology plan (source).

Design and implementation of construction engineering survey management system (name), JCZ20100 (No.).

References

1. Zeng, K.: Analysis on the importance of construction engineering technology management. Guangdong Sci. Technol. (10) (2007)
2. Chen, Y.: Research on integrated management of super large engineering construction project based on modern information technology. Tianjin University, Tianjin (2004)

3. Lee, S.-K., Yu, J.-H.: Success model of project management information system in construction. Autom. Constr. **25**, 82–93 (2012)
4. Guo Fanglong's analysis on the importance of construction engineering technology management. Today Keyuan (16), 86 (2009)
5. Tu, J.: Discussion on construction engineering technology management. Manage. Technol. Small Med. Enterp. **04**, 104–105 (2013)

Research on the Application of Information Technology in Psychological Education of College Students

Qiudi Xing[1(✉)] and Xuefang Chen[2]

[1] Fuyang Preschool Teachers College, Fuyang 236015, Anhui, China
[2] Department of Psychology and Behavioral Sciences, Zhejiang University, Hangzhou 310028, Zhejiang, China

Abstract. The mental health level of college students not only directly affects their own growth, but also affects the stability of the campus, and then affects the social harmony and the improvement of the quality of the whole people. Therefore, the psychological problems of college students have aroused widespread concern in the society. The psychological intervention of college students has become a hot spot in the research of college students' mental health. With the development and maturity of data mining technology and the successful application in all walks of life, the advantages of this technology in discovering hidden rules or patterns in data are incomparable with other technologies. This paper collects and reviews a large number of relevant literature, and discusses the application of data mining technology and data analysis of college students' psychological problems.

Keywords: Data mining · Data preprocessing · Decision tree · Psychological problems of college students

1 Introduction

Data mining, also known as knowledge discovery, is to discover hidden mineral resources knowledge from assive data. It is a comprehensive application of statistics, artificial intelligence, database and other technologies. Using the tools and methods of data mining, valuable knowledge can be extracted from the rich data, otherwise the vast "data ocean" will become the "data grave" of lack of information [1].

Some research shows that most of the students' weariness, dropout, suicide and hurting others are caused by mental health problems, and the number of students with poor mental health has been on the rise. According to a survey of 126000 college students in China, 20.3% of them have psychological problems, mainly manifested as terror, anxiety, obsessive-compulsive disorder, depression and neurasthenia.

According to the survey, the current college students' psychological problems mainly include three aspects: psychological confusion, psychological obstacles and psychological diseases. Among them, the students with psychological confusion are more common.

M. A. Jan and F. Khan (Eds.): BigIoT-EDU 2021, LNICST 392, pp. 487–492, 2021.
https://doi.org/10.1007/978-3-030-87903-7_60

Although they are mild psychological problems, they do not affect their health. However, if the minor problems can not be adjusted and dredged in time, they will develop into mental disorders. If psychological barriers are not timely adjusted and treated, they will develop into mental diseases. Mental illness will seriously affect their physical and mental health and all-round development, and even lead to malignant events.

2   Data Mining Technology

2.1   Cluster Analysis

Clustering is to classify data objects into several classes or clusters according to the principle of "maximizing the similarity within a class and minimizing the similarity between classes". The similarity of objects in the same class is very high, but the differences of objects in different classes are very high. Clustering analysis is the process of classifying according to some similarity of data and analyzing the formed multiple classes. Clustering methods mainly include hierarchical method, partition method, grid based method, density based method, model-based method, etc.

2.2   Classification of Data Mining System

Data mining technology comes from many disciplines, which will have an impact on data mining, as shown in Fig. 1.

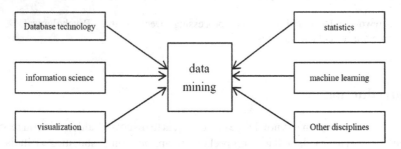

Fig. 1. Multiple disciplines influence data mining

Because data mining is an interdisciplinary subject, data mining will produce many different types of data mining systems [2]. Accurate classification of data mining system can provide scientific basis for users to choose the most suitable data mining system.

2.3   Decision Tree

In data mining, decision tree is mainly used for classification. Each node represents the distribution method of the top-level tree, and each node represents the distribution method of a class, and each node represents the distribution of a class. According to different characteristics, the decision tree uses tree structure to represent the classification, which is used as the basis for generating rules. The main advantages of decision tree are simple description, fast classification speed, easy to understand the generated model, and high precision. It is widely used in all kinds of data mining systems. Its main drawback is that it is difficult to construct a decision tree based on multiple variables (Fig. 2).

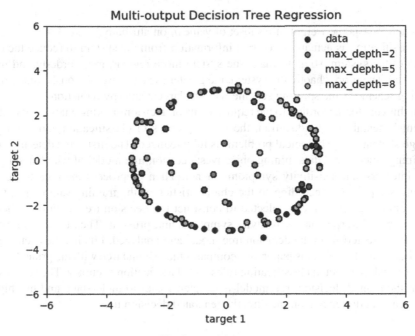

Fig. 2. Decision tree

3 Application of Data Mining in the Analysis of College Students' Psychological Problems

Attribute selection measures, also known as splitting rules, determine how to split samples on a given node. Here are two popular attribute selection metrics: information gain and gain rate [3].

1. Information gain.

Let node n store all samples of data partition D. The expected information required for the classification of samples in D is given by the following formula:

$$Info(D) = -\sum_{i=1}^{m} p_i \log_2(p_i) \tag{1}$$

Where p_i is the probability that any sample in D belongs to C_i.

The expected information required for sample classification of D based on attribute A can be obtained as follows:

$$Info_A(D) = \sum_{j=1}^{v} \frac{|D_j|}{|D|} \times Info(D_j) \tag{2}$$

Where $\frac{|D_j|}{|D|}$ is the weight of a subset of value a_j on attribute A.

Classification is actually to extract information from the system to reduce the confusion of the system, so as to make the system more regular, more orderly and more organized. The more chaotic the system, the greater the entropy. Obviously, the optimal splitting scheme is the splitting scheme with the largest entropy reduction.

In this chapter, according to the requirements of decision-making analysis of college students' mental health education, the whole process of classification and mining of college students' psychological problems is fully realized. The first is the determination of mining objects and data mining objectives: decision tree model of whether students have interpersonal sensitivity symptoms or not. Then preprocess the data to get the training sample set. According to the characteristics of the training sample set, C4.5 algorithm of decision tree is selected to construct the decision tree model of whether students have interpersonal sensitivity symptoms and prune it. Then the classification rules are extracted from the decision tree model and analyzed. Finally, the accuracy of the model is evaluated. This paper also compares the original tree with the pruned tree in terms of scale, extracted classification rules and classification accuracy. The conclusion is that the pruned decision tree model is simpler, easier to understand and has higher classification efficiency than the directly generated decision tree.

4 Design and Implementation of College Students' Psychological Data Management System

The collection and analysis of students' psychological evaluation data is a necessary basic work for colleges and universities to carry out mental health education. With the rapid increase of enrollment and the improvement of the connotation of psychological data analysis, more and more psychological data need to be analyzed and processed more deeply. Although some well-known and powerful psychological assessment software has appeared in China, these software are expensive and have not applied data mining technology. Therefore, it is necessary to develop a college students' psychological data management system based on data mining technology and BS mode, so as to improve the work efficiency of psychological evaluation data collection and increase the depth of psychological data analysis [4].

The student function module is oriented to students, which mainly realizes the collection of students' basic information and psychological evaluation information, and establishes psychological files for students. After students enter the system, the system generates a dynamic psychological file for them. Students can modify their personal password by modifying the password sub module; modify the personal basic information by modifying the basic information sub module; at the same time, the system can collect the students' basic information; through the psychological evaluation sub module, online psychological self-assessment can be realized, and the evaluation results can be viewed, and the system can collect the psychological evaluation data (Fig. 3).

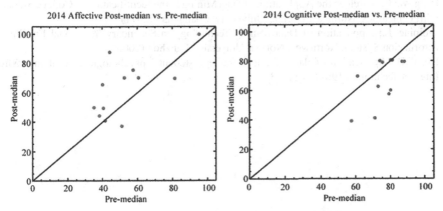

Fig. 3. Design and implementation of college students' psychological data management system

The administrator function module is used to retrieve the administrator, mainly to modify personal password, customize psychological questionnaire, information management, data mining, psychological prediction and other functions. The administrator realizes the management and import of the questionnaire through the self-defined psychological questionnaire sub module; realizes the self-determination of the students' basic attributes, the management of the students' basic information and the management of the psychological evaluation information through the information management sub module; realizes the data preparation, the generation of the decision tree and the generation of the classification rules through the data mining sub module; through the psychological prediction sub module, students' psychological problems can be predicted.

5 Conclusion

This paper analyzes the key technologies of data mining, deeply studies the classification in data mining, and analyzes and compares several commonly used classification algorithms, which provides the basis for the application of decision tree algorithm in the analysis of college students' psychological problems. The classification rules are extracted from the optimal decision tree model, which provides an important reference for the school psychological consultation work, and realizes the classification and prediction of new data by using the model, which provides a scientific basis for the early warning and intervention research of college students' psychological problems.

Acknowledgements. This is the achievement of Anhui Province top talent cultivation funding project excellent young backbone talents domestic visit project (NO. gxgnfx2019111).

References

1. Guojun, M., Lijuan, D., Shi, W.: Principle and Algorithm of Data Mining. Publishing House of Tsinghua University, Beijing (2005)
2. Peng, W.: Research on the Application of Data Mining in the Identification of College Students with Difficulties. Northeast Normal University, Changchun (2011)
3. Weiping, L.: Application of Data Mining Technology in Secondary Vocational Enrollment Information System. Northwest Normal University, Lanzhou (2008)
4. He, G.D.: Application of data mining in college students' psychological problems. Wirel. Internet Technol. **2**, 196–197 (2013)

A Lexical Analysis Algorithm for the Translation System of Germany and China Under Information Technology Education

Haopin Luo[✉]

School of International Education, Wuhan Business University, Wuhan 430056, China

Abstract. This paper proposes a rule-based lexical analysis algorithm for German Chinese machine translation. The algorithm can not only effectively restore the original morphemes of various deformed words, but also provide useful part of speech and various grammatical features for the subsequent parsing mechanism in the system, Through the part of speech information of specific morphological changes, we can only extract the part of speech information in the original dictionary definition of the deformed word and its corresponding dictionary entry definition, so as to facilitate the analysis and processing of the word.

Keyword: Lexical analysis of machine translation

1 Introduction

If only the original words are stored in the dictionary of the system, the lexical analysis algorithm not only needs to segment all kinds of sentence building units, but also needs to distinguish the root, affixes and their morphological features according to the rules of morphological changes. Through comparison in the program, the deformed words in the input string can be returned to the original words. This method can greatly reduce the number of dictionary entries and the storage space of the dictionary However, because the program is directly related to the affixes of specific natural languages, the algorithm is not easy to modify and maintain [1].

In this paper, a rule-based lexical analysis algorithm for German Chinese machine translation is proposed. By designing a lexical rule representation and corresponding rule processing mechanism, the dictionary only needs to include the definition of original words, and the morphemed words are reduced by using lexical rules. The algorithm realizes the independence of program and data, In order to provide deep grammatical information for the subsequent syntactic analysis mechanism, we add the corresponding deformed part of speech and morphological feature information to the lexical rules. Below, we will first summarize and summarize the rules of various morphological changes of German words, then give the computer internal representation of these rules, and finally give the specific lexical analysis algorithm [2].

© ICST Institute for Computer Sciences, Social Informatics and Telecommunications Engineering 2021
Published by Springer Nature Switzerland AG 2021. All Rights Reserved
M. A. Jan and F. Khan (Eds.): BigIoT-EDU 2021, LNICST 392, pp. 493–502, 2021.
https://doi.org/10.1007/978-3-030-87903-7_61

2 The Choice of Translation Methods

2.1 Literal Translation and Free Translation

In the 19th century, translators preferred free translation (meaning to meaning) rather than literal translation (word to word). Many translators and translation theorists supported the idea of free translation because if literal translation was too close to the original text, it would produce a strange translation, which would make the readers unable to understand the true meaning of the original text; while free translation could reproduce the content of the original text. Jerome once used a metaphor to describe the relationship between the source text and the Translation: the source text is like a prisoner, who is taken into the translation by the conqueror. Different text types should be classified and appropriate translation methods should be selected. Different translation methods can even be used in the same text to make the translation more "faithfulness, expressiveness and elegance" [3].

Free translation may be more fluent in sentences, but it may be different from the original in terms of content. For general literary works, these do not affect reading, but for scientific and technological literature, there is no tolerance for a single error, which is also the inevitable result of the accuracy, objectivity and preciseness of scientific and technological literature. Therefore, in general, literal translation (word to word) is the first choice for scientific and technological literature translation [4].

2.2 Semantic Translation and Communicative Translation

Peter Newmark, a British translation theorist, first introduced the concepts of semantic translation and communicative translation in his book "approaches to translation" published in 1981. Semantic translation means that the translator expresses the meaning of the original text as accurately as possible under the permission of the semantic and syntactic structure of the target text, so the target text is closer to the original text in form and style. Communicative translation, on the other hand, tries to make readers get the same effect as the original readers. It advocates adopting different translation methods according to different types of texts. On this basis, Newmark divides text types according to language functions. In his two books, approaches to translation and a textbook of translation [5], Newmark divides them into six types: expressive function, informative function, vocative function and aesthetic function (aesthetic function), phatic function and metalingual function [6].

Among them, information text emphasizes the information function of text, its content often involves a wide range of knowledge fields, and its form is generally very standard [7]. The core of informative text is authenticity, where language comes second. Scientific and technological literature text is basically information type text. After comparing semantic translation with communicative translation, Newmark believes that communicative translation is more suitable for information texts (including technical texts, publicity texts and various standard texts).

3   The Rules if German Word Form Change

3.1   Lexical Analysis

The change of case is mainly the change of case such as noun, determiner, adjective and pronoun; the upgrade is the change of comparative and superlative of adjective and adverb; the change of morpheme refers to the variety of verb; the change of fusion form two prepositions merge into one preposition [8].

1) The change of rules is generally reflected by the change of regular affixes. It can be divided into the following categories: (1) the change of suffixes forms words with different lexical features through the regular change of suffixes, For example, Leib → Leiber (2) inflection + suffix change forms words with different lexical features through regular suffix change and inflection Bache (3) infix + suffix changes form words with different morphological characteristics through regular infix and suffix changes, such as aufmachen − aufgemacht (4) prefix + suffix changes form words with different morphological characteristics through regular prefix and suffix changes. For example, machen → gemacht (5) preposition fusion forms a new word through the fusion of two prepositions, such as an DEM bank [9].

2) Irregular changes the irregular morphological changes of German words are not regular. Generally, they can not be summarized by the rules of morphological changes. They can only be listed one by one. For example, essen-a 3. Wenig ± minder. In addition, sometimes a German prototype needs to be deformed many times to produce a specific word. Generally, it needs to be changed several times. Typical examples are (1) adjective upgrading and then changing case, such as EB liebst -Liebsten (2) verbs change irregularly, ten people call the ending change. For example, sprechen → sprach → sprach (3) verb first participle + case change. For example, machen → machender (4) verb second participle + case change. The translation process of German Chinese translation system is shown in Fig. 1 below:

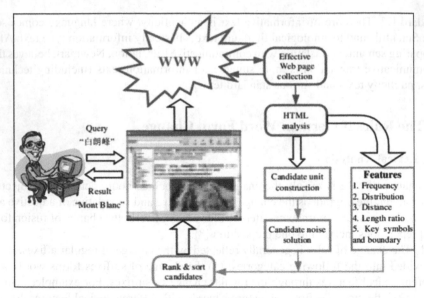

Fig. 1. Translation process of German Chinese translation system

3.2 Apriori Algorithm

Apriori algorithm accounts for a large proportion in association rules algorithm. Apriori algorithm is usually divided into two steps: the first step is to select the frequent itemsets that meet the user's requirements, and select the frequent itemsets that are larger than the user's agreed minimum support threshold in all databases. The second step is to summarize the expected association rules in the process of using frequent itemsets. The confidence level must be greater than or equal to the total critical value of the minimum confidence level agreed by the user, which is the most fundamental condition to find out the association rules [10].

$$\sup port(X \Rightarrow Y) = P(X \cup Y) \tag{1}$$

$$confidence(X \Rightarrow Y) = P(Y|X) \tag{2}$$

4 The Construction of a Small Parallel Corpus Between German and Chinese

Because of the difficulty in collecting and aligning bilingual corpus and the special characters different from English in German, the construction of German Chinese parallel corpus is much harder than that of English-Chinese parallel corpus. Therefore, it is necessary to solve the technical problem that the retrieval software can not recognize both German and Chinese bilingual. After trying to overcome the technical problems, the author has built a small German Chinese one-way parallel corpus with a scale of

600000 words. Parallel corpus consists of two parts: parallel corpus stored in computer in electronic text form and positioning retrieval software for managing and retrieving these corpora, Corpus construction is also carried out from these two aspects [11].

4.1 Selection and Collection of German and Chinese Bilingual Corpus

First, we should select the corpus according to the research purpose and research needs. Scherer pointed out that when creating a corpus, four principles should be considered, namely representativeness (Germany]repr $\Box$ sentimental $\Box$ T), persistence ([Germany] best $\Box$ ndigkeit), scale ([Germany]gr9 $\Box$ E) and content ([de] inhalt), among which representativeness is the highest goal of creating a corpus. Therefore, the constructors of corpus should take representativeness as the basis for corpus selection, and usually choose random sampling or stratified sampling. Of course, some scholars believe that all corpora can not be completely objective and not subject to the subjective influence of researchers. Therefore, the most common method of corpus is to choose corpus appropriately, which is also the development trend. Under the balance of the two, the author believes that as long as the selection of corpus can achieve the predetermined research purpose and meet the predetermined research needs, this is the proper process of corpus selection [12].

In the process of collecting data, the builders of corpus can use different ways, such as downloading German and Chinese reference materials through the Internet. In order to collect specific text and improve the quality of the corpus, scanners can also be used. Because most of the corpus needs to be saved as TXT format at last, the corpus builder should master different format conversion methods [13].

4.2 The Arrangement of German and Chinese Bilingual Corpus

Before using paraconc to search the German and Chinese bilingual corpus, we must deal with the corpus to meet the requirements of the software. The author briefly describes the process of corpus arrangement according to the experience of self-construction of German Chinese parallel corpus.

First, the preprocessing of corpus includes the unification of format and the removal of various impurities. Especially, the files downloaded from the Internet have different text formats such as font, paragraph arrangement and document format. Sometimes there are redundant spaces, broken lines, random codes, unnecessary or unrecognized graphics and symbols, which have no significance for the research. Therefore, it can be used in Microsoft Word and powergrep or "text finisher", Sometimes it is necessary to assist manual proofreading again [14].

Next, the segmentation of bilingual corpus, parallel alignment between Chinese and German corpus are followed. Corpus alignment is the key to deal with bilingual or multilingual parallel corpus by using paraconc software. It refers to the source language corpus and its target language corpus being stored in different texts respectively, and the corpus in two texts is aligned according to the relationship between paragraphs, paragraphs or sentences, sentences or words and words. At present, parallel alignment at the lexical level is almost difficult to achieve, and sentence level alignment needs to be combined with software application and manual intervention. The author has realized sentence level alignment in the corpus built by myself. Before parallel alignment of the corpus, the German and Chinese materials are divided into sentence units. The steps are as follows: create a new office word document, copy and paste the text file, select all the text, and then "in the Chinese corpus". Replace all with ". ^P^P", to avoid four newline characters between some sentences, you need to set the retrieval item to "^p^p^p^p", and replace it with "^p^p". Similarly, the German text is also processed accordingly, only the Chinese period is replaced by the period in the German text. However, considering that there are some abbreviations in the original German, such as bzw., etc., it is still necessary to manually check after segmentation. Because Chinese is a word-based writing unit, there is no obvious distinguishing mark between words, and paraconc can not recognize and calculate Chinese as English. Therefore, it is necessary to divide Chinese text before aligning the corpus. The author suggests that the Chinese word segmentation system ictcola of the Institute of Sciences be used to segment Chinese. Although paraconc software has align format drop-down menu, there are four alignment options: not aligned, new line delimiter, delimiter, starter/stop tags, but the author finds that manual adjustment is still needed after automatic alignment through paraconc, which is very difficult. Therefore, it is recommended to use Excel table to manually align text (sentence alignment), and then load it into paraconc software. The advantage of this operation is that the corpus can be automatically aligned after loading, and the actual operation is more convenient and practical than relying on software. The specific operation is to copy the corresponding German and Chinese bilingual corpus to an excel file as required, adjust it to align the sentence and sentence, and then copy and paste them into different word files after the alignment is completed. In this process, the builders of corpus need to pay special attention to the fact that there may be less translation and additional translation in the target language corpus [15].

5   Regular Expression of Lexical Rules

In order to realize the independence of lexical analysis algorithm and specific data, and at the same time, it is not necessary to provide dictionary definition for each morpheme, we propose a rule-based lexical analysis algorithm, in which rules are used to represent various morphological changes and their corresponding morphological features.

On the other hand, although the complex morphological changes of German words bring complexity to German lexical analysis, these rich morphological changes can also provide very useful deep grammatical information for German syntactic analysis, We should not only consider how to recognize each word string in the input sentence, but also consider how to transfer the feature attributes of the word string obtained in the process of recognition to the later syntactic analysis module, It can provide useful grammatical information for syntactic analysis [16–19]. Considering that the morphological change of the original word only corresponds to one part of speech, the corresponding deformed part of speech information is also given in the lexical rules. When looking up the dictionary, only the dictionary entry definition of the original word with the corresponding part of speech features is taken, so as to facilitate the processing of word classification [20].

6   Lexical Analysis Algorithm

Based on the above lexical rule representation, In order to save the result of lexical analysis, we design a rule-based lexical analysis algorithm. For each word, we use a lexical analysis information table to save the result of lexical analysis. Each element is represented by the structure of (prototype word string, part of speech/feature mark, morphology feature table) [21]. The prototype word string is analyzed by lexical analysis mechanism and lexical rule library The string representation of the original word after deformation processing: the part of speech/feature mark is used to represent the part of speech mark corresponding to the corresponding deformation of the analyzed word or the mark of the component of the special sentence (such as punctuation, number, etc.); the form feature table is used to store all kinds of form feature information corresponding to the deformation of the word, Then the original word string is set to empty [22].

6.1   Algorithm Flow

The flow chart of lexical analysis algorithm is shown in Fig. 1. This algorithm first divides the input of source text into sentences and their components, and then carries out different processing according to different types of sentence components. The specific algorithm is described as follows.

(1) The input sentences of the source text are segmented by components (such as words, punctuation, numbers, etc.)

(2) Take a component in a sentence and deal with it differently according to its type: if it is a special symbol or expression, turn it.

(3) If it's a word, turn to (4) (3) if it is a special symbol, the punctuation mark and the string representation of the special symbol are directly added to the lexical analysis information table; if it is a number, date and other special representation forms, such as 321.3112194, which are not easy to be defined directly in the dictionary, the number mark and its string representation are added to the lexical analysis information table [23].

4) Search the dictionary with candidate morpheme string and transfer.

The flow of lexical analysis algorithm is shown in Fig. 2 below.

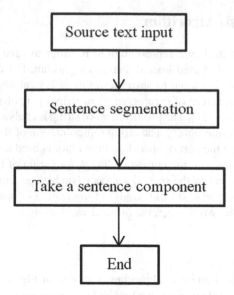

Fig. 2. Lexical analysis algorithm flow.

6.2 Lexical Analysis Information

When querying the dictionary according to these lexical analysis information, we will extract the corresponding part of speech information in the lexical analysis information table of each word, and only extract the entry definition corresponding to the part of speech information in the dictionary meaning and the part of speech information in the lexical analysis information table [24], In this way, the morphological features are only associated with the dictionary definition of the corresponding part of speech. Therefore, this lexical rule representation and the corresponding lexical analysis algorithm can comprehensively and accurately analyze the word class/features and morphological features of each sentence component, In the lexical analysis stage [25], it provides useful information for the following translation processing mechanism to deal with the multi category of words. From the above algorithm, we can see that all kinds of morphological deformation and feature information related to specific languages are expressed by the rules in the lexical rule base. The lexical analysis algorithm is only related to the expression form of lexical rules, but has nothing to do with the specific content of lexical rules, As long as we use the same rule form to represent the morpheme rules in different languages, this algorithm is applicable (see Fig. 3).

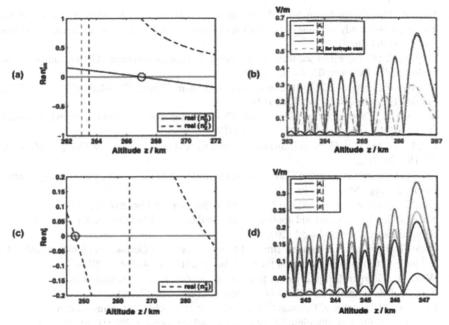

Fig. 3. Simulation for analysis information

7   Concluding Remarks

We present a rule-based lexical analysis algorithm for German Chinese machine translation. By designing a lexical rule representation and the corresponding rule processing mechanism, the algorithm solves the problems existing in the traditional descriptive lexical analysis algorithm and procedural lexical analysis algorithm, and also helps to deal with the dual category of words in machine translation processing, so as to achieve the high efficiency of the lexical analysis algorithm.

Acknowledgements. Metaphor in the Language of Science and Technology——A Comparison between Chinese and German.

References

1. Gao, T.: Design and application of web based collaborative translation system. Electron. Des. Eng. **28**(19), 85–88+92 (2020)
2. Hanji, L., Haiqing, C.: Philosophical reflection on the dilemma of machine translation technology. J. Dalian Univ. Technol. (Soc. Sci. Ed.) **41**(06), 122–128 (2020)
3. Bin Bin, G., Huang, Z.S.: Design and application of web based collaborative translation system in Colleges and universities. Comput. Program. Skills Maintenance **2020**(08), 23–25 (2020)
4. Qiang, H., Ruili, H.: Translation studies from the perspective of modern science and technology: an introduction to the future of translation technology: towards a world without Babel Tower. Orient. Translation **04**, 92–95 (2020)

5. Jing, Z.: Poetry translation from the perspective of Benjamin's translation Transcendence Theory: a case study of Pound's translation of Chinese poetry into English. J. Chizhou Univ. **34**(01), 91–94 (2020)
6. Tingting, L.: Comparison and translation of animal metaphors between German and Chinese. J. Jilin Radio TV Univ. **02**, 90–92 (2020)
7. Zhiyong, Z., Fenglan, G.: On the basic unit of German Chinese Translation: Yuyuan. Fudan Foreign Lang. Lit. **01**, 125–130 (2017)
8. Shangke, G.: Comparison of German and Chinese idioms in cross cultural translation mode [J]. Anhui Lit. (Second Half) **08**, 76–78 (2017)
9. Lei, L.: Research on German translation of science and technology. New Campus (Reading) **05**, 184 (2017)
10. Ge, N.: Research on German Chinese translation of patent documents supported by corpus. Beijing Foreign Studies University (2017)
11. Lan, G.: A study on the course of Dehan. Shanghai Normal University (2017)
12. Nannan, G.: Construction and application of small German Chinese parallel corpus. German Humanities Res. **4**(02), 44–52 (2016)
13. Zhou, T.: Translation of cultural image differences between German and Chinese under the guidance of functional translation theory. Chin. J. **2016**(11), 4–5+61 (2016)
14. Wang, Y.: German Chinese patent translation practice report - Taking word and sentence translation as an example. Xi'an Foreign Studies University (2016)
15. Li Xiang, Li Dongliang. German Chinese science and technology translation and auxiliary translation in the new situation. China Sci. Technol. Translation **29**(02), 30–31+29 (2016)
16. Li, D., Li, X., Wu, M.: Research on the course of German Chinese general practical translation. Teach. Chin. Univ. **11**, 66–69 (2015)
17. Liyan, H.: Application of functional translation theory in German Chinese translation. Test Weekly **93**, 11–12 (2014)
18. Yin, W.: Corpus assisted contrastive study on the translational accents of Jude's nameless Chinese versions. Dalian Maritime University (2014)
19. Bo, L.: The application of structural adaptation in German Chinese translation. J. Jilin Radio TV Univ. **2013**(05), 80–82+111 (2013)
20. Jing, J.: Grasp the cultural factors of professional text categories and improve the translation skills of German and Chinese science and technology – Taking the instructions of German and Chinese household appliances as an example. Innov. Appl. Sci. Technol. **08**, 281–282 (2013)
21. Gao, F., Huang, X.: Translation of synonyms in German scientific articles: a case study of standards of German automobile industry association. J. Tonghua Normal Univ. **34**(01), 83–87 (2013)
22. Xiaoqing, W.: On the curriculum design of German Chinese translation. Xueyuan (Educ. Sci. Res.) **19**, 69–70 (2012)
23. Juan, S., Chao, Y.: Discussion on the interpretation of German neologisms: a case study of German Chinese neologisms DICTIONARY. J. Neijiang Normal Univ. **27**(05), 60–62 (2012)
24. Xingzhou, H.: Reflections on the current German translation textbooks in colleges and universities in China – comments on Professor Wang Jingping's new German Chinese translation course. Orient. Translation **02**, 22–27 (2012)
25. Hui, C.: An analysis of German English appellation of some linguistic terms. Chin. Terminology Sci. Technol. **14**(01), 7–13 (2012)

Analysis and Research of Data Encryption Technology in Network Communication Security

Weiqiang Qi$^{(\boxtimes)}$, Peng Zhou, and Wei Ye

State Grid Zhejiang Electric Power Corporation Information and Telecommunication Branch, Hangzhou 310016, China

Abstract. With the rapid development of information technology, especially with the wide application of the Internet, the communication security in the network has become an important problem we have to face. In order to solve this problem, this paper takes the network report system of a mining enterprise as the research object, comprehensively expounds the design of network reporting system and data encryption in network communication from the aspects of demand investigation, demand analysis, system design and software system development, and develops and realizes it by using BS architecture + independent client. This paper first briefly introduces the background and significance of computer network communication security research, explains the importance of data encryption in network communication, and the development process and status quo at home and abroad.

Keywords: Network report system · File encryption · RSA algorithm · Design and development · System testing

1 Introduction

The development of information technology in today's world is changing with each passing day, especially after the emergence and popularization of the Internet. Nowadays, almost everyone is inseparable from the Internet, and they are always in contact with the Internet [1]. The number of Internet users in China is also increasing. Network anti-corruption, that is to expose the corruption behavior in life through network public opinion and network report, provides a new mode for government and enterprise anti-corruption. Under the background of the rapid development of Internet technology and anti-corruption, many governments and enterprise affairs groups give the greatest support and recognition to the report and the emerging network reporting system. In recent years, some staff of Beijing Donghua Yishi Technology Co., Ltd. have seen the great application prospect of this reporting system, and have designed and developed an online reporting system for some governments and enterprises. Here, I choose the construction project of the network reporting system of the grass-roots employees of a mining enterprise in Tengzhou that I participated in. In order to let the scientific, legal and democratic governance accept the comprehensive supervision of grass-roots employees, Tengzhou beixulou coal mine insists on keeping pace with the times and introduces the network security reporting system. The effective use of this system can play a great role in anti-corruption and daily production.

© ICST Institute for Computer Sciences, Social Informatics and Telecommunications Engineering 2021
Published by Springer Nature Switzerland AG 2021. All Rights Reserved
M. A. Jan and F. Khan (Eds.): BigIoT-EDU 2021, LNICST 392, pp. 503–511, 2021.
https://doi.org/10.1007/978-3-030-87903-7_62

2　Analysis of Encryption Technology

2.1　The Importance of Data Encryption

With the progress of science and technology and the rapid development of informa-tion technology, people have enhanced the awareness of the protection of information, information security has become increasingly important, now people are paying more and more attention to information security, and began to explore the relevant process-ing methods [2]. Information is a macro concept, which is composed of data. In other words, data appears as a carrier. Therefore, we must take certain measures to protect the data security, to avoid the data being stolen or destroyed or intentionally modified. The best way to solve the problem of data security is to prevent data leakage through file encryption.

2.2　Principle of Data Encryption

We call the original information which has not been transformed into plaintext (P), and the information after transformation is called ciphertext (C). This transformation from plaintext to ciphertext is called encryption (E), and is usually implemented by some encryption algorithm. The process of recovery transformation from ciphertext to plaintext is called decryption (d), which is usually implemented by some decryption algorithm. As shown in Fig. 1 below.

Fig. 1.　Data encryption principle model

　　The sender encrypts the plaintext P into ciphertext and sends it to the receiver. After receiving ciphertext C, the receiver uses the corresponding key for decryption to restore ciphertext C to the original plaintext P [3]. In this way, even if the information in the transmission process is stolen by others, he can only get ciphertext C. without the decryption key, he can't understand it, which plays a role in protecting the information. When encrypting, the encryption key we use is the parameter K, where k is only one randomly selected from the key space, and we can also select any other value. If only one key is used, then this is symmetric encryption technology using symmetric key, and the shared key is K.

　　The purpose of adding activation function to the network is to add nonlinear mapping with nonlinear factors to enhance the nonlinear expression and fitting ability of network model. The main expressions of the activation function are 1, 2 and 3 of the activation function of CNN.

$$sigmoid : f(x) = \frac{1}{1 + e^{-x}} \qquad (1)$$

$$\tanh : f(x) = \tanh(x) \qquad (2)$$

$$ReLU : f(x) = \max(x, 0) \qquad (3)$$

3   Analysis of Classical Encryption Algorithm

3.1   Overview of Encryption Algorithm

The encryption algorithm obviously is to process the data to be transmitted into the data that can't be understood by outsiders, and then transmit it to the correct receiver, and then the receiver will restore the ciphertext processing and read the information with the agreed processing method in advance. In this way, even if the interceptor is intercepted on the way, the information cannot be read after intercepting because the interceptor has no corresponding decryption algorithm.

There are two major classes of encryption algorithms. One is the earlier one, which is not based on key. This premise is that the algorithm is confidential, just like the connection code. The disadvantage is obvious. If the algorithm is leaked or cracked, it will not be available. The other type is naturally based on key, which is commonly used by us now. Key is generated by the algorithm, the algorithm is public, but the key is confidential, so only need to change the key, no need to change the algorithm, so it is much more flexible. The guarantee of security falls on the key, and the length of key determines the security. Symmetric encryption and asymmetric encryption are two kinds of key based encryption algorithms. The previous chapters have already been described, so I will not repeat them here. The following introduces and analyzes several commonly used encryption algorithms, and selects the encryption algorithm adopted by the encryption and decryption client.

3.2   DES Algorithm

DES, also known as data encryption standard, is a very classic symmetric encryption algorithm. It was first developed by IBM company, and then developed as data encryption standard by the United States and ISO. DES is a group symmetric encryption and decryption algorithm. Before encrypting the plaintext, all plaintexts are divided into multiple groups, and the length of each group is set to 64 bits. Then the binary data encryption operation of each group is performed. After encryption, a group of 64 bit ciphertexts are generated. Finally, the ciphertexts of each group are spliced together to get the whole ciphertext. The length of the key used is 64 bits. It should be noted that 8 bits are used for parity check (see Fig. 2).

DES algorithm involves, of course, the three parameters that are usually involved in the algorithm: the original or processed data or information, the suitable key generated according to the algorithm, and the running stage mode [4].

The specific operation process of DES is as follows: when the working mode is switched to the encryption mode, the plaintext is encrypted with the preset key, and then the ciphertext is generated and the output is generated; when the working mode is switched to the decryption mode, it is processed with the preset key, and then the correct information is recovered and the output is generated. Before data transmission, both parties agreed in advance obtain the same key according to the corresponding method. Before data transmission, DES algorithm is used to encrypt the data, and then the encrypted information is transmitted to the place where it needs to be reached. After the data arrives at the correct place, the data receiver decrypts the data with the same secret key [5]. Obviously, for outsiders who do not know the key, it is necessary to use the same secret key to decrypt the data, This is enough to ensure the security of data transmission (see Fig. 2).

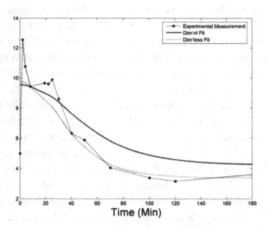

Fig. 2. Simulation with DES algorithm.

3.3 MD5 Algorithm

The main purpose of MD5 is to ensure the correct transmission of information [6, 7]. The specific implementation process is as follows: for example, I generated a text document, and then generated an MD5 value. I release this file for others to download and use. When the download person is not sure whether the file is safe or it is suspected to contain a virus, MD5 can be used to verify. If the value is the same, then it is safe. If it is not the same, it is modified and dangerous! There are many MD5 small programs on the network, mainly used to verify data integrity. Thus, MD5 is widely used, and its security and reliability are also relatively mature [8].

The principle of MD5 algorithm can be summarized as follows: MD5 does not require the length of input information, but the output must be fixed to 128 bits. The basic methods of its implementation include finding the remainder, taking the remainder and adjusting its length, and performing cyclic operation with linked variables [9].

Through the above analysis, we conclude that the main purpose of MD5 is not consistent with the project we are going to do. Because the most important thing we need to do is to ensure that the reported information is not stolen by others, not just to ensure that the reported information is not tampered with. In addition, MD5 is not enough to ensure the security of information. As early as 2004, a professor from Shandong University had decoded MD5, so we rejected the MD5 algorithm. Therefore, it is no longer necessary to elaborate, and only the next section describes the flow of the algorithm [10].

Although RSA is the first proposed public key algorithm, it is also a very comprehensive public key algorithm. It can not only be used to encrypt all kinds of information, but also has great use in signature and authentication. This algorithm is the most widely used and trusted algorithm. Since the date of release, it has been attacked and cracked by scholars and hackers, but no one can crack it successfully [11]. Of course, it is undeniable that we can not use scientific means to prove that it can not be cracked, but countless facts have proved that it is reliable and its security is guaranteed. Therefore, RSA today is very important, can be said to be indispensable, has been adopted to a high degree (see Fig. 3).

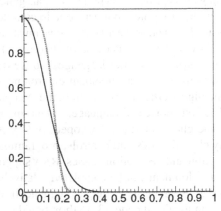

Fig. 3. Simulation with MD5 algorithm.

4   Design and Implementation of RSA File Encryption in the System

4.1   The Client Function of Encrypting and Decrypting Report Files

(1) The key can be generated and decrypted for the client only. 2) The client can load any public key or private key. Both the employee client and the administrator client have this function. (3) The client can load the txt file with the report content encrypted by the public key, and then the ciphertext can also be saved as the txt file. Only the employee client has the encryption function. (4) The client can load the private key to decrypt the ciphertext TXT file, and then save the plaintext as the txt file. Only the administrator client has the decryption function. (5) To emphasize one point, ordinary employees use the encryption client. Generally, the public key loaded by them is released by the administrator (who also owns the decryption client) and downloaded by ordinary employees. In this way, the key pair can be updated regularly to ensure more security! To sum up, the employee client can load the public key (download the public key from the server) to encrypt the file, And save the public key and the encrypted ciphertext. The administrator client can generate and save the key pair, upload the public key to the designated server, decrypt the ciphertext of the report file with the corresponding private key, and save the decrypted plaintext [12].

4.2 Technical Route Selection

Based on the actual situation of client development, this paper analyzes and compares several common technical means, and combined with their own familiarity, finally selects the best technical route, and tries to use easy to learn and easy to use Python language to write the client [13]. (1) Development based on Java platform the workload of client development based on Java is not large, because the Java class library provides complete RSA tool classes, including encryption, decryption, key pair generation and other methods. Therefore, it does not need to write too much code to complete, and one advantage is that it can cross platform. However, as we all know, Java must have its virtual machine, so the efficiency will be discounted, and because of its cross platform nature, it can't be as close to the operating system as other languages, so we don't choose it. (2) Net platform is a relatively new integrated development environment, which is convenient for rapid development and highly efficient. In addition, the. Net platform provides a very powerful class library. The first choice is C language. C # has many desktop applications on Windows platform. The client interface developed by C # is delicate and easy to use. Because of its many class libraries and controls, programming becomes relatively easy. Net provides encryption and decryption classes, RSA support class is RSA crypto service provider, so the development can be completed relatively quickly [14]. One of the disadvantages is that the system must be equipped with net framework. The other one is that it is less efficient than localized code. (3) Python based development language is simple and easy to use, with simple implementation logic and enough performance. As we all know, in design, the simpler the logic is, the higher the reliability is, and the less likely it is to make mistakes! Python is an interpretive language, which can be executed directly without compilation, which can improve the speed of development and testing, and facilitate subsequent modification and function improvement. Secondly, Python is very convenient for the large number storage operation in RSA algorithm compared with C++, because it is a dynamic type and can automatically adjust the storage type according to the needs. In addition, python can package programs directly without the limitation of additional support like Java virtual machine or net framework, so it has good cross platform performance [15].

Considering the software's executability, maintainability, reusability and development workload, the software adopts layered implementation, and the underlying RSA algorithm is encapsulated into a reusable RSA encryption and decryption library by python, which is called by the upper layer. WxPython, a GUI library, is used in the upper graphical interface. The library is based on wxWidgets cross platform GUI tool library developed by C++. It can run in Microsoft Windows system, almost all uni "X" and similar systems without changing the code. Of course, Mac OS can also be used. The advantages of this development are: the main functions are at the bottom and can be easily improved, extended and modified. As mentioned above, this client development is divided into three parts: using Python to write RSA algorithm, encapsulating it into a reusable RSA encryption and decryption library, and using wxPython, the Gu graphics library, to realize the graphical interface of the basic operation of the client [16].

4.3   Design and Implementation of Encryption Algorithm

The main function modules that need to be developed are: (1) generation of key pair (only administrator client has this function); (2) encryption and decryption of file by loading key (appearing separately in employee and administrator client); (3) opening and saving of text file and uploading and downloading operation (common); (4) encryption and decryption of file by loading key; (4) Design and production of graphical operation interface (common). This section focuses on the generation of key pair, the encryption and decryption operation of loading key pair file. From the analysis in Sect. 3, we can know that the basis of RSA is large number and its operation. Therefore, the design and implementation of encryption algorithm can be summed up in the following three aspects: the first is the search and test of large prime number, the second is the generation of public key and private key, and the last is the encryption and decryption of TXT report file. The search and test of large prime number is the most fundamental and the first. If there is no large prime number, there will be no RSA algorithm. Generally, we search and constantly test to find out and determine the appropriate large prime numbers P and Q in RSA algorithm. The generation of RSA key pair is actually the generation of public key for encryption and private key for decryption, and whether the key is reliable and secure. This fundamentally defines the security of this security system, We need methods to generate extremely secure and reliable trusted key pairs. Another important operation is encryption and decryption. The main operation involved is modular exponentiation, which fundamentally defines whether this algorithm is effective and fast. Obviously, we need an efficient algorithm, but at the same time, it is used for small TXT text encryption and decryption, so this requirement is not too high, and sufficiency is the most basic [17].

5   Software Implementation of Generating Large Prime Number

Here is an important point to declare. So far, it is very difficult for computers to generate a large prime number randomly. If we want to ensure that we can get a relatively accurate large prime number, we need and must get it by looking up the prime table. But this way is dangerous, because if this vital prime table is stolen, then the reliability and security of RSA algorithm in this way is questionable. Originally, we wanted to adopt this method. Later, considering the possible security problems, we gave up. Now we use random calculation to generate large prime numbers. In this way, the disadvantage is that it greatly increases the complexity of the algorithm, and the advantage is that it greatly improves the security of the system. This method can't guarantee 100% generating primes in a short time, just try to find the right P and Q by searching and testing.

Prime numbers are infinite, that is to say, the number of prime numbers is countless, which has been proved. Although a lot of efforts have been made, there is still no solution for factoring large integers! This proves that a method of producing prime numbers is not feasible, that is, by trying to factorize large integers. So far, the most effective way to find large prime is to judge whether a large integer can pass the test of some prime detection algorithm.

6 System Test

6.1 Test Purpose

The purpose of system testing: 1. System as a whole: all functions of the reporting system are correctly implemented in accordance with the requirements of customers, without defects affecting its normal operation, and the performance indicators have reached the standards set by the industry and customers. 2. Function: ensure that the reporting system can operate correctly and realize various functions correctly when it is used on a centralized scale in the company. 3. Performance: ensure that the reporting system can meet the large amount of data inquiry, multi-user concurrent operation, the number of failures or errors and the maximum or average response time of the system are within the normal range. 4. Security: the report must have a complete user rights management function, all files storing passwords must be encrypted, the system must transmit key data in the network environment, and the transmitted data must be encrypted twice.

6.2 Test Scope

The whole system test includes the following process: 1 module test: each functional module should be tested several times to see whether each module is normal. If not, the statistics of error probability and error reason should be done for subsequent modification. 2. Integration test: after the establishment of the reporting system, test whether the business links between different modules are normal. 3. Confirmation test: check whether the reporting system can meet all the functional and non functional requirements mentioned in the customer requirements. 4. System test: check whether the reporting system and the corresponding client can operate normally in the real environment and have good compatibility. 5. Acceptance test: after the system is tested by us, it needs to be tested by the customer to let the customer judge whether it meets the customer's needs.

This test includes functional test and non functional test, the details are as follows: (1) functional test (2) performance test (3) user interface (UI) test (4) interface test (5) security and access control test (6) failure transition and recovery test (7) fault tolerance and exception test.

7 Conclusion

In a word, in the new era, network communication data encryption technology has obtained rapid development, and data encryption technology has been widely used in various industries and fields. The relevant technical personnel must strengthen the research, further optimize and improve the data encryption technology, create a good network communication environment for users, and promote the sustainable and healthy development of computer technology.

References

1. Xiaosong, Z.: Application of data encryption technology in computer network security. J. Jiamusi Vocat. Coll. **07**, 254–255 (2019)

2. Xiangna, C.: Exploring the application of data encryption technology in computer network communication security. Netw. Secur. Technol. Appl. **6**, 23–24 (2019)
3. Xue, H.: Research on data communication network maintenance and network security. Electron. Components Inf. Technol. **2**(08), 6–8+17 (2018)
4. Xu, Z.: Analysis of data encryption technology in network communication security. Comput. Prod. Circ. **2019**(11), 28+30 (2019)
5. Zou, H., Xu, P.: Introduction to Cryptography. People's Posts and Telecommunications Press, Beijing (2004)
6. Lecai, C.: Applied cryptography. China Electric Power Press, Beijing (2005)
7. Wei, R.: Modern cryptography. Beijing University of Posts and Telecommunications Press, Beijing (2011)
8. Chen, Z.: Implementation of RSA public key cryptography software package. Master's thesis of Guangzhou University (2002)
9. Xie, X., Wei, B.: Analysis of network information encryption technology. Science and Technology Plaza (2007)
10. Jian, Z.: Cryptography Principle and Application Technology. Tsinghua University Press, Beijing (2011)
11. Yuefei, Z., Yajuan, Z.: Public Key Cryptography Design Principles and Provable Security. Higher Education Press, Beijing (2010)
12. Report on the Development of Cryptography in China, Group of Chinese Cryptography Society. Electronic Industry Press, Beijing (2011)
13. Xueli, W., Dingyi, P.: Theory and Implementation of Elliptic and Hyperelliptic Curve Public Key Cryptography. Science Press, Beijing (2006)
14. Xiaoyong, G., Yang, F.: Network Security Operation and Maintenance. Higher Education Press, Beijing (2011)
15. Xiaohua, Z.: Computer Network Security Technology and Solutions. Zhejiang University Press, Zhejiang (2008)
16. Ma, G., Bai, Y.: Analysis of RSA public key system technology and design of related algorithms. Comput. Sci. **33**(8) (2006)
17. Shi, X., Dong, P.: Design of a new encryption core based on RSA algorithm. Microcomput. Inf. **12**, 3 (2005)

Application Analysis of Data Technology in Computer Information Security Education

Xiaopeng Qiu[✉]

School of Mathematics and Statistics, Central South University, Hunan 410083, China
8201180315@csu.edu.cn

Abstract. With the development of information technology, big data technology is widely used to improve people's work efficiency. At present, computers are widely used in various industries, and information security is of great practical significance for enterprises. In order to further improve security, big data technology must be effectively applied to computer information security protection. This paper discusses the big data technology in computer information security to a certain extent. On this basis, the in-depth analysis and Research on the application of big data technology provides a reliable guarantee for computer information security, and has certain reference significance for technical personnel engaged in related business.

Keywords: Big data · Computer · Information security

1 Introduction

With the development of information technology, people's daily life has brought serious changes, and the work efficiency has been continuously improved, which has brought a lot of convenience to people's life. However, in the process of information technology application, its information security is affected by many adverse factors of flexible publicity and application of technology. As a relatively advanced processing technology, big data can improve a certain level of information security in the safe operation of computer information through effective transmission, so as to provide reliable guarantee for computer information security.

2 Application Status of Big Data Technology in Computer Information Security

2.1 Overview of Big Data Technology

The core technologies of big data technology include big data acquisition, big data preprocessing, big data storage and big data analysis. The specific processing process is shown in Fig. 1. First collect the data, then collect the original evidence, as shown in the

M. A. Jan and F. Khan (Eds.): BigIoT-EDU 2021, LNICST 392, pp. 512–520, 2021.
https://doi.org/10.1007/978-3-030-87903-7_63

figure, through data collation, data integration, data conversion and protocol operation, and then storage and data analysis to obtain high-quality data [1]. With the development of big data technology, the application of computer science is becoming more and more important. It is worth noting that the big data technology encountered difficulties in many data analysis, which is also the research direction of this paper.

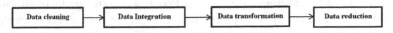

Fig. 1. Big data preprocessing

2.2 Computer Information Security in Big Data Environment

Data technology has penetrated into the industry, and the development of computer information has also brought the security of computer information. The application process of big data technology has been integrated with the needs of various industries. The impact of big data technology on computer information security is in the process of big data technology, computer information storage, information security management equipment, hacker attack, virus penetration, computer information stolen in the process of transmission and storage, so as to ensure the security of information.

3   Big Data Technology of Computer Information Security

3.1 Cloud Computing Technology

Computing technology is a typical representative of the era of big data, and also a key technology to promote the rapid development of big data technology. The flexible promotion and application of big data technology based on parallel computing method and the systematic and comprehensive sorting and statistics of relevant computer data are of great practical significance. With abundant information resources on the Internet, parallel grid computing is carried out, resources are integrated and reasonable planning and arrangement are made [2]. At present, with the development of cloud computing technology, big data technology can combine with the security requirements of computer information system, and take corresponding safeguard measures. The effective application of computing technology is a huge application of computer data processing capacity. In the process of cloud computing technology, a large amount of information needs to be stored. The continuous improvement of data storage capacity and the expansion of data information space have played a certain role in promoting the penetration of computer information security based on data technology into various fields.

3.2 Data Backup Technology

Although information technology has brought good interaction with economic life, it still challenges the information security of computer. The effective application of big data technology can improve the level of information security and provide reliable guarantee for important information security. As an important technology of big data technology, data backup plays an important role in information security. In the process of technology application, especially in the process of introducing big data technology into enterprise information system, data backup technology can maximize the information security of enterprises and provide safe and reliable storage space for important information of enterprises. This is the age of information explosion. People are enjoying a lot of information, accompanied by the risk of privacy information leakage. For individuals or enterprises, information security has very important practical significance. Data backup technology can fundamentally solve the important data loss caused by various emergencies, reduce the probability of data loss, and ensure information security as far as possible.

3.3 Analysis of Network Security Behavior

In order to ensure that the effect of computer information security management can be expected, it is necessary to apply security evaluation mode to evaluate various network security behaviors, such as confirming whether there is intrusion in the network system. In order to achieve this goal, it is necessary to evaluate the security behavior of network packets by relying on the relevant content of matching algorithm If the evaluation results show that the behavior of intrusion features is found, it can rely on the corresponding command retrieval and identification, and group various potential or existing system security risks through matching rules [3]. When the final result shows that the current network system has been in an unsafe state.

In view of the above problems, BM algorithm can be used to evaluate the network intrusion behavior in the comprehensive grey correlation clustering method. If the evaluation results show that the characters in the network intrusion string are different from the characters in the network, the correct parameters of the network intrusion character string can be determined by this function. The specific calculation specification is:

$$delta(x) = m(distance) \tag{1}$$

$$delta(x) = m - max \tag{2}$$

In the above formula, when there is no X at the end of the detected string, formula 1 can be used, while formula 2 can be applied to other types. However, in this case, the relevant personnel also need to consider a special case, that is, large character space or short character, and the whole average displacement of delta (x) is m, The BM method can be used to evaluate the network security behavior, At present, the content of network information security assessment is very extensive, mainly focusing on the temporary network security assessment, chat favorite security assessment, comprehensive network security assessment, and this type of assessment will be further refined and classified. Therefore, for the relevant technical personnel, when studying related issues, they can extract multiple time periods of security management information, Form a comprehensive information security strategy, in this process, through file transfer package and other methods, a large number of security information is directly transmitted to the target, which can not only ensure the effect of information transmission to the greatest extent, but also fit all kinds of security assessment system with grey correlation clustering. From the Perspective of security assessment, all kinds of security incidents are evaluated and solutions are found.

4 Big Data Application

4.1 Application of Cloud Computing Technology

Cloud computing can combine the network and cloud platform organically, build a model with the formed technology, store all information and data in the cloud, and significantly use computers to process various data information, bringing good economic benefits to enterprises. In the cloud computing process, the technical service software of the package can be transmitted through the relatively stable data transmission network of the local alkali network, and the information processing of the big data can be upgraded to the locking feedback information processing mode. In the process of using cloud computing to process the data, the whole calculation process runs through the non cyclic data flow, and the system is comprehensive [4]. On this basis, combined with the parallel computing characteristics of cloud computing, a cloud computing service model is formed through a series of groups and suggested channels for remote data transmission [5].

4.2 Application of Data Backup Technology

With the continuous development of enterprise scale, a large number of production information data will be produced in the process of production and operation [6]. This should give them enough storage space. However, there are many adverse factors in the actual data storage process, so the safe storage of data may bring great risk. Therefore, the application of data backup technology to computer information, in order to avoid the loss of important data in safe operation, can minimize the loss of enterprises. In the process of data backup and storage, in order to ensure the information security of enterprises, in the traditional process of data backup, it is necessary to carry out conventional important production, generally using U disk and mobile hard disk to store data [7]. However, with the increase of the total amount of data information, these traditional storage devices used to store data can not meet the requirements of data backup. In order to provide a broader storage space, cloud platform can have a large amount of storage space and has high security (see Fig. 2).

4.3 Application of Hado

At present, the number of data types is increasing gradually, showing a variety of development trends. For some data with special storage requirements, the traditional storage method is no longer applicable, which will bring computer information security risks [8]. Applied to the system, can meet a variety of data storage and centralized management has good advantages, the whole process of data management has systematic and effective operation to coordinate with each other, support the function equivalent to computer data security processing. At the same time, hado system is also in the stage of continuous improvement, the function division of each component is clearer, and the application system of big data information analysis platform is more detailed and functional (see Figs. 2 and 3).

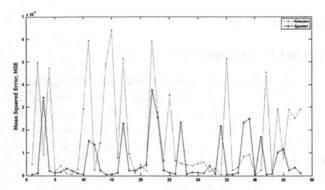

Fig. 2. Simulation of MSE for data backup

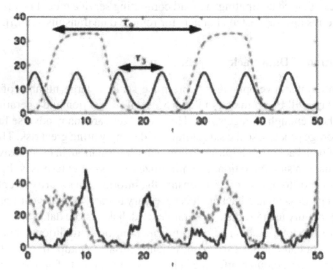

Fig. 3. Simulation for hado

5 Analysis of the Problems in the Network Security Education of College Students

5.1 The Virtual Nature of Network Alienates the Main Body of Network Security Education

With the rapid development of Internet technology, today's society has entered the era of information and digitization, "virtual" has become the most iconic feature of the network society. As the scholar Zhang Mingcang said: "virtual in the contemporary context refers to the way of expression, composition and Transcendence of digitization". These ways enrich the network life of contemporary college students and meet their diversified network needs [9]. However, the virtuality of the network is a virtual environment created with the help of the network, which is different from the real physical environment [10]. In the virtual network space, the identity, behavior and image of college students are digitized. The daily network activities of college students are actually symbolic activities, The communication between college students has evolved into the interaction between symbols. In the virtual world, people have become slaves of symbols, which deviates from the nature of human beings. People lose themselves and become the other, resulting in the alienation of people, which is not conducive to the all-round development of people. According to Marxism, "alienation, as a social phenomenon, comes into being together with class. It is a social phenomenon in which people's material production, spiritual production and their products become alien forces and in turn dominate people". In the process of alienation, people lose their subjective initiative and are enslaved by alien material or spiritual forces, which leads to the fact that people's personality can not develop in an all-round way, can only develop one sidedly or even abnormally. "New York Times" commented: "like most people in the network society, I am deeply immersed in the media life, which is a state of excitement and fatigue. I can't distinguish between the virtual world and the real world. I feel that my personality is submerged in the vast network world, my own memory disappears in the fog of the virtual world, and I become a lifeless bystander [11].

5.2   The Openness of Network Leads to the Complex Environment of Network Security Education

In 1978, China carried out reform and opening up and took the initiative to open the door to the outside world, making today's open, inclusive, shared and innovative China [12]. Today in the 21st century, opening up has become the result of the times. As long as you have an electronic device connected to the network, no matter where you are in the world, you can share the rich information resources on the network. Marxism believes that: contradiction is the source and power of the development of things, things exist in the form of contradiction unity, in short, contradiction is the unity of opposites, so we should treat the openness of the network in two. The Internet is an open platform and a public domain. Anyone can receive and transmit information in this open area. According to the same preferences and in different time and space, Internet users all over the world rely on the Internet as a medium to freely express and freely exchange information. Due to the openness of the Internet, it is difficult to monitor the network information and resources in real time, Some individuals and organizations who have other plans have spread the bad information of violence, pornography gambling and cults, which are not in line with socialist core values. Recently, the data of the SF customers have been leaked. More than 300000000 customer data is selling two bitcoins on the dark Internet. The "dark net" has again appeared in the public's vision. In early 18, it was pushed by the official account of the public. Let the majority of young people, especially college students, know that "dark net" is a foreign website full of crime, violence and metamorphosis. Since then, "dark net" has been registered in China, which has seriously polluted the network environment of our country, and also affected the environment of network security education, which is not conducive to the smooth development of network security education [13].

5.3   The Fragmentation of Network Information Weakens the Effectiveness of Network Security Education

"Fragmentation appeared in the relevant literature of" postmodernism "in the 1980s". Fragmentation "originally means that the complete things become broken and scattered things". Nowadays, fragmentation has been widely used in politics, education, sociology, communication and other fields. From the relevant discussion of communication, we can learn that "media fragmentation has become a new trend of mass media. Compared with newspapers, radio and other traditional media, the network is a new carrier of mass media, through the network in the form of numbers or symbols, relying on the terminal of electronic equipment to provide information resources for Internet users [14]. With the advent of the era of big data, massive information and resources provide convenience for the public, but also present a fragmented state in the communication content. As the media of information release, many traditional paper media have strict review standards, and full-time personnel carry out reasonable and legal processing based on the authenticity and reliability of information, carefully proofread, and finally summarize and bind. Different from the traditional paper media, the Internet media has not yet formed a set of perfect censorship standards for information release. In the new media era, information publishers have a wide range of discourse power. Information publishers come from different social strata and social groups, and spread information with the help of different apps. College students browse all kinds of information through microblog, wechat, post bar and forum, Due to the imperfection of the network accountability system, in order to attract attention and get hits, APP reports hot events out of context. "Through a text, a video, and a picture, it is difficult for college students to piece together a comprehensive and complete event. For the incomplete and incomplete event, college students express their views actively combined with what they have learned, and the conclusions also present a fragmented state. Fragmented, good and bad information flooding the network, polluting the network environment, threatening network security. In the long run, college students are used to the fragmented way of information acquisition, which will have an impact on the comprehensive and systematic network security education classroom to a certain extent, weaken the leading role of the network security education classroom, and is not conducive to the construction of a complete network security knowledge system, the cultivation of logical thinking and divergent thinking of college students, and increase the difficulty of network security education, As a result, network security education is difficult to achieve the desired effect [15].

6   There is a Contradiction Between the Diversity of College Students' Network Needs and the Singleness of Network Security Education

With the deepening of reform and opening up, the continuous improvement of market economy and the continuous improvement of China's productivity, China's main contradictions are also changing with the times and the situation. The 19th National Congress of the Communist Party of China has a new discussion on the main contradiction of our country: "the contradiction between the growing needs of the people for a better life and the unbalanced and inadequate development". today, the needs of a better life are more focused on the needs of a better spiritual life. The important way for contemporary college students to enrich their spiritual life is to rely on the Internet, The Internet meets the diverse network needs of college students in a subtle way, but the network world is complex, and college students are not deeply involved in the world. At this time, colleges and universities need to carry out a variety of network security education, provide positive spiritual resources for enriching college students' spiritual life, and build a rich and colorful spiritual world for college students. However, the current situation of network security education in Colleges and universities can not effectively meet the needs of college students, there are contradictions and deviations [16].

After the heavy schoolwork burden of high school and the baptism of the silent war of college entrance examination, we have entered the ivory tower of our dreams. The education mode of "strict in and broad out" has left enough free time for college students. Different from the high school step-by-step, busy life, university life is more idle and empty, idle. With the advent of the Internet age and the popularity of electronic devices, college students have found the carrier of emotional attachment. The post-95 college students are the aborigines of the Internet world. For them, the Internet is not only the existence of tools, but also a way of life. College students are the pillars of our country. At the same time, they are also at the forefront of the society in the operation of the network. While meeting the diversified needs of college students, the network is constantly occupying the world of college students. The network has become a new carrier of contemporary college students' lifestyle. Through the network platform, college students can carry out digital symbolic social activities such as online shopping, online games, online love, online chat, etc., With the continuous improvement of living standards and the degree of network socialization, the network needs of college students are not limited to entertainment and social communication. College students are fundamentally "born for learning" groups, and education is also an important part of college life. However, the network security education in Colleges and universities is not flexible and sufficient, and the supply of unsystematic resources is difficult to meet the network needs of college students. As an important way to meet the network needs of college students, network security education in Colleges and universities should carry out information supply side reform to meet the network needs of college students in an all-round and wide range. However, in reality, we find that the supply of network security education resources is not flexible, sufficient and systematic.

7 Conclusion

In short, with the continuous development of information technology, people's daily life has brought new changes, improved work efficiency and brought many conveniences to life. However, the complexity of the network environment has a negative impact on the information security of computers and the application of information technology. Therefore, we should effectively use various computer security technologies to ensure the security of big data.

References

1. Xiaoyan, X.: Taking big data analysis as an example to analyze the application of computer technology in information security. Build. Eng. Technol. Des. **22**, 263 (2019)
2. Lin, Z.: Analysis of the application of computer technology in information security – taking big data analysis as an example. Heihe J. **4**, 29–30 (2019)
3. Qiming, C.: Analysis of computer information data security and encryption technology. Enterp. Technol. Dev. **06**, 68–69 (2015)
4. Di, W., Dengguo, F., Yifeng, L., et al.: A utility evaluation model of security measures in a given vulnerability environment. Acta Softw. **7**, 1880–1898 (2018)
5. Chen, M.: Network security education for college students based on 5W theory. J. Minnan Norm. Univ. **2** (2018)
6. Zhang, Z.: On the value of family school interaction from the perspective of Ideological and political education in Colleges and universities. J. Changchun Inst. Educ. **2** (2034)
7. Han, A.: Research on the optimization of Ideological and political education environment in Colleges and universities. J. Jilin Educ. Coll. **3**(4) (2018)
8. Li, C.: Research on the path of improving the ideological and political education in Colleges and universities in the era of network information fragmentation. J. Hubei Corresp. Univ. **31**(10) (2018)
9. Lin, H.: Problems and Countermeasures of network security education for college students in the new era. J. Hangzhou Univ. Electron. Sci. Technol. **14**(4) (208)
10. Chen, W.: Network security education for college students in the new media era. J. Jilin Radio Telev. Univ. **9** (2018)
11. Feng, G.: Review and Reflection on the quality evaluation of Ideological and political education in Colleges and universities since the reform and opening up. Teach. Res. **3** (2018)
12. Liu, Q.: Dilemma and strategy of network security education in colleges and universities in the new era of mobile Internet. J. Chongqing Univ. **24**(5) (2018)
13. Zhang H.: Strategy research on network security education for college students. Dec. Making Forum (2017)
14. Tan, Y.: International comparison and enlightenment of network security education for college students. E-government **2** (2017)
15. Liu, X.: Thoughts on strengthening college students' network security education. Jiangnan Forum. **10** (2017)
16. Pan, Z.: Ten questions and ten answers on network security education. China Inf. Sec. **10** (2017)

Computer Network Security and Effective Measures for the Era of Big Data

Changliang Zheng[✉]

Beijing Polytechnic, Beijing 100176, China

Abstract. The era of big data has come. At present, the application of big data is more and more, which has a certain impact on people's life. With the use of big data, computer network security problems also appear. This problem is not conducive to development, so it is urgent to find out preventive measures to solve this problem. Therefore, this paper focuses on the analysis of computer network security problems, and studies the effective measures to prevent computer network security problems.

Keywords: Computer network · Research on preventive measures · Big data · Network security

1 Introduction

The advent of the era of big data has brought great convenience to mankind, and it is also the inevitable trend of the development of modern science and technology. In such a background, the computer network has developed rapidly, and there are still some problems to be solved in the process of development. If we can solve the problem of computer network security, the development of computer network will move forward.

Network security, usually refers to the security of computer network, in fact, it can also refer to the security of computer communication network. Computer communication network is a system that connects several computers with independent functions through communication equipment and transmission media, and realizes information transmission and exchange between computers with the support of communication software. Computer network is a system that uses communication means to connect several independent computer systems, terminal devices and data devices which are relatively scattered in the region for the purpose of sharing resources, and exchange data under the control of protocol. The fundamental purpose of computer network lies in resource sharing, and communication network is the way to realize network resource sharing. Therefore, computer network is safe, and corresponding computer communication network must also be safe. It should be able to realize information exchange and resource sharing for network users. Below, network security refers to both computer network security and computer communication network security.

The basic meaning of security: objectively there is no threat, subjectively there is no fear. That is, the object does not worry about its normal state being affected. Network

M. A. Jan and F. Khan (Eds.): BigIoT-EDU 2021, LNICST 392, pp. 521–529, 2021.
https://doi.org/10.1007/978-3-030-87903-7_64

security can be defined as: a network system is free from any threat and infringement, and can normally realize the function of resource sharing. In order to make the network realize the function of resource sharing, the hardware and software of the network should run normally, and then the security of data and information exchange should be guaranteed. As can be seen from the previous two sections, the abuse of resource sharing leads to network security problems. Therefore, the technical way of network security is to implement limited sharing, its structure is shown in Fig. 1.

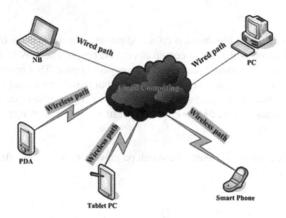

Fig. 1. Network security structure

2 Computer Network Security in the Era of Big Data

2.1 There Are Loopholes in Computer Network System

Most of the enterprise computers use Windows system, the utilization rate of windows system is very high, but there are some problems in the system. Whether it is the invention and update of windows system or the process of downloading windows system, there are some bugs, which will bring security risks to the computer network. In order to solve these bugs, the designers of windows system will constantly upgrade and maintain them, and timely launch new systems to solve these bugs. Nevertheless, there are still some loopholes in the new windows system that need to be solved [1–3]. Therefore, if we want to develop windows without any loopholes, it cannot be realized at present. We can only improve the windows system in the process of continuous attempt and development. In addition to Windows system vulnerabilities, there are many other system vulnerabilities. In the long run, these vulnerabilities will affect the user's information security, and may lead to user data theft and other problems.

Maximum weight matching can be defined as:

$$\text{max} - value = \max \left\{ \sum_{c_i \in R f_j \in S} w_{ij} \right\} \tag{1}$$

The description of matching similarity between services can be defined as:

$$\sin RS = \frac{\max -value}{R}, 0 < \sin RS \leq 1 \tag{2}$$

2.2 Safety Problems Caused by Improper Human Operation

Some improper actions of users in the process of using computer will cause safety accidents. First of all, there may be problems in the user's operation technology, some unintentional actions will bring security risks to the computer network, such as the user has no intention to close the firewall or computer information and data security protection function; secondly, there are many users deliberately attacking the computer network, enterprises should pay attention to this phenomenon, and allow such staff to be dismissed.

With the development of computer technology, most enterprises have installed computers, but the technical level of computer operators is still in the primary stage. These operators can only operate some simple operations, and they are not proficient in some basic software of the computer, so there are often phenomena such as improper human operation causing security problems [4]. In addition, some intentional people attack the computer network and steal the important data in the computer network.

2.3 Network Virus Infection

Users often watch the network virus when they operate the computer. The emergence of the network and work with convenience, many enterprises through the founder of the website to obtain profits, attracted a large number of users to register information. For example, the user base of Taobao Mall and other websites is large, and these users will pay for online shopping, so they must bind their own silver speed hackers to specially develop some kind of sick users' computers infected with these viruses. If they have been infected with network viruses, it is difficult to completely eliminate them [5–7]. In addition to the threat to the user's property security, the computer infected with network virus also affects the computer's protection system and hardware equipment, thus reducing the life of the computer.

3 Effective Preventive Measures for Computer Network Security Problems in the Era of Big Data

3.1 Application Firewall

With the advent of the era of big data, there are more and more kinds of viruses, and the number is also increasing. If we do not prevent and control viruses, it will affect the security of computer networks. At present, the best way to deal with viruses is to use firewalls. The function of firewall is to filter the data in the computer network, it will not affect the normal transmission of computer network information data, will leave safe information data, for the malicious virus processing, these viruses and some junk information block out. So the installation of firewall in the computer can provide security for the computer network. In addition, the firewall function can also effectively distinguish the company and personal information, and process the relevant information separately, so both enterprises and individuals can use the technology safely. In addition, the firewall has been monitoring the data in the computer network to review the internal and external of the system. Firewall can block the virus invasion in time to ensure the security of computer network.

3.2   Application Safety Monitoring System

In order to ensure network security, users can set up security monitoring system in the computer. The system can detect the operation status of the computer system at all times, and ensure the user to use the computer safely [8–10]. At present, there are many security monitoring systems, including 360, computer security manager and so on. These safety monitoring systems are free of charge, and have many internal functions, including virus detection, comprehensive physical examination, garbage cleaning and other functions. In addition, there are some charging security monitoring systems with higher security performance. In order to avoid the internal important data being stolen, it is necessary to pay to use these security monitoring systems. After using the security monitoring system, the security of computer network will be greatly improved, providing users with a more healthy and safe network environment.

3.3   Using Antivirus Software

The emergence of virus directly threatens the security of computer network, and virus is also one of the biggest problems affecting the security of computer network. Now the network virus is more and more powerful, more and more kinds, such as Trojans, worms and other viruses seriously threaten the user's computer data security. In order to avoid computer being invaded by virus, enterprises and individuals can install anti-virus software in the computer. Anti virus software can help users find and remove viruses. Now there are many anti-virus software in the computer market, some anti-virus software specifically for a virus, some anti-virus software can kill common viruses. Now most of the anti-virus software functions are more and more complete, which can basically meet the needs of users [11]. At present, the more commonly used anti-virus software are computer security manager, Jinshan drug bully and so on. Enterprises and individuals can install some anti-virus software according to their own needs, and set relevant settings in the software, such as setting automatic detection of virus or killing virus, etc. All in all, the use of antivirus software can help users deal with the computer virus, so that users in a safe state of the Internet.

3.4   Ensure the Security of Information Storage and Transmission

After enterprises use big data, the network is transmitting data and information all the time. For enterprises, some data is very important and must not be lost. Therefore, in order to avoid data loss or theft, the security of information transmission and storage should be guaranteed. There are many ways to deal with this problem. The most common way is to use encryption technology to encrypt the transmitted data, so that the data can be more safely transmitted to the other party's computer. After receiving the data, the other party needs to decrypt the data information, and only the two parties know the key. Taking this way to transport the data information improves the security of transmission [12]. The lawless person can't crack the key of data transmission in a short time. In addition to ensuring the security of information transmission, but also to protect the security of information storage, so the enterprise can also encrypt the important data inside the computer, to avoid intentional intrusion into the computer to steal data. In fact, enterprises can employ excellent computer talents to deal with the problems of computer network security.

4 Simulation Analysis of the Importance of Solving Computer Network Security Problems in the Era of Big Data

Let's first understand the meaning of big data. From the three words of big data, big data refers to a large amount of data. At present, big data is characterized by diversification and transmission through computer network technology. Compared with the previous local network transmission mode, this kind of transmission mode is faster and more convenient, and the data processing speed is faster. At present, the era of big data has arrived, and has promoted the development of computer network technology. According to understanding, most enterprises in China are currently applying big data, such as Jingdong tiktok, Taobao, jitter and so on, all of which have used big data [13–15]. The application of big data solves the core problems of enterprise development, changes the survival mode of enterprises, and promotes the development of enterprises.

There are many methods of data information transmission, and the media used are extremely complex. Therefore, there are some problems in the current network data dissemination. The biggest problem is the security problem. The security factors are mainly divided into: first, it may be caused by human activities; second, with the development of big data, the requirements for computer network technology are higher and higher, But the development speed of computer network can not keep up with the pace of the times. Finally, because of the openness of the network, many viruses invade the network, causing network security accidents (see Fig. 2 and Fig. 3).

In this case, we must take some measures to prevent and solve the problem of computer network security. Only when the problem of computer network security is solved, can it be more conducive to the development of national enterprises and promote the development of national economy [16]. So it is very important to solve the problem of network security.

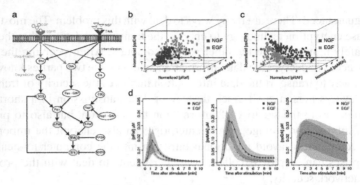

Fig. 2. Network security with big data

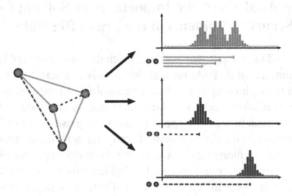

Fig. 3. Attack frequency and effectiveness

5 Effective Response to Computer Network Security Problems Under the New Normal

Under the new normal, in order to effectively deal with the security threats of computer network, ensure the security of computer network, create convenience for users to better use computer network, and prevent unnecessary attacks, the following effective countermeasures can be taken.

5.1 Application of Network Intrusion Detection Technology to Effectively Prevent Hacker Attacks

Using network intrusion detection technology, we can find and eliminate the hidden security in time, and the computer network is safe and reliable. Intrusion detection can also improve the foresight of security management, prevent and control security problems in advance, and take technical measures to intercept and block security risks in time for suspicious activities, which is conducive to the security operation of computer network [17]. The management personnel should find the security problems existing in the network in time, and take measures to calculate and manage the network. "Rukou" attaches great importance to the daily tracing of computer tears leakage, and timely repair when double leakage occurs. As shown in Fig. 4. Add computer system management and daily push. To ensure the safety of software and hardware of jizhunge, we will ensure the safety and effectiveness of delivery limit setting, user setting and code setting, and timely access and recovery of safe delivery. Focus on the sweeping measures of the database, update and upgrade the pushed data laws, improve the security performance of the database, effectively ensure that the user's right of assistance, user settings, secret penalties and orders meet the requirements of the Taiwan security specification, and perform the patch according to the requirements to effectively ensure the network security of the gambling computing machine.

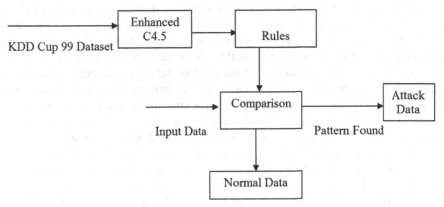

Fig. 4. Application of network intrusion

5.2 Strengthen the Network Security Management, Effectively Guarantee the Computer Network Security

Managers should understand the importance of computer network security management, and make comprehensive use of management rules and regulations, technical measures, etc. to do a good job of security. On the one hand, they should pay attention to computer network information encryption, avoid data loss or be improved. There are two methods of network security management: water code and public key code, which have different characteristics and should be reasonably selected according to the needs of security management. Conventional cipher means that both the sender and the receiver use the same key, which effectively guarantees the information security, has strong confidentiality, and can withstand the test of time [18]. When conveying information, it must be delivered in a secure way to avoid password leakage. Public key cryptography is quite different, and it is difficult to derive the decryption key in the application process. Simple management, meet the requirements of network development, can achieve digital visa and character verification, has strong confidentiality, can make the information transfer and delivery. But its algorithm is complex and not easy to use, which restricts its promotion and application. In order to improve the level of computer network security management, we can combine the two, which is conducive to enhancing the level of computer network security management. On the other hand, improve the computer network security management rules and regulations 61. Establish a strict responsibility system, clear the responsibilities and authority of relevant personnel [19–24]. To effectively guarantee the computer network security, we should formulate the network security use system, formulate the access management system, and establish the network security maintenance and emergency response system.

5.3   Reasonable Application of Encryption Technology to Ensure the Security of Network Information Resources

There are many sources of computer network security risks, not only including websites, hacker attacks, Trojans and so on, but also from the internal database of the network [25]. For example, criminals illegally steal user accounts and passwords, use the database beyond their authority, arbitrarily and steal the information resources of the database. In order to make up for this deficiency, ensure database security, and create conditions for people to make rational use of computer network information resources, it is necessary to adopt encryption technology in management. The specific method is to set the password properly according to the needs of database and computer network resources. Users can access the corresponding database and obtain the required information only by virtue of the password and after passing the system kernel. In order to avoid unauthorized access, illegal intrusion, fundamentally prevent the occurrence of data change phenomenon, ensure the security of computer network [26]. Pay attention to the backup and recovery of teaching data to ensure the integrity of data. In order to prevent malicious invasion, data loss and other problems, computer network management should pay attention to the backup of data. Staff should fully understand the importance of data, make rational use of micro cloud, hard disk and other tools to backup and store important data, and take it as a habit, Protect important data [27]. Through data backup, even if there are security problems in the computer network, it can also restore the data in a short time, ensure the integrity of the data, and provide convenience for the use of information.

6   Conclusions

At present, the era of big data has come, people's lives have undergone tremendous changes, computer network technology has brought convenience to personal life and work, at the same time, computer network security issues threaten people's security. Therefore, enterprises and individuals should pay attention to the problem of computer network security. This paper is divided into three parts, respectively, for everyone to analyze the importance of computer network technology in the era of big data, computer network security problems and effective preventive measures.

References

1. Huang, G.: Analysis of computer network security risks and preventive measures. Popular Sci. Technol. **18**(12), 4-5+23 (2016)

2. Xinyue, H.: On computer network security technology. Heilongjiang Sci. **7**(22), 44–45 (2016)
3. Yuanbo, T.: Discussion on computer network information and network security and its protection strategy. Electron. Technol. Softw. Eng. **18**, 234 (2016)
4. Xiang, S.: Analysis of computer network application security. Mod. Econ. Inf. **21**, 360 (2015)
5. Huang, B.: Computer network information security and protection measures. Ind. Des. **09**, 179+185 (2015)
6. Juxi, X.: Research based on computer network security. Exam. Wkly **58**, 120 (2015)
7. Song, F.: Research on computer network security and firewall technology. Enterp. Guide **12**, 146+148 (2015)
8. Jun, Z.: discussion on internal computer network security maintenance. Inf. Comput. **10**, 84–85 (2015)
9. Li, Y.: Computer communication network security and protection countermeasures. Digit. Technol. Appl. **04**, 177 (2015)
10. Wei, W.: Research on computer network security problems and effective preventive measures. Comput. CD Softw. Appl. **18**(02), 164–165 (2015)
11. Xinyu, H.: Research on computer network security. Wireless Internet Technol. **12**, 62 (2014)
12. Wenjie, L.: Brief analysis of factors affecting computer network security and preventive measures. Comput. CD Softw. Appl. **17**(21), 194–195 (2014)
13. Xu, T.: Computer network information security analysis in civil aviation air traffic control. Silicon Valley **7**(20), 206+198 (2014)
14. Sun, M., Sun, P.: On the main hidden dangers and avoidance measures of computer network security. Inf. Technol. Inf. **09**, 189–190 (2014)
15. Guoyun, S., Wei, Z., Ping, D., Yuanlong, Z.: Effective improvement of computer network security and preventive measures. Comput. Knowl. Technol. **10**(04), 721–723 (2014)
16. Lin, L.: on preventive measures of computer network security. Electron. Technol. Softw. Eng. **23**, 243 (2013)
17. Xueqing, B.: Effective measures to improve the security of WLAN. Comput. Netw. **39**(22), 55–57 (2013)
18. Fengyan, S., Shimin, X.: Explore the threat and prevention of computer network security. Comput. Knowl. Technol. **9**(31), 6975–6977 (2013)
19. Meng, Z.: Analysis of computer information network security management construction. Wireless Internet Technol. **06**, 20 (2013)
20. Jianwu, N.: Effective measures of computer network security maintenance and management. China New Commun. **15**(08), 18 (2013)
21. Jun, H.: current situation of computer security prevention and control. Inf. Comput. **06**, 82–83 (2013)
22. Chuan, J.: On computer network security and prevention. Dig. Technol. Appl. **01**, 164 (2013)
23. Peng, L.: Discussion on computer network security and preventive measures. Silicon Valley **5**(01), 143+155 (2013)
24. Liu, Y.: Internet network database security management measures. Comput. CD Softw. Appl. **16**(01), 92+97 (2013)
25. Hailing, S.: Computer network security risks and effective maintenance measures analysis. Inf. Comput. **18**, 6–7 (2012)
26. Bokai, H., Ling, L.: On the harm and Countermeasures of computer network attack. Manag. Technol. Small Medium-Sized Enterp. **10**, 222 (2011)
27. Yongjian, D.: Analysis of preventive measures for computer network risks. Inf. Comput. **16**, 49–50 (2011)

Construction and Application of Employment Platform for Intelligent Poor College Students Under Information-Based Education

Peng Zhang[✉]

Harbin Finance University Heilongjiang, Harbin 150030, China

Abstract. The wide application of artificial intelligence and big data has given birth to a new employment mode and a new ecology of vocational education. In order to solve the problems of lagging behind and slow employment, asymmetric employment informatization and lack of individualized employment guidance, this paper, based on the thinking and technology of big data, constructs a model with student training as the guidance and accurate employment as the core, The University intelligent employment service mode and employment platform, which integrates employment, recruitment, education, evaluation, monitoring, research and judgment, realizes online vocational course learning, accurate job recommendation, career planning guidance, student difference analysis, student career portrait, employer job portrait, enterprise recruitment, Graduate portrait and other services, and helps to upgrade and reform the employment mode, Improve the quality of personnel training and employment competitiveness of college students.

Keywords: Information education · Poor college students · Platform construction

1 Introduction

Employment is an eternal topic in the world, and the employment of college graduates is the most special one. By 2019, there are 2956 colleges and universities in China. The total number of students in higher education is more than 38.3 million. In 2020, there will be about 8.74 million college graduates, an increase of 400000 over the previous year [1]. At the same time, the complex and severe situation is prominent. The employment of college graduates is not only related to the personal development and quality of life of graduates, but also related to the quality of higher education reform and development, and even the reform and development of the whole national education. At present, there are some problems in college graduates, such as lack of professional consciousness, inaccurate job-hunting orientation, improper methods and so on, which lead to poor job-hunting results. The imbalance between the supply of talents in Colleges and universities and the demand of talents in the market, and the imbalance between the supply of labor market and the demand of graduates' employment create three major difficulties: the difficulty of graduates' employment, the difficulty of employers' recruitment, and the

© ICST Institute for Computer Sciences, Social Informatics and Telecommunications Engineering 2021
Published by Springer Nature Switzerland AG 2021. All Rights Reserved
M. A. Jan and F. Khan (Eds.): BigIoT-EDU 2021, LNICST 392, pp. 530–539, 2021.
https://doi.org/10.1007/978-3-030-87903-7_65

difficulty of school education. In recent years, Internet plus employment has become one of the key measures to improve the quality of employment for graduates. The widespread application of AI and big data has created new employment patterns and new ecological education for occupation, which helps to meet the quality expectations and effective docking of graduates' occupation expectations and jobs [2]. At present, how to use big data technology to promote precision employment, improve the employment competitiveness of college students, reduce the difficulty of employment, improve the efficiency of employer recruitment, and improve the quality of school personnel training need to be solved.

2   General Detail Feature Extraction Algorithm

As shown in Fig. 1, set point P as a target pixel (image point to be processed):

$$T_{Sum}(P) = \sum_{i=1}^{8} P_i \tag{1}$$

$$T_{Sub}(P) = \sum_{i=1}^{8} |P_{i+1} - P_i| \tag{2}$$

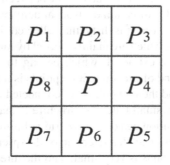

Fig. 1. Definition of 8 adjacent pixels.

The conventional detail feature extraction algorithm is to refine the image first, then repair the ridge, and finally extract the detail feature points by using formula (1) or formula (2).

Among them, formula (1) can work correctly only when the fingerprint ridge is strictly refined to a single pixel width, while formula (2) can correctly extract the detail feature information for the fingerprint image that is not fully refined, But when the quality of fingerprint image is bad and the noise is serious, it is very difficult to repair the fingerprint lines, and the feature effect of conventional algorithms will be seriously affected.

3 Improved Algorithm

There are a lot of noises such as burr, ridge discontinuity and ridge cross before repairing the thinned fingerprint image. The detail features extracted directly from this image by using formula (1) or formula (2) often contain a lot of pseudo feature information [3]. However, if the characteristics of various kinds of noises in fingerprint image are analyzed in depth, the formation causes and distribution rules of pseudo feature points are summarized. We can design the corresponding algorithm to eliminate the false and retain the true, and screen out the real feature point set.

3.1 Classification and Characteristic Analysis of Noise in Thinning Fingerprint Image

(1) When the finger is relatively dry, the fingerprint image collected often has a large number of ridge discontinuities. Where the ridge is discontinuous, the detail feature extraction algorithm will detect two ridge endpoints, which belong to pseudo feature points. The characteristic of this kind of pseudo feature points is that the distance between two points is very small, and there is no ridge along the local ridge direction.

(2) When the finger is wet or dirty, the collected fingerprint image will often have more crossed lines, that is, the lines that should not be connected are glued together. In this position, the detail feature extraction algorithm will extract two points, which belong to pseudo feature points. The distance between two points is approximately equal to the average ridge distance, and the line between two points is approximately perpendicular to the ridge direction of its local neighborhood.

(3) Short line when the fingerprint is dirty, the collected fingerprint image is prone to appear more short lines, which are mainly caused by random noise. In this position, two ridge endpoints are extracted, which belong to pseudo feature points. The characteristic of this kind of pseudo feature points is that the distance between two points is very small, and the two points are connected by a ridge.

(4) The appearance of very small hole like structure is mainly due to the influence of random noise. This position can detect two ridge bifurcation points, which belong to pseudo feature points. The characteristic of this pseudo feature point is that the distance between two points is very small, and the direction of the connecting line between two points and its local neighborhood ridge is approximately parallel.

(5) The appearance of burr is also due to the influence of random noise. This position can detect a ridge endpoint and a ridge bifurcation point, which belongs to the pseudo feature point. The feature of this pseudo feature point is that a pair of endpoints and the bifurcation point are connected by ridges, and the distance between the two points is relatively close.

3.2 Analysis of the Inherent Distribution Law of the Detail Feature Points of the Motif

According to the relevant data and experimental observation, the fingerprint lines and detail feature points have the following characteristics: (1) the change trend of fingerprint lines is gentle except for individual areas such as pattern area, and the width between two adjacent lines is roughly equal. (2) at the resolution of 500 dpi, Generally, there are no detail feature points whose distance is less than 8 pixels in the fingerprint image. (3) there is basically no burr line crossing structure with mutation property in the fingerprint image, such as the structure of two lines with very close ends and two lines sense points or burr line sense connection (see Fig. 2).

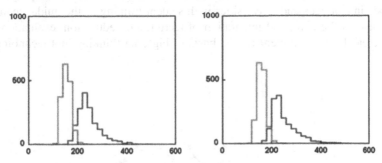

Fig. 2. Distribution law of the detail feature points of the motif.

4   Construction of Intelligent Employment Service Mode in Colleges and Universities Based on Big Data

Big data, artificial intelligence and blockchain technology are triggering a new round of educational informatization reform, providing important support for improving the quality of educational decision-making and educational governance ability, and promoting education to be precise, personalized and intelligent. As a driving force for the development and transformation of education in the era of Internet plus, its thinking and technology have been promoting the scientific decision-making of education, the intellectualization of management and the individualization of teaching [4–7]. Big data technology will reconstruct the education ecosystem, provide scientific support for education decision-making, provide innovation and Practice for education management and evaluation, and provide accurate support for personalized teaching. The application of big data in college employment will break through the limit of employment recommendation method and bring new opportunities for college students' employment reform.

The intelligent employment service in Colleges and Universities Based on big data will fully realize and deeply integrate the employment information services such as online employment market demand survey, students' online counseling and vocational ability evaluation, paperless online recruitment, etc., so as to meet the needs of college students for employment information and personalized counseling, realize the accuracy of employment education, and solve the problems of insufficient teachers and weak pertinence. At the same time, we use big data technology to analyze and process the collected massive employment and entrepreneurship information data, excavate the hidden correlation and regularity between the data, obtain the application value, and strive to improve the school teaching methods, improve the employment competitiveness of graduates, so as to solve the structural contradiction between talent training and market employment demand.

The quality of personnel training will directly affect the quality of graduates' employment, and the quality of employment will directly reflect the quality of personnel training [8]. Based on big data, the intelligent employment service in Colleges and universities needs to contact all departments of the University, build an intelligent employment service mode in Colleges and universities with student training as the guidance, accurate employment as the core, and integration of recruitment, education, evaluation, monitoring, research and judgment on the basis of big data thinking and technology (see Fig. 3).

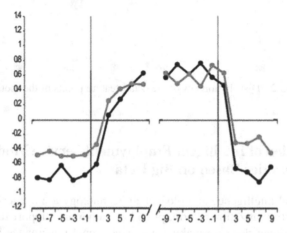

Fig. 3. Simulation with employment service mode in Universities Based on big data

5   Research Background

With the development of computer technology and application, the demand for software engineering professionals at home and abroad is increasing. How to effectively cultivate high-quality software engineering talents and how to meet the requirements of new engineering construction. Strengthening the ability training of software engineering talents is an important challenge for the current computer professional education. Software engineering is a highly comprehensive and practical discipline, and one of the fastest developing branches in the information field. It requires mastering the basic theory of software engineering, software development technology and software engineering testing and management technology; it cultivates engineering high-quality practical talents with the ability of computer software research and development, software engineering management, and the ability to engage in demand analysis, design, development, testing, implementation and management. Software engineering course group includes a series of courses, such as introduction to software engineering, system analysis and design, software architecture, design pattern, unified modeling language and tools, unified software process, software quality assurance and testing, software project management and software engineering case analysis and practice, It plays an important role in improving the comprehensive ability of students [9–12]. At the beginning of its establishment, the National Model Software Institute set up a software engineering course group. From the teaching situation for many years, there are many problems, such as more theoretical knowledge, insufficient practical ability, limited innovation ability, and learning content lagging behind the latest international research and application hot spots, The training objectives are out of line with the actual needs of enterprises.

6 Problems of Software Engineering Course Group

6.1 Unreasonable Curriculum System

The courses in the software engineering curriculum group are relatively independent and have complicated relationships [13]. The existing curriculum system does not consider the connection and correlation between these courses. Although most colleges and universities will set up an introduction to software engineering, but the content of the course is too much scattered, lack of systematicness, and does not really play the role of connecting other courses in the course group. There are three situations in the software engineering curriculum group: ① the knowledge points are not completely covered, many courses mistakenly think that they have been taught in other courses and ignore them for a specific knowledge point, which eventually leads to the omission of knowledge points; ② the order of courses is not reasonable, and the semester of courses with application relationship is reversed; ③ knowledge points are repeatedly taught by multiple courses, Due to the redundancy of knowledge points in each course, students need to repeat the knowledge points they have mastered in valuable classroom time.

6.2 The Teaching Content is Out of Date

Nowadays, with the rapid development of software industry, new concepts, technologies and knowledge continue to emerge, which increases the difficulty of education [14–16]. With the development of big data, cloud computing and artificial intelligence, higher requirements are put forward for software engineering. The teaching content of the school is relatively old and backward, mainly teaching life cycle methodology and object-oriented methodology. Although some teaching materials briefly introduce the agile process such as scion, which is widely used in Internet applications, However, JRA, GitHub and other development methods based on cloud and services, which are widely used in engineering practice at home and abroad, are rarely mentioned in textbooks at home and abroad, and professional education has lagged behind engineering practice.

6.3 The Combination of Curriculum Group Teaching and Industrial Demand Is Not Close

With the continuous progress of computer science and technology, important changes have taken place in the form of information industry, the application of new computing systems is deepening day by day, and the training of computer professionals must keep pace with the times. However, it is difficult for school teachers to grasp the latest development information, popular methods and tools of enterprises, as well as the latest requirements of enterprises for talent training in time, and the teaching is out of line with the actual social needs, It limits the effective connection between the talents trained by the school and the market demand.

6.4   The Evaluation Method of Curriculum Group Is Not Scientific Enough

The close correlation of courses in the course group will be reflected in the students' scores of each course to a certain extent. However, teachers usually evaluate students' performance independently, and do not consider the relevance of courses in the curriculum group, and then do not refer to students' performance evaluation results in other related courses [17]. Therefore, in the course evaluation, especially in the performance evaluation of subjective problems, it is likely to give contradictory scores, which can not truly reflect the ability of students to solve complex engineering problems. In addition, for the practical courses related to software engineering, teachers usually evaluate them at the end of the practice session according to the project documents and system demonstrations submitted by the student project team. Although this method can comprehensively investigate the quality of software process (project documents) and software product (system demonstrations), it often lacks objective standards and has strong subjectivity, Without the support of effective tools, instructors are usually unable to effectively supervise and control the individual work situation in the multi person collaborative project team, and thus cannot objectively evaluate the students' individual course performance.

7   Construction and Teaching of Industry University Research Platform for Software Engineering Course Group

7.1   Promoting Education

In order to promote the engineering education and improve the teaching effect of software engineering curriculum group, the construction plan of production, learning and research platform for software engineering curriculum group includes two meanings: ① defining a set of training mechanism combining production, learning and research oriented to software engineering curriculum group, teachers can complete teaching and research and development demonstration based on this mechanism, and students can obtain basic theory based on this mechanism, Master basic skills and train its practical and innovative ability in software engineering discipline; ② an operational system integrating Tencent tapd and other tools can be accessed through PC and mobile phone, and it has the functions of automatic support for knowledge point synchronization, coverage calculation, scientific and technological literature reading recommendation, case base use monitoring and so on.

7.2 Analysis of Logical Relationship of Software Engineering Curriculum Group

The upper and lower edges of the course indicate the process flow involved in the course, and the left and right edges restrict the semester of the course. As shown in Fig. 4. The scheme divides the software engineering course group into three levels: the basic level, the detailed level and the application level (correspondingly, the courses in these three levels are represented by dotted lines, thin solid lines and thick solid lines respectively) [18, 19]. Three core courses are set in each level. Basic layer: introduction to software engineering (SE), unified modeling language and tools (UML and software analysis and design (SAD); refinement layer: unified software development process (RUP), design pattern (DP) and software architecture (SA); application layer: Software Project Management (PM), software quality assurance and testing (SQA) and software engineering case analysis and Practice (SECs). Both SA and DP courses cover the workflow of analysis and design, which can be taught in parallel with different emphasis on software design technology [20–24]. Because testing is the most important means of SQA, SQA covers the testing workflow.

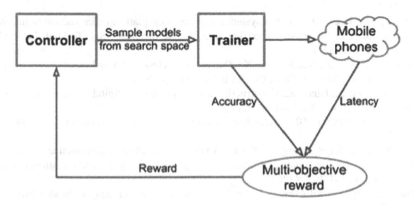

Fig. 4. Analysis of logical relationship of software engineering course group

8 Conclusion

On the basis of the conventional minutiae feature extraction method, the fingerprint minutiae feature extraction algorithm is improved, that is, all minutiae feature points are directly extracted from the thinning fingerprint image without repairing the lines. Then, the pseudo feature points are deleted by using the distribution law of pseudo feature points in mathematical morphology, The algorithm greatly improves the speed of feature extraction, and the accuracy of feature extraction can basically meet the needs of practical application.

Acknowledgements. A New Characteristics of the Ideological Development of Poverty University Students in Universities and Colleges by the Provincial Basic Business Fee Project of Harbin Institute of Finance (2018-KYWF-E017).

References

1. Wang, T.: Analysis of the current situation of employment and entrepreneurship of college graduates and countermeasures. Educ. Occup. **21**, 74–76 (2015)
2. Mo, Y.: The implementation of career guidance in the employment of college graduates. Adult Educ. China **9**, 73–75 (2015)
3. Yang, X., Guo, L., Jin, X., et al.: Big data help the reform of the new college entrance examination: framework design and implementation path. Res. Audio Vis. Educ.(2), 30–37 (2019)
4. Li, Z., Zhou, D., Liu, N., et al.: Modern educational technology (1), 100–106 (2018)
5. Qiao, J., Zhang, B., Yang, Y.: Research on construction mode and application of production dispatching command platform of engineering company. Project Manag. Technol. **19**(03), 127–131 (2021)
6. Zhao, R., Li, J.: Construction and application of power grid visual intelligent management platform. Electr. Era (03), 19–21 + 27 (2021)
7. Fu, Q., Ren, F., Zhao, Z., Liu, Q., Wei, Z.: Exploration on construction mode of Navigation Engineering Laboratory. J. High. Educ. (09), 50–53 + 57 (2021)
8. Jinli, H.: Practice points of urban gas Internet of things. Urban Manag. Technol. **22**(02), 60–63 (2021)
9. Lei, Y., Yao, J., Liu, J.: Support system for information platform construction of innovation and entrepreneurship under the background of "Internet plus". Sci. Technol. Econ. Guide **29**(07), 35–36 (2021)
10. Ning, W., Zhang, J., Wei, S., Hou, X., Zhu, H.: A wireless individual video system integrated into enterprise platform. China Constr. Inf. **04**, 71–73 (2021)
11. Peng, Q., Liu, L., Duan, X.: Construction and application of digital horse academy. Beijing Educ. (Moral Educ.) **01**, 13–15 (2021)
12. Li, M.: Application of BIM technology in intelligent building construction. Jushe **06**, 32–33 (2021)
13. Xu, Q., Liu, C., Su, P., Huang, Y., Xie, H.: Analysis on the construction practice of "multi planning integration" application platform under the goal of sharing and collaboration. Geospatial Inf. **19**(02), 125–130 + 8 (2021)
14. The application of information technology in reservoir engineering in Shanxi Province [J, 198,47]
15. Xu, N.: Construction and application of real estate management platform based on GIS. J. Changchun Normal University **40**(02), 132–136 (2021)
16. Ouyang, S., et al.: Construction and application of 3D virtual simulation experiment platform for waste tire thermal conversion to biofuel process. Internet of Things Technol. **11**(02), 106–109 (2021)
17. Li, C., Guo, W., Wang, S.: Research on the construction and application of HRP based economic management data platform for public hospitals. Chinese General Accountant (02), 26–28 (2021)
18. Teng, J.: Application of cloud computing in information construction of secondary vocational colleges. China Arab Sci. Technol. Forum (Chin. Engl.) **02**, 130–132 (2021)
19. Feng, T., Che, H., Kang, H., Mei, F.: Construction and application of industry university research platform for software engineering curriculum group. Comput. Educ. (02), 144–148 (2021)
20. Wang, L.: Application of big data technology in smart city. Comput. Inf. Technol. **29**(01), 64–67 (2021)
21. Ma, H., Wang, Y., Yang, J.: Online teaching course construction and practice exploration based on Intelligent vocational education cloud platform – taking "Engineering Mechanics" course as an example. Sci. Technol. Wind **04**, 95–96 (2021)

22. Wu, Z., Zhang, X., Ma, W., Wang, X.: Research on the construction of practical teaching platform in and out of application-oriented colleges and universities – building a practical platform for the cultivation of innovative talents. J. Sci. Technol. Econ. **29**(04), 127–128 (2021)
23. Shan, K., Yuan, S., Xu, F., Zhang, H., Wang, J.: Research on the application of blockchain in university informatization construction. China Educ. Inf. (03), 36–39 (2021)
24. Zhang, J., Zhang, T.: Research on platform construction of mechanical engineering training center for applied talents training. Fortune Today **04**, 190–191 (2021)

Design and Realization of Interactive Learning System for Art Teaching in Pre-school Education of Artificial Intelligence Equipment

Lijuan Zhong[✉]

XianYang Normal University, Xianyang 712000, China

Abstract. This paper analyzes the needs of art teaching and research interactive study of preschool education major, and designs related functions. It can realize various functions of art teaching and research interactive learning system through the network. Through the design of access rights and the control of user roles, the security of the system is fully guaranteed. Through the design and implementation of the system, online learning, online homework, online examination and other functions are realized. Finally, the interactive system of art teaching and research is tested. The results show that the system can achieve the functions of landing, online learning, forum communication, online homework and online testing in art teaching, and achieve the goal of the system.

Keywords: Art teaching system · Design · Implementation

1 Introduction

Its main task is to maximize the use of modern computer and network communication technology to strengthen enterprise information management, through the investigation and understanding of the human, material, financial, equipment, technology and other resources owned by the enterprise, establish correct data, process and compile various information materials, and provide them to the management personnel in time, so as to make correct decisions, Continuously improve the management level and economic benefits of enterprises. At present, the enterprise computer network has become an important means for enterprises to carry out technological transformation and improve the level of enterprise management. University information system is one of the main tools of university management in recent years, and it is also the development direction of university management in China [1]. Nowadays, with the rapid development of information technology, computer network is no longer unfamiliar to most people. People's dependence on the network has become more and more. It can be said that people have been closely linked with the network. The network has penetrated into people's study, work and life, and played an important role, causing major changes in the field of education. In recent years, with the rapid development of software engineering technology, information communication technology and other related technologies, network education

M. A. Jan and F. Khan (Eds.): BigIoT-EDU 2021, LNICST 392, pp. 540–548, 2021.
https://doi.org/10.1007/978-3-030-87903-7_66

has been gradually popularized in people's educational activities [2, 3]. In these related education, network testing has become an indispensable part of network education, is an important part of network education.

2 System Analysis and Main Functions

2.1 System Setting Principle

The advent of information technology makes the rapid development of modern digital education and digital campus. The information age of school management has arrived. The network has become the most important technical basis and the carrier of resource acquisition in modern education. Important changes have taken place in the form of modern education [4]. The rapid development of distance education and network education has put forward new challenges for China's education. And after making clear the design principles and functions of the system, the system finally adopts the three-tier BS design mode based on Web technology. In the development process, it strictly follows the development principles of practicability, universality, integrity and openness, and the development platform is selected VS.NET The background database is SQL Server 2005, And using the web forn technology of aspnet and adonet database connection technology, a brand-new art teaching and research system is successfully developed and implemented. The system has the following functions: Students' basic information management, automatic examination database management, students' online test, automatic evaluation of test results and management of teachers and students [5, 6]. The establishment of such a course website can achieve low-cost operation, and the website has good functions at the same time.

2.2 System Setting Purpose

It is a powerful breakthrough to help students realize their achievement motivation and get a sense of achievement in teaching. If there is a result, there must be a cause. "Students' sense of achievement" is a result, so what is the cause? The author summarizes it into the following two aspects. 2. Guide students to find their own value. Every student has unique personalized value, which sometimes appears dominant and sometimes recessive. Sometimes in the process, sometimes in the result. This requires teachers to actively guide different students to present diversified values with an attitude of respect, understanding and tolerance [7, 8]. For example, every student sees and thinks different things about the same piece of wood art. Some students will analyze it from the perspective of culture, some from the perspective of emotion, some from the perspective of society, and some from their own experience. Therefore, teachers should pay attention to observe students, find out the unique value of their thinking, guide and affirm them, and then stimulate their intrinsic achievement motivation.

3 System Performance Analysis

3.1 System Objectives

The mutual aid system of teaching and research requires students to realize the purpose of self-education and learning. Through the platform of art teaching and research system, we can realize the interaction of art teaching and learning, and stimulate students' self-learning potential. The system mainly includes students' autonomous learning, online homework, forum communication, online self-test and other functions. Through the analysis of students' learning and testing, the system can help students master their own learning effect, and can learn the relevant knowledge in time. The design and development of BS (Browser / server) mode art teaching and research system realizes the instant learning, and can carry out interactive learning without any time and geographical restrictions: teachers upload learning materials, which can be easily used by students and other users; they can independently select test questions information resources, and combine these resources together, And these resources can be assigned to students [9–11]. The student's test machine should install a browser, and then enter the user name and password to log in to the art teaching and research mutual aid system.

3.2 System Function Analysis

The art teaching and research system comprehensively considers three types of users: teachers, students and administrators. Through the use of the system, it is very convenient to carry out the whole education and teaching activities: examinees log in to the test system to participate in learning through identity verification, while students can inquire about all kinds of relevant information; teachers manage test questions through identity verification, At the same time, realize the management of students' information, learning situation and other related information, teachers can easily release the relevant information about education and teaching in the information system, while administrators can manage the whole process of the whole test.

1. Teacher users
(1) Release information: after passing the identity authentication, manage the test database, including uploading, modifying and deleting [12]. The test questions are selected by the teacher. Each teacher user authorized by the administrator has the right to publish assignments and enter test questions. The information entered can be displayed on the management test page.
(2) Student management: students manage their information through identity verification, mainly including viewing relevant information, viewing courseware, viewing homework assigned by teachers, online testing, forum communication, etc.
(3) Forum management: forum management is mainly for teachers to realize the function of news management through the release and management of learning related information.

$$s, t \sum_{j=0}^{r} \lambda_i, x_i \leq x_j \tag{1}$$

$$\sum_{j=0}^{n} \lambda_i = 1, \forall \lambda_i \geq 0 \tag{2}$$

2. Student users.
(1) Participate in learning: select students to log in and jump to the main page directly. Check the teacher's courseware, homework forum, communication and online test, and master the learning content in time. Through online self-test, we can master our own learning situation. Students can achieve the combination of learning and practice through the submission of online homework.
(2) Forum View: search and browse the learning materials and related information released by teachers, and realize interaction between students and teachers.
3. Administrator user.
(1) Teacher management: management of teacher authentication information, that is, user name and password.
(2) Student management: manage the teacher's identity authentication information, that is, user name and password.

4 System Overall Design

B/S structure has some advantages in system design, development and implementation. Information system based on B/S structure has good openness and expansibility. B/S structure is easy to realize in modular design management. B/S structure (Browser/server, browser / server mode) is widely used in the development and utilization of modern information technology. It is a network structure mode emerging after the development of web. After the installation of web browser and other related database software, it can easily operate the information in the system. Its main features are: simple maintenance and upgrading and low development cost. There are many kinds of implementation technologies of art teaching and research system [13, 14]. The most common network test system is the traditional architecture based on client / service model, In this mode, the smooth progress of each test must be related to the installation and configuration of each client, so the work efficiency is relatively low, and the operation is more cumbersome.

B/S structure features: it has strong operability, and is conducive to the maintenance of the system. The opening of the system is more prominent. In addition, aspnet technology has the characteristics of flexibility, security, expandability, friendly access and browser independence [15]. This test adopts browser server structure, which is composed of client and database server. Because the system adopts Internet related technology, the fine arts teaching and research system based on B/S mode adopts the web solution based on aspnet technology proposed by Microsoft. The solution is shown in Fig. 1.

The implementation process is as shown in. Users can send requests to the server to call the dynamic page through is server. The system will return the dynamic browsing page to customers through the automatic execution of foot code. Users can operate the static page through their own needs ADO.NET Directly access the background database, query the learning information, and then generate the HTML page, The system will return the corresponding query results to the user, the client department needs to install any application, only need to install the relevant application in the server; the maintenance and upgrade of the system are concentrated in the server.

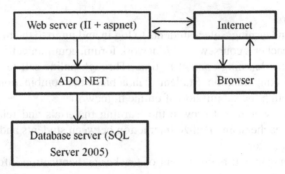

Fig. 1. General scheme

5   Detailed Design of Art Teaching Interactive Study System

The design and implementation of the art teaching and research system must be able to meet the needs of teachers and students in all aspects of education and teaching. At the same time, the school educational administrators should also be able to manage the system. The main purpose of the system website is to realize the combination of art teaching and research and information system, which can provide strong support for teachers and students [16–18]. Using this website, teachers can answer questions online, upload courseware and assign tasks; students can learn more information and knowledge by logging on the website; school educational administrators can have both the authority of teachers and students, and manage the information of teachers and students as administrators (see Fig. 2).

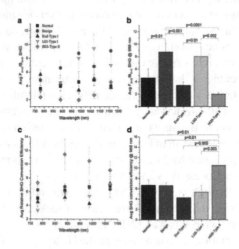

Fig. 2. Simulation result with art teaching interactive study system

5.1 The Design of the Main Function Modules of the System

In the learning system, it can provide certain navigation and search functions, such as arranging the location of network courseware in a certain order, and providing the keyword search function according to certain knowledge points, so that the system users can easily and quickly find the learning resources they need. The learning system can also record the individual learning situation of learners at any time, so that learners and teachers can master the relevant learning progress and make teaching assistant arrangements [19]. The main design of online learning function module is the upload of course resource function, and the main technology of course resource production function is aspne. Aspnet is a programming framework. It is a new framework based on the operation of common language. It can realize the establishment and generation of Web programs. Web programs have powerful functions. Compared with the traditional web program, the web program based on aspnet has great progress in function and operability.

5.2 Design of Online Operation Function

The design of online homework function requires teachers to arrange homework according to the progress of learning. As far as the art teaching and research mutual aid system is concerned, the assignment of homework is the same as other subjects, but most of the assignments are submitted by students who upload their own works. Teachers comment on students' works, and then realize the online homework function of the system, There is basically no standard answer, because the automatic evaluation of art painting is still very difficult to achieve, so it is basically through the teacher's evaluation of the work to achieve the scoring of the homework, the students' achievements will be recorded in the art teaching and research mutual aid system database, and put forward opinions, convenient for students to consult their homework.

6 Promotion Method

6.1 Interest Stimulation Method

Psychological research has proved that the biggest and most lasting learning motivation is students' internal interest in subject content. Only this kind of inner interest can stimulate students' thirst for knowledge. There are many ways to stimulate students' interest, from the following aspects can mobilize the enthusiasm of students () the use of physical import is the most intuitive, but also one of the most effective teaching tools. For example: in the lesson "perception and emotion of color", what kind of visual effect will a white ball of the same size compare with a black ball? If there is no real object, let the students think out of thin air, it will make the teaching content boring and too abstract, which will make the students lose interest in learning [20]. On the contrary, if the same size of white stone ball and Black shot put are used to import the real object, it will be clear to the students at a glance, It's intuitive and easy to understand knowledge, and the classroom atmosphere will also be active. 2) using riddles to lead into life is boring without humor, and it's boring without humor in art class. In art teaching, if we can use humorous language, the teaching process will be more relaxed and effective. It makes the distance between teachers and students closer, makes them form psychological compatibility and join in the new teaching situation.

6.2 Imagination Stimulation

In art teaching, we can recite all kinds of methods, such as vivid grammar description, music camp, face-to-face display, students' performance and so on, to create situations for students, so as to improve their imagination. In addition, a good art class in addition to a good lead-in, teachers in the process of carrying out art activities also need to fully mobilize the students' creativity (-) display excellent students' art works, stimulate creative consciousness, in the art classroom homework, collect some excellent students' works, in the process of students' hands-on production, in order to motivate students. -In the process of students' creation, teachers should be good at discovering the talents of each word, and guide them to express their innovative consciousness in their works. In classroom teaching, we should know their ideas in time and encourage their creativity. When evaluating students' works, we should try our best to find some shining points and praise them to make them have a sense of success [21]. Observation is very important in painting. As shown in Fig. 3. Without it as the basis of drawing circles, there would be no artistic expression. People, things and things in life can be regarded as objects of observation. Clear observation, correct observation methods and good observation habits can gradually improve students' observation ability. In the process of observing sketching, constantly improve their own strength.

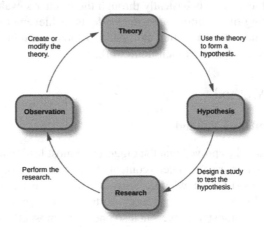

Fig. 3. The importance of observation.

6.3 Integration of Teaching, Learning and Doing

In the art classroom teaching, in addition to the teacher's vivid explanation, more time is the process of students' hands-on practice. This requires students to combine the teacher's knowledge and hands-on production ability, that is, the integration of teaching, learning and doing. For example: in the lesson "basic knowledge of color", the teacher uses the hue ring to let students understand the three elements and deployment methods of color, but the teacher's explanation is not as intuitive and solid as the students' hands-on practice [22–24]. For example, the teacher explained that Yellow + blue = green, but the result of students' practical operation is that each student's green is not the same, some students' green is light, some students' green is dark. The reason for this is the different proportions of yellow and blue. This kind of practice makes students not only grasp the knowledge, but also find new problems in practice, and finally sum up the reasons and solutions [25]. In a word, in the future life, we will continue to sum up the experience, and make full use of the interactive teaching method of teaching, learning and doing, which is the integration of teaching, learning and doing, And keep their painting naive, simple, full of interesting imagination art style, improve the quality of students' painting.

7 Concluding Remarks

In the development process of this system, a number of technologies are comprehensively applied, which are: asp net page technology, object-oriented programming based on C#, object-oriented programming based on C# ADO.NET SQL database query technology, XML technology. When the system is running, there are three modules: administrator module, teacher module and student module. In the development process of this system, the role of teachers in education and teaching has been greatly improved. Through this system, teachers can upload information related to learning information. For students, they can participate in learning, take part in tests, and check their scores at the same time. The function of the system administrator is responsible for the management of teachers and the system. Because of the advantages of this system, it can be used in other subjects. Through the design and implementation of this paper, the design and implementation of art teaching and research mutual aid system is realized in a low-cost environment. For art teaching, it has the advantages of simple operation. However, in the process of research, there are still great deficiencies, which need to be further improved.

Acknowledgements. Education and Teaching Reform Research Project of Xianyang normal University 2019 "Research on the Reform and Innovation of Art Curriculum system of Preschool Education Specialty under the background of excellent Preschool Teachers training" Project No.: Y07; 2019.
"Young Backbone Teachers" No. XSYGG201907, Xianyang Normal University, 2019.
Xiamen University from 2018 to 2019.

References

1. Zhang, H.: Practical research on Interactive Teaching in primary school art teaching. Popular Sc. Fairy Tale **22**, 154 (2020)
2. Ma, Y.: Interactive learning: an exploration of creating multiple learning spaces for high school fine arts. College Entrance Exam. **16**, 192 (2020)

3. Wu, X.: Construction of interactive teaching mode in primary school art classroom. In: Proceedings of Academic Conference on School Management And Teaching Innovation in 2020, Intelligent Learning and Innovation Research Committee of China Intelligent Engineering Research Association (2020)
4. Fan, R.: Teaching reform of landscape art course in secondary vocational landscape technology specialty. Tianjin Polytechnic Normal University (2020)
5. Li, Y.: Research on interactive teaching practice. Primary School Sci. (Teach. Ed.) **05**, 135 (2020)
6. Wang, T.: Practical analysis of interactive teaching mode in primary school art classroom. Acad. Weekly **16**, 149–150 (2020)
7. Ren, Q.: Interaction between teachers and students - a new teaching mode of high school art appreciation. Sci. Fiction Pictorial **05**, 262 (2020)
8. Yihui: Art observation (05), 40 (2020)
9. Cai, X.: On the application of "mutual assistance and interaction" teaching in art class. Navigation Arts Sci. (First Ten Days) **05**, 94 (2020)
10. Ma, L.: On the practice of interactive teaching mode in primary school art classroom. Huaxia Teach. **12**, 31–32 (2020)
11. Huang, Y., Bi, X.: Measures to cultivate students' free creativity in primary school art education. Chinese J. Multimedia Netw. Teach. (Next Issue) (04), 251–252 + 254 (2020)
12. Tang, H.: Practical research on art interactive teaching strategy in primary school under information technology environment. School Educ. **08**, 94 (2020)
13. Zhu, Y.: Discussion on art teaching and operation mode of Su fan Vocational College. J. Jiamusi Vocat. Coll. **36**(04), 115–116 (2020)
14. Zang, H., Wang, G.: Analysis on the connotation and strategy of interactive teaching in art appreciation class. Chinese Youth (11), 257 + 260 (2020)
15. Huang, Z.: Research on interactive teaching mode of secondary vocational art class. Chin. Foreign Entrepreneurs **12**, 225 (2020)
16. Zhou, T.: The use of interest groups in primary school art education and teaching. New Curriculum **13**, 69 (2020)
17. Zhu, X.: Practical research on interactive teaching mode in primary school art classroom. Art Educ. Res. **05**, 174–175 (2020)
18. Zhu, J.: Classroom "live" means students "live" – construction of interactive classroom teaching mode of junior high school art. Art Eval. **05**, 137–138 (2020)
19. Cao, H.: Practice and reflection of primary school art teaching in the era of "Internet plus". New Wisdom (07), 21 (2020)
20. Zhu, Q.: Art teaching strategies for home study in special period. Future Educator (z1), 50–54 (2020)
21. Yan, W.: Strengthening the interaction between teachers and students to promote art teaching and learning in primary schools. Primary School Era **07**, 36–37 (2020)
22. Ma, Z.: Research on the interaction between teachers and children in art teaching activities in middle class of kindergarten. Shanghai Normal University (2020)
23. Zheng, Y.: Exploration on the implementation of interactive teaching method in primary school art classroom. Inner Mongolia Educ. **06**, 93–94 (2020)
24. Guan, Y.: Tracing the origin of life and promoting regional culture. Navigation Arts Sci. (02), 74 + 76 (2020)
25. Zeng, L., Liu, J.: Multiple interaction and cross-border integration: a preliminary study on the teaching mode and method based on the connotation of public art in Sichuan Academy of Fine Arts. Public Art **01**, 28–35 (2020)

Design of Online Vocal Teaching Auxiliary System in Colleges and Universities Under the Background of Big Data

Meng Ning[✉] and Xun Luo

Huainan Normal University, Huainan 232038, Anhui, China

Abstract. In today's society, the Internet has penetrated into all aspects of daily life, and has become an indispensable part of people's life. In the teaching of vocal music, teachers and students constantly put forward problems that need to be solved in teaching. For example, due to the lack of teaching time, teachers do not have enough classroom time to teach comprehensive vocal music skills for students. The homework assigned by teachers needs to be checked and scored uniformly. The workload is heavy, and there is no communication channel between teachers and students except classroom time. In this paper, the application of software engineering, computer network, web site architecture and other aspects of knowledge, the requirements of the system are analyzed, and on this basis, the overall architecture of the system and its internal modules are designed in detail. In the process of implementation, with the help of ThinkPHP, the mature software framework, on the basis of which, the data persistence layer and presentation layer of the system are realized, and the business logic layer of the system is set up between the two layers to realize the hierarchy and module structure of the system design. After the implementation of the system, we also use the system testing method to test the system in an all-round way to ensure the correct use of the system.

Keywords: Vocal music teaching · Auxiliary system · MVC website architecture

1 Introduction

Information management system based on the Internet has always been an important part of the Internet, shouldering the responsibility of solving various practical problems. After long-term development, the industry has accumulated a lot of successful experience in building information management system. In the face of practical problems in the actual work and life, we can appropriately learn from the industry's previous successful experience to solve [1].

Music (especially vocal music) teaching and learning because of its own particularity determines that a large number of live demonstration teaching methods must be used between teachers and students to teach singing skills and practice skills guidance. However, in the actual teaching process, according to the schedule of the course, there

M. A. Jan and F. Khan (Eds.): BigIoT-EDU 2021, LNICST 392, pp. 549–558, 2021.
https://doi.org/10.1007/978-3-030-87903-7_67

is often not enough time for teachers and students to have face-to-face interaction after the necessary theoretical study.

For a long time, in vocal music teaching practice, this problem is solved by the way of transmitting and watching videos between teachers and students: teachers record the singing skills they need to teach students into videos, and then distribute them to students. Students who need to learn singing skills by downloading and watching videos. In the process of practice, if students have any questions or difficulties that are difficult to solve, they can record their own practice videos and send them to relevant teachers for guidance (see Fig. 1).

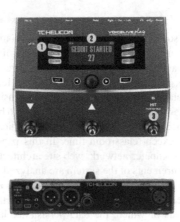

Fig. 1. Online music system

2   Overall Design

Before the overall design, it is necessary to clarify some basic design principles to be followed in this stage. In general, the overall design should make the designed system simple, flexible, stable and reliable, and easy to maintain [2]. Specifically, the following design principles will be followed to achieve this purpose:

At present, the common design methods are top-down method and bottom-up method. The latter is a more traditional method, which is suitable for building new systems with existing underlying designs. This method can quickly complete the construction of the system by combining the underlying modules when the underlying design is determined. At present, this method is not suitable for use, because this research project is to gradually design and implement a new system from scratch, and there is no ready-made underlying design for use. The top-down design method is more in line with the actual situation of the project. Because the core idea of the top-down method is to gradually divide complex and macro problems into simple and micro problems. After several steps of abstract processing, the problems can be described qualitatively and quantitatively, thus making the problems controllable. In addition, the top-down design method can gradually refine the design of the system in the process of design, or find the deficiencies in the previous design and make up for them, so as to make the design of the system more unified and coordinated.

The entity connection of teaching video management module is shown in Fig. 2.

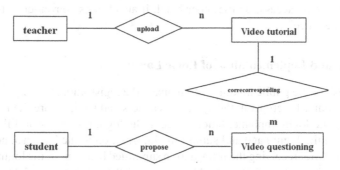

Fig. 2. ER diagram of teaching video

Group fitness F:

$$F = \frac{\sum_{i=1}^{n} f_i}{N} \tag{1}$$

Where: n is the number of individuals in the group, and N is the total number of elements in the training set that should be matched by all individuals in the group.

Coverage rate P:

$$P = \frac{\sum_{i=1}^{m} f_i}{N} \tag{2}$$

Where: n, N have the same meaning as formula (1), m < n, that is, the fitness of the subgroup composed of the first m individuals of the group.

3   Detailed Design and Implementation of the System

3.1   Design and Implementation of Data Persistence Layer

In the system, under ideal circumstances, each part of the system performs its own duties and has a clear division of labor. However, many times, data and system logic are closely coupled [3]. This will leave hidden trouble for the future maintenance of the system, because in each module of the system, data access operation will be carried out. If each module of the system does its own work and there is no unified data access scheduling center, it is easy to fill each module of the logic layer with similar data access representation. If the data changes in the future, there will be more than one place to be modified, which will easily cause omission or inconsistency after modification.

In order to solve this problem, it is necessary to add a data persistence layer in the system. The main task of the data persistence layer is to unify the data access operations to this layer. The modules outside the data persistence layer no longer implement their own data access operations. They only call the data access interface provided by the data persistence layer for data operations when necessary. On the one hand, it ensures the consistency of data access, on the other hand, it also brings convenience to the future system maintenance.

3.2   Design and Implementation of Logic Layer

The main task of the public module is to manage the registration, login, password modification and other functions in the system. Teachers and students are the main users of the module. The login implementation process: the login user needs to fill in the login information in the login page and submit the information to the background login processing page. After receiving the information submitted by the user, the login processing page encrypts the password submitted by the user with the same rules as the registration process, and queries the user's information in the system database together with the submitted user name. If the corresponding information does not exist, an error message will be returned. Otherwise, the system login page will set the session for the user on the server side.

3.3   Design and Implementation of Presentation Layer

The presentation layer is located on the outermost layer (the top layer) of the system and is closest to the user. As an interface for interaction between users and systems, presentation layer is often used to display data or receive data input from users.

The presentation layer has two important aspects to consider: interface style and data display. These are two different fields. The interface style needs the participation of special art staff and web front-end engineers to determine the best layout and presentation style of the user interface, while the background programmer is generally responsible for accessing the background data to the user interface. This leads to a problem: background programmers usually don't know the technology of the front end of the page, so when embedding background data on the page, the layout or style of the page may change inadvertently.

The communication mechanism between logic layer and presentation layer is realized by filling template parameters in logic layer. The template parameters and their types that need to be filled in when the template is used to display data will be described in the template instructions. The logic layer only needs to fill in the specified data according to the requirements, and does not need to care about the expansion of data in the template. Generally, there are two ways to fill template parameters: using the variables in the language to fill in the parameters, or returning the files in JSON or XML format by the logic layer, and then filling the template parameters by JavaScript operation data on the client side.

3.4 Summary of this Chapter

This chapter mainly carries on the detailed design of each level of the system. In this stage, the internal representation of each layer and the communication between layers are designed in detail, and the internal structure of each layer in the division of the overall design stage is refined. The logical rationality and necessity of the hierarchy and its internal module division are preliminarily verified, so that the outline of the system is more and more clearly visible [4]. On this basis, this chapter also briefly introduces the specific implementation of each layer of the system, and introduces the technical selection and technical difficulties in the system implementation one by one. So far, the preliminary implementation of the system has been completed, and it can be put into use after testing. The simulation of teaching effect is shown in Fig. 3.

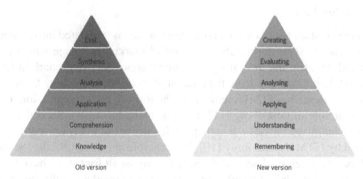

Fig. 3. Teaching effect simulation

4 Overall Design of System Technical Framework

Framework is an application skeleton that can be customized by developers according to their needs, which makes the whole or part of the system reusable [5]. It is represented as a group of abstract components and the method of interaction between component instances. In the application system, the biggest advantage of applying mature framework is reuse, which can reduce the burden of developers to establish solutions to complex problems [6]. With the increasing social needs and the rapid development of software technology, today's software system is very complex, especially the design of server-side application software, which will involve a lot of knowledge, content and problems. Using mature and robust framework in the development of complex application system can greatly simplify the development difficulty and workload, and make system designers concentrate on the business logic design of the system. An application system based on framework usually includes one or more frameworks, component classes related to the framework and function extensions related to the application system [7].

4.1 Web Design Technology

XHTML (the Extensible HyperText Markup Language) is a basic web page design language. XHTML is not only the successor of HTML language, but also a markup language based on XML. The syntax requirements of HTML are relatively loose. For web designers, it can be operated more freely and conveniently. However, for machines, the looser the syntax is, the more difficult it is to process, and the worse its compatibility with various browsers is. XHTML standardizes and rewrites HTML language through XML syntax, which is more strict. Therefore, XHTML is essentially a transitional technology, which combines the powerful functions of some XML and inherits the simple features of most HTML, showing good usability and compatibility [8].

Xhtmil language combined with CSS (cascading style sheets) can better reflect its advantages. The combination of the two can separate the style and content of the web page, combine the web page code organically, and even mix all kinds of XML in a single file, such as SVG, MathML and so on [9].

4.2   CSS Technology

CSS is an abbreviation for cascading style sheet, which is translated into cascading style form and cascading style sheet. CSS is a kind of markup language which is designed to control and enhance the design style of web pages, and it is a markup language to separate the style information from the content of web pages. By using CSS technology, the browser can describe HTML elements in the form of preset expression. CSS can be regarded as the rule of describing HTML elements by browser. CSS is the best performance design language based on text display. It can control many attributes that can not be controlled by HTML technology [10]. For example, it can control the location layout of Web objects to pixel level accurately, support almost all font style, have the ability to edit the image and model styles of web pages and can conduct preliminary interactive design. CSS language can be regarded as a breakthrough in web design. Through CSS technology, designers can control the style and layout of multiple pages at the same time. Website developers can define the unique style for each HTML element and apply it to any number of pages [11]. If you need to update the page globally, simply change the cascading style sheet, all elements in the website will be updated automatically. With the help of the powerful function of CSS, web designers can not only release the design inspiration fully, and beautify the web page as much as possible. CSS can simplify and optimize the code according to the understanding ability and programming style of different designers, so as to reduce the design difficulty and improve the work efficiency. In the development of Web site, CSS technology can also cooperate with struts tags to realize the design of web pages. The performance layer of the Java intelligent assistant teaching system designed in this paper is based on Struts2 framework. Struts2 provides a very rich tag to help the design of the performance layer. Many pages in this system use tags provided by Struts2. When calling struts tags, it is important to note that when defining a presentation style in the tag body, you cannot use the style keyword, but instead use cssstype keyword [12].

5   Design and Implementation of the Main Function of the System

5.1   Design and Implementation of Media Learning Environment

Media learning environment consists of three important functional modules: basic learning materials module, training and testing module and in-depth learning module. Among them, the basic learning materials module is a visual and interactive narrative e-book based on the syllabus and teachers' courseware, referring to the classic java books and network materials, and then designed through a variety of RA methods. The visual narrative e-book in the media learning module is completely realized by the background, and does not provide editing function. The training and testing module provides links to other functions of the system, which can be edited by teachers or administrators. The in-depth learning module provides links to network resources, which can be added by teachers or students according to their needs [13].

5.2 Design and Implementation of Test Exercise Environment

There are two modules involved in the system: the teacher oriented test resource management module and the student oriented test exercise module. The test resource management module is located in the website background management center, which is mainly responsible for providing teachers with visual test entry, classification, editing and other functions; the test exercise module is located in the java learning system, which is mainly composed of test exercise and homework system.. According to the actual needs of Java teaching, the question bank designed in this paper supports three types of test questions, which are judgment questions, multiple-choice questions and other objective questions with standard answers, subjective questions and programming exercises. The system can automatically mark the objective questions with standard answers and the programming exercises with test cases; for the subjective questions, due to the lack of marking standards, the system can not mark the subjective questions, so it can only hand over the answers submitted by the students to the teachers, and then the teachers can complete the review and feedback [14].

5.3 Question Resource Management

System for teachers and system administrators to provide two main ways of test management: manual editing and formatting test file entry. Teachers can edit the test questions manually according to their needs. The main operation functions are search, add, modify and delete. But the amount of test questions in the question bank is very large, especially for the objective questions [15]. Manual editing will add a huge workload and teaching burden to teachers, so the system provides the format file entry function for teachers and system administrators. Teachers and system administrators can upload specific format of Excel test questions and test paper files to batch add test questions and test papers, which can not only greatly reduce the burden of teachers, but also export the test questions in the Java question bank currently used by the school to excel file, and then import them into the Java Teaching system, that is, realize the data migration of the question bank through Excel.

5.4 Design and Implementation of Programming Training Environment

The background implementation of programming training environment is mainly composed of four parts: Java source program compilation error detection, logic error detection, foreground editing environment and class structure analysis. Compiler error detection is to dynamically compile java source program to detect whether there are syntax errors in the code submitted by students. Logical error detection means that when the source code submitted by students passes the compilation error detection, the system preset test cases are used to check whether the running results of Java code are correct in all aspects. In order to facilitate students to write code in the browser, the system not only provides a design friendly java source program editing interface, but also highlights Java keywords, and outputs members in the Java source program in the form of a list.

Generally, the compilation of Java source program is implemented by using javac command after the code is written, whether it is directly calling the command line or using the configured ide. However, the situation of programming training environment in this paper is special. It needs to compile java source code online (browser side) and get the results, which requires the system to have the ability to compile, load and run java source code on the server side. There are many ways to compile, load and run java source programs by calling Java API, the most common of which are: (1) to compile java source code by calling native javac commands. (2) The application program receives the data, generates java files and saves them to the hard disk, and then calls API to perform compilation operation. (3) Call Java API to compile and execute string data, and complete Java dynamic compilation, loading and execution in memory.

6   Research on the Key Technology of the System

6.1   The Necessity of RIA in Java Teaching Assistant System

RIA is the abbreviation of rich Internet application, which can be translated as rich Internet program. It has rich user experience, high interactivity and page design ability. The traditional web model is based on HTML page design, the page design is simple and needs frequent refresh, lack of browser side intelligent mechanism, almost unable to complete complex user interaction. With the development of the times, people's requirements for the complexity of network applications are increasing. The traditional web model has been difficult to meet the needs of users. The emergence of RIA solves this problem well. RIA absorbs the advantages of desktop applications, such as fast response and strong interactivity, and can provide a richer, more interactive and responsive user experience. RIA architecture can be understood as a CS application running on B/S structure. It not only has the advantages of strong interaction, fast response, rich pages and so on, but also has Internet, which is very convenient for users. The Java teaching assistant system designed in this paper has very high requirements for user experience, especially the design of media learning environment. Media learning environment not only provides users with visual and interactive narrative learning resources, but also embeds code running process demonstration and a small amount of programming training functions. These functions need strong page display ability, rapid response ability and complex interaction ability to achieve good, and the traditional web architecture is powerless for such a high user experience requirements, so RIA is the cornerstone of Java intelligent teaching assistant system design, which is necessary for the development and design of the system.

6.2 Overview of Ajax Technology

The full name of ax is asynchronous JavaScript and XML (asynchronous JavaScript and XML), which is a web development technology to create interactive web applications. Ajax technology is one of the most important technologies in Web2.0. It is a framework of U I function and concept driven by request and response server call model based on XML. AAX application is essentially a kind of RIA. Compared with other RIA technology frameworks, AJAX has obvious advantages: the application based on Ajax does not need to download any plug-ins from the browser, and has excellent compatibility, and can run well on almost any browser.

7 Conclusion

According to the characteristics of vocal music teaching field, this system studies and discusses how to use modern information technology to help teachers and students in art colleges to carry out vocal music education better and more efficiently. At the same time, it also reflects the humanization of students' education, so as to promote the development of modern technology combined with traditional education. This system is committed to the use of modern software engineering methods to build and develop a practical online vocal music teaching interactive platform, and in the development process, comprehensive use of various methods in software engineering as a guide, for vocal music education provides more possibilities, but also for students learning vocal music to bring more efficient learning methods.

References

1. Huang, T.: Management Information System, 4th edn. Higher Education Press, Beijing (2009)
2. Xu, R., Yu, R.: A new approach to software engineering object oriented methodology. Opt. Precis. Eng. **8**(6) (2000)
3. Zhiying, Z.: Basic Methods of Modern Software Engineering (Volume II). Science Press, Beijing (2002)
4. Chen, H., Wang, J., Dong, W.: High trust software engineering technology. Acta Elect. Sin., Beijing **S1** (2003)
5. Gang, L.: Integration of Struts + Hibernate + Spring Application Development Details. Tsinghua University Press, Beijing (2007)
6. Cay, S., Wang, H.: Java Core Technology Volume, 9th edn. China Machine Press, Beijing (2014)
7. Yin, S., Fu, L.: Java reflection mechanism research. Educ. Technol. Guide **11** (2008)
8. Sun, Y.: Compiler Principle and Implementation, vol. 4, pp. 4–17. Tsinghua University Press, Beijing (2005)
9. Xi, L., Column, Q.: On double subject teaching and personality education. J. Inner Mongolia Normal Univ. **15**(1), 21–23 (2002)
10. Gang, L.: Lightweight Java EE Enterprise Application Practice, pp. 173–733. Electronic Industry Press, Beijing (2007)
11. Piersol, K.: Building an open doc part handler. Mactech Magazine 10 (2005). Feng, X.: Teaching interactive platform based on ax and jQuery. Comput. Knowl. Technol. (2011)
12. Zhang, X.: Combination of EJ and O/R mapping. J. UESTC **35**, 559–563 (2004)

13. Zhang, T.: Research and Application of Agent in Modern Distance Education. Hunan University, Changsha (2006)
14. Konstantinos, S.: Nikoaos Avouris learning from Coabo rating Intelligent
15. Shoham, Y.: Agent-oriented programming. Artif. Intell. **60**(1), 51–92 (1993)

Discussion on the Innovation of Computer Virtual Reality Technology in the Training and Teaching of Aerobics in Vocational Colleges

Yongfu Wu(✉)

Sichuan Vocational and Technical College Suining, Sichuan 629000, China

Abstract. Virtual reality technology is a new technology which is based on the Internet and integrates a variety of information technologies. In only ten years, it has penetrated into the daily life of human beings. Based on the analysis of the principle of virtual reality technology and the actual characteristics of aerobics training, this paper expounds the application of virtual reality technology in the field of aerobics training. On this basis, through the analysis of examples, this paper further constructs the aerobics training innovation in vocational college education and teaching, and prospects the future development of virtual reality technology in the field of physical training.

Keywords: Computer virtual technology · Vocational colleges · Aerobics training

1 Introduction

Ancient Greeks are famous for their respect for human beauty. They believe that in all things in the world, only the body building is the most symmetrical, harmonious, solemn and most vigorous. Ancient Greeks like to use running, throwing, soft gymnastics and bodybuilding dance to exercise the beauty of human body. They put forward the idea of "Gymnastics exercises, music edifies the spirit". The appearance of gymnastics is an important factor in the formation of aerobics.

In ancient India, a yoga technique was popular for a long time. It combined posture, breath and mind closely. It adjusted body (posture), breathing (adjusting breath), adjusting heart (meaning to keep the dandian quiet), self-regulation of body and body by using consciousness, and body building, and reaching the life extension. Yoga fitness movement includes standing, kneeling, sitting, lying, lunge and other basic posture. These postures are consistent with the basic postures commonly used in the current popular aerobics. The pursuit of fitness and fitness by ancient people and the idea of combining gymnastics with music are the foundation of the formation and development of modern Aerobics.

At the end of 19th century and early 20th century, there were many gymnastics schools in Europe. Their innovation in theory and practice played a role in promoting

© ICST Institute for Computer Sciences, Social Informatics and Telecommunications Engineering 2021
Published by Springer Nature Switzerland AG 2021. All Rights Reserved
M. A. Jan and F. Khan (Eds.): BigIoT-EDU 2021, LNICST 392, pp. 559–568, 2021.
https://doi.org/10.1007/978-3-030-87903-7_68

the development of aerobics. In the early 1960s, it was the germination period of aerobics. It was the first physical training content designed by Dr. Cooper, a doctor of NASA, for astronauts. In the early 1980s, with the development of fitness fever and entertainment sports all over the world, aerobics has become popular in the world with its strong vitality. The United States is a country that has an important influence on the development of World Aerobics. Its representative, film star Jane Fonda, wrote a book "Jane fonta health art" according to his own fitness experience and experience. The book has caused a stir in the world since it was published in 1981. She has promoted the promotion of Aerobics worldwide by her present example. At the same time, since 1985, the United States officially held the annual Aerobics Championships, and determined the competition items and rules, so that aerobics development into competitive sports.

Aerobics not only develops rapidly in the United States, Britain, France and other countries, but also has been carried out in some developing countries and regions to different extent. The former Soviet Union has already included Aerobics in the physical education syllabus of the major, middle and primary schools. In Asia, Japan, Philippines, Singapore and other countries have also built many aerobics activity centers and fitness clubs. People begin to take aerobics as their main fitness mode, thus forming a worldwide "aerobics fever".

Aerobics originated in 1968. In 1983, the first aerobics competition was held in the United States and the first far east aerobics competition in 1984 was held in Japan. Therefore, aerobics has been widely used in all over the world. The annual international events are: World Championships of aerobics, world cup, world championships and world tour.

In September 1992, China aerobics association was established, with its headquarter in Beijing. In 1987, Beijing held the first national aerobics Invitational Competition, followed by four Invitational competitions in Beijing, Guiyang, Kunming and Beijing in 1988, 1989, 1990 and 1991. Since 1992, it has been renamed the national championship, which has become a traditional event held every year.

In addition, in 1992 and 1995, two National Aerobics Championships were held in Beijing. In 1998, the national championship and national aerobics games were held.

With the continuous improvement of people's living standard, more and more people pay attention to the practical value of health care, medical treatment, fitness, bodybuilding and entertainment. It attracts fans of different ages to participate in it and forms a certain scale of consumer group. TV stations at all levels have produced special programs with aerobics competition and popularization as their content, and their video reception rate is far higher than other programs.

Because aerobics competition can be held in the gymnasium and stage, and the characteristics of the field use concentration in aerobics, it creates opportunities for enterprises to carry out advertising and publicity in combination with the competition. Aerobics project is favored by more and more enterprises.

2 Definition of Concept

2.1 Virtual Reality

Virtual reality, or virtual reality, was put forward by Jaron Lanier, a famous computer scientist and founder of VPL research in 1989. Its basic meaning is as follows: (1) use computer graphics system to unify various display and management interface instruments. (2) It can be produced on the computer and can produce immersion feeling in the simulation environment of human-computer interaction. In other words, virtual reality is the description of the real world, with the framework of the real world, but also in the virtual environment for human-computer interaction.

2.2 Sports System Simulation Platform

Sports system simulation platform, namely sports simulation support environment, is a set of software [1]. The platform consists of assistant modeling, model debugging and operation support system. Through the organic combination of these systems, we can provide users with a simulated real sports environment in the whole simulation cycle.

2.3 Principle of Motion Attitude Detection Based on Nine Axis Sensor

Generally, Euler angle, quaternion, moment or axis angle are used to express attitude. Different expressions are used in different fields, and each expression has its own advantages, Euler angle and sicai onhard Euler are the representation methods used in this project. Euler angle was first proposed to describe the orientation of rigid body in three-dimensional Euclidean space. The following mathematical model can be used to describe the mobility.

For a reference system in three-dimensional space, the orientation of any coordinate system can be expressed by three Euler angles [2]. The reference system is also known as the laboratory reference system. It is a static coordinate system, which is determined by the rigid body and rotates with the rotation of the rigid body, as shown in Fig. 1.

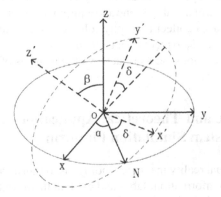

Fig. 1. Euler angle rotation coordinate system

The following mathematical model can be used to describe the mobility:

$$y = A - Tanh(B - K * x) \tag{1}$$

3 Discussion on the Application Principle of Virtual Reality Technology in the Field of Aerobics Training

3.1 The Basic Architecture of Aerobics System Simulation Platform Based on Virtual Reality Technology

Aerobics system simulation platform is a computer simulation system which uses virtual reality technology to simulate aerobics training. The core of the system is data collection, establishment of three-dimensional model and construction of virtual reality environment. The core of aerobics training simulation platform is the common basic structure. That is, data acquisition relies on position and direction tracking, data glove, conversion equipment, glove input conversion instrument and other components of the input system. The establishment of three-dimensional model depends on the generation of effect and the conversion of signal to build the defined output system to complete the aerobics virtual reality training environment. The main components are as follows: first, the user's use system; second, the three-dimensional space disposal instrument; third, the three-dimensional model database; fourth, the simulation manager; fifth, the computer and all kinds of feedback equipment.

3.2 Working Principle of Sports System Simulation Platform Based on Virtual Reality Technology

Data acquisition and analysis virtual reality technology essentially relies on the big data generation under cloud computing, that is, a large number of Aerobics data is input into the virtual environment generator, through the system software to analyze these data, and then use signal converter, raster display and other processing, so as to achieve the desired virtual effect [3]. The main purpose of data analysis is to realize the following functions, that is, to build training virtual reality scene and equipment in aerobics training, to capture Aerobics Athletes' data, to collect their physiological, biochemical and psychological data, so as to analyze the data, to reconstruct after deconstruction, to realize the repetition and display of aerobics actions, so as to achieve the analysis of training effect and the role of scientific selection.

4 Basic Principle and Theoretical Application Environment of Aerobics System Simulation Platform

The application of virtual reality media technology is to transform the basic movements of Aerobics into video information, label and explain them. In the process of training, according to the actual situation of teaching, the prescribed movements are repeatedly broadcast and taught, and then through the correct demonstration of the coach, the students can show clear Aerobics movements in their brains, so that they can intuitively grasp the essentials of the movements and speed up their learning, Promote the enthusiasm and initiative of students in learning and training. At the same time, it can also effectively find the error, and through the discussion of the causes of the error action, it can effectively correct the problem. In theory, virtual reality technology in the field of aerobics can be applied to the following three aspects.

(1) The virtual simulation training model and virtual reality training environment are established.
(2) Simulate the key actions.
(3) The recreation of Aerobics action.

5 The Application of Virtual Reality Technology in the Field of Aerobics Training

5.1 Characteristics of Aerobics Training

Aerobics combines dance art and sports, so in aerobics, the training content system of both has its own characteristics. Aerobics training includes three major contents, each of which can be decomposed. The three major items include comprehensive physical training, including strength, flexibility and endurance; action skill training, including basic action skills, standard action and difficult action; aerobics overall quality training, including aerobics integration ability, music experience and performance ability, aerobics creativity and psychological quality training.

5.2 Data Acquisition and 3D Model Construction of Virtual Reality Technology

Using virtual reality technology to carry out aerobics training, the first step is to capture the trainer's movement data. In the construction of human skeleton, quaternion representation is widely used in the field of virtual reality to calculate the rotation of joint points. By obtaining the rotation of all joint points, the root joint point is located, and the human motion data is collected through the relative position. After obtaining a large number of relevant data, the aerobics training database is established, and the three-dimensional model of Aerobics athletes is established by reorganizing the data. Generally speaking, the model is an athlete in virtual reality. In the performance of aerobics, it can be said that the model in virtual reality is equal to the real athlete. Through the analysis of virtual athletes, aerobics athletes will greatly improve their awareness of their own state (see Fig. 2).

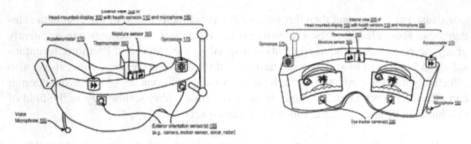

Fig. 2. VR simulation and use

5.3 Construction of Aerobics System Simulation Platform

In the field of aerobics training, in order to improve the training efficiency, we can use virtual reality technology to build aerobics system simulation platform. The platform consists of three independent and co operating sub platforms, that is, action generation sub platform. Its function is to rearrange the aerobics action based on the data captured by sensors on the scientific basis. The sub platform of action mode design is to design the action mode of aerobics. The group formation mode change simulation sub platform is a simulation model based on virtual reality technology [4–6]. Through data input, the aerobics formation will be changed from the initial state to the final state. Under the virtual reality environment, the mode of model team change can provide help for the aerobics formation mode design (see Fig. 3).

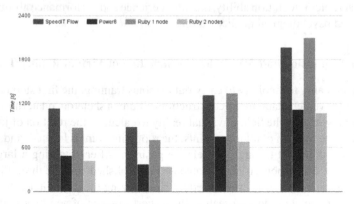

Fig. 3. Simulation time for VR

6 The Dilemma of Improving Aerobics Teaching in College Physical Education

With the continuous development of curriculum reform in Colleges and universities, aerobics is no longer a fixed traditional teaching method. Under the guidance of the idea of health first and lifelong education, our College Aerobics teachers have been able to choose the main content of aerobics teaching according to students' interests and hobbies. At the same time, it can also stimulate the independent exercise of contemporary college students and cultivate their awareness of exercise. However, with the further deepening of the curriculum reform in Colleges and universities, there are still many drawbacks in our aerobics teaching, which form a dilemma for Contemporary College Physical Education and hinder the normal development of aerobics teaching.

6.1 Under the New Curriculum Reform, the New Goal of Teaching Idea is Combined with the Modern Teaching Mode

Now our College Aerobics Teaching Reform, although has been fully spread out, but our traditional teaching methods can not be completely abandoned. In order to set the goal, we need to realize the reform of teaching, improve our syllabus and teaching mode, which has become an important content in the process of college physical education reform [7]. However, in the process of the teaching of health exercises, the demonstration teaching of our university teachers still plays a leading role. The teaching of university teachers can not be completely wiped out, and the whole classroom should be handed over to the students. There must be a big difference between College Students' active learning and passive learning, but without listening and imitating the learning process, we can not achieve the set goal of teaching [8, 9]. After we give the classroom to college students, although there are more exchanges between college students, such exchanges are not for cooperation in aerobics, there will be many students in the process of laziness and chatting. Therefore, the aerobics teaching in Colleges and universities should be combined with the new teaching goal and teaching mode. Let the students in this environment really aerobics exercise, further improve the creative ability.

6.2 We use the Aerobics Teaching Material is not Targeted, the Content of Theory Teaching is not Enough

Through the analysis of many versions of aerobics teaching materials, we will find that the order is not targeted enough, and the service object is to face the students of Public Physical Education in Colleges and universities or the students of Aerobics major in physical education colleges and universities, whether it is the teaching materials used by teachers or students' learning, and whether it is the teaching materials of compulsory courses or elective courses, Almost all teaching materials are the same, so the result is that it is difficult to reflect the special nature of physical education [10–12]. At present, the teaching content of Aerobics theory course in most universities only includes the sports value, overview, characteristics, significance of aerobics and a brief introduction to the methods of competitive aerobics, There are only a few sports colleges and universities involved in the scientific exercise of aerobics, as well as the principles, sports evaluation, methods, physiological health and sports health knowledge creation methods. We should advocate fitness. The basic theory of sports fitness is these insiders. Only by mastering the scientific principles and methods of fitness, can we better integrate theory with practice and get healthy development in practice.

7   Analysis on the Strategy of Innovative Education in College Physical Aerobics Teaching

Aerobics teaching must combine theory with practice, and the theoretical framework can not be ignored. We use multimedia to carry out the teaching process of practice course and theory course, and combine theoretical knowledge with practical action. In addition, our university teachers can widely use the social media such as microblog, QQ and wechat, which are widely used in college students, to learn the theory and experience of mutual exchange. We should strengthen the knowledge construction and promote the continuous development of Aerobics Teaching in Colleges and universities. Just as colleges and universities can adopt campus forum, we need to discuss the rules of competition, the cooperation and innovation of techniques and tactics, and the results of competition.

7.1   To Build a Matching Aerobics Teaching Mode with the Goal of Students' Ability Training

In the classroom teaching of Aerobics in Colleges and universities, we should establish a healthy educational concept, advocate students to actively learn aerobics, and carry out with various teaching methods. Through improving personal ability training as the basis of aerobics teaching. In the process of physical education and teaching in Colleges and universities, we should take the ability training of college students as the basis to build a modern aerobics teaching method in line with the aerobics syllabus [13, 14]. As shown in Fig. 4. We should use scientific teaching methods and teaching means to reflect the theory and practice of aerobics. We must let students understand the connotation of aerobics. We should build on the basis of basic movements and exercise essentials to promote students' autonomous learning and practice their own ability.

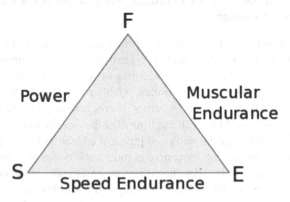

Fig. 4. Goal of students' ability training

7.2 Scientific and Reasonable Construction of Aerobics Evaluation System

Aerobics in university courses should be assessed according to the contents of the semester, and the grades should be used to evaluate the results [15]. However, this kind of evaluation mode will have certain limitations and compulsion, which is separated from all aspects of education. We should respect students' subjective consciousness and needs, and promote the healthy development of College Students' personality. One of the basic ideas of the setting of aerobics teaching content is that the assessment of aerobics course is flexible [16–18]. According to this teaching concept, the evaluation of aerobics class should pay attention to the mastery of College Students' sports skills and the improvement of College Students' comprehensive quality, including college students' sense of social responsibility, lifelong exercise desire and ability, innovation and practice ability. Only by making our evaluation results more reasonable and objective, can it be helpful for our university teachers to use feedback information to guide teaching.

8 Epilogue

In a word, virtual reality is creating a new world with the trend of destroying the past. To borrow a sentence from the Goldman Sachs report, fundamentally speaking, virtual technology has created a new and even more intuitive way to interact with computers [19]. Virtual reality technology will provide us with a greater vision. As far as aerobics training is concerned, people will no longer be bound by the original training system, which means that the easy-to-use property and wide application potential of virtual reality technology, coupled with the broad development [20, 21]. Promotion and development prospects of sports simulation training system, this system and technology will be greatly developed in the field of sports training.

References

1. Wen, H.: Practice of virtual reality technology in aerobics training. Sports Fashion **4**, 48–104 (2020)
2. Qi, J.: Discussion on the application of virtual reality technology in aerobics training. Sports Fashion **8**, 57 (2019)
3. Song, M.: Research on the application of virtual reality technology in aerobics training. J. Chifeng Univ. (Nat. Sci. Edn) **35**(5), 139–141 (2019)
4. Wang, J.: Discussion on the application of virtual reality technology in aerobics training. Contemp. Sports Sci. Technol. **7**(24), 46–48 (2017)
5. Yang, H.: Research on the influence of aerobics training on College Students' psychological quality. Sports Culture Guide **11**, 114–116 (2013)
6. Chen, Y., Men, Y.: The influence of Aerobics Teaching on College Students' body self-concept. J. Shenyang Inst. Phys. Educ. **1**, 112–115 (2014)
7. Zhang, L., Zhang, L., Yang, N., et al.: Effects of long-term fitness activities on impression management and mental health of college students. J. Beijing Sport Univ. **9**, 114–118 (2013)
8. Lu, X.: Research and Analysis on the influence of Aerobics on female college students' psychological quality. Hubei Sports Sci. Technol. **6**, 91–92 (2010)
9. Liu, J., Zhao, S.: The influence of different class types of physique and aerobics on female college students' body shape, function and quality. J. Beijing Sport Univ. **12**, 87–89 (2006)

10. Zhou, Y., Zeng, Q.: Effects of aerobic exercise on female college students' physique, anxiety and mood. J. Guangzhou Inst. Phys. Educ. **5**, 47–49 (2004)
11. Ge, Z.: Practical research on Innovative Education in College Aerobics Teaching. Sports **2**, 103–104 (2017)
12. Chang, H., Cao, X.: Analysis on the application of happy sports teaching mode in College Aerobics Education. Sports Suppl. Technol. **8**, 73–74 (2015)
13. Wang, C., Zhai, F.: Analysis on the value and current situation of offering aerobics elective course in Colleges and Universities. J. Shijiazhuang Univ. **5** (2005)
14. Wang, C.: The integration and development of college sports and community sports in China from the perspective of harmony. J. Nanjing Inst. Phys. Educ. **22**(3) (2008)
15. Li, S., Gao, J.: On the qualities of physical education teachers. J. Heilongjiang Inst. Educ. **6** (2008)
16. Wei, Y.: Factors influencing the reform of physical education curriculum in Colleges and Universities. J. Heilongjiang Inst. Educ. 6 (2008)
17. Lin, D.: On the harmonious development of teacher student relationship in College Physical Education. Shijiqiao **12** (2007)
18. Liu, Y.: Application of cooperative learning mode in College Aerobics Teaching. Contemp. Sports Sci. Technol. **3**(4), 1–3 (2013)
19. Li, L., Ji, L.: The influence of Aerobics on the personality development of primary and secondary school students. China Sports Sci. Technol. **45**(5), 116–121 (2009)
20. Wu, Y.: Current situation and Countermeasures of Aerobics Research in Colleges and universities in China. J. Nanjing Inst. Phys. Educ. **16**(1), 35–37 (2002)
21. Yang, L.: Practice of innovative education in College Aerobics Teaching. Contemp. Sports Sci. Technol. **4**(31), 90–92 (2014)

Empirical Analysis on the Effect of Actual Combat Elements in Simulated Taekwondo Teaching Based on VR Technology

Yuan Song[⊠] and Jian Liu

Department of Physical Education,
Chongqing University of Technology, Chongqing 400054, China

Abstract. Based on the competitive, ornamental and practical features of VR technology, most students in Colleges and universities choose Taekwondo Courses because they want to learn some practical fighting skills through Taekwondo Courses, constantly improve their own accomplishments, and advance to higher levels and sections. At present, the teaching contents of Taekwondo Courses in Colleges and universities are guided by WTF competitive Taekwondo, and the teaching contents are compiled according to the technical specifications and competition rules of WTF Taiwan Association. The teaching mainly focuses on the teaching of competitive Taekwondo technology and character, and its content has certain limitations. It can not fully meet the students' expectations of selecting courses, so that students can get the Taekwondo skills they want in class. In view of this, this paper uses the methods of literature review, investigation, interview and comparative research to analyze the teaching content of Taekwondo Courses in Colleges and universities, and to compare with TF Taekwondo technology.

Keywords: VR technology · Taekwondo teaching · Integration · TF actual combat elements

1 Introduction

Among the three applications of visc, virtual reality is one of them, which has been developed rapidly in recent years. Virtual reality is a kind of simulation environment which can replace the real world by comprehensively using computer system and various special software and hardware. This environment is real and credible for the user's sense. Virtual reality has the characteristics of immersion, real-time and interaction, so it has been widely used in manufacturing industry. Virtual reality can realize interactive visual simulation and information exchange. It is an advanced digital man-machine interface technology.

Virtual reality, with the virtual nature beyond reality, is called the three most promising technologies in the 21st century, together with network and multimedia technology. It is changing and influencing our life. At present, the most widely used in this field is the

M. A. Jan and F. Khan (Eds.): BigIoT-EDU 2021, LNICST 392, pp. 569–578, 2021.
https://doi.org/10.1007/978-3-030-87903-7_69

special workstation produced by SGI sun and other manufacturers. Image display equipment is the key peripheral device used to produce stereo vision effect. At present, the common products include light valve glasses, three-dimensional projector and helmet mounted display.

2 Related Work

The establishment of virtual scene is the core content of virtual reality technology. It is the condition of producing immersion and Realism: the scene is too simple, it will make the user feel false; and the complex and realistic scene will increase the difficulty, reduce the speed of rendering, and affect the real-time. At present, there are three main ways to construct virtual scene: 3D geometric model modeling and rendering based on computer graphics, also known as graphics based modeling and rendering, which is a traditional virtual scene construction technology; advanced modeling technology, image-based modeling and drawing technology and image-based rendering technology.

Many phenomena in nature have the characteristics of self similarity. M Andel BMT points out that the main characteristic of a fractal surface is its self similarity and steady increase. Natural terrain is likely to be a typical fractal surface. In the terrain description model established by berry and hanayca, the famous nanogram function is given:

$$E[X(x) - X(x+d)]^2 = k(d)^{2h} \tag{1}$$

And the power spectral density of the terrain profile:

$$G(w) = 2\pi k w^{-\delta} \tag{2}$$

In the above two formulas, X (x) is terrain elevation, E is statistical expectation, and H is statistical parameter describing terrain change. Formulas (1) and (2) have been widely used in geosciences. They are a classical model and are mainly used to represent Taekwondo.

3 Analysis on the Teaching Contents and Causes of Taekwondo

Taekwondo in college physical education has experienced the promotion and maturity stage, and is now in a stable stage. According to the survey of students, many students have different views on the teaching content of Taekwondo due to the influence of we media. Students no longer consider the performance attribute of Taekwondo, but pay more attention to the actual combat role of Taekwondo, This is also related to the fact that students classify taekwondo as martial arts at the beginning of course selection. In the course selection motivation, there are the plots of strengthening body and practicing martial arts to defend oneself.

At present, the teaching content of Taekwondo in Colleges and universities is mainly in accordance with the competition technical requirements of the World Taekwondo Federation, as well as the technical and action tests needed to be completed in the promotion level and stage level in Shanxi [1]. On the other hand, the teaching of Taekwondo is also constantly reforming and enhancing the technical movements that students are willing to accept and are willing to learn, especially those high difficulty turning technology, rotating technology and flying technology. However, these high difficulty technologies not only have higher requirements for the physical quality of practitioners, but also have certain risks. With the physical requirements of zero accident management of physical education courses in Colleges and universities, No teacher is willing to take the risk of safety to teach this stunt, so Taekwondo has lost its appeal to students..

4 Simulation Analysis of Adding TF Practical Skills to Taekwondo Teaching Based on VR Technology

4.1 ITF and WTF

World Taekwondo Federation (WTF) and International Taekwondo Federation (TF) belong to different system schools of Taekwondo. The World Taekwondo Federation (WTF) was founded in Seoul in May 1973. Jin Yuncheng was elected as the president. At that time, more than 20 countries from all continents joined the organization. In 2000, the Sydney Olympic Games included taekwondo as an official competition item. The entry of WTF into the Olympic Games had a great impact on Taekwondo [2]. In 1980, the International Olympic Committee officially recognized WTF. In order to make Taekwondo develop faster and better, and provide the exchange and exchange of martial arts skills among Taekwondo practitioners, with the continuous efforts and advocacy of general Cui Hongxi, taekwondo practitioners from military to civilian have developed rapidly in South Korea, which has become an indispensable competitive event in all major competitions in South Korea. Since Taekwondo entered the Olympic Games, WTF rules have been widely spread and applied in the world through media propaganda, and practitioners have sprung up all over the world. The International Taekwondo Union (TF) was founded by general Cui Hongxi, a two-star major general of the South Korean army, in Seoul, South Korea on March 22, 1966. The name of the 24 sets of routines is called ter, which symbolizes 24 h of a day from heaven and earth to unity. The name of each group is taken from the outstanding historical figures and national heroes who have never invaded other countries in Korean history for nearly 5000 years; In terms of technology, it emphasizes the control and exertion of force to the limbs. It is allowed to use the combination of boxing and leg techniques to obtain the scoring points in the match. During the competition, the players should wear boxing and foot covers to protect the opponent and reduce the damage. ITF competition is divided into five parts: Tel, duel, stunt, power and body protection.

4.2 Analysis of the Current TF Taekwondo Teaching Content and Competition Action

Tae (TAE), which means kicking and bumping with feet, Kwon with fist and do, is an artistic principle and method. Taekwondo is an artistic method of kicking with boxing and feet. It is mainly footwork, with 70% footwork and 30% boxing. In the TF system, there are many differences between the competition requirements and WTF. First of all, in terms of boxing, WTF only allows forward boxing, that is, PA run Ju Mok is a technique that uses the front of the clenched fist to attack the front of the opponent's trunk in a straight line with speed and power. Although the technical attack of boxing is allowed, it limits the position and method of boxing, and in the process of Taekwondo competition, few competitors use boxing to score. However, in the process of I f competition, it is quite different. In the process of competition, not only can you use the fist attack, but also can use the hand knife attack to score. Moreover, the score of the fist is the main scoring technique, and it can also hit the face, with the score ranging from 1 to 2. Secondly, in the two systems, the use requirements of leg techniques are similar, but there are great differences in the process of competition. In professional WTF competitions, such as the Olympic Games, we can hardly see fancy leg techniques. Instead, we use the advantages of height and leg length to win the competition by swinging and hitting each other's head with hook and loop as much as possible. However, in the process of TF competition, due to the more use of boxing, its ornamental value is better than that of WTF, but the intensity is different from that of WTF. By watching the competition of TF and WTF, and looking for relevant information, it can be concluded that ITF is more practical and practical than WTF [3].

4.3 Feasibility Analysis of Adding TF Techniques to VR Taekwondo Teaching

Taekwondo belongs to the same field of competitive sports, with a lot of advantages, such as large amount of physical contact exercise, in college sports elective courses have a wide range of student groups, many colleges and universities Taekwondo elective courses are full of thought, due to the strength of teachers and venues, finally have to through flexibility and flexibility and other special qualities will return some students, let them choose other special subjects. Therefore, it can be seen that Taekwondo elective course in Colleges and universities has a mature foundation. In Colleges and universities, physical education courses are usually offered in freshmen and grades. Different colleges and universities have different requirements [3, 4]. However, in order to maintain the continuity of physical education courses, most colleges and universities encourage students to keep the same physical education curriculum options in the first and second year. However, in recent years, due to the monotonous teaching content of WTF, students are encouraged to keep the same options, Many students have lost their interest in Taekwondo elective courses in their sophomore year, which makes it different from the freshman full situation in the process of setting up taekwondo in the second year of University. Therefore, the necessity and urgency of teaching content reform of taekwondo course are obvious. Through the author's questionnaire survey on 2017 undergraduate and 2017 graduate students of Xi'an University of Electronic Science and technology, the results show that more than 50% of the students hope that taekwondo course can increase combat skills and antagonism, which can also meet the students' motivation to choose courses, continue to keep Taekwondo Courses better carried out in Colleges and universities, and become a sports event that college students are willing to practice. It can be seen from the above that it is feasible to integrate ITF practical Taekwondo elements into college Taekwondo curriculum. The simulation software is shown in Figs. 1 and 2 [4].

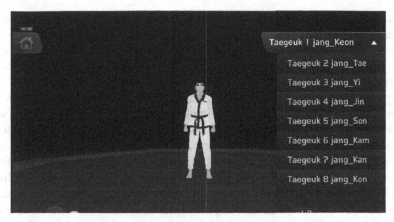

Fig. 1. Simulation software for Taekwondo with VR

Fig. 2. Simulation software for Kundo with VR

5   Research Results and Analysis

5.1   Significance of Training Students in Taekwondo Teaching of Physical Education Major

In order to study the major of physical education, we should have some understanding of its connotation. In 1988, the State Education Commission put forward in the "explanation of the revision of the catalogue of PE Majors in national colleges and universities": the division of majors should be based on the discipline system, with due consideration given to the work needs of the business departments, each major must have its own theoretical system and Curriculum system, and the professional direction and business scope should be obviously different from other majors [5]. Since the National Education Commission changed the original "physical education major" to "physical education major" in 1988, it has put more emphasis on the normal nature of physical education major. From the relative independence of Physical Education (in a broad sense) and the normal requirements of physical education major, physical education major has two attributes of "physical education" and "education". Therefore, how to deal with the relationship between the two is the physical education curriculum reform must be considered. The training goal of physical education major is to cultivate high-quality comprehensive talents, not only to learn knowledge, but also to know how to apply it, so as to achieve the purpose of teaching and educating people. From the perspective of physical education specialty, physical education specialty itself is comprehensive. It is not only a physical education specialty, but also a normal education specialty of higher education. In essence, physical education belongs to the normal major, and its normal nature is reflected by "professional" courses, and it is the requirement of the times to reflect a certain degree of comprehensiveness in professional courses and professional basic courses. Students majoring in physical education are responsible for teaching and educating people. It is the goal of the school to set up reserve talents for the development of physical education. To cultivate students majoring in physical education is to transmit physical education work to the society, spread all kinds of sports activities, and let people know and understand all kinds of sports [6].

5.2 Characteristics of Taekwondo

Taekwondo belongs to an open, multiple variation combination of action technology structure, is a "fast" to win, to "smart" to win the sport. The uncertain factors often appear in the competition, so in the actual competition, any action of the athletes will be constantly interfered by the opponents and fierce confrontation, which runs through the struggle of restriction and anti restriction, exertion and anti exertion [7]. The unpredictable factors always surround both sides of the competition, and the technical actions are difficult to implement according to the usual training program, we must constantly adjust our technical actions according to the changes of the situation on the spot, and take appropriate countermeasures and action methods [8]. In other words, in a real competition, technical elements include: vision, body center of gravity, sense of distance, timing, judgment, speed, hitting strength and adaptability. Taekwondo belongs to the category of fighting events. In terms of its technical essence, its purpose and function is to attack the opponent without being attacked, or to resolve the opponent's attack and counterattack. As far as its function is concerned, taekwondo technology is mainly divided into seven parts: attack technology, counterattack technology, defense technology, connection technology, fake action technology, combination technology and comprehensive technology. Taekwondo competition is a comprehensive competition project of intelligence, physical strength, technology, skill and will quality of both sides of the fighters, which has a high degree of actual combat and fierce antagonism. In the application process of Taekwondo technology, only to grasp its inherent laws and characteristics, is the premise of success [9]. From the movement structure and characteristics of Taekwondo, whether it is attack or defense, the basic elements of its movement are nothing more than the point, line, distance and speed, weight, accuracy in attack and slow, stable and flexible in defense. Only by paying attention to these basic elements and applying them to the practice of training can we achieve the goal of conquering the enemy. We should see that: action is controlled by consciousness, to act according to circumstances, we must have conscious control, otherwise, any action we want to implement is impossible. Therefore, consciousness forerunner becomes the connotation and core of technical action. It includes observation, judgment, foresight and special sense and perception of

Taekwondo (sense of foot, sense of distance, sense of opportunity, sense of space, sense of rhythm, etc.) [10].

5.3 Moral Cultivation for the Body, Etiquette for the Use

Taekwondo sports through physical practice and training to achieve self-cultivation, improve the noble purpose of personality [11]. Physical behavior is the external performance, and its connotation is not only the skills and skills of Taekwondo, but also the reflection of people's inner world, including emotion, sentiment, behavior and quality. The teaching and training of Taekwondo is not only a process of strengthening the body and cultivating certain self-defense skills, but also an effective means of cultivating Tao's sentiment and noble moral character. As a taekwondo Professor, he should not only "teach" and "solve doubts", but also "preach"; he should not only have a good professional and technical level, but also strictly require himself in the code of conduct and be a model for students; he should be a good teacher and a good friend to create a harmonious and lively teaching environment, in the process of imperceptibly cultivating and improving students' good moral quality and indomitable fighting spirit [12].

6 The Role of Multiple Intelligences in Taekwondo Teaching of Physical Education Major

Traditional Taekwondo Teaching exaggerates the role of sports intelligence and overemphasizes the importance of double Basics (basic knowledge and basic skills). According to Gardner's multiple intelligences, it is necessary to re integrate the teaching content of Taekwondo to connect it with more intelligence. From the perspective of the essential attribute of Taekwondo sports technology, we can think that taekwondo course belongs to skill learning courses, belongs to the field of physical cognition; at the same time, most of Taekwondo learning is completed in the practice of mutual cooperation between students, which needs the interaction between people. To sum up, it can be concluded that the nature of taekwondo course is a practical course based on competitiveness and multiple intelligences, which makes it necessary and feasible to use multiple intelligences to realize the diversified construction of taekwondo course. In physical education teaching, taekwondo course has a very high demand on students' comprehensive intelligence. On the contrary, reasonable and effective teaching methods can help students improve their multiple intelligence levels, mainly in the following four aspects [13].

(1) Sports intelligence, taekwondo and other sports technology, as long as we provide students with reasonable teaching means, promote students to practice effectively, we can improve students' physical ability to a certain extent. (2) Space intelligence, taekwondo technology, in addition to a single technical action, more is the effective completion of the combination action, especially in the actual competition process, athletes need to make judgments at any time, control the body space feeling. (3) Interpersonal intelligence, taekwondo most of the technical action learning needs to be done in pairs or even more people with practice, so the cultivation of students' interpersonal intelligence is indispensable. (4) Self cognitive intelligence, after a day of hard training or the end of a game, the teacher asked the students to make training notes or summary of the game,

record their own feelings, analyze the key to success or failure, and formulate further goals. This link focuses on training the students' self cognitive intelligence. Therefore, Taekwondo Teaching is not only a project that can develop students' physical and sports intelligence, but also can cultivate students' spatial intelligence, self cognitive intelligence and interpersonal intelligence. Therefore, taekwondo course can well develop students' multiple intelligences [14].

7   Teaching Design of Cultivating Students' Multiple Intelligences in Taekwondo Teaching of Physical Education Major

The traditional teaching of Taekwondo only pays attention to the improvement of students' skills and tactics, but ignores the development of students' multiple intelligences. Taekwondo itself requires students to have multiple intelligences to promote the improvement of their skills and tactics. The two complement each other. Therefore, this research is to combine Taekwondo Teaching with multiple intelligences theory, cultivate students' technical and tactical ability, improve students' multiple intelligences ability, and make students become comprehensive talents. According to the characteristics of multiple intelligences theory, scientific research theory and Taekwondo Teaching, after widely soliciting the opinions of experienced teachers, the author tries to find the connection point between the two and set the effect index of this teaching experiment [15].

Taekwondo classroom teaching is a multi sensory learning experience course, which provides students with rich learning opportunities, including visual, auditory, tactile, discussion, cooperation, reflection and other ways. Teachers should choose the best teaching strategies according to their teaching objectives. Of course, even if we choose four different teaching methods to develop students' four intelligences in Taekwondo class, it does not mean that we need to teach in four steps. Some activities are carried out separately, but some activities can happen at the same time. The following seven teaching methods are often used in the teaching of Taekwondo combined with the theory of multiple intelligences. No matter what kind of teaching method is, it will reflect the cultivation of students' one or several intelligent abilities [16]. The teaching design of cultivating students' Multiple Intelligences in Taekwondo Teaching is shown in Fig. 3.

Demonstration teaching is one of the commonly used methods in Taekwondo Teaching. It refers to the learning method that students accept technology through their own visual perception by observing the teacher's correct demonstration actions. Through the demonstration of technical movements, teachers can let students understand the structural characteristics, technical essentials and skills of the learned technical movements, and quickly establish the movement representation. The specific requirements are as follows: 1. The demonstration should be standardized and concise: teachers should ensure the quality of demonstration actions, be standardized, coordinated and fluent, and be accurate from the order of action, the track of movement to the coordination of all parts of the body. 2. The demonstration should highlight the diversity and pertinence: first, it should be based on the actual needs of students, not blind demonstration; second, it should be based on different teaching stages, teaching objectives and tasks to carry out targeted key demonstration; third, the action demonstration should be conducive to students' observation: the teacher's action demonstration should let all students see as

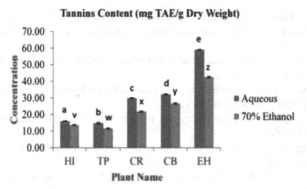

Fig. 3. Teaching design of cultivating students' Multiple Intelligences in Taekwondo Teaching

the criterion, Fully consider the speed, angle and other factors. 3. Demonstration teaching should be combined with explanation and enlightening students' Thinking: for new teaching technology, it can be explained first and then demonstrated, or demonstrated first and then explained, or demonstrated while explaining; at the same time, some small questions should be designed for students to answer, so as to inspire students to think about technical actions.

8   Conclusion

Compared with the traditional martial arts as a new school of martial arts system, KUNDO has been highly valued and favored by different education and teaching circles since it was introduced into China. Through the development of Taekwondo, it has continuously absorbed the cultural essence of Chinese traditional martial arts. Under the new historical development background, it has gradually become the mainstream sports in the world. The world competitive sports pattern is constantly changing. The presidents and presidents of ITF and WTF have repeatedly expressed their good wishes and have the intention of merging. This is also a great joy of Taekwondo. Through the merger, the gap brought by historical reasons can be filled, laying the foundation for the development of Taekwondo, paving the way for the historical inheritance and continuous innovation and reform of Taekwondo. Therefore, for practitioners of different Taekwondo systems, we should abandon the common sense, actively carry forward the techniques and tactics of Taekwondo, better serve the development of human sports, better guide the people to keep fit, and promote the peaceful, friendly, healthy and orderly development of all mankind.

References

1. Sun, H., Sun, W.: Research on the effectiveness of VR technology in helping students overcome psychological barriers in actual combat learning of Taekwondo. J. Shandong Agricult. Eng. Univ. **37**(4), 25-26+29 (2020)

2. Sun, H.: Practical research on immersive learning based on VR technology in Physical Education Teaching – Taking taekwondo course as an example. J. Dali Univ. **4**(6), 95–100 (2019)
3. Lin, H.: Analysis of Taekwondo teaching methods in Colleges and universities. Contemp. Sports Sci. Technol. **27** (2014)
4. Li, P.: Problems and countermeasures of Taekwondo teaching in higher vocational colleges. Fighting (Martial Arts Science). **11** (2015)
5. Huo, L.: Multiple intelligence theory and multiple intelligence curriculum research. Educ. Sci. 142–143 (2003)
6. Gu, M., Meng, H.: New Concept of International Education, pp. 106–119. Hainan Publishing House, Haikou (2001)
7. Du, Q.Y.: Modern Taekwondo Course, vol. 4, pp. 131–132. Hubei Science and Technology Press (2007)
8. Qu, Z., Gu, Y.: Curriculum Reform of Physical Education Teachers. People's Sports Publishing House (2006)
9. Gardner, translated by Shen Zhilong. Multiple Intelligences, pp. 214–216. Xinhua Press (1999)
10. Wang, J.: Development and Reform of Physical Education Curriculum. Central China Normal University Press (2003)
11. Wang, Z.: Teaching and Training of Modern Taekwondo, p. 132. Beijing Sports Publishing House (2007)
12. Zeng, Q.: Modern Taekwondo Course, pp. 234–235. Jinan University Press (2006)
13. Campbell, et al.: Trans. Wang, C. Strategies of Multiple Intelligences Teaching, p. 9. China Light Industry Press, Beijing (2001)
14. Zhong, Z.: Interpretation of Multiple Intelligences Theory. Kaiming Publishing House (2003)
15. Zhong, Z., Wu, F.: Interpretation of Multiple Intelligences Theory. Kaiming Publishing House, Beijing (2003)
16. Liu, B.: Multiple Intelligences and Teachers, p. 1–12. Shanghuan Education Press, Shanghai (2005)

On the Influence of Ganpo Elements on Oil Painting Creation in Education and Teaching Under the Background of Ant Colony Algorithm

Hui Luo[✉]

Jiangxi Normal University Science and Technology College, Gongqing 332020, Jiangxi, China

Abstract. "Image" is the core category of Chinese traditional aesthetics and the core spirit of Chinese art. Image landscape oil painting is the product of the fusion of "oil painting", a Western media, into China since it was introduced into China. It not only contains the influence of western literary concepts and technical language on Chinese oil painting, but also contains the impact of national literary spirit, especially aesthetic feelings and landscape spirit on landscape oil painting. This study takes the image landscape oil painting as the research object. The basic idea of the research is to trace and analyze the related concepts of the image landscape oil painting in the context of Chinese traditional aesthetics; elaborate the self-consciousness and the "image" characteristics of the image landscape oil painting; explore the influence of Chinese traditional aesthetics on the image landscape oil painting and its internal logic; through comparative study, exchange interview, self exploration and other ways, this paper analyzes the development status and problems of image landscape oil painting, and puts forward personal opinions. On this basis, this study makes a further cultural consideration and Prospect of image landscape oil painting. Ant colony algorithm, which has some advantages such as positive feedback, heuristic search and distributed computing, is better used to solve the optimization problem of the solution in the path planning.

Keywords: Ant colony algorithm · Chinese traditional aesthetics · Context · Image landscape oil painting

1 Introduction

Oil painting is a kind of painting originated from the West. "Image landscape oil painting" is the product of oil painting blending with Chinese local literature and art since it was introduced into China. It has experienced a process of exploration from unconsciousness to consciousness [1]. The study of "image landscape oil painting" should be based on three dimensions, one is "oil painting", oil painting is its form language and painting medium. In the exploration process of image landscape painting (especially in the early stage), the development and evolution of Western landscape oil painting provide a strong language basis for it. The other is "scenery", which is the subject matter and the object

M. A. Jan and F. Khan (Eds.): BigIoT-EDU 2021, LNICST 392, pp. 579–588, 2021.
https://doi.org/10.1007/978-3-030-87903-7_70

of expression. Different from the subject matter of characters, the theme of "landscape" has its "particularity" and "openness". In addition, due to the different understanding of nature, the expression of "landscape" is different in the process of artistic creation. For example, the western countries speak more about "landscape", and the Chinese side speak more about "landscape" [2]. There are differences between the performance objects themselves. For a single ant, its behavior movement is relatively simple, while the biological groups formed by a large number of ants show very complex species information. Ants in the process of movement will leave a chemical substance on the path, which is called pheromone. This substance can lead other ants to higher concentrations. Ants are evenly distributed at the beginning [3]. When they encounter obstacles, they choose the next feasible path according to the principle of equal probability. Later, with the increase of the number of iterations, ants tend to move to places with high pheromone concentration, which makes the number of ants on the shorter route increase, while the number of ants on the longer route decreases. The ants choose the shortest route to transport food through pheromone communication.

2   Research on the Formal Beauty in the Teaching of Oil Painting Landscape Painting

The connotation and embodiment of formal beauty are different in different historical periods. It is an important carrier and expression of aesthetic feeling in artistic works. The theme of artistic creation determines the basic framework and connotation of formal beauty. In the teaching and research of landscape painting in oil painting, the content and form of painting depend on each other and cannot be separated. How to raise the rich natural image elements to the form of painting with artistic beauty, form a unified and harmonious picture relationship, and create a unique painting language is a problem that teachers and students face and need to solve. In the teaching and research of landscape painting in oil painting, we should gradually guide students to reasonably use the modeling means such as lines, modeling, light and shade, color, space and perspective, follow the artistic principle of "from life, higher than life", and create paintings that meet the inner needs and are different from others through various painting languages and forms. In the abstract painting art, the simple use of point, line, surface and abstract form and color structure for artistic creation can also create a harmonious, unified and unpredictable painting mood. It can be seen that the constituent factors play an important role in the creation of formal beauty [4]. The formal beauty of painting language also depends on the artist's performance skills. On the basis of the rules of modeling and formal beauty, we can reasonably summarize and utilize the picture factors such as ups and downs, changes and unity, contrast and harmony, size and number, height and density, so as to form a well ordered picture, achieve the unity of form and content, and convey the artistic beauty. The rule of formal beauty is that after the factors of point, line, surface, shape, color, texture and their combination are creatively recombined and sublimated by the painter, a picture relationship of mutual coordination and restriction is formed. The painter will make reasonable and orderly treatment and adjustment of these form factors from life according to the law of formal beauty, so that the picture has a unified and harmonious aesthetic feeling [5]. The law of formal beauty originates from the laws of nature and human physiology and psychology. It is the experience of artists based on long-term life accumulation and artistic practice. Through the teaching practice of landscape sketching, we can know that many form factors of an excellent painting are interrelated and interdependent organic whole, which is not the replication and accumulation of naturalism. In reality, the images are changeable, and the beauty of form is hidden in them. It needs painters to explore and discover [6]. In this point, students are facing greater challenges. Therefore, in the teaching practice of oil paint-ing landscape sketch, teachers should constantly infiltrate the objective law of formal beauty creation into the teaching of oil painting landscape sketch. The natural scenery is complicated and complicated. Teachers should cultivate students' ability to analyze, select and reconstruct the pictures, establish the correct concept of formal aesthetics, and guide students to explore the formal beauty of art in the performance of natural shapes. After long-term practice and training, the law of creating formal beauty has gradually transformed into subconscious instinct in the practice of sketching. Figure 1 below is an oil color picture of the snow cloth in the East and north of Renji [7].

Fig. 1. The oil color picture of the snow cloth in the East and north of Renji

The emotional expression, image building and artistic conception of an excellent work can not exist in isolation from form. In the teaching and research of oil painting landscape painting, we should gradually guide students to recast the content and spirit of the work through art forms, so as to obtain spiritual perception. Without the formal beauty, the image performance and artistic conception building have the risk of falling back into the description of nature. Art form is higher than genre and real image, and also higher than the narrative expression of the picture, which is the artist's understanding and perception of the aesthetic law of all things in nature. The teaching research of formal beauty in oil painting landscape painting needs to guide students to use the rules of formal beauty to make a reasonable but detached expression of the real representation of nature and convey aesthetic interest [8–10].

3 Research on the Formal Beauty in the Teaching of Oil Painting Landscape Creation

Wu Guanzhong thinks: "art teachers mainly teach the art of beauty, and teach the laws and rules of formal beauty Formal beauty is the main content of art teaching...." Therefore, in the teaching practice of oil painting landscape towards creation, we should pay attention to guiding students to understand and comprehend the beauty of painting form, and take the expression ability of reorganizing nature as an important basis for creating the beauty of painting form. From the stage of landscape painting to the stage of creation, we need to gradually abandon the description of pure naturalism, and make the picture tend to the expression of formal beauty through reasonable subjective adjustment, so that the work can achieve the improvement of aesthetic level [11]. The law of formal beauty originates from nature, so we should guide students to extract the formal beauty under the real image, and use the complex and changeable abstract beauty law to connect and expand with the real scenery properly, so that students can constantly summarize and understand the law of formal beauty from the aesthetic experience, and lead students to a higher level of oil painting landscape creation stage. In the teaching and research of oil painting landscape creation, the guidance of form and style should be different from person to person, because each individual has different feelings and understandings of nature. Teachers should pay special attention to the self feelings of students with different temperament to ensure the uniqueness of individual life in artistic expression. In the teaching process of oil painting landscape creation, the development of color personality is particularly important. Teachers should fully guide and protect students' color personality in oil painting landscape learning [12]. In the scene selection of oil painting landscape painting and creation, we need to change different regions, guide students to exercise their observation ability in the flowing and changing colors of scenery, and constantly cultivate students' ability to learn from objective and natural colors the ability to transform into the subjective beauty of form and the color of the picture. Figure 2 is the fifth of Ren Jidong Wuyuan painting series.

Fig. 2. The fifth of Ren Jidong Wuyuan painting series

In the field of contemporary painting, painters pay more and more attention to the pursuit of formal beauty. It has become an important means to express the feelings and personality of painters and to express their personal feelings. Therefore, in the teaching of oil painting landscape creation, we should pay attention to guiding students to understand and grasp the rules of formal beauty, summarize the rules of formal beauty and the exact aesthetic connotation, emphasize that the creation practice should have a distinct personality and sense of the times, avoid stylization and moaning without illness in expression, pay attention to the perception of life and the true appeal of emotion, integrate into the painter's thoughts and emotions, and actively explore the differences between art and art In the past, different from other people's art form, the picture has rich connotation and strong artistic appeal. The excellent landscape painting works of all ages do not take the reproduction of nature as the ultimate goal, but look for the unique art form, internal structure, framework and the relationship with life in the landscape. The application of the rule of formal beauty should be based on the painter's emotional intention, follow the principle of artistic creation, and create a unified picture form of external image and connotation [13–15].

In the teaching of oil painting landscape creation, we should train students to pay attention to the structure and painting language of the picture, help them gradually get rid of the dependence on the objective description of the natural scenery, and use various rules of forming the formal beauty to realize the reconstruction of the picture order and the creation of the artistic conception. The factors of formal beauty in oil painting landscape creation have the characteristics of diversification, and nature is originally a rich and changeable whole [16]. The pursuit of oil painting language and form also needs diversification, in order to constantly enrich the connotation of formal beauty of works. However, diversity needs to be based on harmony and unity. It should be closely related to the theme of the picture and form an organic whole with different shape languages. In the teaching of oil painting landscape creation, students can be guided to integrate the proposition of society and humanity with the thinking of culture, highlight the theme and aesthetic taste of landscape painting through the unique perspective of the times, stimulate students' interest in landscape painting, constantly experience the artistic height that landscape painting can achieve, and show the unremitting pursuit of painting form and language [17].

4 Ant Colony Algorithm

4.1 Mathematical Model

Ants make a choice between the paths, each time only one path selection, then use the tabu list to place the selected city. When the ant traversal completes a search for all nodes, that is, all nodes are placed in the tabu table, and then the corresponding tabu table will be cleared to facilitate subsequent ants to do the next search. In a certain period of time, the pheromone concentration in ants' path will be evaporated as time goes on. In order to ensure that the probability of state transition is affected, it is necessary to control the pheromone concentration effectively. After N seconds, the pheromone concentration on the path (i,j) is updated as follows.

$$\tau_{ij}(t+n) = (1-\rho) * \tau_{ij}(t), \rho \in (0, 1) \tag{1}$$

$$\Delta\tau_{ij}(t) = \sum_{k=1}^{m} \Delta\tau_{ij}^{k}(t) \tag{2}$$

In the above formula, $\Delta\tau_{ij}(t)$ represents the total pheromone amount left by all ants on the link between nodes after traversing the complete path node ij: the size of ρ fluctuates between 0 and 1.

4.2 Update of Pheromone Volatilization Factor

The factors that affect the global convergence ability and optimal solution are the volatility coefficient of pheromone, which also affects the pheromone content of the path to a certain extent. The size of pheromone volatilization factor indicates the change of pheromone over time. Generally speaking, the setting of this factor cannot be too large or too small. If it is set too large, it can speed up the convergence of the algorithm, but it will also make the pheromone volatilization faster on the path that has not been passed. If the setting is too small, the ant will fall into the current local optimal state when searching for the path, so that the ant can not jump out to find more solutions.

After improving the ρ value of pheromone volatilization factor A, the formula 3 is obtained.

$$\rho(x = k) = \frac{1}{\sqrt{2\pi}\sigma} e^{-\frac{(k-\mu)^2}{2\sigma^2}} \tag{3}$$

Where k is the number of the outstanding ants.

$$\mu = \frac{\sum_{k=1}^{m} \tau_k}{m} \tag{4}$$

$$\sigma = \frac{\sum_{k=1}^{m} (\tau_{ij} - \mu)^2}{m} \tag{5}$$

In the above formula, τ_k is the pheromone quantity of ant excellent solution in the current state; μ is the search expectation of ant excellent solution after the completion of iteration; σ is the variance value of optimal ant k and poor ant pheromone quantity of traversing path after the ant completes an iteration; and M represents the iteration times of ants.

4.3 Advantages and Disadvantages of Ant Colony Algorithm

1) The advantages of ant colony algorithm are as follows:
   Because of its strong ability to find the optimal path and its own positive feedback mechanism. The algorithm can quickly find the optimal solution. Through pheromone communication and feedback, ants can find feasible solutions in a short time. In addition, ant colony algorithm can also absorb some advantages of other algorithms and fuse them, so as to improve the overall performance of the algorithm to a certain extent [18].

2) The disadvantages of ant colony algorithm are as follows:
    Ant colony algorithm is prone to premature stagnation in convergence speed and search performance, and when the number of ants is too large or the search space changes greatly, the selectivity of ants in a short time is relatively low. In addition, because of the positive feedback phenomenon of the algorithm itself, the quality of the solution is enhanced and the selectivity of the solution is weakened. At this time, the amount of path pheromone is also continuously enhanced. In addition, the pheromone is constantly volatilizing with the change of time, which leads to less chances for the previously unselected path to be selected later, As a result, it is impossible to find better path information, and the search scope of the solution is shrinking and even the search is stagnant. In order to slow down the occurrence of this phenomenon, we must measure the proportion between the search speed and the search space, and use the correct and effective improvement method to avoid the algorithm falling into the local optimal solution (see Fig. 3).

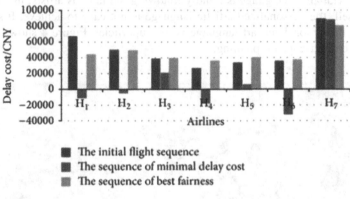

Fig. 3. Simulation for colony algorithm

5   Exploration Trend and Problem Analysis of Image Landscape Oil Painting

The language form of art is always closely related to the creator. In the exploration process of image landscape oil painting, due to regional factors, the aesthetic tendency of subject and object, as well as the different technical language and aesthetic pursuit, image landscape oil painting also presents different exploration trends. Based on this, this chapter will explore the current situation of the development of image landscape oil painting combined with cases. Things always develop in contradiction. In the development process of image landscape oil painting (which can be extended to the current Chinese landscape oil painting to a certain extent), there are also some problems that appear or appear. These problems need us to make calm thinking and rational judgment. This chapter will discuss and analyze these problems, and put forward some humble opinions, hoping to arouse the thinking of landscape painting creators and lovers on related issues [19].

British critic Richard Richards once said: "a poet's creation not only stems from his life experience, but also comes from his language which he can use freely." The same is true for art creation. Personalized artistic language is often a sign to measure whether an artist is mature or not. In the development of landscape oil painting, the jumping light and color language in Monet's works, the gorgeous color expression in Van Gogh's works, the ink charm in Wu Guanzhong's works, and the Chinese style brush and ink flavor of Zhang Dongfeng's works have attracted much attention due to their strong recognition. Because of this, the pursuit of personalized artistic language has become an upsurge in the exploration process of image landscape oil painting. However, as far as the "meaning" of landscape oil painting is concerned, the formation of personalized art language is not achieved overnight. It needs a long-term exploration process [20, 21]. At present, the blind pursuit and even imitation of personalized art language also leads to the "convergence of artistic language": in terms of individuals, it is shown as blindly following famous artists; in terms of creation itself, it is shown that the individual art language is blindly following the famous artists; in terms of the creation itself, the personalized artistic language is not easy to achieve, It shows that the pursuit of personalized art language is too superficial. Figure 4 shows the image simulation of landscape oil painting.

Fig. 4. Simulation for image landscape oil painting

6   Conclusion

Image landscape oil painting is an exploration of Chinese oil painters based on the reality of Chinese society, practicing the spirit of Chinese literature and art, taking landscape as the theme and oil painting as the media. Although its artistic media was introduced from the west, its creative and aesthetic subjects are more Oriental. Image landscape oil painting is not a whim to convey the spirit of Chinese literature and art, nor is it a superficial expression. Image landscape oil painting has a self-knowledge of its culture, and also has a profound understanding of the state, role and status of its culture. Image landscape oil painting is based on the integration of Chinese and Western literature and art, and the embodiment of cultural consciousness and cultural confidence in the process of Chinese oil painting localization.

Acknowledgements. Jiangxi Province University Humanities and Social Sciences Research 2018 Project: Ganpo regional elements in oil painting application research, project approval number: YS18232.

References

1. Du, W., Gu, F., Meng, X., et al.: Environment modeling method for ground robot based on robust elevation boundary. High Tech Commun. **29**(10), 985–994 (2019)
2. Chang, N.: The Meaning of Art. Jiangsu Fine Arts Publishing House, Nanjing (2007)
3. Fan, B.: Outline of Aesthetic History of Chinese Painting and Calligraphy. Jilin Fine Arts Publishing House, Changchun (1998)
4. Shao, Y.: History of Chinese Literary Criticism. Baihua Literature and Art Publishing House, Tianjin (2008)
5. Xia, W.: Teaching resources integration of contemporary oil painting creation based on wechat auxiliary platform. Shenhua **2**(4), 93–94 (2021)
6. Liu, S.: Research on the effective integration of Chinese food culture and oil painting creation teaching. Food Res. Dev. **42**(4), 231–232 (2021)
7. Yang, L.: Application analysis of freehand brushwork and its techniques in oil painting creation. Art Educ. Res. **3**, 122–123 (2021)
8. Huang, C.: Ways to cultivate students' innovative thinking in oil painting teaching in Colleges and universities. Beauty Times (Middle) **2**, 85–86 (2021)
9. Chen, H.: The use of Guangxi folk art resources in the teaching of oil painting creation. Beauty Times (Middle) **2**, 91–92 (2021)
10. Li, J.: Why oil painting materials and techniques are very important in oil painting creation and Teaching. Qingfeng **2**, 45–47 (2021)
11. Xiong, H.: Thinking and creation of contemporary landscape oil painting: a review of landscape oil painting. J. Trop. Crops **4**(1), 330 (2021)
12. Tang, L.: Research on the teaching of urban landscape oil painting for art majors in Colleges and universities. Contemp. Educ. Theory Pract. **13**(1), 87–91 (2021)
13. Wang, J., Zheng, B., Wang, M.: The influence of post impressionism art style on oil painting creation in Colleges and universities. Masterpieces **1**, 114–115 (2021)
14. Liu, J.: Persistence and development – Research on Yao Yongmao's oil painting. J. Yunnan Acad. Arts **04**, 69–74 (2020)
15. Lin, F.: The cultivation of unique painting language and feeling in the teaching of oil painting creation in Colleges and universities. Changjiang Ser. **35**, 16–52 (2020)
16. Ren, J.: Research on the formal beauty in the creation and teaching of landscape painting in oil painting. Art Observ. **11**, 138–139 (2020)
17. Ji, Y.: Research on the acceptance of Russian oil painting by Heilongjiang oil painting creation. Harbin Normal University (2020)

18. Yang, H. (Dawa Zaxi). Analysis of basic oil painting teaching and creative practice. Modern Voc. Educ. **44**, 160–161 (2020)
19. Wu, L.: Reflections on oil painting teaching strategies in Colleges and Universities under the background of diversified art culture development. China National Expo **18**, 37–38 (2020)
20. Zeng, Y.: Impressionism style oil painting landscape painting creation teaching practice strategy. J. Chengdu Normal Univ. **36**(9), 110–117 (2020)
21. Yang, D.: Strategies for cultivating students' innovative thinking in oil painting teaching. Grand View (Forum) **9**, 156–157 (2020)

Practice and Exploration of Completely Online Network Teaching with ID3 Algorithm

Jingjing Gao[⊠]

College of Education, Xi'an FanYi University, Xi'an 710105, China

Abstract. Using big data information technology can extract rules from massive data. Through the information collation, collation and analysis of the massive data of online teaching, it can provide effective decision-making reference, as well as the education industry. Introducing information technology into network teaching platform can effectively improve students' learning effect and teaching management level. Using information technology to explore the inherent law of online teaching can provide reference for the decision-making level of education and teaching, and can also provide guidance for the overall teaching task and teaching plan.

Keywords: Big data · Data mining · Association rules · Clustering algorithm · Personalized education

1 Introduction

In recent years, China has vigorously promoted educational reform and the application of educational technology. The application of computer technology, information technology, network technology and other new technologies has made the prospect of higher education more positive development. Through the survey, it is found that the application of most university network platform is relatively simple, and most of the network platform has become the educational administration platform or network library platform of each university, and the teaching task is not satisfactory. From the current technical point of view and teaching task requirements, it is not feasible to completely use the internet teaching platform to replace manual teaching, but it is feasible to use the internet teaching platform as an auxiliary means of artificial teaching or even as an elective course teaching platform. After observing the network teaching system of several colleges and universities, it is found that the current network teaching platform is the carrier of students' course selection, score inquiry, registration information and other functions. From the perspective of functionality, the network teaching platform is more inclined to the educational administration system [1]. Many students' learning information, student status information, course selection information, grades and other contents are not related to each other, but in fact, the connotation law information has not been used.

M. A. Jan and F. Khan (Eds.): BigIoT-EDU 2021, LNICST 392, pp. 589–598, 2021.
https://doi.org/10.1007/978-3-030-87903-7_71

2 Integration, Optimization and Quantity Construction of Existing Resources

2.1 Further Enrich and Improve All Aspects of Resources, Optimize the Existing Courseware

Although there is a complete set of teaching software in the teaching resource library of the drawing course, and there are more than 1000 test questions stored in the computer system, and the content covers the existing content of the whole course, in practice, we try our best to make more dynamic renderings, so as to make the two-dimensional image into a vivid three-dimensional effect, and make the students understand the knowledge of the textbook more intuitive and easier, At the same time, the teacher guides the students to do it by themselves, enriches the online resources, and selectively puts the students' exercises into the resource database, so that the later students can have more resources to read, draw and test constantly, so that the students can consolidate their knowledge in a very relaxed environment. The objectives of teaching reform are as follows: (1) based on the advantages of network resources, students can quickly establish more abstract spatial thinking and spatial imagination ability, integrate information technology and graphics organically, and promote students from passive and abstract learning thinking to actively watch realistic objects on the network, so as to build a three-dimensional world to two-dimensional plane.

2.2 Use the Existing Network Platform to Build a Comprehensive Network Teaching System

In order to give full play to the advantages of network resources, our school selects "hero Shuangyi" as the teacher resource database, and builds it on this platform. We have made five modules, They are: (1) the specific content of the online course content module has been described earlier; (2) the content of the student learning guidance module includes the syllabus, experimental syllabus, teaching schedule and learning objectives of each chapter; and (3) the tutoring and answering module, which mainly provides online discussion, topic forum, and so on, Questions and answers 16 edification, etc.: Practical Exploration of drawing teaching reform based on network resources (4) learning evaluation module can check the answers of exercise questions online, test online, submit homework, query scores, etc. The module also provides thinking questions, noun explanation, judgment questions, multiple-choice questions, and blank filling questions related to the learning content. (5) The related resource module provides students with links to similar and related courses, as well as references and additional learning materials, the latest scientific research achievements of engineering graphics, and the latest CAD software, so as to expand students' horizons and understand the most cutting-edge technology of graphics science and technology.

3   Making Full Use of the Existing Multimedia Courseware and Question Bank to Carry Out Online Teaching

3.1   Master Feedback Information and Develop Skills

As early as 2001, computer multimedia teaching has been implemented in the drawing course of our school. Then the school encouraged teachers to carry out the teaching reform of network course. We moved the original multimedia courseware of drawing to the Internet by categories. According to the key and difficult points of the course, we integrated the courseware into three parts (descriptive geometry, mechanical drawing, computer drawing) to make use of the image and vivid animation effect of network resources, Reappear the content explained in class, help students to establish the ability of spatial thinking and spatial imagination, and solve the contradiction of less class hours and more learning content. Organize students to study creatively in the network environment, enhance students' information inquiry ability, self-study ability and speculative ability. At the same time, use the existing question bank, let students evaluate their study on the Internet, master the feedback information in time, and finally cultivate their own skills.

3.2   Teaching Method

Based on the established network teaching system, the teaching reform of drawing teaching based on the utilization of online resources is carried out. In the three major modules of teaching, there are 60 class hours to arrange classroom situation teaching, 30 class hours to carry out personalized online autonomous learning, and 20 class hours to combine classroom situation teaching with personalized autonomous learning. Based on the classroom situation teaching: each module in the beginning of the process is in the multimedia network classroom situation teaching, the teaching requirements of this module, the teaching progress, the focus of this section, the difficulties and the problems to be discussed, the questions to be answered to the students a comprehensive introduction. According to the difficulty of the teaching content and the students' acceptance ability, the teacher makes the more abstract and difficult content concrete, which makes the students easy to accept and disperse the difficulties in the cognitive process. The teaching time of each module is about 3/4 of the class hours of the content, and 2/5 of the time is for the students to browse the content of the module online.

4   Decision Tree Analysis Algorithm

4.1   Basic Algorithm of Decision Tree

Decision tree is a common and important data mining method. The implementation of the algorithm is to use the top-down greedy algorithm to sum up the given data samples, extract classification rules from the unordered data tuples, and recursively generate a tree structure from the top root node. Each branch node of the tree structure represents test or selection results, Through the reasonable classification of each selection result, the process continues until all the attributes are traversed, and finally the decision tree is generated. Decision tree algorithm mainly includes two processes: constructing tree and pruning decision tree. The former means that the input training data is taken as the function value of the established algorithm, the output different attribute values are generated into each branch, and each branch continues to carry out recursive operation to the lower level, and finally forms the decision tree; for the newly established decision tree, a considerable number of branch nodes are generated because the input training sample data contains abnormal content, This is why the decision tree must be pruned. The whole decision tree process is shown in Fig. 1. At present, the typical decision tree algorithms are cart, ID3, CHAID and so on [2].

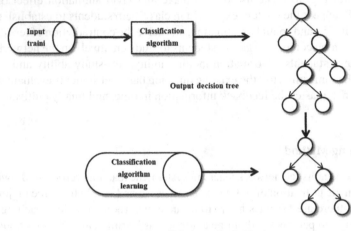

Fig. 1. Decision tree generation process

4.2 ID3 Algorithm

Among many decision tree algorithms, D3 is a basic algorithm formed earlier. It is a greedy algorithm, which uses the top-down recursive classification structure to generate the decision tree. The model generation method of the algorithm is relatively simple and robust, and the classification accuracy is high. It has good classification and statistical ability for the non incremental data sample set in the process of network learning, which is more suitable for application in the field of educational data mining [3]. The following is a simple discussion of ID3 algorithm.

ID3 uses the information gain as the measurement unit when selecting the branch node attributes. For data set s, the expected sample information is as follows:

$$I(n_1, n_2, \cdots, n_m) = \sum_{i=1}^{m} P(y_i) \log_2 P(y_i) \tag{1}$$

In the formula of information gain, we need to weighted average the information quantity of system samples, and the parameter obtained is information entropy. The entropy of data set s divided by description attribute d is as follows:

$$E(D_f) = \sum_{s=1}^{q} \frac{n_{1s} + \cdots + n_{ms}}{cout} I(n_{1s}, \cdots, n_{ms}) \tag{2}$$

Among them:

$$I(n_{1s}, \cdots, n_{ms}) = - \sum_{i=1}^{m} p_{is} \log_2(p_{is}) \tag{3}$$

By calculating the information gain of all the attributes, D3 algorithm forms the test attribute in the data sample set s with the largest gain, and then generates the branch node. The branch node is also marked as index attribute and classified into the given sample set.

5 Exploration of Teaching Methods

5.1 Pay Attention to Combine with Practice

Secondly, due to the concentration of teaching in different engineering master's class, students are prone to fatigue and inattention in the process of continuous listening. Therefore, how to ensure the quality of students' listening and stimulate students' interest in class is the main content of teaching method exploration. For example, when teaching the chat program based on wins α K, first ask the students to think about the basic functions of the client-side and server-side programs, the functions to be called, and the call process, and encourage the students to demonstrate in class, On the basis of the students' program, the teacher points out the mistakes or improvements; (2) pay attention to the combination with the reality of life, and try to use simple language to express the theoretical principles. Improving teaching quality.

Most of the part-time software engineering postgraduates are on-the-job students. In order to make students apply what they have learned, improve their learning initiative and ability to solve practical problems independently. The course group encourages students to put forward the problems and needs encountered in the actual work and discuss with teachers, and independently design their own comprehensive experiments. For example, the students working in the hospital designed and implemented the medical information management system according to the medical treatment process; the students working in the telecom designed and implemented the WAP Portal background management system according to their own work needs; the students working in the website designed and implemented the streaming media live broadcast system based on mode according to the customer needs [4–6]. The course group first constructs the advanced content system from the perspective of network software design and development; then designs the online experimental platform in strict accordance with the software development specifications; finally provides rich teaching resources and communication environment to students from all over the world through the resource sharing platform. In addition, some teaching methods such as "basic teaching links", "interactive teaching mechanism" and "students' participation in the course construction" are put forward to ensure the teaching quality of the course and improve students' interest in learning.

6   Pay Attention to Combine with Practice

Secondly, due to the concentration of teaching in different engineering master's class, students are prone to fatigue and inattention in the process of continuous listening. Therefore, how to ensure the quality of students' listening and stimulate students' interest in class is the main content of teaching method exploration. For example, when teaching the chat program based on wins α K, first ask the students to think about the basic functions of the client-side and server-side programs, the functions to be called, and the call process, and encourage the students to demonstrate in class, On the basis of the students' program, the teacher points out the mistakes or improvements; (2) pay attention to the combination with the reality of life, and try to use simple language to express the theoretical principles. For example, when introducing the CS mode, the students can first list the applications of CS mode in daily life (dining room, bank, etc.), and then explain to the students the advantages of using the CS mode in the network program by analogy, (3) emphasis on students' hands-on practice, requiring students to complete the corresponding necessary experiments in time after the completion of the theoretical lectures.

7   The Importance of Summary

In addition to the construction of hardware resources, the course team also explored the teaching methods of this course in combination with the characteristics of software engineering postgraduates, including determining the basic teaching links, ensuring the teaching quality, using heuristic, interactive and other teaching methods to stimulate students' interest in the classroom, The first step is to determine the basic teaching links. The course group divides the teaching content into seven relatively independent lectures. Each lecture includes such basic links as "review of the previous lecture content, introduction of learning objectives and learning methods, introduction of main contents and examples, and summary of the lecture content".

8   The Practice of MOOC Teaching in the Internet Age

According to the exploration and practice of programming class, we find that the effect of MOOC teaching is closely related to the perfection of online platform. In order to better realize MOOC teaching, the following characteristics should be provided. First, it can provide learning resources and realize online real-time interactive communication. This platform needs not only complete learning resources, but also the communication between teachers and students and teachers in the same school. On the basis of school internal communication, we can further realize cross school communication and foreign school communication. Teachers and students have different permissions of the system, teachers can push the outline of the class, homework layout and correction, while students mainly submit homework. Students first learn micro video resources online and complete relevant exercises. Any problems they encounter can be published on the platform. Classmates and teachers can answer them. In addition to supporting web browsing, they can also learn through the corresponding mobile app. In the Internet age, the use of smart phones has been very common, and the common characteristics of smart phones are large screen, fast running and convenient networking. In modern society, many functions of field brain have been replaced by portable smart phones to a certain extent. Therefore, compared with computers, learning through smart phones is a more popular way for students. Students' timetable is usually full and time is tight. Using mobile app can make students' learning and homework time more flexible.

9 Simulation Analysis for Course Recommendation

9.1 Overview of Recommendation System Based on Clustering Collaborative Filtering

Clustering is a method to classify objects with physical form or abstract form according to some characteristics, which is very suitable for personalized course recommendation. Because through the above data mining, we can get some inherent problems of students' learning rules. After finding the clusters of students' groups, similar cluster matching is carried out, and then Personalized Course recommendation is made for students.

The basic process of personalized recommendation system based on clustering collaborative filtering is as follows: firstly, the interaction process between students and the course system is determined and information is accumulated; secondly, the database is established by using the accumulated information; secondly, the information data in the database is preprocessed; thirdly, users are clustered according to the processing results; thirdly, similar matching is performed for different clusters after clustering; thirdly, curriculum recommendation is made according to the matching results. In short, the process of recommendation is formed by processing the relevant information learned by students in the network teaching platform (or network educational administration system), which is the basic principle of clustering collaborative filtering personalized recommendation system.

9.2 Data Acquisition

The first step is to get personalized courses and students' preferences. Before operation, the core meaning of data acquisition must be clear, and the students' preference for some courses can be determined through the interaction between students and online course system. The background of the network teaching system can query the students' course subscription, and the degree of students' interest in the course can be found by observing the subscription frequency of some courses; in addition, for some necessary courses, the degree of students' preference for the courses can be determined according to the students' learning frequency, access frequency, review frequency after learning and other information. In addition, the network teaching system is also embedded in the curriculum evaluation system, through the students' subjective and objective evaluation of the course, the students' satisfaction degree feedback can be determined. As the feedback of the adoption criteria, it can be determined by objective scoring, and the subjective evaluation is used as the reference for curriculum improvement and curriculum arrangement. Before and after the course selection, the students will conduct an objective questionnaire survey. The questionnaire includes the students' interests, professional course quality, infrastructure literacy and other information, as well as the satisfaction degree of the course, class arrangement and content arrangement. The comprehensive mapping is carried out through implicit data and image data [7].

The standard differential evolution algorithm is a random search algorithm. The basic idea of the algorithm is to calculate the vector difference of two individuals in different populations from a random initial population, and then sum with the third individual according to some specific rules to generate a new individual, and compare the new individual with the single individual determined at present, If the fitness of the new individual is better than that of the determined one, the new individual will be replaced [8–11]; otherwise, the new individual will be deleted, which is similar to the elimination algorithm of survival of the fittest, which gradually guides the search process to the most favorable result. In addition, the convergence rate of the algorithm will be greatly affected by the convergence rate of the algorithm in the later stage, and even the convergence rate of the algorithm will be improved. By introducing a more reasonable shrinkage factor and cross probability parameter, the algorithm is adaptively adjusted, and the termination condition of the original algorithm is improved, because the termination condition required by the calculation is not the optimal solution in the standard algorithm.

We know that some recommended courses are not willing to be learned by students, but they are compulsory, as shown in Fig. 2. From Fig. 1, we can see that the willingness of teachers is very high, but the willingness of students is very low. As a result, students sometimes do not learn through the courses recommended by teachers, which is also the reason for students' low learning efficiency.

From Fig. 3, we can also see that if it is a compulsory recommended course, students' willingness to learn will reach the peak. After the peak, students' willingness to learn will decline rapidly.

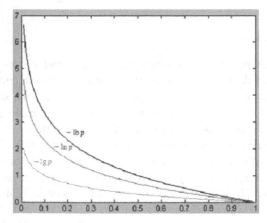

Fig. 2. Decesion Algorithm for Course recommendation

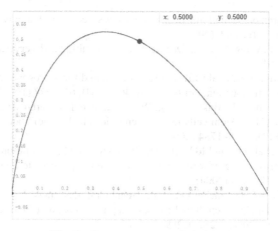

Fig. 3. Course recommendation rate

10 Conclusion

Network teaching requires teachers to extract reliable and useful learning feedback information from a large number of redundant and complicated teaching process data, and timely update and adjust teaching strategies, methods and contents on this basis, so as to solve the problem that online teaching can't carry out on-site interaction, realize personalized guidance for individual teaching, truly teach students in accordance with their aptitude, and improve the effect of online learning. Therefore, educational data mining technology plays an important role in online learning environment. Among many data mining technologies, ID3 decision tree algorithm is especially suitable for large-scale online learning because of its strong learning ability and easy implementation. In addition, the generated decision tree can express the classification rules corresponding to different branches vividly, and the algorithm is easy to read and use, especially suitable for the technical tool of educational data mining (EDM). With the rapid development of Internet technology and the popularity of big data technology and cloud technology in recent years, it has triggered a series of changes in university teaching. The introduction of MOOC teaching is a very obvious performance.

Acknowledgements. "Online course Construction Project of Kindergarten and Primary School Class Management" of Xi'an FanYi University (Project No.: ZK2019).

References

1. Huang, W.: Data mining technology and application research. Netw. Secur. Technol. Appl. **7** (2018)
2. Sun, J., Zhao, P., Lei, M.: Research on the application of data mining technology in university teaching evaluation. Small Technol. Inf. **17** (2014).
3. Yang, J.: Yanxia's web-based application platform. Sci. Technol. Entrep. Monthly **2** (2013)

4. Zhu, W.: Research on data mining decision tree classification technology and its application. South China Univ. Technol. (2004)
5. Sun, Y., Liang, D., Wang, X., et al.: DeepID3: face recognition with very deep neural networks. Comput. Sci. (2015)
6. Lyden, D., Young, A.Z., et al.: Id1 and Id3 are required for neurogenesis, angiogenesis and vascularization of tumour xenografts. Nature **401**(6754), 670–677 (1999)
7. Spits, H., Couwenberg, F., Bakker, A.Q., Weijer, K., Uittenbogaart, C.H.: Id2 and Id3 inhibit development of Cd34+ stem cells into predendritic cell (Pre-Dc)2 but not into Pre-Dc1. J. Exp. Med. **192**(12), 1775–1784 (2000)
8. Kowanetz, M., et al.: Id2 and Id3 define the potency of cell proliferation and differentiation responses to transforming growth factor β and bone morphogenetic protein. Molecul. Cell. Biol. **24**(10), 4241–4254 (2004)
9. Saika, S., Ikeda, K., Yamanaka, O., et al.: Adenoviral gene transfer of BMP-7, Id2, or Id3 suppresses injury-induced epithelial-to-mesenchymal transition of lens epithelium in mice. AJP Cell Physiol. **290**(1), C282–C289 (2006)
10. Shalekhah, Z., Sholihin, M., Rohman, M.G.: Penerapan algoritma ID3 untuk penentuan penerima bantuan program keluarga harapan (Studi Kasus: Desa Kelolarum Kabupaten Lamongan). Cybernetics **4**(2) (2021)
11. Han, J., Ma, Y., Ma, L., et al.: Id3 and Bcl6 promote the development of long-term immune memory induced by tuberculosis subunit vaccine. Vaccines **9**(2), 126 (2021)

Research on the Application of Cross-Industry Information Education in Agricultural Electronic Commerce

Yang Luo and Xiaohui Wang[✉]

Zhejiang Ocean University, 1 Ocean University S.Rd, Lincheng New District, Zhoushan 316022, Zhejiang, People's Republic of China

Abstract. The current application of cross industry information data mining technology can fully organize and analyze all kinds of information in agricultural e-commerce. In addition, the relevant fields and personnel can use the data mining model to calculate the data information reasonably. In this analysis, we can make targeted marketing plan according to the behavior characteristics of agricultural products. This paper analyzes the application of cross industry data mining in agricultural e-commerce.

Keywords: Cross-industry data mining · Agricultural products · e-commerce · Data acquisition

1 Introduction

At present, cross-industry data mining is a very important mainstream data mining. First used in finance, health insurance, marketing and retail. However, under the background of the rapid development of the Internet at this stage, the marketing process of agricultural products has been further developed through e-commerce, and the way of marketing has changed. For its formation of various data information, the value of in-depth mining.

2 Application Base of E-commerce Website CRISP-DM Agricultural Products

The effective application of data mining technology to business can effectively solve many technical problems, such as database marketing, customer group division, background analysis and so on. During the development of Internet technology, its information technology presents interactive attributes, and a large amount of data information is stored in the Web page, or there is a database established. In the current use of information, people need to take different technical means to extract the corresponding information internal information. For the e-commerce website of agricultural products, in the process of customer operation, will leave a lot of information data, at the same time by the website comprehensive collection [1]. As shown in Table 1. For this reason, we need to

M. A. Jan and F. Khan (Eds.): BigIoT-EDU 2021, LNICST 392, pp. 599–606, 2021.
https://doi.org/10.1007/978-3-030-87903-7_72

use a special way to dig into the documents in the Web and some valuable information in network activities. After processing and analyzing these information effectively, a large amount of valuable information can be obtained. Below is the CRISP-DM process.

Table 1. CRISP-DM process

Process	Process 1	Process 2	Process 3	Process 4	Process 5	Process 6
Content	Business understanding	Data understanding	Data preparation	Modeling	Assessment	Deployment

Besides analyzing the information in the database, the e-commerce of agricultural products in the process of development, its website also has various types of customer groups, brokers, growers information, with the help of the above information, can be very good application of CRISP-DM technology. After mining cross-industry data, we can analyze the problem of customer loss in detail, and evaluate the credit of customers. After realizing these functions, the network marketing of agricultural products can be realized on the basis of the structure of farmers' actual purchase habits, so as to optimize and adjust the existing websites. Let the user in the process of browsing the website has a high degree of comfort. In the current understanding of customer consumption habits, can be very good sales information adjustment, in order to carry out precision marketing. Only when we fully understand the actual needs of users can we carry out targeted marketing services and make users produce a certain consumption viscosity. In addition, we can make good use of the results of data mining, optimize and adjust the current website, and further improve the operation level of e-commerce website.

3   Preparation of Application CRISP-DM for E-commerce Websites on Agricultural Products

3.1   Basic Data from a Business Perspective

In the development of e-commerce, its essence is that it has always been a business type. E-commerce is based on the carrier of network. In the analysis of e-commerce, it is necessary to collect and organize all kinds of data information in electronic website in detail. In general, in the operation of e-commerce websites, it can be basically divided into two different types of information data, namely, traffic and volume. The main content of the inspection of the number of visits is to investigate the situation of a customer visiting many times or the same customer, and to analyze the degree and depth of the visit. In terms of turnover, users analyze the consumption habits and the internal relationship between products in the process of purchasing product combination [2, 3]. Agricultural e-commerce is similar to other types of websites and has obvious commercial attributes.

3.2   Data Acquisition Pathways

In the current process of data mining, the e-commerce data under various paths are analyzed in detail, and the source of e-commerce data is mainly in the e-commerce web page, such as click flow, results, research and competitive data these four different data types.

For example, click stream data is all the information data types formed in the operation of agricultural e-commerce website, which can record the user's website access behavior in detail. In the result data, the sales of agricultural products in the website to achieve data records. Therefore, in the data mining behavior of agricultural products e-commerce, the result data is the main mining object.

3.3 Data Mining Content

In the current network log, there is a lot of information content. However, for the current e-commerce data analysis behavior, because a lot of data can not be directly analyzed and processed, it is necessary to be able to exchange and process this type of data information in advance. In order to process, mining potential information value. In addition, it is necessary to synchronize the data mining of individual data sets.

4 Application of CRISP-DM of E-commerce Website for Agricultural Products

4.1 Projections of User Purchase Behavior

In the actual data mining, it is necessary to establish the prediction model of the user's purchase behavior. In the use of the model, the decision tree model is basically used, which can be well based on rule division to ensure the classification of data and good prediction. In the agricultural products e-commerce website, the user needs to confirm the order information in the formed payment interface. For the establishment of this model, the user's consumption behavior can be well predicted and analyzed. CART, CHAID and other calculation formulas are basically used in the analysis of decision tree model. At present, the C5.0 model analysis is mainly used in the data mining of agricultural products. In this way, in the process of model analysis, it can effectively act in the big data set.

After establishing the C5.0 model, it is necessary to select the internal information gain rate effectively. In addition, the maximum information gain is also needed to realize the sample split analysis of the field. However, it should be noted that before applying this model algorithm, it is also necessary to ensure that the access record analysis in the network log can guarantee that a single access behavior contains only one record entry. In addition, we also need to let users access the web page, reference web page, web page top-level directory, and so on, need to carry out separate modeling analysis. However, in the behavior of user ordering determination, and the low frequency of use of payment interface, the balance of variables is needed in modeling.

4.2 Accurate Agricultural Delivery

In the process of accurate marketing of agricultural products, the establishment of its model mainly uses clustering analysis algorithm, which can scientifically and reasonably divide the purchase behavior of users. In the actual calculation, several agricultural products can be calculated well for each cluster formed. After applying the C5.0 decision tree model, its clustering function can be effectively based on the actual behavior of users visiting the website, as the information entropy of the calculation link, so as to realize the classification of users. In addition, according to the way of accessing users, we can analyze their access behavior scientifically and reasonably [4–6]. After the model is established, it can effectively provide some products of interest to users according to the design of user's access page. For example, for some farmers to recommend some chemical fertilizers and pesticides, so that the formation of precision marketing behavior.

4.3 Humanized Web Page Recommendation Model

In the current construction of agricultural products website, we need to effectively analyze the user's actual usage habits, search habits and user's access records. And for these three aspects, the collection of variables to ensure that in the process of C5.0 model modeling, users can achieve customary clustering analysis. As shown in Fig. 1. After establishing a reasonable model, users can push accurately after successful access to three different pages. In addition, in the actual analysis process, can make good use of the value of various depths in the data, for the user's page personalized design, to ensure in the marketing process, Can realize the scope of marketing and the depth of the promotion [7–9]. This marketing method can also tap some potential consumers and push some information about agricultural products.

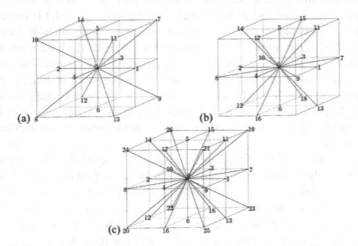

Fig. 1. Simulation of CRISP-DM of e-commerce website for agricultural products

Problems in the selection of teaching materials in Colleges and Universities.

At present, the selection of teaching materials in Colleges and universities is fundamentally the problem of how teachers choose teaching materials. The main reasons for teachers' inability to select suitable teaching materials are that the responsibility of teaching material management is not fully implemented and there is not enough information support for teaching material selection and evaluation.

5 Problems in the Selection of Teaching Materials in Colleges and Universities

At present, the selection of teaching materials in Colleges and universities is fundamentally the problem of how teachers choose teaching materials. The main reasons for teachers' inability to select suitable teaching materials are that the responsibility of teaching material management is not fully implemented and there is not enough information support for teaching material selection and evaluation.

5.1 The Implementation of the Responsibility of Teaching Material Management

There are some problems in the implementation of textbook management responsibility, such as the national macro management function and the local and school management.

Since the reform and opening up, the state has explored the management mechanism of teaching materials, from the teaching material management mechanism dominated by the plan to the teaching material management system under the central macro management, and then to the three-level system under the macro management, gradually clarifying the management responsibilities of all levels [10–13]. We will introduce various rules and regulations for the compilation, examination and selection of teaching materials in Colleges and universities, promote the construction of teaching materials classification by taking engineering teaching materials as a breakthrough point, and implement the bibliographic system. In 2017, the Ministry of education set up the Teaching Materials Bureau. In 2019, the management measures emphasizes the three-level management system, and emphasizes that "the Party committee of colleges and universities is responsible for the teaching materials work of the University". The state attaches great importance to the management mechanism of university teaching materials, improves the quality of university teaching materials, and brings prosperity to the publication of university teaching materials.

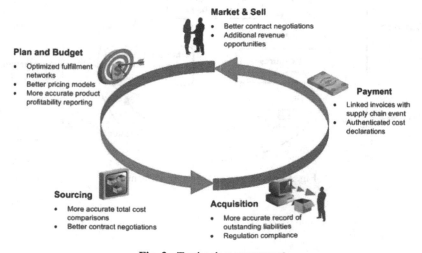

Fig. 2. Textbook management

However, the implementation of the National Textbook Policy is not in place in local and universities. As shown in Fig. 2. From the organizational point of view, local and university teaching material management institutions are weak.

5.2 Textbook Management Responsibility

There are some problems in the implementation of textbook management responsibility, such as the national macro management function and the local and school management. Since the reform and opening up, the state has explored the management mechanism of teaching materials, from the teaching material management mechanism dominated by the plan to the teaching material management system under the central macro management, and then to the three-level management system under the macro management, gradually clarifying the management responsibilities of all levels. We should promote the compilation and review of teaching materials, and introduce the rules and regulations for the selection of teaching materials. In 2017, the Ministry of education set up the Teaching Materials Bureau. the management measures emphasizes the three-level management system, and emphasizes that "the Party committee of colleges and universities is responsible for the teaching materials work of the University". As shown in Fig. 3. The state attaches great importance to the management mechanism of university teaching materials, improves the quality of university teaching materials, and brings prosperity to the publication of university teaching materials. However, the implementation of the National Textbook Policy is not in place in local and universities [14–16]. From the organizational point of view, local and university teaching material management institutions are weak.

Fig. 3. Quality management process model

5.3 Highlight the Operator's Own Advantages

In the design of responsibility distribution system, responsibility distribution is carried out according to the advantages of operators in each link. Educational administration department is the Department in charge of educational administration and construction in Colleges and universities. It is at the core of the system and plays the role of communication, coordination and management. In the relationship with teachers, because it belongs to the management department of textbook selection, we can formulate relevant systems to regulate or reward teachers' textbook selection behavior. To guide the teacher's teaching material selection behavior macroscopically or microcosmically. In the relationship with the supplier, there is no management relationship between the supplier and the supplier, but it can sign a contract with the supplier to ensure that the supplier provides teaching material services (including teaching material information, teaching material distribution, etc.) in strict accordance with the contract and supplies high-quality teaching materials. The advantage of suppliers is that they can obtain the publication information and authoritative evaluation information of teaching materials at a lower cost, and organize large-scale teaching material exhibitions and promotion meetings. Based on the fact that many textbook publishers have settled in MOOC platform, it is also the responsibility of suppliers to provide feedback information of MOOC platform students on the use of textbooks [17–20]. The main task of teachers is to use teaching materials to prepare lessons and organize teaching. The effective implementation of the responsibilities of educational administration departments and the timely provision of textbook information by suppliers can save a lot of time for teachers to choose textbooks and make them put more energy in the process of textbook research and teaching. The main responsibility of students is to feed back the use information of teaching materials, because the big data analysis technology of MOOC platform can obtain effective evaluation information of students' teaching materials by capturing the learning situation of students' teaching materials in real time.

6 Summary

To sum up, in the analysis of this paper, this paper mainly expounds the value and practical application of cross-industry data mining technology in the development of agricultural products e-commerce. After the application of this technology, the accurate marketing of agricultural products is realized effectively, the development of electronic commerce of agricultural products is greatly promoted, and more economic benefits are created.

References

1. Tall: Analysis and prediction of gas card customer churn -- based on "cross industry data mining standard process". Int. Petrol. Econ. **27**(10), 99–105 (2019)
2. It's always bright: Apply cross industry data mining model to standardize the data development and utilization strategy of aerospace manufacturing enterprises. China Equipment Engineering Press. (06), 217–218 (2019)

3. Gao, W., Kang, F., Zhong, L.: Yes. Data mining process improvement and model application microelectronics and computer science **28**(07), 9–12 + 16 (2019)
4. Liu, S., Liu, X.: Discussion on the development of forest health care industry in Dangchang assisted by under forest economic products. Gansu Forestry (02), 29–31 (2021)
5. Sun, J., Yang, J.: Exploring the UI design of mobile shopping platform – taking Xiaomi Youpin as an example. Art Educ. Res. **06**, 43–45 (2021)
6. Lu, B., Wu, H., Zhu, L.: Research on opening up the blue ocean of postal rural e-commerce with system intelligence and process standardization. Postal Res. **37**(02), 46–48 (2021)
7. Wang, H.: Achievements and prospects of Zunhua's e-commerce economic development. Tangshan Labor Daily (005) (2021)
8. Cai, J.: Current situation and operation mode of cross border e-commerce logistics in China. Investment Cooperation **03**, 74–75 (2021)
9. Ding, W., Zhan, J.: business model innovation of e-commerce platform from the perspective of value co creation: a case study of Jingdong Mall and Suning yunshang. China Bus. Theory **06**, 23–25 (2021)
10. Jia, S., Jia, Z., Cha, L., Deng, Q.: The marketing mode of Sichuan Baijiu liquor brand. China Bus. Theory **06**, 52–54 (2021)
11. Xue, J.: Providing clearer guidance and institutional support for the scientific and effective supervision of online transactions. China Market Supervision J. (003) (2021)
12. Shang, L.: On the standard of "small amount" in the measures for the supervision and administration of online transactions from the perspective of taxation. China Market Supervision J. (003) (2021)
13. Gong, C., Xu, F.: Problem analysis and regulation suggestions on "live delivery" of e-commerce. Commercial Econ. Res. (06), 83–86 (2021)
14. Wu, M., Zou, L.: Innovative development path of cross border e-commerce between China and Russia: from the perspective of heterogeneity of e-commerce development. Commercial Econ. Res. **06**, 142–145 (2021)
15. Li, X., Lin, C., Dong, J.: Thinking and practice of new business specialty construction under the background of Digital Economy – Taking International Business School of Sichuan Foreign Studies University as an example. J. Higher Educ. **10**, 86–89 (2021)
16. Meng, G.: Study on the rules of electronic transport records to be established in the revision of maritime law. Maritime Law China **32**(01), 16–22 (2021)
17. Fang, S.: New ideas of community marketing for e-commerce of agricultural products. J. Nuclear Agriculture **35**(05), 1252 (2021)
18. Han, Z., Huang, J.: Establishment of coordination and linkage mechanism of anti unfair competition department in Anhui Province. China Price Regul. Anti Monopoly **03**, 20 (2021)
19. Wu, K.W.: The possibility and limitation of platform neutrality Obligation -- starting from the legal regulation of non neutrality behavior of e-commerce platform. China's Price Regul. Antitrust (03), 26–29 (2021)
20. Zhang, J.: Application of problem-based teaching method in E-commerce teaching reform of vocational colleges. Sci. Educ. Wenhui (zhongxunjiao) **03**, 156–157 (2021)

Research on the Design of Internet Plus Home Care Service Platform for Intelligent Elderly

Yanping Xu[⊠] and Minhang Qiu

Nanchang Institute of Technology, Nanchang 330044, China

Abstract. With the development of aging population, the concept of "smart pension" is proposed, which can alleviate the contradiction between the growing demand of better life for the elderly and the unbalanced and inadequate development of the elderly service, and meet the diversified and multi-level needs of the elderly. Through the Internet information system, a digital information platform including emergency assistance, life service and home service is established. The elderly who meet the standards of the elderly can enjoy the corresponding elderly care service through the platform. Therefore, this paper designs and studies such a platform, which has both the support of the technology of the elderly and the innovation of advanced pension management. In addition, the Internet as an information integration system is therefore called "smart pension" service system platform, and the security mechanism of the pension system is improved to realize the intelligent elderly care.

Keywords: Internet plus · Service system platform · Intelligent pension · Security mechanism

1 Introduction

The increasing demand of the elderly population and its services has brought great pressure to the pension services in China. Since the implementation of China's family planning policy for more than 30 years, it has made a great contribution to restraining the rapid growth of population, but it also makes the social population structure become very unreasonable, the aging speed is accelerating, and there are more and more one-child families, which makes the social function of family pension gradually lose. The National Bureau of statistics released the latest population data on January 21, 2019: China's elderly population aged 60 and above has reached 249 million, accounting for 17.9% of the total population. Compared with 2017, the proportion increased by 0.6%, as shown in Fig. 1. It is predicted that the population aged 60 and above will soar to 329 million in the next 20 years. The above data can show that: entering the new era, China's aging problem is aggravating.

With the rapid development of global information technology and the gradual acceleration of population aging, smart pension is a new way of pension in the experimental stage. From the national level, smart pension service is very beneficial to China's pension system in terms of both quality and cost. On the one hand, it can greatly alleviate

© ICST Institute for Computer Sciences, Social Informatics and Telecommunications Engineering 2021
Published by Springer Nature Switzerland AG 2021. All Rights Reserved
M. A. Jan and F. Khan (Eds.): BigIoT-EDU 2021, LNICST 392, pp. 607–612, 2021.
https://doi.org/10.1007/978-3-030-87903-7_73

the huge pension pressure of the country, on the other hand, it provides data support for the government's scientific and reasonable allocation of pension resources and pension planning. From the perspective of each family, intelligent pension service has its unique advantages in emergency alarm, disease prevention and relieving economic pressure. On the one hand, it improves the quality of life of the elderly, on the other hand, it can also reduce the burden of pension for many young people and reduce children's worries about their parents' pension problems [1]. As shown in Fig. 1 below.

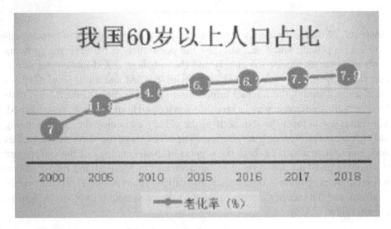

Fig. 1. Struts frame structure

2   Internet Plus Intelligent Endowment Mode

2.1   Defects of Pension Service

(1)   Information exchange is not timely and flexibility is poor

The unique advantage of home-based care service mode lies in the integration of family and community care resources, providing one-to-one or many to one door-to-door service, which is recognized by the society and the elderly. However, there are some problems in practice. At present, the process to meet the needs of the elderly is as follows: the elderly first put forward the service needs to the community day care center for the elderly or the community neighborhood committee, and then the two departments send people to provide door-to-door services, and then the social workers provide services and return to the community care center or the neighborhood committee for record. Not to mention that it may be inconvenient for the elderly to put forward their service demands to the above two institutions. Even if the elderly can put forward their service demands smoothly, the day care center for the elderly must have enough manpower and appropriate time to arrange the door-to-door service of social workers, which is a common problem in reality. Similarly, the on-site service of nursing workers and elderly volunteers also needs the detailed data of elderly care needs. If the day care center for the elderly can not provide the detailed service needs of the elderly within its jurisdiction, it is difficult for them to carry out their work effectively [2].

(2) Immature service mechanism

At present, China's smart pension service is in the stage of crossing the river by feeling the stone. Our city only carries out demonstration sites in some communities, and has not formed a complete and mature smart pension service system, let alone mature guidance of relevant laws and regulations. Because of this, without certain legal supervision and restriction [3], smart pension service will encounter various problems, such as different service labels and chaotic service market. On the other hand, from the perspective of the protection of the rights and interests of the elderly, the only authoritative document with national legal significance issued by the government is the law on the protection of the rights and interests of the elderly, and other documents are administrative provisions and notification reports of local governments. However, these local documents do not have the authority and force of law, and can not achieve good results in the process of implementation.

2.2 Core Technology of Smart Pension

There is no accurate service delivery strategy in the traditional way of providing for the aged, so the actual needs of the elderly can not be met. In order to get rid of the disease, we must collect data to understand the actual needs of the elderly. However, these data are laborious, time-consuming, costly and not comprehensive. In the era of big data, intelligent collection of pension data is no longer far away. In addition, the traditional way of providing for the aged is a kind of "cause and effect", which is characterized by top-down. Firstly, the demand model is given, then the target object is located, and finally the service is delivered. However, big data analysis is a scientific search and analysis in massive data to find the connection between things, which breaks the traditional pension "from cause to result" supply mode, and its characteristics are bottom-up.

2.3 Internet Plus Home Care Concept

As a new thing, "Internet plus home care" is relatively late in China. The author found through CNKI search that "Internet plus home care for the aged" was first put forward by Hu Liming in 2007 by the concept of "digital pension". After 2010, the concept has been widely recognized and used in academic circles. In 2011, Ma Feng Ling published a thesis on "accelerating technological innovation to promote technology endowment" at the sixth Beijing International Rehabilitation forum, and clearly put forward the concept of "technology endowment". In 2012, Shi Yuntong put forward the concept of "network endowment", and then developed into "intelligent endowment" and "intelligent endowment". As shown in Fig. 2.

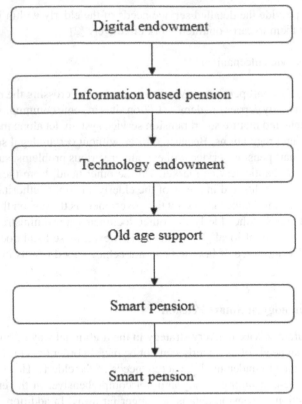

Fig. 2. Conceptual evolution

3 Design of Intelligent Pension Service System Platform

Under the background of "Internet plus", we should establish a daily monitoring and management mechanism for health information of the elderly. The monitoring of health information mainly adopts the method of mobile self-service detection equipment or fixed self-service detection equipment and medical staff detection to monitor the health information of the elderly under different conditions and different scenes. The elderly health information management is the data obtained through different monitoring methods, which are processed and integrated, and uploaded to the data center in a unified data format. The data center needs to analyze and process the acquired data, and at the same time, combine with the developed intelligent medical system, The elderly community information service platform is seamlessly connected with the medical information system of the remote hospital and the information exchange is realized to realize the real-time remote consultation or monitoring [4]. The elderly health information monitoring terminal can use the elderly intelligent health detection terminal device, including portable biochemical analyzer, 3G blood pressure meter, 3G weight scale and 3G motion detector to meet the needs of elderly health monitoring. Secondly, we should monitor the abnormal behavior of the elderly. The elderly's daily behavior habits are tracked and analyzed, and more effective service schemes are provided according to the analysis results, and the abnormal situation is reported in time, and corresponding emergency treatment is carried out. RFID, video monitoring and GPS positioning technology can be used in the monitoring mode. The platform architecture of smart elderly care service system is shown in Fig. 3 below.

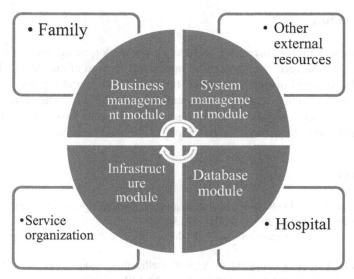

Fig. 3. Platform architecture of smart pension service system

The system architecture is shown in Fig. 3. The overall architecture of the system mainly includes infrastructure module, database module, business management module and system management module. Families, service institutions, hospitals and volunteers monitor the life of the elderly, provide consultation, carry out psychological comfort and emergency assistance after selecting relevant functions in the corresponding modules through the network.

4   Internet Plus Home Care for the Elderly

The government should adhere to its leading position and improve the top-level design, so as to optimize the allocation of existing pension service elements and form a good environment for the development of pension industry. The government should work in the formulation of relevant laws and regulations, infrastructure construction, service platform construction and other aspects.

The development of smart pension is inseparable from the strong support of the government, enterprises, social institutions and other aspects, so all parties need to work together. First of all, the government should introduce a variety of preferential policies, such as financial support, infrastructure support, human resources support, to attract social forces to participate in the construction of smart pension services. In addition, improving the relevant laws and regulations of smart pension industry can create a good external environment for the development of smart pension. The government should reasonably divide the duties of functional departments, guide enterprises and social organizations to participate in smart pension through various incentive measures, and form a pattern of multi-agent joint participation and collaborative pension service.

5   Conclusion

In the new era, with the increasing problem of population aging, the traditional way of providing for the aged cannot meet the increasing demand of the elderly. Therefore, under the background of mobile Internet, Internet of things and big data technology, intelligent pension emerges and gradually enters people's vision. Therefore, this paper analyzes how to reform supply to meet the demand from the two aspects of "supply" and "demand", so as to promote the development of smart pension.

References

1. Liu, M., Zuo, M., Li, Q.: Research on informatization demand of home care for the aged based on community service. J. Inf. Syst. (01) (2013)
2. Wang, L., Rao, P.: Design criteria of IT products for Chinese elderly users. Ergonomics (03) (2013)
3. Lu, Z.: Research on development strategy of intelligent pension service industry under the background of "Internet plus". J. Ningbo Polytechnic (02) (2017)
4. Yuan, B.: CNNIC: smart pension perspective, Ding. Service outsourcing (04) (2015)

Research on the Process of Informatization Education in Publicity and Translation on Government Website

Yanwei Jiao[✉]

Xi'an Fanyi University, Chang'an District, Xi'an 710105, China

Abstract. Under the guidance of information education, this paper discusses the process of publicity translation of government websites by using the research approach of information theory. It is found that in the process of publicity translation of government websites, the translator directly interprets the full clarity of the original text and clarifies the implied context of the text in the process of cognitive interpretation. In the process of the generation of the target language, the translator dynamically chooses the language that conforms to the social and cultural context, language context and aesthetic level of the target readers.

Keywords: Informatization · Government website publicity · Translation process

1 Introduction

As an important channel for the government to translate and introduce China, the importance of government website translation is self-evident, because it directly affects the construction of China's foreign image. At present, many scholars pay attention to the translation strategies, translation status and interdisciplinary theoretical guidance of government website publicity. Mao Donghui found that the quality of English translation of many government websites is not satisfactory, and there are many mistakes in the language level alone. Based on the theory of translation norms, Yao Yanbo,analyzes the problems and causes of Zhoushan government's English website, and puts forward that the English version of the government website should meet the standard expectations of the target language readers [1]. From the perspective of relevance adaptation in cognitive pragmatics, this paper will explore the whole process from reading the original text to the output of the translated text with the English website texts of local governments in Hebei Province as the corpus, so as to provide a new theoretical perspective for the English translation and introduction of government websites.

2 Data Mining Algorithm

At present, data mining algorithms are mainly divided into two categories: one is supervised algorithm, the other is unsupervised algorithm. Supervised algorithm mainly

M. A. Jan and F. Khan (Eds.): BigIoT-EDU 2021, LNICST 392, pp. 613–622, 2021.
https://doi.org/10.1007/978-3-030-87903-7_74

includes support regression, BP neural network, decision tree random forest, etc.; unsupervised algorithm mainly includes clustering analysis and association rule analysis. Unsupervised algorithm is a kind of machine learning based on specific rules. It uses some data evaluation indexes to judge some rules existing in the database, that is to find specific rules in the known database, but it does not play a predictive role. Therefore, this paper selects three data mining algorithms including regression, BP neural network and decision tree to predict the transformer hot spot temperature.

2.1 Support Vector Regression

Support vector regression (SVR) is widely used in power load forecasting because of its good measurement accuracy. For a specific historical data set $\{(x_i, y_i), i = 1, 2, \ldots, N\}$ Where x is an input vector and Y is its class label (output value). The support vector regression algorithm uses the nonlinear mapping $\varphi : x_i \leftarrow \varphi(x_i)$ to map the data to the multidimensional feature space. The general regression equation of the feature space is expressed as:

$$f(x) = \omega^T \phi(x) + b \tag{1}$$

In order to calculate ω and b in the regression equation and minimize the error between prediction and reality, the following objective functions are established:

$$\min_{\frac{1}{2}} \|\omega\|^2 + C \frac{1}{N} \sum_{i=1}^{N} (\zeta_i + \zeta_i^*) \tag{2}$$

2.2 Decision Tree

The decision tree algorithm spreads big data from the root node to the leaf node in turn by imitating the tree shape results, forming different types as the basis for making decisions [2–4]. Among them, the decision tree computing model is one of the most widely used models at present. It can not only deal with the logical data set, but also solve the incomplete problem in big data. For the regression problem in this paper, we define the calculation rules of each branch node of the decision tree to minimize the quadratic variance of the node:

$$\min RE(d) = \sum_{l=0}^{L} (y_1 + y_L)^2 + \sum_{r=0}^{R} (y_r + y_R)^2 \tag{3}$$

Where y_r and y_1 are the left and right branches of the decision tree node respectively; y_L and y_R represent the sample size of the left and right branches respectively; y_L and y_R are the average output values of the left and right branches. For all the values, the minimum value of quadratic variance is set as the parent node, and the recursive method is used to establish multiple child nodes. Each child node uses the variance size relationship to generate the parent node again, and so on, until no new node is generated in the model.

2.3 Innovate the Way of Propaganda and Improve the New Ability of Propaganda Work

The emergence of emerging media has changed the way for the masses to obtain information. Social media such as forum and post bar have gathered a large number of users, and deep interaction among users has formed a wide and profound influence. Compared with traditional media, the pervasive communication power of this social network has brought innovative inspiration for our propaganda work. Mobile Internet has changed people's daily reading habits. Traditional print media and TV broadcasting need deliberate and focused audition space. On the contrary, the "fragmented" shallow reading mode of wechat, microblog, micro video and client is more suitable for people's fast-food reading habits. In this case, relying entirely on traditional media publicity will be far less than the expected effect [5]. Only to cater to the public taste, the content is humorous, the text is short and concise, the comprehensive use of pictures and short videos, the mainstream ideas and core values into it, make it both entertaining and knowledge, teaching in fun. Make full use of the social function of "three micro terminals", strengthen the interaction with Internet users, and pay attention to and guide the network public opinion while disseminating and publicizing.

3 Theoretical Review

3.1 Relevance Adaptation Theory

Relevance adaptation theory is an organic combination of Sperber and Wilson's relevance theory and Verschuere η's adaptation theory to explain language communication and translation. According to relevance theory, language communication, including translation, is a process of ostensive reasoning between the two sides of communication. The speaker expresses the information intention to be conveyed, and the listener infers

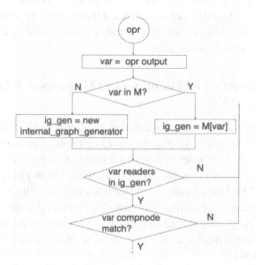

Fig. 1. The algorithm flow chart

the implied meaning of the speaker, that is, communicative intention [3]. Only when the hearer makes unnecessary cognitive efforts to infer the speaker's communicative intention, the maximum contextual effect can be achieved [6–8]. As shown in Fig. 1. According to adaptation theory, the process of language use is a continuous process of language selection, because language has the characteristics of variability, negotiability and adaptability. Li Zhanxi, a scholar, suggests that the two should be combined organically to explain and describe the translation process from the perspective of relevance adaptation, thus providing a new theoretical framework for the study of translation process. The algorithm flow chart is following:

3.2 The Specific Requirements of Government Propaganda in the New Media Era

The arrival of the new media era not only brings convenience to the publicity of school archives, but also brings new challenges to the management of school archives. First of all, the use of new media to promote school archives requires file managers to have higher quality. The traditional methods of file publicity and management are not applicable in the new media era. School archives managers must master the use of new media, ways of communication, use the Internet and other information platforms to publicize school archives, change the original way of work, actively and seriously learn the relevant knowledge of new media, keep pace with the times, and enhance their own ability, so as to keep up with the pace of the information age and make new media more convenient for school archives publicity service. Secondly, the new media era requires the publicity of school archives to be more interactive [9–12]. Although the traditional forms of propaganda such as news, newspaper and radio are convenient, they lack the interaction with the information recipients, so the propaganda effect is not ideal. In the new media era, archives publicity can actively mobilize the enthusiasm of the participants, let the information users participate in the archives publicity work in the form of message, discussion and voting, or make the school's development information files into videos and put them on the Internet to publicize the school, which greatly promotes the development of the school's archives publicity work and makes it more vitality and vitality, It is also more conducive to people's acceptance.

3.3 The Current Situation and Improvement Measures of School Case Publicity in the New Media Era

In the new media era, if the school's archives publicity work wants to be effectively carried out, it is necessary to give play to the advantages of new media and traditional media, and combine the two to learn from each other, so as to achieve the best publicity effect. At present, the school's archives publicity work has not played the best effect in the new media era [13]. The utilization rate of new media is still low. Generally affected by the traditional concept, the school's archives demanders get the archives information in the traditional way of looking up materials and reading newspapers. Some demanders go directly to the school's archives to look up the relevant information, this not only results in the low efficiency of file management, sometimes due to poor data management, information users may not get relevant information. Therefore, the archives propaganda work of the university must combine the traditional propaganda media and give full play to the advantages of the new media. The school should strengthen the training of archives management personnel, master the information transmission channels in the new media era, follow the way of government microblog, actively do a good job in archives publicity, learn more advanced multimedia related knowledge at home and abroad, learn from each other, and enrich their knowledge and ability. In the new media era, information operations such as website operation and microblog operation have to have backstage maintenance and need to constantly update the content, which need special information management personnel to operate [14]. This part of the cost needs special budget investment from the school to ensure the normal operation of the website and microblog. In short, the arrival of the new media era represented by the rapid popularization of the Internet and mobile phones and the popularization of information and communication technology have broadened the communication channels and forms for the school's archives publicity work, and also brought convenience for the popularization of archives information. School archives managers should make full use of the communication advantages of new media to maximize the resource sharing of school archives information, We should create characteristic archives resources, better publicize the school, serve the school, and improve the level of propaganda and management of school archives.

4 Analysis on the Current Situation and Problems of Propaganda Work Under the Condition of Informatization

4.1 The Cadres do not Grasp the Change of the Current Concept of Information Dissemination

The traditional way of communication belongs to "mass communication", which refers to the one-way communication of the media or other propaganda departments for the broadest masses of the people. One content is for all different types of audiences, and there is only one voice in the whole society. However, with the progress of science and technology and the continuous transformation of information society, a single way of Ideological and cultural propaganda has been marginalized by the rich and colorful cultural life of the people. "Does mass communication no longer monopolize the way of information dissemination? In the information age when ordinary people are constantly seeking the right to speak, can the government better use high-tech in Ideological and cultural propaganda, whether we can publicize the party's principles and policies in a way that people like to hear and see, so as to grasp the right to speak and guide the mainstream public opinion, has become one of the criteria to measure the quality of propaganda work under the condition of informatization [15, 16]. This requires the propaganda department to accurately grasp the new changes of the current information communication phenomenon, change the old communication concept, establish a more scientific and more advanced working idea, and adapt to the new requirements of the information age.

4.2   Under the Condition of Informatization, the Function of Public Opinion Supervision of Traditional Propaganda Methods is Gradually Weakening

Under the condition of information technology, the function of public opinion supervision of traditional propaganda methods is gradually weakening. To do a good job in Ideological and cultural propaganda, we must focus on consolidating and strengthening the mainstream ideological and public opinion, and enhance the ability of public opinion guidance. In a country, the government usually dominates the discourse power, and establishes a set of systematic core value system. In recent years, traditional media has been gradually integrated and replaced by new media, and "three micro - end" has become the new focus of public opinion in China, replacing traditional paper media and radio and television, Gradually weakened the government's use of the traditional way of public opinion supervision function [17]. Mobile Internet has the characteristics of openness and virtuality, which makes the environment for netizens to express various voices more relaxed. The timeliness and portability of traditional newspapers, magazines, television and even the traditional Internet are far behind the mobile Internet, which gradually leads to the weakening of the traditional propaganda mode in the function of public opinion supervision.

4.3   The Unidirectional Communication of Traditional Propaganda Lags Behind the Development of the Times

The traditional ideological and cultural propaganda is a kind of one-way information transmission mode, which spreads from top to bottom. In the era of mobile Internet, everyone is "We Media". More and more people no longer accept it passively, but talk and express it actively. The openness and inclusiveness of mobile information network provide a new platform for netizens. In the past, the self talk of government agencies has not adapted to the current complex Internet format. Ideological and cultural propaganda is a highly operational work. In order to achieve a new breakthrough in Ideological and cultural propaganda, we must start with the reform of propaganda methods and the use of high-tech means.

5   Two Stages in the Process of Government Publicity Translation

5.1   The Process of Interpretative Communication

According to relevance theory, there are both explicit and implicit utterances. Explicit speaking is the speaker's information intention which can be easily obtained by the hearer, while implicit is to convey a communicative intention, that is, to let the listener understand that the speaker has an intention to transmit information. The listener should not only understand the utterance according to the semantic representation of the utterance, but also infer the implied meaning according to the speaker's contextual hypothesis.

When the author of the original text delivers completely explicit utterance, the translator can obtain the maximum contextual effect at the cost of the least cognitive effort in the process of reasoning. There are many completely explicit words in the translated texts of the website. For example, Changli has a long history and rich cultural heritage. There are 45 intangible cultural heritage projects at or above the county level, including 3 national intangible cultural heritage projects, 9 provincial intangible cultural heritage projects, 11 municipal intangible cultural heritage projects and 37 county-level intangible cultural heritage projects.

5.2 The Process of Discourse Productive Communication

The process of discourse production and communication from the pragmatic perspective is a process in which the translator dynamically chooses and adapts to the cognitive context of the target language. The translator infers the information intention and communicative intention of the original text in the process of text interpretative communication. Then, the translator will choose the language from different language levels to convey the intention of the original text [18–20]. According to the three major attributes of language in the theory of adaptation, translators will consider both internal and external factors when making language choices in the process of discourse communication. They should not only flexibly choose the language form of the translation, but also choose pragmatic strategies. In the process of translation and translation, the translator should take into account both the reader's and the reader's communicative competence.

As a cross-cultural communication activity, translation in accordance with social and cultural context often encounters cultural differences or cultural conflicts. It is these differences that pose great obstacles or challenges to translators in translation communication, because they will directly affect or determine the translator's Choice of translation in the process of translation.

In the process of discourse production and communication, translators need to judge whether the contextual assumptions that the original author is trying to convey exist in the cognitive context of the target readers [21]. If there is, the translator should consider the aesthetic expectation and acceptance ability of the target language readers when choosing the language, so as to achieve the maximum harmony between the real world presented by the original text and the horizon of the target culture.

6 Simulation Analysis

All the government online translation process is not only a process of information transmission, but also a process of external publicity, so no matter which language, whether it is English, French or other languages, mutual translation is actually a process of mutual understanding. In fact, sometimes the expression is not very accurate in translation. But just convey the basic meaning correctly. Figure 2 shows that the accuracy of various languages can be used as a basic reference for translation.

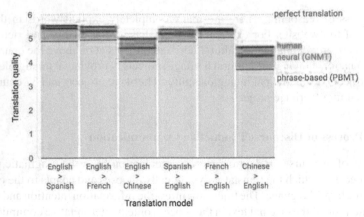

Fig. 2. Translation model

After the emergence of government translation, when it was just launched, people's attention rate was relatively high due to its complexity. However, with the passage of time, people's attention rate decreased year by year [22–24]. Therefore, if we want to obtain a relatively high attention rate, we should simplify it in the process of translation, as shown in Fig. 3.

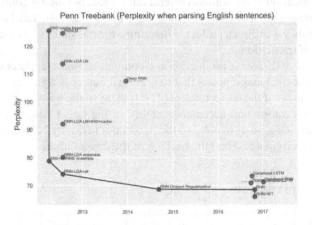

Fig. 3. Time translation comparison chart

7   Conclusion

The publicity translation of government websites is a complex cross-cultural communication activity. The whole process is a process in which the translator interprets the original text and re produces the original text information and communicative intention in the target text. In this process, the translator is not only the reader and researcher of the original text, but also the creator and negotiator of the target language. Based on the research results of relevance adaptation theory, this paper argues that the process of government website translation from the pragmatic perspective is such a process: in the process of seeking the best relevance of the original text's communicative intention, the translator interprets the explicit meaning of the original text and clarifies the implied context of the original text, so as to fully appreciate the artistic conception effect of the original text, Then, under the condition of confirming the cognitive harmony of the target language readers, the translator can give full play to the subjective consciousness and make a dynamic choice of language so as to realize the adaptation in the social and cultural context, language context and the aesthetic level of the target readers.

Acknowledgements. 2018 Special Scientific Research Project of Education Department of Shaanxi Provincial Govrnment "Analysis of the Status Quo and Countermeasures of the Translation of the Websites of the People's Governments at Various Levels in Shaanxi Province from the Perspective of Eco-translatology" (Project NO. 18JK0997).

References

1. Mao, D.: Discussion on some translation problems of English version of government website. China Sci. Technol. Transl. **2**, 18–21 (2008)
2. Yao, Y.: Discussion on the standardization of publicity translation of government websites – Taking the English version of Zhoushan Municipal government website as an example. J. Zhejiang Ocean Univ. (Human. Soc. Sci. Ed.) **3**, 81–85 (2012)
3. He, Z.: A summary of the principles of inference and relevance cognitive pragmatics. Foreign Lang. Teach. **4**, 1–10 (1997)
4. Li, Z.: Reader centered principle of cognitive harmony in translation. Foreign Lang. Teach. **1**, 101–104 (2012)
5. Industry trends. China Inf. Secur. **5**, 18–19 (2015)
6. Current political news. China Municip. Newsp. **5**, 5 (2015)
7. Wan, S.: Optimizing government websites to promote economic and trade development and cultural exchanges – Taking Anhui Province as an example. Northern Econ. Trade **3**, 8-9+12 (2015)
8. Li, P.: On the thematic construction of government website propaganda content – Taking the portal website of the State Bureau of surveying, mapping and geographic information as an example. News World **2**, 97–98 (2015)
9. Liu, P.: Attitude is More Outrageous than Literal Errors. China Press and Publication, October 23 (003) (2014)
10. Dong, X.: Research on Provincial Sports Government Websites in China. Shanghai Institute of Physical Education (2014)
11. Qin, M.: Persisting in reform and innovation to promote the work of ethnic propaganda in Guizhou Province. Guizhou Ethnic Daily (2014) (A03)
12. Lu, H.: The development of Chinese government websites from the perspective of US government website construction. News World **1**, 77–78 (2014)
13. Chen, L., Lai, M.: Discussion on the construction of service oriented government website. News Commun. **10**, 121 (2013)
14. Xu, Y.: We must do a better job in public opinion propaganda. People's Daily, October 15 (007) (2013)

15. Jiang, C.: Innovation of propaganda ideological and cultural work under the network environment. Theoret. Learn. **3**, 37–39 (2013)
16. Che, W.: Research on the English translation of government portals. Overseas English **1**, 129–130 (2013)
17. Sang, S.: Improving working mechanism of China Railway 11th bureau to deal with news crisis scientifically. Architecture **1**, 49–50 (2013)
18. Cheng, L., Chen, Y.: Strategies for optimizing the network publicity scheme of the city's international image – Taking the optimization of the English interface of the portal website of Chengdu municipal government as an example. People's Forum **32**, 230–231 (2012)
19. Sha, B.: An attempt to promote the propaganda work of earthquake prevention and disaster reduction at district level by using the Internet platform. In: Shenyang Municipal Party Committee and Shenyang Municipal People's Government. Proceedings of the 9th Shenyang Science Annual Conference (Agricultural Science and Environmental Protection Volume). Shenyang Municipal Party Committee and Shenyang Municipal People's Government: Shenyang Science and Technology Association (2012), p. 3
20. Liu, Y.: On the transformation of social communication mode and the challenge of government microblog. GUI Hai Lun Cong **28**(5), 94–97 (2012)
21. Xu, X.: From "news propaganda" to "topic construction". Fudan University (2012)
22. Huang, Y.: Discussion on operation and maintenance management of foreign language version of government website. Dongfangcheng Township News (2012) (A05)
23. Cao, J., Xie, M., Ye, X.: The portal website of Ministry of land and resources innovates the information service mode of government website around the propaganda activity of "ten thousand miles of rural land consolidation." Land Resour. Informat. **1**, 70-73+58 (2012)
24. Gao, L.: Image building with publicity strategy. China Quality News, 13 Feb 2012 (003) (2012)

Teaching Team of Accounting Professional Training and Guidance Teachers for Multi-interactive Vocational Colleges Under the Background of Artificial Intelligence

Xiaona Guo(✉)

Shandong Xiehe University of Jinan, Jinan, Shandong, China

Abstract. This paper summarizes and analyzes the background of multiple interactive era under the background of artificial intelligence; the problems and weak links in the construction of accounting professional training teachers in higher vocational colleges. Based on the successful experience of the construction of vocational teachers at home and abroad, this paper explores the construction of accounting teachers in Higher Vocational Colleges under the background of artificial intelligence, and realizes the goal of "diversified interaction of teachers, diversified training structure, diversified training methods, alternation of work and study, and synchronous teaching and learning ability".

Keywords: Artificial intelligence · Multiple interaction · Accounting major · Instructor

1 Introduction

The accounting major of Higher Vocational Colleges trains high-quality applied talents facing the front line of accounting industry. The training of students' application ability and technical ability is mainly completed by the professional training instructor (hereinafter referred to as the training teacher's teaching). Practical teachers play an important role in Higher Vocational Colleges and play an important role in teaching.

According to the relevant regulations of the Ministry of education, non normal and non-medical colleges and universities should gradually standardize their school name suffixes as "Vocational and Technical College" or "Vocational College", while normal and medical colleges and universities should standardize their school name suffixes as "College" [1]. Higher vocational education includes two levels of academic education: undergraduate education and junior college education, while in other countries and regions, the higher vocational education system completely includes both undergraduate education and junior college education.

M. A. Jan and F. Khan (Eds.): BigIoT-EDU 2021, LNICST 392, pp. 623–632, 2021.
https://doi.org/10.1007/978-3-030-87903-7_75

2   Related Work at Home and Abroad

In the 21st century, higher vocational education is booming in China. CNKI and VIP have found that in recent years, there are more than 20 papers about the construction of professional training instructor team, and almost all of them come from higher vocational colleges, especially in the fields of engineering, automobile maintenance and mechanical processing. There is no special research on the construction of accounting practical training instructors in higher vocational colleges, Only in the research on the construction of professional training base, the problem of "double qualification structure" teaching staff is mentioned, which is not suitable for the urgent demand of our country's sustainable development of accounting industry for the cultivation of advanced application-oriented talents [2–4]. As the lifeline of school running, it is urgent for Higher Vocational Colleges to find the correct orientation and strengthen the construction in the increasingly fierce competition of accounting education.

3   The Teaching Function of Three Dimensional Integrated and Multiple Interactive Learning Platform

3.1   It has Online Learning Function

In the application of multiple interactive learning platform, teachers can make a short video of 35 min, which is sent to the class students through QR code scanning. The micro lesson video integrates the key points and key points of classroom teaching, and visualizes the classroom teaching content to students, so that students can watch and learn in fragmented time, and change the original boring and boring learning method. At the same time, the learning platform has the functions of exercise, lottery, interactive game or test [5]. Students can make targeted learning plans according to their own learning needs, and the learning platform will also receive students' learning feedback.

3.2   It has Intelligent Teaching Function

In the application of three-dimensional integrated and multiple interactive learning platform, teachers implement diversified hierarchical teaching methods according to the difficulty of teaching materials, guide students to form cooperative groups, and formulate corresponding teaching objectives and teaching plans for students at all levels, so as to determine the requirements of position and professional ability, and guide students to complete corresponding post tasks [6, 7]. As shown in Fig. 1. In order to master the accounting knowledge and skills. In addition, teachers can organize online teaching activities on the learning platform to strengthen the communication and cooperation between teachers and students, while teachers mainly guide students as consultants, tutors or evaluators to strengthen the dominant position of students.

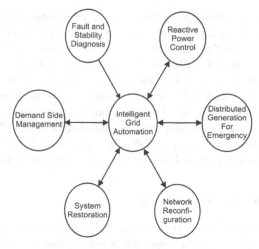

Fig. 1. Intelligent teaching function

3.3 It has the Function of Autonomous Learning

In the three-dimensional integration, multiple interactive learning, the effective integration of personalized learning and ability training, in the construction of student-centered learning system, cultivate students' habit of autonomous learning, so that students can discuss, ask questions, share in interactive learning, make learning break through the limitations of the classroom [8, 9]. Achieve the effect of learning anytime and anywhere.

4 Implementation and Construction of Teaching Mode of Three-Dimensional and Multiple Interactive Learning Platform Under the Background of Informatization

4.1 Design of Classroom Teaching Link

In the context of information technology, with the help of three-dimensional integrated, multi interactive learning platform, teachers can design the whole teaching process, play the role of 0 learning platform, and optimize the classroom teaching system. First, learning tasks. Before class, teachers should design learning tasks according to teaching content and teaching objectives, make theme accounting micro Lesson Videos, and upload learning task list and learning materials to the learning platform. The implementation of the principle of teaching students in accordance with their aptitude, the design of multi-level learning task list, different difficulties, so that all levels of students can complete the learning task, improve the students' sense of achievement, through the way of independent inquiry learning and understanding of accounting knowledge, to achieve the purpose of Preview, independent inquiry learning. Before class, students log into wechat group or QQ group to communicate with students and teachers online through mobile devices such as laptops, tablets and smart phones. Third, classroom interactive teaching [10–12]. In the class, teachers should change the original indoctrination teaching method, design multiple interactive tasks, help students consolidate and internalize the key knowledge of this class, guide students to carry out the study of accounting tasks and key topics, cultivate students' habit of independent exploration and group cooperation, so that teachers and students can explore knowledge together and make progress together. Fourth, evaluation of learning effect. After class, teachers should design the content of after class test according to students' feedback. On the one hand, it is to consolidate students' learning content, on the other hand, it is to evaluate the learning effect, so as to understand the shortcomings of the current teaching system, so as to improve and optimize. In addition to the individual test task, the test content also includes group test task, which can give full play to the strength of the team, strengthen students' autonomous learning, and cultivate students' innovative thinking ability.

4.2 Classroom Teaching Process Design

Three dimensional integration, multiple interactive learning platform classroom teaching process. Before class, the teacher releases the self-study task list, guides the students to carry on the inquiry independent learning through the case design or the topic design, and carries on the repeated thinking with the question, strengthens the practice; in the class, the teacher plays the micro lesson video which has been made in advance, according to the difficulty of the teaching content, sets a_ The teaching content of level B, from simple to deep, is in line with the students' cognitive law, and the difficulty increases gradually from a to B. As shown in Fig. 2 after learning a, enter the advanced learning of level B, implement the principle of teaching students in accordance with their aptitude, and realize the students' personalized learning; after class, timely feedback the students' learning situation, so as to adjust the teaching system.

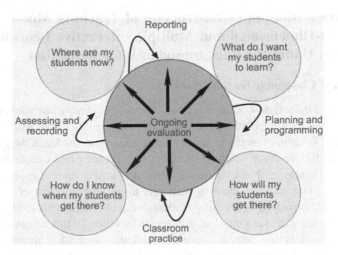

Fig. 2. Classroom teaching process design

4.3 Case Analysis

Taking intangible assets in financial accounting as an example, this paper selects umu learning platform software to design three interactive links before class [13]. One is to take the initial measurement of intangible assets as the main content, design questionnaire interactive activities to review the knowledge of last class; the other is to take the follow-up calculation of intangible assets as the main content, design exam interactive activities to investigate the key and difficult points of this class, The third is to design interactive games, follow the principle of combining work and rest, fit the teaching content, and relieve students' learning pressure and tension. These interactive links mainly rely on the way of scanning the QR code. Students can read the QR code, log on to the platform, enter the game link and questionnaire link, stimulate their interest in learning and meet their learning needs [14–16]. In class, according to the interactive results and answers of students' questionnaires, they can find out the missing and make up for the missing, and grasp the digestion of old knowledge and new knowledge, The platform will automatically display students' achievements and time, students can carry out reinforcement learning according to their own situation. After class, students can understand their own autonomous learning achievements through the answer results, while teachers can adjust the teaching progress and brake the targeted learning strategies and learning plans through students' learning feedback.

5 Correlation Analysis of Management Accounting Indicators

The traditional method of analyzing the correlation coefficient of variables is used to analyze the correlation of the selected financial indicators. The calculation formula of the correlation coefficient is as follows:

$$r = \frac{\sum_{i=1}^{n} (x_i - x)(y_i - y)}{\sqrt{\left(\sum_{i=1}^{n} (x_i - x)^2 \sum_{i=1}^{n} (y_i - y)^2\right)}} \tag{1}$$

Where: R is the correlation coefficient of the two variables, and X and y are the two variables. If R is equal to 1, it means that X and y are completely positive correlated; if R is equal to 0, it means that X and y are completely uncorrelated; if R is equal to $-1 \leq r \leq 1$, it means that X and y are completely negative correlated. In general, 1R "1, which means that there is a definite linear relationship between X and y.

6 Analysis on the Current Situation of the Construction of Accounting Training Teachers in Vocational Colleges

This paper summarizes and analyzes the existing problems and weak links in the construction of the current practical training teacher team of accounting major in higher vocational colleges, which can be summarized into five aspects. Firstly, many of the "double qualified" teachers in full-time teachers belong to "certificate type", and lack of practice opportunities and practical ability; secondly, the current personnel management system of higher vocational college teachers still applies to the teachers of ordinary colleges and universities Third, according to the feedback of teaching evaluation of external teachers, although their operation ability is strong, their comprehensive teaching ability is still not as good as full-time teachers [17]. For a long time, the traditional inertia of attaching importance to theoretical teaching and neglecting practical teaching often makes practical teaching more laissez faire, lacking effective evaluation and supervision, and even becoming the scene and ornament of professional teaching; fifthly, at present, most of the new teachers in Colleges and universities require master's degree, and lack of practical experience in the industry, so they are generally not competent for the teaching of practical training courses.

7 Theory and Method of Multi Objective Robust Optimization Design

This chapter mainly describes the general situation of robust optimization and the basic process of robust optimization. In order to deal with the situation that the interaction between multi-objective responses is not fully considered in the process of robust optimization of compliant structural products, based on the reliability and robustness theory, the factor analysis optimization method considering the interaction between multi-objective responses is introduced, which provides the corresponding theoretical support for the following chapters.

7.1 Basic Theory of Robust Optimization Design

7.1.1 Basic Concepts of Robust Optimization Design

Robust design is also known as robust design because of its different English translation. Its purpose is to find the optimal parameter combination under different conditions by adjusting the size and level of technical parameters, so as to reduce the performance fluctuation of products when they leave the factory. The fluctuation degree of product specific quality characteristics will be reduced as much as possible after robust optimization design, so as to ensure that all performance indicators of the product in its life cycle can reach the design value [18–20]. This method comes from the new quality supervision technology created by the famous Japanese quality management expert Dr. Xuanyi Taguchi. This technology is not only the essence of modern enterprise design and production at the beginning of its establishment, but also famous for its high efficiency, science, objectivity and low cost.

7.1.2 Common Robust Optimization Design Methods

(1) Taguchi robust design method this method mainly carries out robust optimization design in the form of experimental design, and analyzes the variance and mean value of the specific performance of the product, so that the mean value tends to be stable and the variance is reduced as much as possible, The purpose is to minimize the impact of external noise, temperature, corrosion and other factors on

product performance, as well as changes in internal structure, natural frequency and vibration mode, resulting in large fluctuations in product quality characteristics, so as to improve the robustness of the product [21]. Taguchi method usually uses SNR formula and orthogonal test to test and process data and results. Dr. Taguchi Xuanyi proposed more than 70 SNR formulas under different conditions. According to the different properties of each product quality response objective function, it can be divided into three types.

(2) Response surface method

The response surface method can express the function relationship by graphic technology, so that we can choose the optimal parameter combination by intuitive observation. The construction of such response surface needs a lot of data to test the experimental model, and modeling is a very important part. Constructing response surface model can intuitively get the functional relationship between the main parameters and the corresponding response variables, and the linear combination of the parameters and the response variables can be obtained through computer calculation of the functional relationship. This method has been used in many researches since it was proposed, and has been used in electrical engineering, mathematical equations, biochemistry and other aspects for many times.

7.2 A New Method of Optimization Design

7.2.1 The Necessity of Considering the Interaction Between Response Objectives in Multi Response Optimization Design

The main reason for the existence of the interaction between multiple response objectives is that the design parameters of the product seldom consider this influence in the design stage, which makes the product not fully play its performance and leads to a high rate of product return. RSM does not fully consider the interaction of multiple response objectives, which leads to the research results not fully in line with the reality. As a result, the interaction between some response objectives based on equipment quality characteristics is ignored in the design stage, which leads to product quality problems, even in the production stage, use stage and customer feedback stage, there are some product quality problems, such as high rate of inferior products, customer complaints, product recall and redesign [22]. For example, the recall of hundreds of thousands of Japanese Toyota vehicles has also had a negative impact on its market reputation and brand influence. Therefore, the interaction of response objectives in product multi response optimization is indispensable to improve product robustness.

7.2.2 A New Method of Multi Response Optimization Design is Proposed - Factor Analysis Method

Based on the correlation matrix, factor analysis classifies many variables with complex relationships (including independent influence and interactive influence) into a small number of common factors. The name of "factor analysis" originated in 1931. It was first proposed by Thurstone in 1904. The famous statistician C Spearman published an article to complete the concept of this method. In his published article, he analyzed the scores of six courses learned by students in a school, and studied the correlation coefficient, similarity degree and arrangement of these six courses, and obtained a model with only one common factor, which is interpreted as "general intelligence". This common factor has contribution to all six courses, and its contribution to different courses is different, Then Mr. Spielman further studied the common factor model. However, due to the complexity of this calculation model and the lack of effective calculation tools and methods at that time, he had to suspend the research. With the development of the times, Mr. Spearman's research has become history, but it has not been forgotten by mankind. The emergence of high-speed computer is convenient for the calculation of big data, making it more convenient.

8   Exploration and Innovation on the Construction of Accounting Training Teachers in Higher Vocational Colleges

The accounting major of Tourism College actively explores and innovates in the construction of training teachers, improves the double quality of full-time teachers through multiple channels, and strives to introduce excellent teachers with industry experience [23]. We will arrange teachers with insufficient experience in the industry to take temporary posts in the corresponding enterprises for training, investigation and study, introduce industry front-line managers or technical experts as part-time training teachers, pay attention to the improvement of their teaching ability, and constantly optimize the structure of professional training teachers, so as to form a full-time and part-time training teacher team with complementary school and enterprise.

8.1   System Guarantee

The system guarantee team has relatively stable high-quality practical training teachers, which can widely accept the industry's talents. From hotels, travel agencies, high-end business clubs, golf clubs and industry associations, middle and high-level management personnel and line craftsmen are employed as special teachers or visiting professors. The practical skills module in professional courses or full responsibility for vocational skills courses are comprehensively responsible for. The school has issued the "accounting senior" and other majors Policies and systems, such as the Interim Measures for the appointment and management of external teachers in the school, the detailed rules for the management of external teachers in the Department of accounting and sports leisure management, etc., clearly define the employment procedures of part-time teachers in enterprises, the rights and responsibilities of post qualification, the conversion standard of post and occupation grade and course remuneration, etc., so as to form a stable part-time teaching staff in enterprises, which plays a positive role.

8.2   Multiple Linkage

Establish a teaching quality assurance system for training teachers, strive to improve the teaching ability of external teachers and the practical ability of full-time teachers. In addition to the conventional teaching evaluation methods such as students' feedback, supervision and listening, we also take full-time and part-time teachers' teaching observation and experience exchange meetings to hold ppt courseware making skills special lectures for external teachers, and provide technical services to the teachers and external teachers of the enterprise cooperate to learn from each other, and promote the improvement and improvement of the theoretical teaching level and the cultivation of teaching norms of part-time practical training instructors outside the school [24]. Through a series of effective measures such as platform construction, system guarantee training promotion, practice experience and ability evaluation, the teaching quality assurance system of accounting management training teachers in higher vocational colleges is constructed (see Fig. 3).

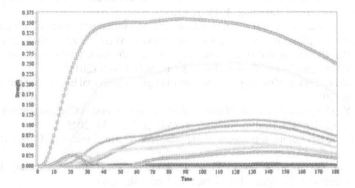

Fig. 3. Simulation of Multiple linkage

9   Conclusion

The construction of accounting teaching staff in higher vocational colleges is crucial to the development of Accounting Specialty in Ningxia. We must find out the problems existing in the construction of accounting teaching staff in each school. We should improve the talent introduction policy, construction planning and construction ideas, and according to the actual needs of the construction of accounting professional teaching staff in our school, it is necessary to expand the construction of accounting teaching staff in Ningxia higher vocational colleges, speed up the construction of accounting teachers with high quality and the combination of theory and practice, so as to better promote the accounting professional education in Ningxia higher vocational colleges.

Acknowledgements. This paper is the research result of the school level teaching reform research project of Shandong Xiehe University in 2020, "Research on the Construction of Instructing Team on Accounting Training in Vocational Colleges by means of Multivariable-Interactive Perspective" (2020gz06).

References

1. Niu, K.: Construction of Ideological and political teachers of business English in Higher Vocational Colleges. Sci. Consult. (Educ. Scientific Res.) **11**, 73 (2020)
2. Shen, Q.: Research on the construction of teaching staff in Higher Vocational Colleges under the background of school enterprise cooperation – Taking Hotel Management Major as an example. Henan Agricult. **24**, 9–10 (2020)
3. Zhang, J.: Research on the construction of accounting teachers in Ningxia higher vocational colleges. Commun. Res. **4**(16), 172–173 (2020)
4. Lan, X.: Discussion on the construction of teaching staff of Intelligent Manufacturing Specialty in Higher Vocational Colleges in Ethnic Areas – Taking Guangxi Modern Vocational and Technical College as an example. Guangxi Educ. **19**, 98–99 (2020)
5. Zhang, R., Fu, G.S.: Research on the application of interactive electronic whiteboard teaching in China. China Educ. Informat. **2**, 13–16 (2016)
6. Ma, C.: Exploration on the construction and sharing mechanism of digital resource management platform in Vocational Colleges. Modern Vocat. Educ. **30**, 6–7 (2015)
7. Su, X., Wu, B.: The application of mobile interactive digital teaching in Higher Vocational Colleges. J. Beijing Vocat. College Labor Secur. **9**(4), 45–49 (2015)
8. Guo, Y.: A brief study of the teaching method of Higher Vocational English. Examinat. Weekly **94**, 79–80 (2015)
9. Cheng, Y.: The feasibility study of interactive teaching method in Chinese teaching of Higher Vocational Colleges. J. Harbin Vocat. Tech. College **5**, 33–34 (2015)
10. Xiao, P.: The development prospect of machine tool electrical teaching. Chizi (Middle Last) **11**, 309 (2015)
11. Hanwenxin: An empirical study of interactive teaching in English Vocabulary Teaching in Higher Vocational Colleges. Minnan Normal University (2015)
12. Shenchaoqun, Li, W.: Discussion on the teaching of "color composition" in Vocational Colleges. Era Educ. **10**, 86 (2015)
13. Wang, K.: The application of interactive teaching mode in public English Teaching in Higher Vocational Colleges. School Park **10**, 66–67 (2015)
14. Zhang, C.: Practice and thinking of interactive electronic whiteboard applied to accounting teaching. China Educ. Technol. Equip. **24**, 29–31 (2014)
15. Li, Y.: Research on interactive 3D teaching system in Higher Vocational Colleges. Heilongjiang Sci. Technol. Inf. **34**, 164 (2014)
16. Zhangjiahe. Curriculum resource construction of CAD/CAM technology. J. Anhui Inst. Elect. Inf. Technol. **13**(3), 66–69 (2014)
17. Fu, W., Zhou, W.: Application of interactive electronic whiteboard in the course of architectural drawing in Higher Vocational Education. Jiangxi Build. Mater. **9**, 286 (2014)
18. Jiang, H.: Discussion on teaching methods of communication principle course in Vocational Colleges. Sci. Educ. Guide (Last Xunjiao) **5**, 124–125 (2014)
19. Zhang, M.: Application of interactive teaching method in special education major teaching in Colleges and universities. J. Suihua Univ. **33**(7), 24–27 (2013)
20. Wang, J.: A study on the interaction of oral English examination model in Maritime Vocational Colleges. J. Qingdao Ocean Crew Vocat. Coll. **34**(2), 67–70 (2013)
21. Yang, N.Y.: An example of the application of interactive virtual operation training system in Vocational Education. Vocat. Educ. Res. **2**, 171–173 (2013)
22. Shi, F.: Application and influence of authorized evaluation in quality development evaluation of Beijing Vocational Colleges. J. Beijing Inst. Technol. **1**, 48–54 (2013)
23. Li, L.: The tool of integrated teaching interactive electronic whiteboard. Henan Educ. (Late Ten Days) **12**, 34 (2012)
24. Zuojiahuai: Exploration of computer curriculum teaching reform in Vocational Colleges. Jintian (Inspirational) **12**, 126+114 (2012)

Design and Research of Network Computer Room Management Platform in Smart Campus Environment

Rui Fu[✉]

Guangdong Industry Polytechnic, Guangzhou 510300, China

Abstract. As one of the important teaching infrastructure of the whole school, intelligent campus network room management is also an important application system of intelligent campus. This paper introduces the research and design of computer room management platform in big data environment.

Keywords: Big data environment · Smart campus · Network room

1 Introduction

The computing center has always been one of the important teaching infrastructure of our school. After more than 10 years of construction, it has completed more than 10 laboratory construction and transformation projects, and has reached the advanced level in computer configuration, network architecture, open management and utilization. At present, our school's computing center has built a three-tier switching LAN, which is equipped with billing server, file server, DHCP, FTP, web and other dedicated servers, and a dedicated gigabit optical fiber direct to the main node of the campus network. The computer center is also equipped with a self-developed computer room management system, which realizes open management, unattended computer room, overall management of computer arrangement, automatic real-time billing and other functions [1]. With the continuous advancement of educational informatization, many high schools have carried out the construction and use of smart campus, and educational informatization has shown its application effect in related scientific research. But relatively speaking, there are still many problems in the construction process of smart campus in high school, such as low utilization rate of information, insufficient platform resources and so on, which affect the process of campus informatization. The arrival of the Internet plus era brings new opportunities and challenges to the construction of the smart campus.

2 Proposed Work for Cloud Computing Technology in the Construction of Smart Campus

2.1 Background of Smart Campus Construction

The rapid development of information technology has promoted the reform of educational concepts and teaching methods [2–4]. With the development of cloud computing,

M. A. Jan and F. Khan (Eds.): BigIoT-EDU 2021, LNICST 392, pp. 633–642, 2021.
https://doi.org/10.1007/978-3-030-87903-7_76

Internet of things, mobile Internet and other new generation information technologies, New "smart campus" teaching management mode has been shown in people's face, such as anytime and anywhere interaction between teachers and students, ubiquitous personalized learning, intelligent teaching management and learning process tracking and evaluation, integrated educational resources and technical services, learning community of home school interaction, and campus culture of teachers and students growing together. Through the electronic campus and digital campus, the traditional campus has gradually stepped into the stage of smart campus, providing a collaborative and comprehensive intelligent perception environment for teachers and students, and providing convenient, personalized and intelligent information services for management, scientific research, teaching and life. However, in the face of many new concepts and concepts of educational informatization, most people do not have a clear understanding of its connotation and characteristics, confusing such concepts as smart campus, digital campus and electronic campus. Therefore, it is necessary to study and sort out the existing theories and concepts when doing a good job in the top-level design of campus informatization [5].Smart campus is shown in Fig. 1. Under this background, we need to work hard to solve the problems encountered in building smart campus, and promote the development of high schools. With the continuous development of information technology, the Internet era has arrived. Under the background of Internet plus, the campus informatization has made great progress, which has greatly enriched the daily learning and life of teachers and students. However, due to the slow construction process of smart campus, there are many problems, so we still need to further strengthen the relevant construction, so that it can really help teachers and students to learn and live, and effectively promote the development of campus.

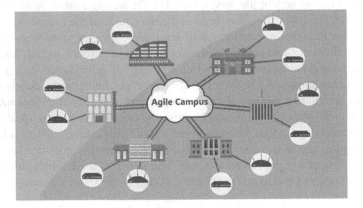

Fig. 1. Smart campus

2.2 Connotation and Characteristics of Smart Campus

The word "wisdom" refers to the ability to understand and solve things quickly, flexibly and correctly. "Wisdom" is a relative concept, which is not limited to human beings. Any system composed of objects has wisdom, but different in height. The characteristics of intelligent system proposed by IBM are more thorough perception and measurement, more comprehensive interconnection, and more in-depth intelligence.

Based on a comprehensive analysis of various definitions and views on smart campus, the author agrees with Jiang Jiafu's view that smart campus has multiple attributes of technology, education and culture [6]. It is a smart campus with high integration of information technology and education, deep integration of information application and education, and wide perception of network and information terminal.

Calculate a document. The calculation formula is as follows:

$$TF_{D_i,w} = \frac{n_{D_i}}{\sum_k n_{D_i,k}} \tag{1}$$

Calculate the inverse document frequency of the document:

$$IDF_{D_i} = \log \frac{|D|}{|p : w \in D_q| + 1} \tag{2}$$

The central node returns it to the system. At this time, the system EA uses formula 3 to decrypt the key K of D, so as to decrypt the ciphertext and get the plaintext document.

$$K' = E_{PKE-A}(E_{ID_A}(K)) \tag{3}$$

$$K = D_{SK_{PKE-A}}(D_{SK_{IBE-A}}(K')) \tag{4}$$

In order to ensure transparent encryption and sharing of encrypted files between different enterprises, we introduce a key management mechanism based on attribute encryption. In encb ox, each user has a master key (PKU, SKU) specially used to protect the file key, where PKU is the public key and SKU is the private key. The encrypted file key is called key lock. In practice, the key owned by the user is stored and managed in the intranet security gateway. For security reasons, encbox stores the key in the trusted platform module.

3 General Design Principles and Simulation

3.1 Standardization Principle

In the early stage of promoting the informatization construction of the school, there are mainly the following problems: the application system of each department is led by each department, lacking the long-term planning of the function and technology implementation scheme; the application system of each department mainly solves the current and local needs, and there is no overall cooperation among the departments, independent construction, maintenance and management, and some even cause the repeated construction of the system, It is disadvantageous to the continuous growth and utilization of school informatization, resulting in a serious waste of resources. Therefore, according to the current national and industrial standards, the school information construction should first formulate and form a unified technical standard system to ensure continuous investment in construction and long-term application [7–10]. The current demand and long-term planning should unify the standard overall planning and construction.

3.2 Data Sharing Principles

Due to the lack of unified data standards, the business system of each department is relatively independent, and the data between the systems is difficult to share, which brings difficulties to the synchronous and unified processing of business cooperation and cross data sharing of each department. In the existing business processes, the systems used by each other can not provide the function of data exchange. When some data need to be exchanged across departments, the work also relies on manual, semi manual or e-mail to transfer information. Today, with the rapid development of information technology, such information interaction mode has lost the effectiveness and convenience of information sharing, and the work efficiency is very low. Therefore, it is necessary to establish data sharing mechanism and specification, realize the co construction and sharing of campus data, collaborative development and workflow reengineering of various business departments.

3.3 Principle of Openness

At present, the development platform, database and running environment of each application system of the school are very different, and there is no unified use and management platform. With the campus network application system and application resources more and more, business system development and maintenance mode is not unified, there are risks in technology upgrading, update and maintenance difficulties continue to increase, the application of the lack of effective organization and unified management, but the school information maintenance management and construction costs continue to increase. Therefore, the construction of smart campus needs an open platform to provide for the future demand change and expansion of the school. Through the open platform, it can continuously improve and flexibly meet the demand, and realize more convenient system maintenance and management [11–13].

3.4 One Stop Service Principle

According to the classification of management information system and teaching information system, the original information construction mode invests and organizes the implementation of education information engineering projects, which is a typical "technology oriented" thinking mode. In the process of information construction in the past, the phenomenon of "emphasis on construction, light application" and "emphasis on management, light service" is serious. It can not provide comprehensive and personalized information service for teachers and students, which not only leads to low investment efficiency of education informatization, but also has little impact on promoting education reform, promoting learning style change and improving education quality. Therefore, the smart campus needs to build a unified information portal, call for one-stop service, respect individual needs and services, and integrate the business processes of various business departments, so as to provide all-round services to teachers and students.

3.5 Construction Objectives

To build a harmonious and unified campus information service platform for all teachers, students and parents to participate. With teachers' professional development, students' comprehensive development and quality growth as the core, and based on the construction of various information resources and application systems of the school, it is oriented to the multi-level information application needs of school education and teaching activities, school education and teaching management, school education and teaching research, campus culture construction, campus life of teachers and students, etc., To provide comprehensive, comprehensive and personalized information resources sharing and business process collaborative services, and build an efficient and scientific educational information resources service and guarantee system [14–16]. We should build advanced information infrastructure, optimize and upgrade the existing campus network environment, computing resources and data storage resources, improve the campus information security system, and build a number of modern digital education and teaching facilities and IOT perception terminals. Build a smart campus basic support software platform to provide stable and efficient support services for various information applications of the school, including portal integration, unified user management, unified identity authentication, organization management, content management and other basic service support, and practice the sustainable and scientific development of school running efficiency.

In the construction of smart campus, our charging project is a big problem, so Fig. 2 and Fig. 3 simulate the state of smart campus from charging.

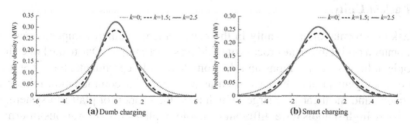

Fig. 2. Charging for smart campus

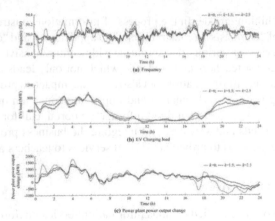

Fig. 3. Power plant for smart campus

4 Analysis Of The Current Situation Of Smart Campus Construction

4.1 The Efficiency of Information Application is not High

At present, although many high schools have established a digital campus which has begun to take shape, but when teachers and students carry out teaching research and learning, there are many problems, such as low rate of information application, lack of centralized information transmission and so on. Data and information are scattered in various departments, which can not achieve effective system collation and application. The way of data transmission is generally one-way transmission, which can not be accurate to the transmission target, and the transmission information is relatively random. At the same time, the high school in the construction of smart campus, can not realize the exchange of information with the outside world, such as teaching resources, curriculum content and other information exchange is still limited to the campus, this situation is not conducive to teachers and students in class time for information sharing and exchange.

4.2 Lack of Unity

The lack of information uniformity in the management of smart campus is one of the most common problems in the process of building smart campus. Due to the large number of people in high school campus and the complexity of departments, there are various kinds of management and services, such as payment of accommodation related fees, score query, and meal card recharge, which need the support of platform system. As a result, many high schools have different technology platforms and various information systems in the construction of smart campus, It makes the platform more chaotic [17]. At the same time, because the information of different platforms are stored in different databases, they can't exchange with each other effectively, resulting in a large number of useless data in the smart campus system, which can't be unified. Therefore, it brings some disadvantages to the normal teaching and research.

4.3 The Infrastructure is Lagging Behind

Because the smart campus needs a lot of capital investment and human investment, as well as technical support, the current high school in the construction of the project, there will be facilities lagging behind. From the current situation of high school smart campus construction, it is unrealistic to only rely on the development of the school itself. Therefore, relevant departments should actively promote the construction of smart campus, provide financial and technical support for it, and ensure that the high school smart campus can achieve more rapid construction.

5 Measures for the Construction of High School Smart Campus Under the Internet Plus Era

In the Internet plus era, building a smart campus is the real need of education. Therefore, how to promote the construction of high school smart campus under the background of Internet plus is the problem facing the high school. The following analysis of several measures on the construction of smart campus.

5.1 Grasp the Direction of Smart Campus Construction Under the Background of Internet Plus

It not only needs technical support, but also needs high school to sort out the idea of effective application of information technology. Therefore, in the construction of smart campus, first of all, we need to grasp the idea of construction, that is, the focus of the current construction of smart campus and what problems need to be paid attention to in the process of construction. Smart campus is a new stage of teaching informatization, which is upgraded from digital campus. Different from digital campus, digital campus pays more attention to the application, sharing and analysis of information, while smart campus pays more attention to how to provide services for teachers and students with the most humanized and intelligent services, so as to improve the use experience of teachers and students. Smart campus uses more advanced and fast computing methods, such as cloud computing, big data and other technical means, which effectively combines high school scientific research, teaching and management, and constantly reflects the person-alized, standardized and comprehensive service [18]. Therefore, the future development direction of smart campus should jump out of the thinking of digital campus period, think about how to use the Internet to make comprehensive use of the information in the campus, so as to provide more quality information services for teachers and students, and realize the ultimate goal of building smart campus.

5.2 Construction of High School Campus Integrated Information Service Platform

In the Internet plus era, high school wants to build smart campus. First, it needs to start from the technical aspect, change the status of all kinds of system applications being restricted to the desktop, take new technology as the main information carrier, and effectively create an integrated information service platform, which includes information interchange and intelligent application, thus changing the state of information dispersion in high school campus. Through the service platform, information can be unified. When teachers and students use information and carry out related management, they can complete it in a database, improve the utilization rate and interaction rate of related data, effectively save costs, and improve teaching efficiency and management efficiency. This is the focus of the construction of high school smart campus in the future.

5.3 Provide Organizational Foundation for Smart Campus by Strengthening Campus Internal Management

At present, there are many reasons for the lag of the construction of many smart campuses, not only in technology, but also in the management of the school itself. For example, many high schools manage teaching, teacher management, scientific research and other information separately. Even if special management departments and agencies are set up, most of their responsibilities are related to the construction of campus infrastructure, lacking the right to dispatch and use relevant information [19]. These reasons lead to the construction of smart campus, All kinds of information can not be effectively transmitted and circulated. Therefore, in the construction of smart campus at the same time, we also need to strengthen the internal management of the campus, to achieve innovative system reform, so as to provide a solid foundation for the promotion of smart campus.

5.4 Increase the Investment in Infrastructure and the Establishment of Guarantee Mechanism in the Process of Construction

Smart campus is a long-term and systematic project, which cannot be completed at one time. Therefore, it is necessary for the school, relevant departments and the whole industry to work together to form a kind of cooperation, and jointly provide some help for the construction of smart campus. At present, many schools are faced with the lack of infrastructure and construction resources, which has become the biggest challenge in the construction process of many campuses. It is not enough to build a smart campus only by the efforts of the school. In the process of construction, we need to actively find partners, encourage relevant enterprises to actively participate in the construction of smart campus, make use of their own advantages and mature services to contribute to the smart campus and speed up its construction process. In addition, the security maintenance of the smart campus system is also very important. In the high school smart campus platform, in addition to teachers can use the system to manage information, the school will also manage some students' personal information and financial information [20–24]. Therefore, it is necessary to protect the security of the information, otherwise, if the information is lost, In view of this situation, it is necessary to employ network professionals in the process of construction to carry out security maintenance on the high school network. The focus of maintenance is to improve the security of information. Therefore, it is not only necessary to implement security maintenance on the system, but also need to carry out maintenance on internal related equipment such as library and canteen, Only in this way can we ensure the security of information and realize the normal operation of smart campus.

6 Conclusions

This paper mainly analyzes and studies the process of smart campus construction based on cloud computing. In order to complete the overall development planning of Kunming No.1 middle school education informatization and the informatization development goal of phased and step-by-step construction, this paper analyzes and studies the connotation and characteristics of smart campus, and the guiding significance and characteristics of cloud computing technology in school informatization construction. Aiming at the construction of education cloud platform with unified data center, unified identity authentication and unified information portal, this paper puts forward the overall planning and design scheme of smart campus based on the concept of cloud computing. According to the idea of construction by stages and in batches, the first phase network construction is designed in detail, and the network construction solutions for the school to implement procurement, system integration and project acceptance are provided. Under the new background of Internet plus, high schools need to focus on cooperation with enterprises to speed up the pace of information management in campus management and services. Meanwhile, we should build a smart campus service platform based on the characteristics of the school itself, so as to provide better standardized and comprehensive services for teachers and students, and realize the new development of the campus under the Internet plus background.

Acknowledgement. A study on the construction and application of network learning space in higher vocational colleges based on Yunzhongtai (Project No .19 JX06240).

References

1. Jiafu, J., Yong, Z., Yulong, W., Li, Z., Huang, M.: Construction of smart campus system based on education cloud. J. Modern Educ. Technol. **23**(2), 109–110 (2013)
2. Ronghuai, H., Zhang Jinbao, H., Yongbin, Y.J.: Smart campus: the inevitable trend of Digital Campus Development. J. Open Educ. Res. **18**(4), 12–17 (2012)
3. Yan, W.: Overall architecture model and typical application analysis of smart campus construction. J. China Audio Vis. educ. **32**, 88 (2014)
4. Qintai, H., Kai, Z., Nanhui, L.: Development and transformation of education informatization: from "Digital Campus" to "smart campus." J. China Audio Vis. Educ. **324**, 38 (2014)
5. Fang, P.: Application of intelligent image technology in optimizing management of network computer room. J. Health Voc. Educ. **16**, 51–53 (2007)
6. Wei, C.: Discussion on the management of school network computer room. J. J. Taiyuan City Polytech. **04**, 147–148 (2007)
7. Guoxun, Y.: Construction and maintenance management of medium and small communication network room. J. Sci. Technol. Consult. Guide **14**, 42–43 (2007)
8. Shujun, W., Yueming, L.: Exploration of network computer room management and maintenance in higher vocational colleges. J. Career Circle **05**, 16–17 (2007)
9. Qunli, L.: Management and maintenance of network computer room in school. J. Sci. Educ. Wenhui First Ten Issues **01**, 185–186 (2007)
10. Xinyao, W., Qiang, Z.: Development and application of paperless examination system. J. J. Chengdu Aviation Voc. Tech. Coll. **04**, 48–50 (2006)

11. Cui, L.S., Li, H.: Application of protection card in network computer room management. J. Hebei Educ. Teach. Ed. **2006**(12), 39–40 (2006)
12. Qunli, L.: Development and design of open automatic network computer room management system. J. Cult. Educ. Mater. **35**, 143–144 (2006)
13. Qunfa, F.: Discussion on network management and maintenance of school computer room. J. Today Sci. **2006**(11), 116 (2006)
14. Computer based automatic network maintenance j. 29–79
15. Zhihui, Z.: Two classics in the management of student network computer room. J. J. Hengshui Univ. **03**, 59–60 (2006)
16. Qingju, H., Chang, L.: Selection and application of teaching and management software for network computer room. J. J. Hubei Univ. Technol. **04**, 148–152 (2006)
17. Aidong, B.: Management and maintenance of network computer room. J. Educ. Inf. **13**, 60–61 (2006)
18. Shen, L.: Thinking and practice of network computer room management J. Information technology education, (06), 72–73 (2006)
19. Juan, L.: Application and analysis of network cloning technology in computer room management. J. Agric. Netw. Inf. **04**, 111–112 (2006)
20. Yuling, S., Minfeng, C.: Management and maintenance of network computer room. J. Sci. Technol. Inf. Acad. Ed. **03**, 311 (2006)
21. Haitao, W., Jidong, M.: Tasks and responsibilities of hospital network manager. J. J. North China Coal Med. Coll. **01**, 112–113 (2006)
22. Hui, C., Dong, P.: Research on the application of intelligent image technology in the management of university network computer room. J. Audio Vis. Educ. Res. **12**, 67–69 (2005)
23. Xiufeng, L.: Discussion on management and maintenance of network computer room. J. China Mod. Educ. Equip. **12**, 5–7 (2005)
24. Shouyu, W., Lingyun, Z.: My opinion on Teaching LAN management. J. Fish. Econ. Res. **06**, 45–47 (2005)

Research on the Application of Information-Based Education in Non-relic Tourism Education

Xiao Wu[✉]

Department of Business Management, Laiwu Vocational and Technical College,
No.1 Shancai Street, Laiwu District, Jinan City 271100, Shandong Province, China

Abstract. Intangible cultural heritage protection education is related to the survival of national culture, and is the historical mission of contemporary people to reshape the national soul. Combined with the core values of relevant international conventions, this paper analyzes the nature of tourism education in intangible cultural heritage sites, probes into the objectives and practical significance of tourism education in intangible cultural heritage sites, and determines the core content of tourism education in intangible cultural heritage sites. In view of the above problems, combined with the role of information education in education, this paper clarifies the social responsibility of tourism education department in the protection of intangible cultural heritage, and provides some ideas for the development of tourism education in intangible cultural heritage areas.

Keywords: Information education · Intangible cultural heritage · Intangible cultural heritage · Tourism education

1 Introduction

The meaning of construction network system reliability of construction engineering project is: in the process of project construction, the ability to safely and effectively reach the predetermined project quality within the required completion period and within the limited cost. Therefore, we can understand the construction reliability of construction engineering project from the four aspects of cost, construction period, quality and safety, which is specifically understood as the construction period reliability, cost reliability, quality reliability and safety reliability of engineering projects. Reliability includes qualitative and quantitative meanings, and is generally described quantitatively by reliability. It is specially pointed out that the meaning of reliability and reliability in this paper are the same, and both represent the size of reliability. The reliability of construction work unit of engineering project refers to the ability of each work unit in the construction project to complete the construction task safely and effectively and achieve the predetermined project quality under the required planned cost in the construction process. The reliability of construction network system is based on the reliability of each work unit. Only when the reliability of each work unit is determined can the reliability of the construction system of the whole construction project be determined.

M. A. Jan and F. Khan (Eds.): BigIoT-EDU 2021, LNICST 392, pp. 643–651, 2021.
https://doi.org/10.1007/978-3-030-87903-7_77

2 Genetic Algorithm

2.1 The Concept of Genetic Algorithm

There are many types of genetic algorithms. Researchers at home and abroad have proposed more improved algorithms on the basis of basic genetic algorithms. Among them, hybrid genetic algorithm is formed by combining genetic algorithm with other intelligent algorithms, including genetic ant colony algorithm, genetic particle swarm optimization hybrid algorithm, genetic worker bee colony hybrid algorithm, genetic simulated annealing hybrid algorithm, genetic neural network hybrid algorithm, fuzzy genetic algorithm, chaotic genetic algorithm, etc.

Map building and population coding:

$$X = \text{int}(N/G_{size}) + 1 \tag{1}$$

$$Y = N\%G_{size} + 1 \tag{2}$$

2.2 Path Planning Steps of Genetic Algorithm

Path planning steps, The general steps of path planning are environment modeling, path searching and path smoothing. The steps of genetic algorithm in robot path planning are shown in Fig. 1.

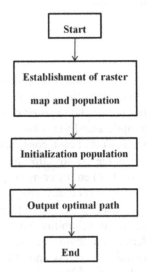

Fig. 1. Flow chart of genetic algorithm path planning

3 The Rise of Foreign Intangible Cultural Heritage Protection Education

In real life, there are many successful examples of tourism education in intangible cultural heritage sites abroad. For example, hula dance and sanweixian, in the form of intangible cultural heritage, have long been the symbols of Hawaii in the United States and Okinawa in Japan. Of course, this is related to the efforts of the two places to develop tourism industry for many years, but one of the most important factors is the development of tourism education in intangible cultural heritage sites [1]. The people of the two places and all stakeholders in tourism industry have considerable investment in tourism education. From the beginning of primary education in Hawaii, students are guided to think about the relationship between regional development and tourism by arranging small papers and mobilizing students to participate in social surveys related to various unique national cultures. In order to ensure the status of the national art in the market, Japanese people who have a higher position in the market are endowed with the status of national art. If it is stipulated that every primary school student must watch a "Neng & quot; play while in school, government officials should entertain them with traditional arts such as Noh opera, Kabuki and rave. In this way, the combined operation of education, cultural market and tourism market has created an excellent environment for the development of intangible cultural property, so that folk art will not lose its equal development right in a single evaluation scale of market economy; school education in Okinawa and Japan pays attention to the explanation and explanation of unique national and regional culture in the details of daily life, Actively encourage children to participate in a variety of national cultural activities or regional unique festivals and celebrations that can attract a large number of tourists from childhood, so as to integrate school education and social education, and realize the goal of intangible cultural heritage education unconsciously.

4 Intangible Cultural Heritage and Tourism Education

The tourism education of intangible cultural heritage refers to the relevant education for all responsible persons and stakeholders, such as residents and tourists of intangible cultural heritage sites, in the development, management and operation of relevant tourism resources of intangible cultural heritage sites, so as to ensure the sustainable development of intangible cultural heritage and tourism. The goal of tourism education in intangible cultural heritage sites is to establish a set of education system with functions of scientific development, management, dissemination and protection of cultural heritage and effective supervision, aiming at different responsibility and interest subjects of intangible cultural heritage, so that different subjects can consciously undertake the social responsibility and obligation of protecting intangible cultural heritage [2]. To jointly promote the sustainable development of intangible cultural heritage sites(see Fig. 2). Its practical significance lies in improving the political, economic and cultural development conditions of intangible cultural heritage sites by improving the quality of the people, so as to urge people to protect the integrity of the world heritage site ecosystem, respect the local community culture, and make the local residents actively share the social and economic benefits.

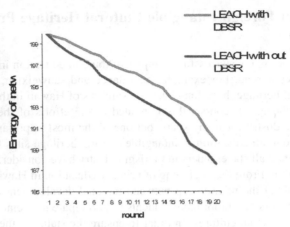

Fig. 2. Simulation of tourism education

5 Tourism Education in Intangible Cultural Heritage Sites

5.1 Universal: Everyone is an Educator and Educatee

The intangible heritage is the legacy of the ancestors' life style and the concentrated expression of their values and wisdom. In a sense, all indigenous people should understand the heritage and be able to choose to inherit it in different forms equally and freely [3]. All indigenous people are responsible and stakeholders of intangible cultural heritage. As indigenous people, when they can perceive, appreciate, carry and express this intangible cultural heritage, they themselves become educators of intangible cultural heritage. When the Aboriginal people or people other than the indigenous people do not understand, feel, appreciate, carry and display the heritage, they have the right to choose whether to accept the relevant intangible cultural heritage education or to choose what form and level of intangible cultural heritage education.

5.2 Public Welfare: Education Reduces the Social Cost of Development and Produces Positive Externality

Intangible cultural heritage education is the selective inheritance and dissemination of national excellent cultural traditions. It plays a very important role in maintaining and maintaining the diversity of cultural ecology, strengthening the cohesion of national culture, improving the quality of people's life and culture, promoting the healthy development of social economy and cultural affairs, coordinating social relations and promoting social harmony. Intangible cultural heritage education can strengthen the sense of national cultural identity, ease ethnic contradictions, reduce the social cost of development and disharmonious factors, and produce positive externalities.

5.3 Pertinence: Adjust Measures to Local Conditions According to the Characteristics of Cultural Heritage Sites

Different intangible cultural heritage sites often have different or the same type of intangible cultural heritage (see Fig. 3). We can see that in the first batch of national intangible cultural heritage list, the same project often becomes the common intangible cultural heritage of different regions and schools at the same time. For example, the folk customs of the Dragon Boat Festival have been selected into the intangible cultural heritage list as follows: the Dragon Boat Festival custom on the Bank of Miluo River in Miluo City, Hunan Province, the Dragon Boat Festival custom in Yichang, Zigui and Huangshi cities in Hubei Province, the Dragon Boat Festival custom in Qu Yuan's hometown, and the Shenzhou festival in Xisai, Suzhou County, Suzhou City, Jiangsu Province. In addition to the common dragon boat race, each place also contains the unique aesthetic consciousness and national emotion of local people. For the education and publicity of similar intangible cultural heritage, we should not take the road of great harmony and great unity, but should combine the unique local conditions and customs to enrich and develop continuously [4]. The Dragon Boat Festival in Jiangling, South Korea, is one of the world's intangible cultural heritages for its unique ancient costume parade, wrestling, masquerade, swing, eating moxa cake and other ethnic activities..

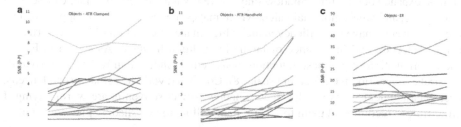

Fig. 3. Simulation of using different methods to protect heritage

6 Research Design

This study uses questionnaire survey method to make quantitative analysis of the target question. At present, there are many kinds of intangible cultural heritage tourist souvenirs in China. This paper chooses shadow puppet tourist souvenirs as the research object to test the theoretical hypothesis, mainly based on three aspects: (1) China's shadow puppet has become a world-class intangible cultural heritage in 2011; (2) Chinese shadow puppet language art, music, drama art and folk cultural psychological model As a whole, it is a comprehensive intangible cultural heritage. (3) shadow play souvenirs have gone out of the concept category [5, 6]. Shadow play hanging pictures, shadow play ornaments and shadow play toys have become popular souvenirs for tourists.

6.1 Scale Design

In the questionnaire design, the latent variables refer to the relevant research of domestic and foreign scholars. There are 19 items using Likert 5-point scale. The consumer participation scale mainly refers to the research scale of Huang Xiaozhi and others. Four items are designed. In the activities related to the experience of shadow puppet souvenirs, I pay physical strength. In the activities related to the experience of shadow puppet souvenirs, I spend time in the experience of shadow puppet souvenirs During the activities, I contributed my wisdom. I fully communicated with shadow puppet intangible cultural heritage artists to exchange consumer knowledge [7]. I learned about the history and characteristics of shadow puppet intangible cultural heritage. I learned about the use of shadow puppet souvenirs. I learned about the characteristics of shadow puppet souvenirs. I learned that the function perception quality of shadow puppet souvenirs is the main reference Wang Haisi et al. Set three test items: I am satisfied with the quality of shadow play souvenirs, I am satisfied with the packaging and other supporting hardware of shadow play souvenirs, I am satisfied with the price of shadow play souvenirs, and I measure the perceived value of consumers, In this paper, we draw lessons from fan Xiucheng and other 34 related researches to design five measurement items. Shadow play souvenirs have educational value, let me acquire intangible cultural heritage knowledge. Shadow play souvenirs have social value, let me enhance the emotion between relatives and friends, expand social circle. Shadow play souvenirs have practical value, can be used in daily life, and have emotional value, which makes me feel novel and happy; Shadow puppet souvenirs have self-efficacy value, which makes me feel that I have contributed to the protection and inheritance of intangible cultural heritage. As shown in Fig. 4. The determination of purchase intention measurement index mainly refers to pavlou's 35 purchase intention measurement scale [8]. Combined with the actual research needs, three measurement items are designed. Shadow puppet souvenirs are worth buying. I am willing to recommend them to others; I am willing to buy shadow play souvenirs.

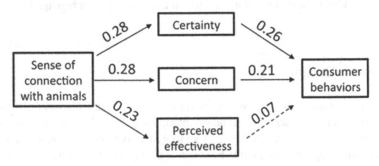

Fig. 4. The impact of consumer participation on consumer knowledge and related hypotheses

6.2 Data Collection

Two professors, five lecturers, three doctoral students, one business person and one intangible cultural heritage inheritor of shadow play were invited to try and answer the initial questionnaire of this research, and they were asked to put forward suggestions for revision after answering the questions, and then the initial scale was adjusted accordingly according to these suggestions, 2) a small-scale questionnaire survey was conducted in Xi'an City, Shaanxi Province, and 20 subjects were randomly selected from the tourists, Try to balance the distribution of gender and age. All the subjects in this round accepted the questionnaire. 3) after the previous two rounds of tests, the paper questionnaire was officially issued [9–11]. The formal survey was conducted in January and June 2018. Taking tourists in Xi'an as the research object, a total of 400 questionnaires were issued, 371 of which were valid, and the effective rate was 928%.

7 Conceptual Model and Related Hypotheses

7.1 The Impact of Consumer Participation on Consumer Knowledge and Related Hypotheses

Consumers can learn more information and knowledge about products or services when they participate in products or services. The degree of participation will affect the amount of information and knowledge that consumers get. It is proposed that consumers can obtain knowledge and details about products and services when they participate in enterprise innovation activities. Liu Wenbo et al.18 pointed out that consumers need certain knowledge when they purchase or use products For example, consumers need to know the function of products and how to extend the life of products [12–14]. Therefore, this paper puts forward the following hypothesis: consumers' participation in intangible cultural heritage souvenir related activities has a significant positive impact on consumer knowledge. In the research on the impact of consumer knowledge on perceived quality, Lu Hongliang and others regard perceived quality as a dimension of brand spillover value. Consumers with different knowledge have different perception of perceived quality. Consumers with low level of product knowledge lack of product cognitive structure, and often can not accurately obtain and interpret product related attribute information, and can not better grasp product quality The following hypotheses are put forward.

7.2 The Influence of Antecedents on Purchase Intention and the Role of Related Hypotheses in the Influence of Consumer Participation on Purchase Intention

At different stages of consumer participation, consumer participation has a positive impact on product purchase intention. Consumers' purchase decision is also affected by their understanding of the product to a large extent. 261 consumers have rich product knowledge, which means that experts will decide whether to buy Swaminathan according to their knowledge. 27 it is found that consumer knowledge can improve the quality of decision-making Cowley et al. 20 research results show that consumer knowledge has a positive impact on consumers\ purchase intention. In addition, consumers usually use their own product perceived quality to make purchase decisions [15]. Qian min et al. 29 in the empirical study of private brand purchase intention, made a regression analysis on the impact of perceived quality on purchase intention, and found that perceived quality can explain 75% of purchase intention, that is, perceived quality is the impact of private brand purchase intention A large number of empirical studies have confirmed that perceived value plays an important role in consumer purchase decision-making. Liu Haoqiang 30 divides perceived value into two dimensions: perceived acquisition value and perceived transaction value, which verifies that perceived acquisition value and perceived transaction value have a mediating effect and can significantly affect tourism consumers' purchase intention.

7.3 Management Inspiration

The conclusions of this study are helpful to enhance the purchase intention of tourism consumers, improve the market competitiveness of intangible cultural heritage souvenirs, and enhance the self survival ability of intangible cultural heritage, It is of great significance to promote the production and sales of the whole tourism industry) to expand the depth and breadth of tourism consumer participation. The more common tourism consumer participation is that tourists invest intelligence, physical strength, emotion and time to participate in the production of tourism souvenirs to obtain a sense of pleasure and satisfaction, Tourism enterprises can learn from LOI's customer participation input model to promote tourists' relationship participation, such as rewarding tourists to publish pictures and words related to intangible cultural heritage souvenirs on friends, microblogs and forums, so as to promote tourists to spread positive word-of-mouth in their personal network. In the extension of participation breadth, tourism enterprises can let tourists participate in intangible cultural heritage performances and feel the charm of intangible cultural heritage; Through interaction and exchange with inheritors of intangible cultural heritage, we can expand our knowledge of intangible cultural heritage, deeply understand the ecological environment of intangible cultural heritage and experience the life of inheritors of intangible cultural heritage in B & B of intangible cultural heritage [16, 17]. Tourists may put forward ideas on the function and design of intangible cultural heritage tourism souvenirs that do not exist at present, so as to help tourism enterprises find the defects of tourism souvenirs and improve product functions.

8   Conclusion

Secondly, it is the combination of intangible cultural heritage development and protection and tourism education system, including tourism education in the process of intangible cultural heritage development, research on cultural heritage resources integration and tourism education, such as the guiding mechanism of scenic spots, museums, exhibitions and intangible cultural heritage protection education, service innovation and intangible cultural heritage protection education. Thirdly, the promotion of intangible cultural heritage protection technology and tourism education research [18]. It mainly includes the formulation and popularization of cultural heritage protection standards, the safety warning knowledge and popularization of cultural heritage protection, the construction of scientific and technological infrastructure and the promotion of tourism education means.

Acknowledgements. Research project of Laiwu vocational and Technical College "Research on the living inheritance of Laiwu Intangible Cultural Heritage Based on research tourism".

References

1. State Council of China. The first batch of national intangible cultural heritage list. GF. 18 (2006)
2. Yang, Y.: Origin, status quo and related issues of the concept of Intangible cultural Heritage. J. Heritage World **2**, 27–31 (2003)
3. Chen, M., Zhang, X.: Summary of the first seminar on intangible cultural heritage education and teaching in Chinese colleges and universities. J. J. Hubei Acad. Fine Arts **4**, 61–62 (2003)
4. Wang, L.: Comparison of Chinese and foreign historical and cultural heritage protection systems. J. Urban Plann. **24**(8), 49–51 (2000)
5. Li, X.: Study on the appropriateness of the teaching content of she minority sports intangible cultural heritage project in rural schools. J. Sci. Educ. Wenhui Next Issue **01**, 129–130 (2021)
6. Zhu, X.: Research on the inheritance and teaching path of intangible cultural heritage folk dance in local universities. J. Media Forum **4**(02), 109–110 (2021)
7. Xiao, C., Liu, D.: Dance education in local colleges and universities and research on intangible cultural heritage dance of food in the context of globalization: a review of pasta dance on the white case. J. Brewed China **40**(01), 230 (2021)
8. Zhou, F.: Infiltration and integration of intangible cultural heritage in school education J. J. Harbin Voc. Tech. Coll. **2021**(01), 42–44 (2021)
9. Wei, L., Zhou, Z.: The education of Chinese cultural identity in the music general course -- from the intangible cultural heritage of Hebei Province J. Overseas Chin. Gard. (01), 68 + 70 (2021)
10. Zhang, Z.: Significance and practical exploration of intangible cultural heritage inheritance in higher education. J. J. Tianjin Radio TV Univ. **24**(04), 67–70 (2020)
11. Wang, H.: Protection and inheritance of folk music intangible cultural heritage. J. J. Hebei Univ. Technol. Soc. Sci. Ed. **37**(04), 116–120 (2020)
12. Wang, X., Yan, M.: Practice and reflection of peer education in the inheritance of intangible cultural heritage skills. J. Cult. Monthly **12**, 46–47 (2020)
13. Cai, T., Chen, Y.: On the campus education inheritance of Dongkeng muyuge in Dongguan J. Lit. Educ. (2), **2020**(12), 60–63 (2020)
14. Cheng, Y.: Process protection of Living Heritage – taking the protection measures of catering intangible cultural heritage items in the list of representative works as an example. J. Natl. Art **06**, 88–98 (2020)
15. Zhang, Y., Yuan, L.: Application of Qiandongnan music intangible cultural heritage in music teaching in primary and secondary schools. J. Contemp. Music **12**, 76–78 (2020)
16. Qian, M.: Experience the interest of bamboo weaving, the root of cultural confidence: Notes on the kindergarten class micro curriculum "the interest of bamboo weaving" C. Beijing Education audio visual press, preschool education magazine. The second Zhang Xuemen education thought seminar. Beijing Education audio visual press, preschool education magazine: Beijing Education audio visual press, preschool education magazine, pp. 623–628 (2020)
17. Hou, L.: Thoughts on the exhibition of intangible cultural heritage. J. Child. Fine Arts **23**, 32–33 (2020)
18. Liu, H., Guo X.: Investigation on the protection of intangible cultural heritage folk songs in Wuling Mountain Area a case study of Wufeng Tujia Autonomous County, Yichang City, Hubei Province, J. China Natl. Expo. **2020**(22), 1–3 + 118 (2020)

Research on the Construction of Mental Health Education System for Higher Vocational Students Based on Computer Software Analysis

Yuting Zhang[✉] and Wenshan Yao

Medical and Nursing Branch of Panjin Vocational and Technical College, Panjin 124000, Liaoning, China

Abstract. From the current overall situation analysis, psychological service has played a positive role in improving students' comprehensive quality. However, due to the unbalanced distribution of resources, imperfect organizational structure, unscientific operation mechanism and unclear process of specific matters, the whole psychological service system has some problems of setting and disability. In order to change this reality, colleges and universities need to establish a sound organizational system, clarify the post responsibilities of all kinds of personnel at all levels, so that they can form a person responsible for their own responsibility, smooth organization and leadership, implementation measures, feedback and correction functions, so as to make the mental health work of vocational college students develop smoothly. This paper studies the mental health of vocational college students by using IP computer analysis software.

Keywords: Computer software analysis · Vocational colleges · Students · Mental health · Service system

1 Introduction

With the diversified needs of society, the psychological state of higher vocational college students has a new development and breakthrough, the whole emotional tone and value orientation have changed, which requires further innovation of mental health education and psychological service research [1]. At present, mental health work is expanding from psychological counseling to mental health education and service with richer connotation and more clear objectives, which is the inevitable trend of the development of mental health service in higher vocational colleges.

2 Analysis on the Current Situation of Mental Health Education for Students in Higher Vocational Colleges in China

2.1 Significance of Mental Health Education

At present, the emergence of psychological counseling institutions or psychological counseling classes in higher vocational colleges is springing up. Based on the theoretical basic knowledge, psychological health education for college students is practiced.

M. A. Jan and F. Khan (Eds.): BigIoT-EDU 2021, LNICST 392, pp. 652–661, 2021.
https://doi.org/10.1007/978-3-030-87903-7_78

However, there are still problems and deficiencies in the mode of mental health education in higher vocational colleges. Mental health education is divided into positive goal education and negative goal education [2]. Negative goal education is to prevent abnormal psychology and treat bad psychological behavior. Positive goal education is to construct the best positive psychological state in an effective environment. Effective development of positive psychological education can improve students' psychological quality, promote the cultivation of sound personality, and build students' psychological quality as a whole.

2.2 The Goal of Mental Health Education is not Clear

From the perspective of positive psychology, it is to cultivate students' positive and optimistic attitude, and have a positive attitude towards life and problems. For a long time, the psychological education in higher vocational colleges is aimed at the special groups, adopting the method of eliminating psychological barriers, aiming at the students with abnormal psychological activities. The goal is to control the psychological crisis, and the main task is to resolve the psychological conflict of students. Generally, it takes the form of a unified deployment in the psychological counseling center, one-time screening, for students' psychological status from the surrounding teachers and students. This form only stays at the superficial stage, which ignores the importance of prevention and psychological guidance, and does not fully understand the psychological development process of students.

3 Construction of Dynamic Monitoring System of Mental Health of Vocational College Students Based on Computer Software Analysis

Computer software technology provides a new research method for mental health education, and provides a realistic basis and feasible guarantee for serving mental health education in Higher Vocational Colleges and realizing dynamic monitoring of students' mental health.

Based on the analysis of computer software, the mental health data monitoring of vocational college students should include traditional manual monitoring and computer system monitoring. The behavior data collected without the students' knowledge is the most real. We need to collect students' personal data based on social media. In the era of network information, a large number of students' social media users publish their daily life behavior records and emotional changes on the Internet, and even some psychological changes that they do not pay attention to will be shown on the network. Social platforms such as SMS, Weibo, wechat, QQ, post bar and other social platforms will generate information, posts, comments, messages and other data every day. All content and usage behavior on the network can be tracked and recorded. This kind of tracking record can provide the most real and comprehensive behavior data source for mental health monitoring.

At the same time, we also need to collect students' social activity behavior records. Nowadays, the development of technology can record the behavior track of individual social activities in detail. In the school, there are traces of students' campus card consumption, book borrowing, attendance in class, examination failure, student loan, participation in campus activities, etc. collecting these information can comprehensively and systematically reflect the behavior rules and characteristics of college students, So as to predict the mental health of students.

The RF propagation loss in free space is shown in Formula 1. Where Lfs is the transmission loss, d is the distance between transceivers (in km), and f is the frequency of wireless signal (in MHz).

$$Lfs = 32.44 + 20\log d + 20\log f \tag{1}$$

The transmitting power range of RF module can be calculated, see formula 2. Among them, Pt is the transmitting power of RF module, Pr is the receiving sensitivity of RF module, and Gt and Gr are the antenna gain of transmitter and receiver respectively [3].

$$P_t \geq P_r + Lfs - G_t - G_r \tag{2}$$

In the actual design, the RF module uses a spring antenna with a gain of 3 dBi, and the maximum receiving sensitivity can reach 124 dBm. According to the formula, the transmitting power of RF module should be guaranteed at 20 dBm.

4   The Effect of Positive Psychology on the Mental Health of Vocational College Students

4.1   Integrating the Function of Mental Health Education

The research direction of positive psychology not only focuses on guiding psychological problems, but also focuses on the prevention of mental health. The positive and healthy education can effectively control and manage their own emotions, so as to solve the students' mental health problems. Positive psychology should be encouraged to find

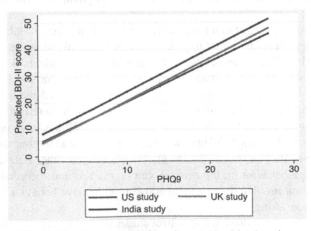

Fig. 1. Prediction simulation of mental health education

positive attitude towards students. Integrating the content of mental health, aiming at the negative factors in mental health, cultivating their own positive quality. The function of integrated mental health education is shown in Fig. 1.

4.2 Deepening the Reform of Mental Health Education

The traditional mental health education is basically a process of diagnosis and correction [4]. Only when there is a problem or a feeling of psychological changes, can we begin to pay attention to mental health [5]. The guidance of positive psychology for the development of mental health is based on a good mentality, through positive psychological counseling to improve students' psychological quality and promote the development of mental health. Positive psychology encourages the cultivation of excellent qualities, such as self-cultivation and temperance, wisdom and knowledge, courage and kindness, selflessness and justice, kindness and empathy, etc., so as to construct the subjective positive psychology of the individual, and to feel the happiness and pleasure from the heart. Positive psychology is the direction of psychological reform. In a positive state, psychology can shape a good personality and better establish the healthy psychology of vocational college students [6, 7].

4.3 Expand the Way of Mental Health Education

The overall quality development of students can not be separated from positive psychological education. However, as an individual, each student comes from different regions, has different family background and different social experience. Therefore, mental health education needs to choose the best method flexibly according to the actual situation. We should popularize the propaganda of mental health education on campus, organize mental health activities regularly, and carry out effective mental health education for students by means of mental health discussion, psychological counseling, psychological training and psychological suggestion [8]. Positive psychology education is not aimed at the treatment of psychological diseases, but focuses on exploring the positive and healthy quality in the individual's heart, giving full play to students' own potential, seeking ways to guide students' mental health from multiple perspectives, and infiltrating positive health concepts. Teachers should not only impart mental health knowledge, but also expand educational approaches in various directions to transmit positive energy to students, Shaping positive psychological quality. Figure 2 shows the prediction simulation of mental health education. Figure 3 shows the statistical simulation of mental health education.

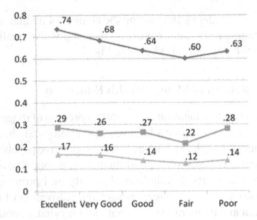

Fig. 2. Prediction simulation of mental health education

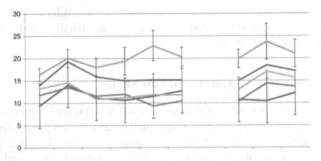

Fig. 3. Statistical simulation of mental health education

5 Countermeasures to Improve the Construction of Mental Health Education System in Higher Vocational Colleges

5.1 Broaden the Channels of Mental Health Education for Vocational College Students

Higher vocational students' mental health education is a systematic project. Although it is included in the system, the development of education should not be formalized because of systematization. In other words, the development of psychological education for higher vocational students should not be limited to classroom, lecture or psychological counseling, but should make full use of various resources to expand the channels of psychological health education, so that education can meet the deep-seated needs of students. The ways of mental health education can be divided into dominant education and recessive education according to the purpose and content of education. Explicit education refers to those education methods with strong purpose, with clear teaching content and curriculum arrangement. And recessive education is more through the creation of environment or conditions, let the educated rely on self-awareness to adjust their mentality. What Hainan Vocational and technical college lacks now is the supplement of recessive education to explicit education. Therefore, the author suggests that we should broaden the coverage of recessive education as far as possible on the basis of existing education methods [9].

In the development of recessive education, campus culture is undoubtedly one of the most important communication channels. The promotion of group effect has twice the result with half the effort for college students' mental health. Hainan Vocational and technical college has taken some measures in the construction of campus mental health culture, but it is far from enough. We should further play the carrier role of campus culture in mental health education, and make full use of campus broadcast, publicity window, school newspaper, classroom blackboard, wall and other media to infiltrate mental health education into all aspects of study and life, Let the positive psychology form a kind of atmosphere and breed an invisible educational force [10].

Network construction is also an important channel for recessive education, which is to use virtual technology to make up for the deficiency of dominant education. If conditions permit, the school can set up a special network platform with the theme of mental health education, and ask professional mental health teachers to be responsible for the management and maintenance. On the one hand, it regularly publishes and updates the mental health knowledge closely related to the life, study and employment of higher vocational college students, integrates the news related to the psychological problems of college students at home and abroad, and puts it in a prominent position as the daily front page headline. In addition, the website also needs to set up an interactive section with students. Students can communicate with teachers by email or online consultation. In this way, it can not only solve the students' inner or emotional confusion, but also better protect the students' privacy. Although the effect of online mental health education is good, the premise of this kind of education is to ensure that online teachers have enough time and energy. The website itself is a one to many open platform, so it is impossible for schools to deploy a large number of people to manage the website. Generally, only one professional psychological teacher answers questions online at a fixed time, or several psychological teachers are on duty in turn. There is no guarantee that all students can answer their questions one by one. In order to solve the contradiction between demand and supply, the school should try its best to deploy full-time and part-time psychological teachers, publish some teachers' email, QQ numbers and telephone numbers on the website, and develop a 24-h hotline for students [11].

5.2 Strengthen the Construction of Teaching Staff

At the beginning of the mental health curriculum, most schools will choose ordinary teachers to take up the role of mental health teaching and counseling. Most of these teachers have not received professional training, and the teaching method is mainly experience teaching, that is, to examine the students' psychology with a "past person" mentality, and hope to help students avoid detours by teaching their own life experience. With the aggravation of students' mental health problems, empiricism has been unable to meet the social requirements of school mental health education. In order to solve the problem of students' mental health, some schools began to try to use external introduction or internal training to obtain professional mental health or psychological counseling teachers. Mental health education has also been widely concerned by the society [12].

The standard to measure the professional level of psychological educators first appeared in the United States in the 1960s. The U.S. government has formulated the training standards for psychological educators and issued them to the whole country in the form of documents. Since then, psychological education has become a hot topic. So far, more than 30000 professional mental health educators have been active in the campus, and have made great contributions to the cultivation of the internal quality of talents in the future. This experience of the United States is gradually spreading in other countries, and the specialization of mental health education team will become a trend [13].

5.3 Strengthen the Mental Health Education and Training of College Counselors

In higher vocational colleges, instructors are not only responsible for the management of students' daily study and life, but also understand students' ideological trends. Many years of experience in Ideological and political work is a favorable condition for counselors to engage in mental health counseling and education. Moreover, compared with professional psychological teachers, counselors are closer to students and understand them better, which is not easy to cause students' rejection. Although the number of full-time psychological teachers in Hainan Vocational and technical college is very limited, the number of counselors is very considerable. Moreover, it is not possible to expand the full-time psychological teachers in a short time, so it is feasible to bring counselors into the team of mental health teachers. The nature of the work of counselors is similar to that of mental health teachers, and the most obvious gap is the lack of professional background of psychology. Schools should provide professional and systematic psychological health education theory and skills training for counselors, so that they can master basic psychological counseling knowledge, form a three-level education counseling network with student associations and professional psychological teachers, and solve the problem of insufficient number of psychological health educators in Hainan Vocational and technical college [14].

5.4 Continuously Improve the Professional Level of Mental Health Education

The development of mental health education is a sustainable circular process, and one of the criteria to judge whether this process is effective is whether the professional level continues to improve. In the process of developing school mental health education, many western countries will continue to put forward new requirements for the professional level [15]. The improvement of professional level is mainly manifested in three aspects, the first is the continuous improvement of professional level of employees. As a pioneer in the development of mental health education, the United States not only takes the lead in formulating the training standards for professional mental health counselors. Subsequently, the United States continued to formulate laws and regulations to strictly enforce the admission qualifications of psychological counseling practitioners. According to statistics, in order to become a qualified school counselor in the United States, we must first have a master's degree or above, and hold the employment certificate of the state government method, but also follow the constraints of thousands of professional regulations, which shows its high requirements for specialization; the improvement of the professional level is reflected in the continuous improvement of the psychological education system. With the highlight of the value of mental health education, many countries have realized that mental health should lay a good foundation, so the psychological counseling course is no longer limited to universities or middle schools, but from primary school, let the mental health education accompany students to complete all their studies. Many European and American countries have put this idea into practice, and may extend the starting point of mental health to the kindergarten stage in the future. The problem of low professional level of mental health teachers in Colleges and universities in China should be solved by improving the industry access threshold.

6 The Training Goal of Positive Psychological Education

Many words in life are related to positivity, such as positive, enterprising and so on. As an adjective, positivity is often used to describe those successful people who have the courage to struggle. But the positive behavior is only external positive, positive behavior may be to obtain money and status for the purpose. Positive mental health education refers to a good psychological quality, internal performance and emotion, mentality and other aspects, and eventually become the guidance of behavior. The formation of this kind of good psychological quality is also the goal of mental health education:

1. Positive cognitive quality can be explained in a positive way. As the saying goes, ruler has advantages, inch has short, different people in different fields have different intellectual advantages, and some of this advantage is dominant, some need to guide and mining. Positive psychological education does not advocate to attack the relatively backward individuals with the difference of intelligence, but to see the talent and potential of different individuals through the difference of intelligence. Positive psychological education should guide individuals to recognize their own inherent advantages, help them establish good self-awareness, and encourage them to position their development direction according to their own advantages.
2. People's emotion comes from life experience, rich and complex, which is also reflected in the content of emotional education. Positive emotional quality requires students to understand good and evil, and love and hate clearly. Take love as an example, love can be divided into small love and big love. It can also be divided into love for people and love for things. Active teaching should cultivate students to have good qualities such as loving parents, enterprising, practical and dedicated; positive emotional quality requires students to have a good sense of obligation and responsibility, attach importance to the sense of achievement and honor, and correctly handle the relationship between people and society, between people and nature, and between people; positive emotional quality requires students to be good at discovering the beauty in life and improving subjective well-being, And can grasp the meaning of life.

7 Conclusion

To sum up, in order to construct the mental health service system in Higher Vocational Colleges from the perspective of positive psychology, we should have a sufficient understanding of students' current psychological development from the perspective of psychology, introduce the concept of positive psychology into health education, unremittingly promote the general education of College Students' mental health, cultivate a peaceful and healthy mentality, strengthen the psychological counseling work for higher vocational students and strengthen humanistic care. Combining theory with practice, this paper explores the reform measures of positive psychology education, guides students to establish positive values, completes students' objective self cognition education, and builds the school into a stable, United, positive and upward Learning Paradise, and promotes the overall and healthy development of students. In the era of big data technology development, if colleges and universities want to do the psychological crisis early warning work of college students in place, it is necessary to integrate the big data technology with the current psychological crisis early warning methods, conduct data analysis in all aspects, so as to realize the dynamic monitoring of College Students' daily behavior activities, and improve the timeliness of psychological crisis intervention on this basis. When carrying out a series of specific work, we should start from many aspects, comprehensively solve the defects of traditional statistical methods, so as to truly do the psychological early warning work in place.

References

1. Li, Y.: The construction of mental health service system in Higher Vocational Colleges from the perspective of positive psychology. Digital Commun. World **3**, 283 (2018)
2. Liao, Z.: Analysis on mental health education of vocational college students from the perspective of positive psychology. Read. Writ. **15**(27), 5–7 (2018)
3. Deng, X.: Investigation and reflection on the teaching needs of students' mental health course in Higher Vocational Colleges. Chiko **34**, 90–91 (2018)
4. Yu, R., Liu, S., Ma, G., et al.: Study on the construction of Higher Vocational Students' mental health education system from the perspective of positive psychology. Occupation **44** (2017)
5. Wu, X.: Analysis on the goal system of mental health education in Colleges and universities. J. Wuhan Univ. Technol. (Soc. Sci. Ed.) **8**, 34–37 (2005)
6. Luo, H.: Construction of higher vocational mental health content education system. Mental Health Educ. **7**, 54–56 (2005)
7. Shen, W.: Research on the content system of mental health education for higher vocational college students. Vocat. Technol. **24**, 35–38 (2006)
8. Qiu, K., Zhou, X.: Research on curriculum system of mental health education in Higher Vocational Colleges. Psychol. Sci. **32**(5), 1259–1261 (2009)
9. Zhao, M.: On the relationship between mental health education and moral education in Colleges and universities. J. Shenyang Normal Univ. (Soc. Sci. Ed.) **5**, 23–26 (2000)
10. Robert, S.: Educational Psychology, pp. 213–239. People's Posts and Telecommunications Press (2004)
11. Zhang, M., Li, G., Li, Y.: Research on the developmental theoretical model and system construction of College Students' mental health education. J. China Inst. Labor Relat. **20**, 4104–4107 (2006)

12. Zhou, W.: Exploration of Mental Health Education for College Students in Higher Vocational and Technical Colleges in China: [Master's Thesis]. Hefei University of Technology, Hefei (2009)
13. Wu, J.: Research on Freshmen's Mental Health Education Based on Positive Psychology Theory: [Master's Thesis]. Dalian University of technology, Dalian (2011)
14. Mi, H., Wang, S., Liu, H.: Shortcomings and countermeasures of mental health education in Higher Vocational Colleges. Teach. Res. **3**, 134–136 (2008)
15. Wang, Y.: Thoughts on implementing positive mental health education. Chinese J. Educ. **4**, 24–37 (2006)

Design of Information Teaching System Platform for University Teachers from TPACK Perspective

Hongyun Du[✉]

Shandong Xiehe University, Jinan, China

Abstract. From the perspective of information technology, this paper analyzes the development of teachers' and students' competency in TPACK The potential direction of development path. Then, from the perspective of TPACK, combined with the existing problems and development path of University Teachers' information teaching ability, this paper constructs a targeted and systematic development model of University Teachers' information teaching ability. Compared with these problems, this development mode can be realized. In the development mode, it puts forward the specific implementation content and method, in order to provide reference for the development of University Teachers' information ability and the construction of University Teachers' team, and establish the design of University Teachers' information teaching system platform.

Keywords: College teachers · Information teaching ability · TPACK

1 Introduction

1.1 Analysis of College Teachers' Information Teaching Ability Based on TPACK

The knowledge of subject teaching method integrating technology is the necessary knowledge for teachers to combine technology with specific subject teaching, which can be used to guide the analysis of University Teachers' information-based teaching ability structure [1]. TPACK knowledge framework includes three basic elements: TK, PK and CK, and four composite elements: TCK, TPK, PCK and TPCK.

For technology, Koehler and Mishra point out that the technology here includes traditional technology and digital technology. Technical knowledge is not unchangeable. With the development of the times, technical knowledge will be constantly updated. Keeping pace with the times is a typical feature of technical knowledge. According to this knowledge, we can get the specific requirements of teachers to master the application of traditional technology and information technology ability, information technology teaching ability is mainly the ability of information technology. For example, teachers can master the courseware making methods and tools, master the use of multimedia equipment, and use the network to search for materials.

M. A. Jan and F. Khan (Eds.): BigIoT-EDU 2021, LNICST 392, pp. 662–670, 2021.
https://doi.org/10.1007/978-3-030-87903-7_79

1.2 Analysis of the Standard of Educational Technology Ability for Primary and Secondary School Teachers

There is no uniform document about the standard of College Teachers' information-based teaching ability. Since educational technology ability is the upper concept of information-based teaching ability, we can refer to educational technology ability to guide the analysis of information-based teaching ability structure. For primary and secondary school teachers, the Ministry of Education promulgated the standard of educational technology ability for primary and secondary school teachers in 2004 [2–4]. The standard specifically describes the teaching staff, management personnel and technical personnel. We mainly analyze the educational technology ability standard of teaching staff. The standard points out that the educational technology ability standard of teaching staff includes four modules, namely consciousness and attitude, knowledge and skills, application and innovation, and social responsibility. Under the first mock exam, there are two levels of indicators under each module. Each index has specific criteria for refinement, and specific capacity requirements are proposed. The educational technology capability indicators of primary school and middle school teachers are systematically and comprehensively defined. Through the specific analysis, we can find that "consciousness and attitude" and "social responsibility" belong to the emotional category, some overlap, and can be integrated.

2 Key Technology of Application

2.1 Wechat Public Account

WeChat public platform is a platform for Tencent Inc to provide services to users. The public platform interface is the basis of services. Developers apply and create public numbers on the website, and get permission to explain the WeChat official account on the WeChat public platform, according to the interface document, which helps developers to develop their needs [5, 6]. Through this account, developers can achieve all-round interactive communication with designated people in text, multimedia and other aspects on wechat platform. It has formed a mainstream mobile communication mode.

2.2 Development Interface

2.2.1 Get Interface

Calling credential interface (1) gets the only credential for access token: user application and official account number. When calling the interface to function, the official account must call access token. For security reasons, access token has only 2 h of validity, refreshing every other time, and the newly acquired access token will cause the old failure. The public number uses appid and appsecrect to call the interface to get access token. The three parameters that must be input are grant type, appid and type respectively, for obtaining the official account of the customer [7]. Appid represents the unique certificate of the third party, and secret is the secret key of the third party. Usually, it sends the data in JON format to the official account access tokenACCESS TOKEN "expires in": 720012) obtaining the official account of WeChat server IP: is considered in the network security and other factors, and needs to get the list of IP addresses of WeChat servers, so as to facilitate the restriction of them.

2.2.2 Receive Event Push

When the official account is exchanged with WeChat users, some operations of the user cause the server to inform the developer server I by way of event delivery, and then get the message. However, whether a developer replies to a message depends on the event description of the push message. Receiving push messages includes cancel and follow events, custom menu events (click menu to pull messages and jump link push events). Cancel and follow the event: wechat sends this event to the URL written by the developer, which is convenient to send binding or unbinding to the user. It includes five parameters: Tour Name: developer wechat account; from user name: sender account; create time: message creation time; msgtype: message type; event: event type (subscribe and unsubscribe). 2) Custom menu event: when the user clicks the custom menu, the developer will receive the user click event sent by wechat server [8–10]. If the secondary menu can pop up, the above situation will not appear. Send and pull message event: including 6 parameters, tour Name: developer wechat account; from user name: sender account; create time message creation time; msgtype: message type; event: event type; eventkey: event key value. Click the menu to jump to the push event of the link: including 6 parameters, tour Name: developer wechat account, from user name: sender account, create time: message creation time, MSG type: message type.

2.2.3 User Group Management Interface

User group management: the interface implements basic operations such as creating, adding, deleting, modifying, and transferring public platform groups. 1) Create group: one account can create up to 100 groups. There are two parameters: oken: call interface voucher; Name: group name. The return value also has two parameter groups ume: group name. Query grouping: only one parameter, access token: call interface voucher. However, the return value has four parameters: groups: group information table group ID; Name: group name; count: number of users in the group. 3) Query user group: query user group ID according to user ID. There are two parameters in total, access to call interface credentials; openid: user openid. The return value only needs group I, groupid: the group ID of the user. 4) Modify group name: three parameters in total, access token: call interface credential ame group name.

2.3 Data Format

XML technology XML (Extensible Markup Language) is a subset of SGML (Standard Generalized Markup Language). Its purpose is to allow general SGML to be served, received and processed in the way of current HTML on the web, and to design ml as a bridge between HTML with limited functions and SGML with complex standards, and it is easy to implement [11]. XML is regarded as a set of rules based on semantic and syntactic tags, through which the document is divided into several parts. Different from

HTM, it is a meta markup language, and developers define tags according to the actual situation. XML tag describes the structure and meaning of text information.

3 JSON Technology

JSON is a lightweight data exchange format, stored in the array, which is conducive to the data exchange between different languages. The text format is completely unrelated to the programming language. JSON is written in JavaScript. Therefore, JSON is defined in the JavaScript file. Because the data format of JSON is simpler than that of XML, there is no need to consider nodes and other issues, so the transmission efficiency is greatly improved. It will also reduce the compatibility problems caused by Xi ml parsing. So far, JSON has supported many mainstream computer programming languages.

3.1 JSON Sequence Mechanism

All kinds of data are stored in the computer, and the transmission time of data in the network is the longest. Therefore, storing the data in JS - n format can avoid the complexity of data conversion, and can be read in many languages. JSON sequence mechanism is divided into forward serialization mechanism and reverse serialization mechanism. Forward serialization mechanism: convert the object stored in the server into JSON format, which is convenient for data transmission to the client [12–14]. If you do not have special permission, you cannot access the data of the instance object. The reverse serialization mechanism is the reverse process of the forward serialization mechanism. It reads the serialized data in the data store and creates a new object. That is to say, after the data in Jon format is parsed by the client, it will be displayed on the web. The combination of forward and backward serialization completes the data transmission in the network. If serialization technology is not used, the client cannot receive the data sent by the server. The purpose of serialization is to store the instance object persistently in a certain format, transfer it between the server and the client, and read the information of the instance object stored in the data cache layer.

3.2 Advantages and Functions of ECS

Compared with physical servers, ECS has incomparable advantages, as follows: (1) disaster recovery backup: each data will save multiple copies, once the original data is damaged, it can be quickly recovered in a short time. Support downtime migration, data snapshot backup and rollback, system performance alarm. It will also enable users to timely alarm when they do not operate the data, so as to avoid losses. (2) Cost saving: according to the needs of the purchase, flexible response to changing business needs. Cloud storage can be flexibly expanded according to the number of users and the storage capacity required. It can start or release 100 ECS instances within 10 min. It supports online bandwidth upgrade without downtime. It can also upgrade CPU and memory within five minutes [15]. At the same time, the enterprise does not need to build an additional platform, and all hardware devices are supported by service providers, thus saving the cost of investment.

4   On the Structure of Information Teaching Ability of University Teachers from the Perspective of TPACK

Through relevant analysis, referring to the connotation of information-based teaching ability, and referring to the opinions of relevant experts, this paper refines the structure of College Teachers' information-based teaching ability from the perspective of TPACK, as shown in Fig. 3.2. The structural model takes information technology as technical support, TPACK knowledge framework as theoretical basis, and information technology as the bottom. The three aspects respectively represent the core elements and basic components of TK, CK and PK, namely technical knowledge, subject content knowledge and teaching method knowledge. The intersection of TK and CK represents TCK, that is, subject content knowledge of integrated technology. Similarly, the intersection of CK and PK represents the PCK formed by the integration of PK and CK, that is, subject pedagogy knowledge. The intersection of TK and PK is TPK, which is the teaching method knowledge of integrated technology. The vertex of the model is the intersection of TK, CK and PK, which is the fusion TPACK of the three, that is, the subject teaching method knowledge of integrated technology [16]. TK, CK, PK and information technology support each other and form a TPACK perspective of University Teachers' information teaching ability structure.

The structure is divided into five layers from bottom to top: attitude and responsibility, foundation and skills, design and development, application and evaluation, integration and innovation. Each module is the fusion of TK, PK and CK, based on these three elements, but each module has its own emphasis [17]. The structure changes step by step from bottom to top, from low to high capacity, and the ability requirements are gradually improved.

5   University Education Informatization Based on Cloud Model Data Mining Algorithm

Among these massive data, there are some dynamic data which have special requirements for the calculation process of judgment criteria, which has higher requirements for the environment of data mining. Data mining is to mine useful information from massive data, find the relationship between data, and promote the application and transmission

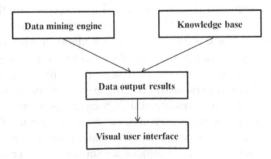

Fig. 1. Cloud model data mining algorithm calculation process

of information. The calculation process of cloud model data mining algorithm is shown in Fig. 1.

In the case of known first mining probability and class conditional probability, the classification result of data samples to be mined depends on all samples in each category. Let the sample set of data mining have m class, which is recorded as $C = \{c_1 \ldots c_i \ldots c_M\}$ The first mining probability of each class is $P(c_i)$.

$$P\left(\frac{c_i}{x}\right) = p\left(\frac{c_i}{x}\right) \cdot \frac{P(c_i)}{P(x)} \tag{1}$$

In all the teaching platforms, we use TPACK to design the teaching platform, as shown in Fig. 2. In this cloud teaching platform, we mainly have the following functions: teaching project and process, how much time is spent, and the error rate of students in this class. These are displayed in the form of curves, which can not only more intuitively see the learning situation of students, but also more intuitively know whether the class is successful or not [18]. If students do not understand in the process of class, they can also suspend real-time learning. In this way, students can achieve the best teaching effect in the shortest time.

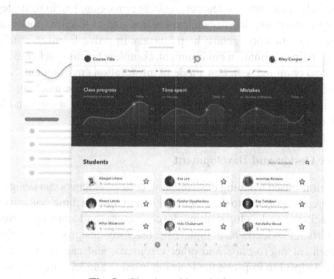

Fig. 2. Cloud teaching platform

6 System Design of University Teachers' Information Teaching Ability Structure from TPACK Perspective

Next, we make a detailed analysis of the various levels of University Teachers' information-based teaching ability structure from the perspective of TPACK. From the perspective of TPACK, University Teachers' information-based teaching ability structure is constructed under the core guidance of TPACK knowledge framework. Each level is the integration of TK, CK and PK, and covers TCK, PCK and TPK composite elements [19]. Combined with relevant standards, it is divided into attitude and responsibility, foundation and skills, design and development, application and evaluation, integration and innovation from bottom to top. The level of ability is constantly increasing, and the requirements for teachers are also gradually improved. Each capability level has specific capability index. In the two levels of application and evaluation, integration and innovation, each capability index is divided into different ability requirements, including basic requirements, advanced requirements and objective requirements, which further refines the ability. Next, we analyze and interpret the specific indicators of University Teachers' information-based teaching ability structure from the perspective of TPACK.

6.1 Information Teaching Design

In the information environment, college teachers can determine and write teaching objectives according to the curriculum standards, and master the compiling methods of common teaching objectives. Be able to analyze and process the teaching materials, determine the teaching content, and clarify the key and difficult points of teaching. It can analyze the characteristics of learners, select appropriate teaching strategies according to teaching content and learner characteristics, select teaching media according to teaching needs and students' needs, help learners improve learning efficiency, design teaching structure in information environment, complete teaching scheme in information environment, and combine with new mode in information environment, Such as flipped classroom, blended teaching, project-based learning, innovative teaching design, restore the essence of learning, optimize teaching, improve learners' ability, and help learners to improve their core literacy.

6.2 Resource Design and Development

In the information-based teaching environment, college teachers can design and develop teaching resources according to the teaching content and teaching objectives. For example, college teachers can develop multimedia courseware and learning website according to the characteristics of learners, complete the teaching design of micro class, complete the development of micro class and other resources, and can modify and update the resources, it is helpful to improve the learning interest and learning efficiency of learners. Teaching evaluation is the evaluation of students' learning process, learning method and learning effect, which can be divided into process evaluation and summative evaluation. In the information environment, on the basic requirements, college teachers can use technical means to collect data and evaluate students. In the advanced requirements, college teachers can complete process evaluation and summative evaluation according to students' learning characteristics and learning content, and provide feedback for teaching [20, 21]. In terms of objectives and requirements, college teachers can integrate technology with classroom teaching, establish electronic portfolio for students, comprehensively record the learning situation of learners, and provide comprehensive evaluation for students.

The specific teaching effect simulation is shown in Fig. 3. From Fig. 3, we can see that both students' satisfaction and recommendation have achieved good results. From another aspect, we can see that our teaching platform has achieved satisfactory results.

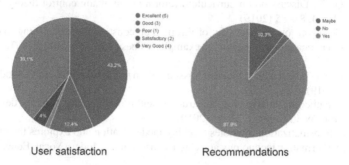

User satisfaction   Recommendations

Fig. 3. Student satisfaction survey

7 Conclusion and Prospect

Based on TPACK knowledge framework, combined with the analysis of relevant competency standards, this paper analyzes the structure of College Teachers' information-based teaching ability from the perspective of TPACK, and compiles the questionnaire of College Teachers' information-based teaching ability, and conducts a survey on students respectively to understand the status quo of teachers' ability from the third perspective, and fills in the questionnaire through expert lectures, From the perspective of experts to understand its development status, and interviews with some university teachers, from the first perspective to understand the status of information-based teaching ability training and teachers' subjective needs. Then through data analysis to understand the current situation of the development of University Teachers' information ability [22–24]. Through the analysis, it is found that college teachers have strong awareness of information-based teaching, good mastery of basic subject knowledge and skills, but in actual teaching, teaching methods are relatively simple, resource development ability is weak, teaching evaluation ability needs to be strengthened, and innovation ability is not enough.

Acknowledgements. Shandong Higher Education Research Project: Research on the development of College Teachers Information Teaching Ability from the perspective of TPACK, Item number: 20HER065.

References

1. Zhengxudong, W.: Higher education informatization and its development trend
2. Jiao, J., Zhong, H.: Research topics and progress of TPACK. J. Dist. Educ. **28**(1), 39–44 (2010)

3. Yan, Z., Xu, F.: TPACK: the knowledge basis of teacher professionalization in the information age. Modern Educ. Technol. **23**(3), 5–9 (2013)
4. Hu, X.: Analysis on the structure of University Teachers' information teaching ability. Modern Dist. Educ. **6**, 67–72 (2012)
5. Ao, Z., Qi, T.: Discussion on curriculum reform of automatic control theory. China Light Indust. Educ. **4**, 81–85 (2019)
6. Sun, X.: Research on the necessity of information construction in teaching management – taking higher vocational colleges as an example. Chinese J. Multimedia Netw. Teach. **8**, 70–71 (2019)
7. Dou, J.: Research on teaching mode of missile system informatization. Educ. Moderniz. **6**(64), 172–174 (2019)
8. Li, J.: Strengthening information construction and promoting high quality development of school. Read. Writ. Comput. **22**, 51 (2019)
9. Zhao, K.: Informatization promotes teaching modernization and explores the effective integration of information technology and classroom teaching. Modern Vocat. Educ. **21**, 262–263 (2019)
10. Wang, A.C., Yao, W.P., Li, J.S.: Exploration and practice of applying information technology to improve the teaching quality of UAV control. Moderniz. Educ. **6**(60), 127–129 (2019)
11. Zhu, M., Shang, X., Zhang, S.: Design and practice of micro lecture in SQL Server database application. Comput. Knowl. Technol. **15**(21), 169–170 (2019)
12. Ningbo, R.: Education Technology Co., Ltd.: leader of education information industry. Basic Educ. Forum **21**, 2 (2019)
13. Guan, Z., Yang, Y., Jia, C., Duan, W., Kang, H.: Innovative research on flipped classroom teaching mode of gun control and electrical system principle. China Educ. Technol. Equip. **13**, 75-76+79 (2019)
14. Wang, X.: Research on the informatization teaching mode of "safety management information system" based on the new constructivism. Sci. Educ. Guide (Last Ten Issues) **19**, 107–108 (2019)
15. Chen, X.: Research on the development of educational informatization in Higher Vocational Colleges – taking a higher vocational school as an example. J. Liaoning Teach. Coll. (Nat. Sci. Ed.) **21**(2): 45–49+83 (2019)
16. Tao, L.: Application of information teaching in Higher Vocational Education. Tiangong **6**, 100 (2019)
17. Zhang, X.: Research on ecological balance of higher vocational information teaching resource system. Indust. Technol. Forum **18**(12), 258–259 (2019)
18. Xie, Y., Yang, L.: Construction of innovative talent cultivation ecosystem from the perspective of ecological thinking – taking applied electronic technology specialty as an example. China Educ. Technol. Equip. **11**, 68–70 (2019)
19. Liu, Y.: Research on the information teaching method of database course under the background of "Internet plus." Comput. Knowl. Technol. **15**(15), 182–183 (2019)
20. Lin, W.: Research on two dimensional expression teaching of product design based on information technology. Art Sci. Technol. **32**(5), 247–248 (2019)
21. Tan, Y.: Teaching practice and thinking of automobile circuit and electrical system detection and maintenance. Southern Agricult. Machin. **50**(9), 210–215 (2019)
22. Yang, L.: Analysis of application mode in primary school English Teaching under e-bag environment. J. Jilin Radio TV Univ. **5**, 51–52 (2019)
23. Fan, L.: Research on Biology Classroom Teaching Behavior in Junior Middle School by Using Interactive Whiteboard and Digital Experiment System. Shaanxi Normal University (2019)
24. Yuan, Q.: Research on innovation of teaching mode based on virtual simulation training system development. Shandong Indust. Technol. **12**, 231-232+235 (2019)

Author Index

Ma, Yong   I-105
Mei, Qionghui   I-76
Meng, Wentao   I-557
Miao, Xiaoqi   I-547
Mo, Dongxiao   I-395
Muqian, Huang   II-258

Na, Xiong   II-436
Nan, Peng   II-25
Ni, Zhili   II-385
Ning, Meng   II-549

Peng, Nan   I-523

Qi, Lili   II-278
Qi, Weiqiang   II-503
Qiu, Deming   I-235
Qiu, Minhang   II-607
Qiu, Shuang   II-268
Qiu, Xiaopeng   II-512

Ren, He   II-13

Shi, Changjuan   II-19
Shu, Jingjun   I-336
Song, Yuan   II-569
Su, Xuelian   II-194
Sun, Mengmeng   II-398
Sun, Xiaofei   II-248
Sun, Yanlou   II-148

Tan, Shu   II-79
Tang, Jiayun   I-282
Tang, Jing   II-454
Tang, Yan   I-225, I-354

Wang, Hailan   II-63
Wang, Hairong   II-175
Wang, Hui   II-43
Wang, Lei   I-53, II-13
Wang, Ping   I-517
Wang, Ronghan   I-59, I-404
Wang, Rui   II-475
Wang, Xiaohui   II-599
Wang, Xiaoli   II-346
Wang, Xinchun   I-216
Wang, Yuan   I-33, II-291
Wang, Zhao   I-354
Wen, Jin   I-26
Wen, Jing   I-345

Wen, Tong   I-105
Wu, Jie   I-494
Wu, Jun   I-133
Wu, Ning   II-3
Wu, Xiao   II-643
Wu, Yongfu   II-559
Wu, Zhaoli   II-366

Xie, Jiayun   II-427
Xie, Juan   I-272
Xing, Qiudi   II-487
Xu, Qian   I-292
Xu, Yanping   II-436, II-607
Xu, Ying   I-506

Ya, Tu   I-105
Yan, Jiaqi   I-395
Yang, Xiaoci   I-377
Yang, Xinyu   I-387
Yang, Yiyuan   I-475
Yang, Yongfen   II-204
Yao, Wenshan   II-652
Ye, Wei   II-503
Yin, Na   II-320
Yin, Xiaoyan   II-48
Yu, Fangyan   II-436
Yu, Haiping   II-469
Yuan, Guoting   II-13
Yuan, Jun   I-181
Yuan, Yuan   II-31

Zeng, Qingtian   II-285
Zeng, Yunan   II-248
Zhang, Donglei   II-391
Zhang, Guihong   II-38
Zhang, Lili   II-185
Zhang, Peng   I-186, II-530
Zhang, Rui   II-238
Zhang, Xiaotian   I-225, I-354
Zhang, Xiaoying   II-155
Zhang, XiuJuan   II-339
Zhang, Yan   II-301
Zhang, Yiping   I-211
Zhang, Yuting   II-652
Zhao, Jing   II-96
Zhao, JingMei   II-175
Zhao, Jinhong   II-330
Zhao, Mingyuan   I-244, I-475
Zhao, Wang   I-225
Zhao, Xinmei   I-438

Printed in the United States
by Baker & Taylor Publisher Services